Biology of the Acanthocephala

Biology of the *ACANTHOCEPHALA*

Edited by

D.W.T. CROMPTON

Department of Parasitology, The Molteno Institute
University of Cambridge

BRENT B. NICKOL

School of Biological Sciences
University of Nebraska – Lincoln

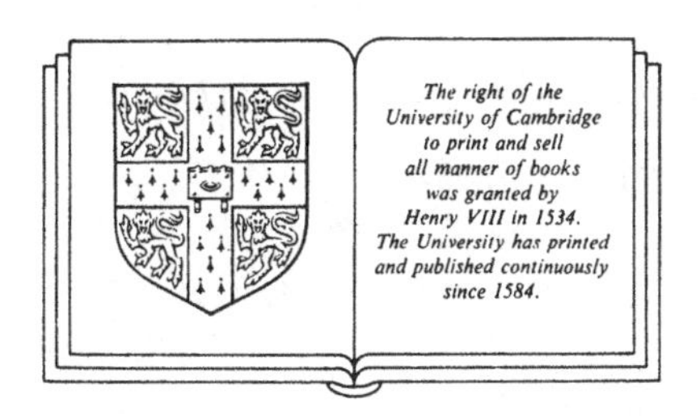

CAMBRIDGE UNIVERSITY PRESS

Cambridge

London New York New Rochelle

Melbourne Sydney

CAMBRIDGE UNIVERSITY PRESS
Cambridge, New York, Melbourne, Madrid, Cape Town, Singapore, São Paulo, Delhi

Cambridge University Press
The Edinburgh Building, Cambridge CB2 8RU, UK

Published in the United States of America by Cambridge University Press, New York

www.cambridge.org
Information on this title: www.cambridge.org/9780521105118

First published 1985
This digitally printed version 2009

A catalogue record for this publication is available from the British Library

Library of Congress Catalogue Card Number: 84-28497

ISBN 978-0-521-24674-3 hardback
ISBN 978-0-521-10511-8 paperback

Contents

Contributors

Omar M. Amin, Division of Science, University of Wisconsin – Parkside, Kenosha, Wisconsin 53141, USA.

Wilbur L. Bullock, Department of Zoology, Spaulding Life Science Building, University of New Hampshire, Durham, New Hampshire 03824, USA.

D.W.T. Crompton, Department of Parasitology, The Molteno Institute, Downing Street, Cambridge CB2 3EE, UK.

*A.P. Dobson, *Department of Pure and Applied Biology, Imperial College, London SW7 2BB, UK.*

T.T. Dunagan, Department of Physiology, Southern Illinois University, Carbondale, Illinois 62901, USA.

C.R. Kennedy, Department of Biological Sciences, University of Exeter, Prince of Wales Road, Exeter EX4 4PS, UK.

Anne Keymer, †Department of Parasitology, The Molteno Institute, Downing Street, Cambridge CB2 3EE, UK.

Donald M. Miller, Department of Physiology, Southern Illinois University, Carbondale, Illinois 62901, USA.

Brent B. Nickol, School of Biological Sciences, University of Nebraska – Lincoln, Lincoln, Nebraska 68588–0118, USA.

Gerald D. Schmidt, Department of Biological Sciences, University of Northern Colorado, Greeley, Colorado 80639, USA.

Jane A. Starling, Department of Biology, University of Missouri – St. Louis, 8001 Natural Bridge Road, St. Louis, Missouri 63121, USA.

* Present address: Department of Biology, Princeton University, Princeton, New Jersey 08544, USA.

† Present address: Department of Zoology, University of Oxford, South Parks Road, Oxford OX1 3PS, UK.

Preface

Nearly 20 years ago, Justus Mueller wrote in a review of parasitic helminths, 'We have neglected the Acanthocephala from our discussion. They are a monotonous group, and few people seem inclined to work with them.' This volume, which may help to correct that view, has developed from the investigations of many workers and particularly from the secure foundations of knowledge set by Anton Meyer and Harley Jones Van Cleave.

Biology of the Acanthocephala has its origins in a symposium held in San Antonio, Texas, in December 1983, under the auspices of the American Society of Parasitologists. We organized the symposium to mark the jubilee of the completion of Anton Meyer's monograph on the *Acanthocephala* in *Dr H.G. Bronn's Klassen und Ordnungen des Tierreichs*. We are deeply indebted to our colleagues whose researches made possible the symposium and volume commemorating the work on acanthocephalan worms by Anton Meyer.

We thank the staff of Cambridge University Press for advice and assistance on editorial matters. Our respective departments in Cambridge, England, and Lincoln, Nebraska, have supported the enterprise; the skilled secretarial help of Mrs Zina Bulstrode and Miss Sarah Johnson has been greatly appreciated, as has help with the Reference List from Miss Virginia Crompton. We also thank Dr William H. Coil, Dr John C. Holmes, Dr Richard D. Lumsden, Dr Richard L. Uznanski, Dr P.F.V. Ward and the authors for their critical reviews of the manuscripts. Finally our special thanks are due to Marion Jowett, whose expertise and care in sub-editing have contributed so much to the production of this book.

D.W.T.C. and B.B.N.
Cambridge, England
and Lincoln, Nebraska

Preface

Nearly [illegible] years ago Justus Mueller wrote [illegible] [illegible] have neglected [illegible] [illegible] They are a monophyletic group [illegible] people [illegible] with them. This [illegible] which [illegible] developed from [illegible] suggestions of [illegible] the [illegible] Appendix [illegible]

[illegible] in December 1983 [illegible]

[illegible] made possible the [illegible]

[illegible] assistance [illegible] Cambridge, England and Lincoln, Nebraska, have supported [illegible] skilled secretarial help of Mrs [illegible] and Mrs [illegible] been greatly appreciated [illegible] has helped with the References [illegible] Dr [illegible] and [illegible] thanks are due [illegible] have contributed so much to the production of this book.

O.W.R. and B.P.N.
Cambridge, England
and Lincoln, Nebraska

Acknowledgements

We acknowledge with thanks the help of the authors, publishers and copyright holders who have agreed to the reproduction of previously published material.

The Harold W. Manter Laboratory, University of Nebraska State Museum (Fig. 3.1)

Parasitology (Figs 5.7; 7.1, 7.2, 7.3, 7.4, 7.5, 7.6, 7.7, 7.8, 7.9, 7.10, 7.14; 11.1, 11.4, 11.5, 11.6)

Journal of Parasitology (Figs 5.5, 5.8, 5.11, 5.13; 7.16; 8.3, 8.4, 8.5, 8.7; 11.3)

Journal of Fish Biology (Figs 11.2, 11.7)

American Journal of Veterinary Research (Fig. 8.6)

Journal of Helminthology (Fig. 8.2)

Transactions of the American Microscopical Society (Fig. 8.1)

Proceedings of the Helminthological Society of Washington (Figs 5.1, 5.3, 5.4, 5.6, 5.12)

Society for Experimental Biology Symposia (Fig. 7.11)

1

Introduction

Brent B. Nickol

Fifty years ago the second part of a two-part monograph on the Acanthocephala was published. The monograph, by Anton Meyer and called simply *Acanthocephala*, comprised the two *Lieferungen* of the *Zweite Buch* of the *Zweite Abteilung* of the *Vierter Band* of *Dr H.G. Bronn's Klassen und Ordnungen des Tierreichs*. Confusion surrounds the year of publication because each *Lieferung* was issued separately, one in 1932 and one in 1933. Afterwards, they were assembled as a complete volume to which the publisher affixed the later date (Van Cleave, 1948). Bound copies of the two parts frequently have a single title page, dated 1933, and a combined table of contents. Covers for the individual parts, however, bear the original 1932 and 1933 dates. This monumental work still stands as one of the principal reference works for the Acanthocephala. To commemorate its fiftieth anniversary, the present volume reviews and interprets many of the discoveries regarding the biology of the Acanthocephala that have been made after 1933.

As the fascinating recent discoveries are read in the following pages, it is appropriate to remember that pioneering work, such as that done by Rudolphi, Hamann, Kaiser, Lühe, Meyer, Travassos, Van Cleave and others, was a prerequisite to the more complete knowledge now amassed. In this era of explosive technology when history is not in vogue, when it is often considered irrelevant, and when ideas of what constitutes science and worthy scientific endeavor are dictated only by the latest issue of *Current Contents*, it is especially important to keep track of the past and to realize where we have been.

Meaningful information about the Acanthocephala is comparatively recent, certainly post-Linnaean and, for the most part, it dates from within the last 100 years. Modern views of the systematic relations of Acanthocephala have the fundamental arrangement of Otto Hamann

(1892) as their basis. Hamann was able to integrate detailed observations by his contemporaries, perhaps most notably those of J.E. Kaiser, into a comparative system. He was the first to evaluate differences at a level above those for separating species, to recognize several genera in place of the all-inclusive *Echinorhynchus*, and to formulate the concepts of acanthocephalan families (Van Cleave, 1948). Meyer added information from his own studies of ontogeny and morphological interpretations to the Hamann scheme to produce his monograph of all the then-known Acanthocephala.

Meyer was not the only early-twentieth-century pioneer to make lasting contributions to knowledge of the Acanthocephala. Beginning in 1913, H.J. Van Cleave made numerous contributions that outlined and defined the acanthocephalan fauna. In 1936 he visualized the concept of Eoacanthocephala, which reconciled incongruent elements in the system advocated by Meyer, and the Meyer–Van Cleave system of classification was formulated. This system provides the basis for most present views of acanthocephalan systematics.

During the three decades following publication of Meyer's monograph, a steady level of investigation continued to accumulate information regarding the diversity of the Acanthocephala and to define the fauna. During this period the first critical studies of life cycles and transmission (Ward, 1940*b*; Moore, 1946*a*, *b*; DeGiusti, 1949*a*) began uncovering some of the ecological facets of acanthocephalan biology. It was soon apparent that the Acanthocephala are ideal animals for the study of many parasitological relationships, especially those of populations and communities. Unlike the Platyhelminthes, they are dioecious, do not multiply within intermediate hosts, and have no free-living stage. The opportunities offered by these attributes were soon exploited.

In the 1970s there was a marked increase in the use of acanthocephalans in parasitological research. Not only did study of them result in new ecological understanding, but it also provided insight into the cellular, molecular, and regulatory biology of helminths and into the host–parasite relationships.

It is the intent of this book to summarize some of the knowledge that has accumulated during the last 50 years. Throughout, the systematic arrangement and nomenclature outlined in Chapter 4 by Omar Amin has been adopted. Synonyms, under which some of the information cited was published, are readily found in that chapter.

Finally, it is impossible to introduce a review such as this without also speculating on the likely content of similar reviews 50 years in the future. Large voids in understanding the Acanthocephala result from the paucity of information on their genetics and their immunological relations with hosts. Advances on these fronts undoubtedly will occur and with them will emerge more enlightened views on pathogenesis, host specificity, host resistance, phylogeny, and many other facets of biology. Increased understanding of cellular and molecular relationships might permit better success in *in vitro* cultivation and with it advances in fields dependent on that technique. A better defined fauna and computer-assisted analysis should produce more detailed and theoretical understanding of community structure and population dynamics. Perhaps puzzling instances of dispersal and distribution among hosts will be resolved.

Whether parasitological study follows these or other paths in future years, present appreciation of the potential held by the Acanthocephala as experimental animals indicates that knowledge of their biology will increase along with their contributions to research in helminthology.

2

Anton Meyer

D.W.T. Crompton

Anton Meyer was born on 28th September 1901 in Mies (now called Stříbro; southwest of Prague, 49.46N, 13.00E) in Bohemia.[1,2] His father, who was probably also called Anton,[1] was a tinsmith from Černošin and his mother, whose maiden name was Buberlová, came from Marienbad.[3] In an application form to the Rockefeller International Education Board,[1] Meyer's citizenship is given as Czechoslovakian and his languages as German, English and Czech in that order. This application form and other sources give no more information about his family and origins. He declared himself as unmarried in 1927[1] and in 1933 wrote 'I am still away from getting married'.[4] Meyer may have died about the time of the outbreak of World War II, possibly in a psychiatric hospital in Dobřany[3] which was destroyed, together with its medical files, by a bombing raid in 1943.[5] Attempts to check this suggestion have yielded nothing of substance, but something may be inferred from the publication in 1958 by another author, C. Spren, of a supplement entitled 'Nachtrag zu Dr Anton Meyer – Klasse: Acanthocephala, Akanthozephalen, Kratzer' in *Die Tierwelt Mitteleuropas*; presumably Meyer must have died.

The record of Anton Meyer's education and academic career is no easier to trace than his personal history. He attended the *Gymnasium*, which we may assume was in Mies, from 1912 to 1920 when he attained his certificate of maturity.[1] He then enrolled in the Faculty of Natural Sciences at the German University of Prague and was awarded his Doctoral Degree on 20th October 1924 for research on the embryology of *Macracanthorhynchus hirudinaceus* (Acanthocephala) under the supervision of Professor C.T. Cori.[1,6] Meyer remained at the German University of Prague as a Demonstrator at the Zoological Institute until he moved to the Department of Zoology at the University of Oxford in England in October 1927 on being awarded a Research Fellowship from the Rockefeller International

Education Board.[1] The Fellowship, which was offered for a year in the first instance, covered the costs of his travel between Czechoslovakia and England, and provided a stipend not to exceed more than US$120 per month.[7] Anton Meyer worked with Professor E.S. Goodrich[8,9] on various aspects of the biology of annelid worms and, after his Rockefeller Fellowship had been extended,[10] he moved in July 1928 to the laboratory of the Marine Biological Association of the United Kingdom at Plymouth where he stayed until February 1929.[11,12]

After his visit to England, Meyer appears to have intended to take up a post as Assistant at the Zoological Institute of the German University of Prague,[13] but these plans may not have been fulfilled. He undoubtedly worked at the Zoological Institute of the University of Leipzig[14] with financial support from the Notgemeinschaft der Deutschen Wissenschaft, an agency acknowledged by Meyer in several of his publications of this period. Anton Meyer next moved, perhaps towards the end of 1931 and apparently with continued financial support from the Notgemeinschaft der Deutschen Wissenschaft, to the Kaiser-Wilhelm-Institut für Biologie, Berlin-Dahlem. He was accommodated in the department headed by Dr Otto Mangold[15] where he worked on his monograph on the Acanthocephala.[16,17]

Between 1933 and 1938, after the publication of his monograph, Meyer does not seem to have had any permanent or even temporary laboratory address. His contribution to *Die Tierwelt Mitteleuropas*, which was published in 1938, gives his address as 'in Mies i Böhmen'. In two letters,[4,18] which mark the beginning and end of this period, it is clear that he had had several arguments with colleagues, editors and publishers about the length of his manuscripts and the significance of his research. For example, he wrote 'Also my monograph of the Acanthocephala became somewhat shortened and deprived of the preface which would have been very necessary, as to secure a right understanding of the disposition of the whole. I had heavily to cross with the publisher...'[4] In the same letter he writes as if recovering from a state of nervous exhaustion and perhaps, although this is merely a speculative suggestion, he was by the end of 1933 beginning to suffer from the illness which may have led to his death in the hospital in Dobřany.

Anton Meyer was remembered by Dr D.P. Wilson[19] as 'a quiet pleasant young man, an earnest and meticulously careful worker'. They corresponded after Meyer's departure from Plymouth[4,19] and, as a token of friendship and respect, Meyer sent Dr Wilson a copy of M.E. Boyle's book *In Search of Our Ancestors* (1927, London, Bombay & Sydney: George G. Harrap & Co. Ltd) in which he wrote 'Dear Mr Wilson. With best wishes for Christmas I send you this book as a small expression of my gratitude. Yours truly Ant. Meyer.' Professor W.E. Ankel[20] recalled Meyer as a typical outsider whom he would have described as a nonconformist. He could not remember where they met, but thought it could well have been at the Zoological Station in Naples. There is no record of Meyer's having worked at Naples,[21] but he could have worked there; he visited the Marine Stations in Helgoland[1] and Messina.[13] Dr H. Kumerloeve[14] clearly remembered Meyer 'as dark-haired, small and with glasses and as a rather shy person who had almost no contact with other scientists' in the laboratory at Leipzig. He gave a seminar there and 'was known as a highly specialized scientist'. From these fragments of direct evidence, Anton Meyer (Fig. 2.1) emerges as an intense, sensitive, dedicated and lonely man. He does not seem to have worked in collaboration with others (see list of publications below), he could not be traced

Fig. 2.1. Anton Meyer. This photograph, which was kindly provided by Dr D.P. Wilson, was accompanied by a handwritten inscription, 'Marienbad, August 1939, Drinking Ambrosias'.

as having been a member of the relevant scientific societies of his day (e.g. Kaiser-Wilhelm-Gesellschaft and Zoologische Gesellschaft), there is no obvious obituary notice, he is not mentioned in the *Biologorum*, there is no record of him in the archives of the Berlin Document Centre and he does not appear to have served as an editor or editorial-board member for any of the prominent scientific journals of the time.[22,23]

During his career (1920–?1939), Anton Meyer achieved much in the fields of acanthocephalan and annelid biology (see list of publications below). In Professor Ankel's view, Meyer's 'most exciting contribution at that time' was his work on flagellated cells in annelids.[20] Professor Cori,[6] whose opinions were endorsed by Professor G.N. Calkins,[24] regarded Meyer as a gifted scholar of unusual maturity and scientific judgement who had already done very creditable work on the cytological aspects of fertilization.[13] Paradoxically, Meyer's researches on annelids, which provided most information about his history, seem to have lost their early impact (for example, his work is given little attention in the relevant volume on *Traité de Zoologie*, edited by P-P. Grassé, Paris: Masson et Cie) while his acanthocephalan contributions, which have so far revealed little about the author, have remained fundamental to our present understanding of the phylum.[25]

Anton Meyer's life and achievements are all the more interesting and remarkable because they took place during a period of major change and instability in Europe. In 1901, when he was born, Bohemia was a province of the Austro-Hungarian Empire.[26] By 1920, when he went to the University of Prague, the fragmentation of Europe during World War I had occurred and the state of Czechoslovakia had been formed from the provinces of Bohemia, Moravia, Silesia, Slovakia and Ruthenia. This amalgamation produced an apparently economically strong country with an important strategic position, but also a population of several languages and divided loyalties.[26] Some looked to Vienna for leadership, others to Budapest and no doubt the German-speaking residents of the Sudetenland,[27] like Meyer, felt drawn towards the rising power of the new Germany as the Weimar Republic (1919–1933) was overtaken by Hitler's National Socialist Workers' Party which advocated anti-semitic, anti-communist and anti-parliamentary policies.[26,28,29] At the end of Anton Meyer's career, and probably of his life, Europe was again on the verge of destruction and terror as World War II approached. Throughout this period, the political uncertainties and power struggles and the collapse of the German economy in the 1920s,[30] when Meyer would have been seeking grants to support his highly academic research, could hardly have helped to create a secure environment for a young and sensitive scientist.

Unless serendipity intervenes, much more research will be needed to elucidate the life history of Anton Meyer and explain how his interest in the Acanthocephala was aroused. Perhaps he should remain as a shadowy figure and perhaps his private life, his hopes and fears, his ambitions and disappointments should not be disclosed. Science is our common heritage and scientists come and go, but I wish we knew more about him.

Time will say nothing but I told you so,
Time only knows the price we have to pay;
If I could tell you I would let you know.

If I could tell you, W.H. Auden,[31] October 1940

Acknowledgements
This account could not have been prepared without the advice, efforts and hospitality of Professor Benno Hess and Frau Ulrike Hess of the Max-Planck-Institute, Dortmund, and the kindness and generosity of Dr D.P. Wilson of Plymouth. I am grateful to all those who helped with enquiries and particularly to Professor W.E. Ankel, Herr Gerhard Baumgarten, Dr J. Chalupský, Mr J.D. Crompton, Mr R. Hughes, Dr H. Kumerloeve and Dr R. Neuhaus. Finally, I thank Cambridge Philosophical Society for a travel grant to visit the Federal Republic of Germany and Dr J. Janovy Jr and the staff of Cedar Point Biological Station, University of Nebraska, for enabling me to study in such delightful surroundings.

Publications of Anton Meyer

1924 Über die Segmentalorgane von *Tomopteris helgolandica* nebst Bemerkungen über das Nervensystem und die rosettenförmigen Organe. *Zoologischer Anzeiger*, **60**, 83–8.

1926 Die Segmentalorgane von *Tomopteris catharina* (Gosse) nebst Bemerkungen über das Nervensystem, die rosettenförmigen organe und die Cölombewimperung. Ein Beitrag zur Theorie der Segmentalorgane. *Zeitschrift für Wissenschaftliche Zoologie*, **127**, 297–402.

1927 Über Cölombewimperung und cölomatische Kreislaufsysteme bei Wirbellosen. Ein Beitrag zur Histophysiologie der secundären Liebeshöhle und ökologischen Bedeutung der Flimmerbewegung. *Zeitschrift für Wissenschaftliche Zoologie*, **129**, 153–212.

Ist *Parergodrilus heideri* (Reisinger) ein Archiannelide? *Zoologischer Anzeiger*, **72**, 19–35.

1928 Die Furchung nebst Eibildung, Reifung und Befruchtung des *Gigantorhynchus gigas* (Ein Beitrag zur Morphologie der Acanthocephalen). *Zoologische Jahrbücher. Abteilung für Anatomie und Ontogenie der Tiere*, **50**, 117–218.

1929 Die Entwicklung der Nephridien und Gonoblasten bei *Tubifex rivulorum* Lam. nebst Bemerkungen zum natürlichen System der

Oligochäten. *Zeitschrift für Wissenschaftliche Zoologie*, **133**, 517–62.
Über Cölombewimperung und cölomatische Kreislaufsysteme bei Wirbellosen. II. Teil. (Sipunculoides, Polychaeta Errantia). Ein Beitrag zur Histophysiologie und Phylogenese des Cölomsystems. *Zeitschrift für Wissenschaftliche Zoologie*, **135**, 495–538.
On the coelomic cilia and circulation of the body-fluid in *Tomopteris helgolandica*. *Journal of the Marine Biological Association of the United Kingdom*, **16**, 271–6.
Ein atavistischer *Tubifex*-Embryo mit überzähligem Gonoblastenpaar und seine Bedeutung für die Theorie der Segmentstauchung bei den Oligochäten. *Zoologischer Anzeiger*, **85**, 321–9.
Zur Segemtierungsanalyse und Stammesgeschichte der Oligochäten. *Zoologischer Anzeiger*, **86**, 1–16.
Tomopteris anadyomene nov. spec. ein Nachweis Phylogenetischer Umwandlung von Nephridialtrichtern in leuchtorgane bei den Polychäten. *Zoologischer Anzeiger*, **86**, 124–33.
1930 Vergleichende Untersuchung der Segmentalorgane von Tomopteriden des Mittelmeeres, ein Nachweis eines Substitutionsprozesses. *Zeitschrift für Wissenschaftliche Zoologie*, **136**, 140–53.
1931 Das urogenitale Organ von *Oligacanthorhynchus taenioides* (Diesing), ein neuer Nephridialtypus bei den Acanthocephalen. *Zeitschrift für Wissenschaftliche Zoologie*, **138**, 88–98.
Urhautzelle, Hautbahn und plasmodiale Entwicklung der Larve von *Neoechinorhynchus rutili* (Acanthocephala). (Ein Beitrag zur Entwicklungsmechanik gebahnter organbildung.) *Zoologische Jahrbücher. Abteilung für Anatomie und Ontogenie der Tiere*, **53**, 103–26.
Infektion, Entwicklung und Wachstum des Riesenkratzers (*Macracanth. hirudinac.*) im Zwischenwirt. *Zoologischer Anzeiger*, **93**, 163–72.
Gordiorhynchus, ein neues Acanthocephalengenus mit innerer ovarialer Pseudosegmentierung. *Zoologische Jahrbücher. Abteiling für Systematik, Ökologie und Geographie der Tiere*, **60**, 457–70.
Neue Acanthocephalen aus dem Berliner Museum. Bergründung eines neuen Acanthocephalensystems auf Grund einer Untersuchung der Berliner Sammlung. *Zoologische Jahrbücher. Abteilung für Systematik, Ökologie und Geographie der Tiere*, **62**, 53–108.
Die Formbildungsmuster und Das Wirbelphänomen in der Cölombewimperung von *Nephthys*. *Zeitschrift für Zellforschung und Mikroskopische Anatomie*, **14**, 222–54.

Das Häutgefasystem von *Neoechinorhynchus rutili* (Formbildung auf plasmodialer Grundlage). *Zeitschrift für Zellforschung und Mikroskopische Anatomie*, **14**, 255–65.

Cytologische Studie über die Gonoblasten und Andere Annliche Zellen in der Entwicklung von *Tubifex*. *Zeitschrift für Morphologie und Ökologie der Tiere*, **22**, 269–86.

Die Stellung des Genus *Heterosentis* Van Cleave 1931 im Acanthocephalensystem. *Zoologischer Anzeiger*, **94**, 258–65.

Die Acanthocephalen d. Arktischen Gebietes. *Fauna Arktica*, **6**, 9–20.

1932 Acanthocephala. In *Dr H.G. Bronn's Klassen und Ordnungen des* Tier–Reichs, Band 4, Abt. 2, Buch 2, Lief. 1, pp. 1–332. Leipzig: Akademische Verlagsgesellschaft MBH.

1933 Acanthocephala. In *Dr H.G. Bronn's Klassen und Ordnungen des Tierreichs*, Band 4, Abt. 2, Buch 2, Lief. 2, pp. 333–582. Leipzig: Akademische Verlagsgesellschaft MBH.

1934 Über Formbildung (Morphodynamik). *Verhandlungen der Deutschen Zoologischen Gesellschaft*, **36**, 218–24.

Die Axiome der Biologie. *Nova Acta Leopoldina Halle*, **1**, 474–551.

1935 Eine regionale Protoplasmazentrierung als Ursache der Formbildung. *Biologia Generalis*, **11**, 122–34.

1936 Dynamese des cölomatischen Wimperfeldes von *Nephthys hombergii*. *Biologischen Zentralblatt*, **56**, 532–48.

Die plasmodiale Entwicklung und Formbildung des Riesenkratzers *Macracanthorhynchus hirudinaceus* (Pallas). I. Teil. *Zoologische Jahrbücher. Abteilung für Anatomie und Ontogenie der Tiere*, **62**, 111–72.

1937 Die plasmodiale Entwicklung und Formbildung des Riesenkratzers *Macracanthorhynchus hirudinaceus* (Pallas). II. Teil. *Zoologische Jahrbücher. Abteilung für Anatomie und Ontogenie der Tiere*, **63**, 1–36.

1938 Die plasmodiale Entwicklung und Formbildung des Riesenkratzers *Macracanthorhynchus hirudinaceus* (Pallas). III. Teil. *Zoologische Jahrbücher. Abteilung für Anatomie und Ontogenie der Tiere*, **64**, 131–97.

Die plasmodiale Entwicklung und Formbildung des Riesenkratzers *Macracanthorhynchus hirudinaceus* (Pallas). IV. Allgemeiner Teil. *Zoologische Jahrbücher. Abteilung für Anatomie und Ontogenie der Tiere*, **64**, 199–242.

Der Rogen (Spawn) und die Entwicklung der Trochophora von *Eulalia viridis* (Phyllodocidae). *Biologia Generalis*, **14**, 334–89.

Die Coelomatischen Wimperfelder, ihre Polaren und Integrätiven Formen bei *Tomopteris catharina. Zoologische Jahrbücher. Abteilung für Anatomie und Ontogenie der Tiere*, **64**, 371–436.
Klasse: Acanthocephala, Acanthozephalen, Kratzer. In *Die Tierwelt Mitteleuropas*, ed. P. Brohmer, P. Ehrmann & G. Ulmer, Band 1, Lief. 6, pp. 1–40. Leipzig: Verlag Von Quelle & Meyer.

Notes and sources

1 Application by Anton Meyer to the Rockefeller International Education Board for a Fellowship. This document, which is handwritten, is dated 1927 and is filed under 'Anton Meyer 1927–1928' with relevant reports and correspondence at the Rockefeller Archive Center, Hillcrest, Pocantico Hills, North Tarrytown, NY 10591, USA.

2 Letter (dated 12th January 1980) from Dr Josef Chalupský, Department of Parasitology, Charles University, 12844 Prague 2, Viničná 7, Czechoslovakia, to Professor Benno Hess, Max-Planck-Institut für Ernährungsphysiologie, D-4600 Dortmund 1, Federal Republic of Germany.

3 Letter (dated 3rd April 1981) from Dr Josef Chalupský to Professor Benno Hess.

4 Letter (dated 18th August 1933) written by Anton Meyer, while staying at Marienbad in Bohemia, to Dr D.P. Wilson, Marine Biological Association of the United Kingdom, The Laboratory, Citadel Hill, Plymouth, England.

5 Letter (dated 15th July 1981) from Dr Josef Chalupský to Professor Benno Hess.

6 Professor C.T. Cori (1864–1954) was an international authority on marine biology. He studied zoology and medicine at the German University of Prague. During World War I he worked on malaria (see Kalmus, *Nature*, *London*, **174**, 774: 1954).

7 Copy of letter (dated 1st August 1927) from W.W. Brierley (no address given) to Professor C. Cori, Zoological Institute, German University, Prague, Czechoslovakia, with copy to Dr Meyer, announcing the award of the Fellowship.

8 Professor E.S. Goodrich FRS (1868–1964), in a report (dated 6th July 1927) to Dr Wickliffe Rose of the Rockefeller International Education Board is quoted as follows: 'I would welcome Dr Meyer and be pleased to afford him every facility for working in this department. It was with great interest that I read his recent paper on the nephridia of *Tomopteris* (Meyer, 1926).' In a publication in 1929, Meyer acknowledged Professor Goodrich's interest in his work.

9 Dr Meyer was registered as a research student in the Department of Zoology at Oxford from October 1927 to June 1928. Letter (dated 12th January 1981) from Mrs J. Loupekine of that department to D.W.T. Crompton.

10 Copy of letter (dated 8th November 1928) from W.E. Tisdale, Rockefeller International Education Board, 20 Rue de la Baume, Paris, to Professor C. Cori, with copy to Dr A. Meyer, announcing the extension of the Fellowship. In his original application to the IEB,[1] Anton Meyer mentioned his interest in working at the Plymouth Laboratory.

11 Dr A. Meyer is listed as having occupied a table at the Plymouth Laboratory in the 'Report of the Council, 1928', *Journal of the Marine Biological Association of the United Kingdom*, **16**, 343.

12 Letter (dated 20th November 1980) from Head of Library and Information Services, Marine Biological Association of the United Kingdom, to D.W.T. Crompton, states that Anton Meyer worked at the Plymouth Laboratory from 21st July 1928 to 8th February 1929.
13 Report (A-25) prepared by the staff of the Rockefeller International Education Board for use during consideration of Anton Meyer's application.[1]
14 Recollections of Dr H. Kumerloeve (Research Assistant, Zoological Institute, University of Leipzig 1931–1935) received by D.W.T. Crompton in a letter from Professor Benno Hess (dated 7th November 1980).
15 *Naturwissenschaften* (1932), Heft 22/24. On page 437, '4. Abteilung MANGOLD. Als wissenschaftliche Gäste arbeiteten:...Dr A. MEYER aus Prag (Tschechoslowakei)'.
16 In addition to the photograph shown in Fig. 2.1, Anton Meyer sent another photograph of himself to Dr D.P. Wilson. This print shows him working on some papers at a table, with a fountain pen in his right hand, and is inscribed in handwriting 'Register-machen: Acanthocephala. Kaiser Wilhelm Institut f Biologie, in der Hundshütte [Doghouse] Mai 1933'. The Kaiser-Wilhelm-Institut für Biologie was a four-storey building with two wings, one for housing work with rodents and small mammals and one for work with insects. Meyer probably studied in the insect wing.
17 Berlin-Dahlem is printed as Anton Meyer's address on the title page of *Acanthocephala*.
18 Letter (dated 4th June 1938) written by Anton Meyer while in Mies to Dr D.P. Wilson[19] at Plymouth, England.
19 Dr Douglas P. Wilson, who was 26 in 1928 when Anton Meyer began work at the Plymouth Laboratory, retired from the scientific staff of the Marine Biological Association of the United Kingdom in 1969.
20 Recollections of emeritus Professor W.E. Ankel, formerly of the Department of Zoology, University of Giessen, received by D.W.T. Crompton in a letter from Professor Benno Hess (dated 10th November 1980).
21 Letter (dated 18th December 1980) from Christiane Groeben, Stazione Zoologica, 80121 Napoli, to Professor Benno Hess.
22 Letter (dated 2nd June 1980) from Dr Rolf Neuhaus, Direktor, Bibliothek und Archiv zur Geschichte der Max-Planck-Gesellschaft, 1 Berlin 33 (Dahlem) to D.W.T. Crompton.
23 Letter (dated 9th June 1981) from Daniel P. Simon, Director, Berlin Document Center, Mission of the USA, Wasserkaefersteig 1, 1000 Berlin 37, to Professor Benno Hess.
24 Professor G.N. Calkins (1869–1943) of Columbia University, New York. Research on cytology, protozoology and systematic zoology.
25 The first sentence in the foreword written by Academician K.I. Skrjabin for V.I. Petrochenko's monograph *Acanthocephala of Domestic and Wild Animals*, vol. 1 (1956, Moscow, Izdatel'stvo Akademii Nauk SSSR), reads, in translation, 'Twenty years have passed since the publication by Meyer of a composite monograph on the Acanthocephala'.
26 Wallace, W.V. (1977) *Czechoslovakia*. London: Ernest Benn.
27 Sudetenland: a region of the Sudeten mountains in Northern Czechoslovakia which was annexed by Hitler's Germany in 1938.
28 Thomson, D. (1966) *Europe since Napoleon*. London: Longman (Penguin edition).
29 Carr, W. (1969) *A History of Germany 1815–1945*. London: Edward Arnold.

30 Childs, D. (1970) *Germany since 1918.* London: B.T. Batsford.
31 W.H. Auden (1907–1973), English poet, educated at Gresham's School and Oxford University where he was an undergraduate while Anton Meyer worked in the Department of Zoology.[9] Auden also lived in Berlin from October 1928 to July 1929 and later, in 1935, married Erika Mann (daughter of Thomas Mann), a declared enemy of the Third Reich, to acquire for her the benefits of British Citizenship.

3

Harley Jones Van Cleave: 40 years with the Acanthocephala

Wilbur L. Bullock

The contributions to this volume demonstrate that there are numerous current investigations of many facets of acanthocephalan biology. Much of this research now in progress in the late twentieth century is based on the work of our predecessors. The pioneering morphological studies of Hamann (1891*a*) and Kaiser (1913) provided a foundation on which Anton Meyer could carry out his morphological, developmental, and systematic studies during the period 1928 through 1938. The most significant of those studies was his 1932–1933 monograph that we commemorate with this volume.

However, Meyer was not the only student of the Acanthocephala in the first half of this century. During the four decades 1913 to 1953, Harley Jones Van Cleave (Fig. 3.1) published more than 125 papers dealing with the spiny-headed worms. Most of these were of a taxonomic–systematic nature and provided us with numerous taxa from species to classes. Professor Van Cleave's publications concerned the Acanthocephala of fish, amphibians, reptiles, birds and mammals; they included specimens from China, India, Venezuela and the Antarctic, as well as from North America and elsewhere. The significant interdependence of the contributions of Anton Meyer and H.J. Van Cleave, two dedicated investigators, can be seen in many ways. Even with the passage of time we are still using what has been aptly described as the Meyer–Van Cleave system. This system is evident in Amin's important updating of acanthocephalan classification elsewhere in this volume. Therefore, as we recognize the fiftieth anniversary of Meyer's epochal monograph, it is fitting that brief recognition be made of the contributions of Van Cleave, contributions that began 70 years ago and continued until his death 30 years ago.

Dr Van, as he was affectionately called by those who knew him, was a native and life-long resident of the state of Illinois. Born in Knoxville,

Illinois, on 5th October 1886, he attended Knox College in the nearby town of Galesburg. After receiving his Bachelor of Science Degree in Biology in 1909 he moved across the state to the University of Illinois in Urbana. Upon earning his Master of Science Degree in 1910 he initiated his studies of the Acanthocephala under the direction of Henry Baldwin Ward and

Fig. 3.1. Harley Jones Van Cleave in 1938. (Kindly provided by M.H. Pritchard, Curator, The Harold W. Manter Laboratory, University of Nebraska State Museum.)

received his doctorate in 1913 on the basis of his dissertation: 'Studies on cell constancy in *Neorhynchus* with descriptions of new species in that genus.' From this thesis came his first publications, which foreshadowed his commitment to acanthocephalan systematics and his special interest in the Neoechinorhynchidae.

Dr Van Cleave remained with the University of Illinois, progressing from Instructor in 1913 to Professor of Zoology in 1929 and Research Professor in 1948. Summer teaching and research took him to numerous localities including Illinois State Normal University (1913–1915); Puget Sound, Washington (1916); Cold Spring Harbor, Long Island, N.Y. (1936); and Isles of Shoals, N.H. (1939). From 1928 to 1934 he, along with Justus Mueller, carried out an intensive study of the parasites of the fishes of Oneida Lake, N.Y. This study was a monumental investigation of the helminths of the fishes of a North American freshwater lake.

In addition to his active research program, Dr Van was also an outstanding teacher and taught a variety of courses including invertebrate zoology, field zoology and helminthology. Many of our present American parasitologists are either students of Dr Van or students of his students. It was to this distinguished group of parasitologists that I had the honor of being added as Dr Van's last doctoral student.

In March 1946, I was a beginning graduate student, recently discharged from the US Army, when I had my first conference with Dr Van Cleave, my assigned advisor. Dr Van seemed to grasp quickly that I was one of those students who was enthusiastic about the study of parasites but who had no idea where to begin or what to work on for my Master's thesis. Therefore, he suggested an ideally limited and precise subject of investigation, a study of the morphology of the praesoma of *Neoechinorhynchus emydis*. Never having heard of either praesoma or *Neoechinorhynchus*, I was obviously unable to respond positively and enthusiastically. Even when he asked if I had ever heard of the Acanthocephala my affirmative answer, though honest, was given in a way that made it all too clear that I did not really appreciate the achievements of my advisor in the realm of acanthocephalan biology. Nevertheless, I proceeded to study the morphology of the praesoma of *Neoechinorhynchus emydis* – or at least that's what we thought it was at the time. My first collection of worms from the turtle hosts turned out to be nematodes (*Camallanus*) and my first set of serial sections of the 'praesoma' of a *Neoechinorhynchus* was actually from the posterior end of a male! Dr Van accepted all this with a patience that I find, in retrospect after 35 years of teaching myself, to be truly outstanding. His major concern was that I was doing something. He calmly and gently pointed out the error of my ways and encouraged me to press on.

What I appreciated most about those early days of our relationship was that, in spite of my inauspicious and stumbling start, I continued to receive Dr Van's encouragement, even when my Master's thesis study began to contradict some of the conclusions in Dr Van's doctoral thesis. He had the truly open mind and humility of a good scientist. He was not easy to convince, but when I assembled my evidence from serial sections, whole mounts and dissections, he accepted the new concepts and encouraged me to continue. Even when I changed my proposed doctoral study from the anatomy of the acanthocephalan nervous system – which he wanted to see studied by someone interested in micromorphology – to acanthocephalan histochemistry, my decision was approved and supported. Dr Van's gentle but firm and demanding guidance was just what I needed to get started on my career as a parasitologist. I, along with his other students, will always be grateful for his dedication to teaching.

In addition to being a thorough and careful worker, an honest and humble scientist, and a patient and gentle teacher, Dr Van also took a personal interest in his students. I have always appreciated the trip on which he took a fellow graduate student and myself so that he could acquaint us with some of the details of the biology and the history of his beloved state of Illinois. We visited the New Salem restoration of Abraham Lincoln's younger days, the State Museum in Springfield, and the Illinois State Biological Laboratory near Havana, Illinois. He also showed a continuing interest in the welfare of my family by taking myself, my wife and our very young twin sons – for Sunday afternoon drives. And when the cumbersome bureaucracy of the Veteran's Administration was slow in processing my 'G.I. Bill of Rights' educational benefits, Dr Van even offered to loan us money until such time as my subsistence checks arrived!

Dr Van Cleave had research interests other than the Acanthocephala. His publications covered a broad area of the natural history of Illinois and included studies on such diverse groups as rotifers, fairy shrimp, jellyfish, snails, fish and snakes. He published several papers on conservation and environmental problems long before such papers were as fashionable as they are today. One of his major interests on the periphery of parasitology and biology was the history of science in North America and the biographies of the important personalities of nineteenth- and early-twentieth-century biology. Those of us involved will never forget our seminar in which we read about and discussed such fascinating individuals as Joseph Leidy, John Sterling Kingsley, Spencer Fullerton Baird, Stephen Alfred Forbes and Charles Atwood Kofoid.

But through it all, from 1913 to 1953, were always the Acanthocephala. Publication after publication added information regarding this unique group of worms and his taxonomic and anatomical investigations provided basic knowledge required for subsequent work on biochemistry, ecology, functional morphology and physiology.

Scientific publications of Harley Jones Van Cleave on the Acanthocephala

1913 The genus *Neorhynchus* in North America. *Zoologischer Anzeiger*, **43**, 177–90.

1914 Studies on cell constancy in the genus *Eorhynchus*. *Journal of Morphology*, **25**, 253–99.

Eorhynchus: a proposed new name for *Neorhynchus* Hamann preoccupied. *Journal of Parasitology*, **1**, 50–1.

1915 Acanthocephala in North American Amphibia. *Journal of Parasitology*, **1**, 175–8.

1916 Seasonal distribution of some Acanthocephala from fresh-water hosts. *Journal of Parasitology*, **2**, 106–10.

Filicollis botulus n. sp. with notes on the characteristics of the genus. *Transactions of the American Microscopical Society*, **35**, 131–4.

A revision of the genus *Arhythmorhynchus*, with descriptions of two new species from North America. *Journal of Parasitology*, **2**, 167–74.

Acanthocephala of the genera *Centrorhynchus* and *Mediorhynchus* (new genus) from North American birds. *Transactions of the American Microscopical Society*, **35**, 221–32.

1917 Observations on seasonal distributions and longevity of some Acanthocephala from fresh water hosts. *Transactions of the Illinois State Academy of Sciences*, **9**, 223–4.

1918 The Acanthocephala of North American birds. (abstract) *Anatomical Record*, **14**, 103.

Acanthocephala of North American birds. *Transactions of the American Microscopical Society*, **37**, 19–48.

Centrorhynchus pinguis, n. sp., from China. *Journal of Parasitology*, **4**, 164–9.

Acanthocephala of the subfamily Rhadinorhychinae from American fish. *Journal of Parasitology*, **5**, 17–24.

1919 Acanthocephala from the Illinois River, with descriptions of species and a synopsis of the family Neoechinorhynchidae. *Bulletin of the Illinois Natural History Survey*, **13**, 225–57.

Acanthocephala from fishes of Douglas Lake, Michigan. *Occasional Papers from the Museum of Zoology of the University of Michigan*, **72**, 1–12.

1920 Two new genera of Acanthocephala from Venezuelan fishes. (abstract) *Anatomical Record*, **17**, 334.

Notes on the life cycle of two species of Acanthocephala from freshwater fishes. (abstract) *Anatomical Record*, **17**, 330.

Acanthocephala of the Canadian Arctic Expedition, 1913–1918. *Report of the Canadian Arctic Expedition 1913–1918*, **9**, E: 1–11.

Notes on the life cycle of two species of Acanthocephala from freshwater fishes. *Journal of Parasitology*, **6**, 167–72.

Two new genera and species of acanthocephalous worms from Venezuelan fishes. *Proceedings of the United States National Museum*, **58**, 455–66.

Sexual dimorphism in the Acanthocephala. *Transactions of the Illinois State Academy of Sciences*, **13**, 280–92.

1921 Acanthocephala parasitic in the dog. *Journal of Parasitology*, **7**, 91–4.

Preliminary survey of the Acanthocephala from fishes of the Illinois River. *Journal of Parasitology*, **12**, 151–6.

Acanthocephala collected by the Swedish Expedition to the Juan Fernandez Islands. In *The Natural History of Juan Fernandez and Easter Island*, vol. 3, ed. C. Skottsberg, pp. 75–80.

Notes on the development and distribution of *Oncicola canis* (Kaupp, 1909). (abstract) *Anatomical Record*, **20**, 206.

Acanthocephala from the American eel. (abstract) *Anatomical Record*, **20**, 209.

Acanthocephala from the eel. *Transactions of the American Microscopical Society*, **40**, 1–13.

1922 A compendium of the hosts of animal parasites contained in Ward and Whipple's Fresh Water Biology. *Transactions of the American Microscopical Society*, **40**, 195–9.

1923 Notes on Acanthocephala from Japan. (abstract) *Anatomical Record*, **24**, 371.

Telosentis, a new genus of Acanthocephala from southern Europe. *Journal of Parasitology*, **9**, 174–5.

Acanthocephala from the fishes of Oneida Lake, New York. *Roosevelt Wild Life Bulletin*, **2**, 73–84.

Acanthocephala from North American mammals. (abstract) *Anatomical Record*, **26**, 355.

A key to the genera of Acanthocephala. *Transactions of the American Microscopical Society*, **42**, 184–91.

1924 Notes on the relationships of Acanthocephala. (abstract) *Anatomical Record*, **29**, 119–20.

A critical study of the Acanthocephala described and identified by Joseph Leidy. *Proceedings of the Academy of Natural Sciences of Philadelphia*, **76**, 279–334.

Additional notes on the Acanthocephala from America described by J.E. Kaiser (1893). *Zentralblatt für Bakteriologie, Parasitenkunde, Infektionskrankheiten und Hygiene*, I, **94**, 57–60.

1925 Acanthocephala from Japan. *Parasitology*, **17**, 149–56.

1927 Morphological changes in the subcuticular nuclei of the Acanthocephala. (abstract) *Anatomical Record*, **37**, 155–6.

An analysis of variability in hook measurements in the Acanthocephala. (abstract) *Journal of Parasitology*, **14**, 132.

1928 Acanthocephala from China. I. New species and new genera from Chinese fishes. *Parasitology*, **20**, 1–9.

Two new genera and species of Acanthocephala from fishes of India. *Record of the Indian Museum*, **30**, 147–9.

Nuclei of the subcuticula in the Acanthocephala. *Zeitschrift für Zellforschung und Mikroskopische Anatomie*, **7**, 109–13.

1929 A new genus and new species of Acanthocephala from the Antarctic. *Annals and Magazine of Natural History*, **4**, 229–31.

1931 Acanthocephala from Japan. II. Two new species of the genus *Acanthocephalus*. *Annotationes Zoologici Japonenses*, **13**, 33–7.

Heterosentis, a new genus of Acanthocephala. *Zoologischer Anzeiger*, **93**, 144–6.

Acanthocephala in North American amphibia. II. A new species of the genus *Acanthocephalus*. *Transactions of the American Microscopical Society*, **50**, 46–7.

New Acanthocephala from fishes of Mississippi and a taxonomic reconsideration of forms with unusual numbers of cement glands. *Transactions of the American Microscopical Society*, **50**, 348–63.

1932 Eutely or cell constancy in its relation to body size. *Quarterly Review of Biology*, **7**, 59–67.

1934 Observations on the status of certain genera of Acanthocephala, chiefly from birds. (abstract) *Journal of Parasitology*, **20**, 324.

Parasites of Oneida Lake fishes. III. A biological and ecological survey of the worm parasites (with Justus F. Mueller). *Roosevelt Wild Life Annual*, 3, 161–334.

1935 The larval stages of Acanthocephala. (abstract) *Journal of Parasitology*, **21**, 435–6.

1936 A tentative survey of the classification of the Acanthocephala (multigraphed).

Tenuisentis a new genus of Acanthocephala, and its taxonomic position. *Parasitology*, **28**, 446–51.

The recognition of a new order in the Acanthocephala. *Journal of Parasitology*, **22**, 202–6.

Acanthocephala from amphibians and reptiles of China. (abstract) *Journal of Parasitology*, **22**, 530.

On the assignment of *Echinorhynchus dirus* to the genus *Acanthocephalus*. *Proceedings of the Helminthological Society of Washington*, **3**, 63 (with Lee H. Townsend).

1937 Acanthocephala of the genus *Corynosoma* from birds of Dyer Island, South Africa. *Goleborgs Kungl Vetershapsoch Vitterhets – Sanhalles Handlinjar*, (*B*), **5**, 1–6.

Developmental stages in acanthocephalan life histories. In *Papers in Helminthology Published in Commemoration of the 30 Year Jubileum of the Scientific, Educational, and Social Activities of the Honoured Worker of Science K.I. Skrjabin, M.Ac.Sci. and of the Fifteenth Anniversary of the All-Union Institute of Helminthology*, pp. 739–43. Moscow: The All-Union Lenin Academy of Agricultural Science.

Acanthocephala from China. II. Two new species of the genus *Acanthocephalus* from Amphibia. *Parasitology*, **29**, 395–8.

Status of a generic name *Profilicollis* of A. Meyer. (abstract) *Journal of Parasitology*, **23**, 563.

1938 Variability in hook measurements in the Acanthocephala. *Journal of Parasitology*, **24**, 25.

1939 A new species of the acanthocephalan genus *Polymorphus* and notes on the status of the name *Profilicollis*. *Journal of Parasitology*, **25**, 129–31.

An analysis of hook measurements in the Acanthocephala. *Vol. Jubilar por Prof. Sadas Yoshido*, **II**, 331–7.

The phylogenetic relations of the Acanthocephala. (abstract) *Abstracts of Communications, Third International Congress of Microbiology, New York, September 2–9, 1939*, 168.

Systems for designating hook patterns in the Acanthocephala. (abstract) *Journal of Parasitology*, **25**, 10.

A reconsideration of the acanthocephalan family Rhadinorhynchidae. (abstract) *Journal of Parasitology*, **25** (suppl.), 1 (with D.R. Lincicome).

On a new genus and species of Rhadinorhynchidae (Acanthocephala). *Parasitology*, **31**, 413–16 (with D.R. Lincicome).

1940 Relationships of the Acanthocephala. (abstract) *Report of Proceedings, Third International Congress of Microbiology, New York, September 2–9, 1939*, 431–53.

Some comparisons of Acanthocephala of marine fishes of the Atlantic and Pacific coasts. (abstract) *Journal of Parasitology*, **26**, 40–1.

The Acanthocephala collected by the Allan Hancock Pacific Expedition. *Allan Hancock Foundation Publications*, series 1, **2**, 501–27.

A reconsideration of the acanthocephalan family Rhadinorhynchidae. *Journal of Parasitology*, **26**, 75–81 (with David R. Lincicome).

A new species of the genus *Centrorhynchus* (Acanthocephala) from the barred owl. *Journal of Parasitology*, **26**, 297–9 (with Edna M. Pratt).

The Acanthocephala of wild ducks in central Illinois, with descriptions of two new species. *Transactions of the American Microscopical Society*, **59**, 348–53 (with William C. Starrett).

1941 Relationships of the Acanthocephala. *American Naturalist*, **75**, 31–47.

Hook patterns on the acanthocephalan proboscis. *Quarterly Review of Biology*, **16**, 157–72.

1942 A reconsideration of *Plagiorhynchus formosus* and observations on Acanthocephala with atypical lemnisci. *Transactions of the American Microscopical Society*, **61**, 206–10.

1944 Physiological responses of *Neoechinorhynchus emydis* (Acanthocephala) to various solutions. *Journal of Parasitology*, **30**, 369–72 (with Elizabeth L. Ross).

1945 A new species of the acanthocephalan genus *Illiosentis* (Rhadinorhynchidae). *Journal of Parasitology*, **31**, 57–60.

A new species of the acanthocephalan genus *Polymorphus* from the American coot. *Journal of Parasitology*, **31**, 128–30.

The genital vestibule and its significance in the morphology and taxonomy of the Acanthocephala, with particular reference to the genus *Corynosoma*. *Journal of Morphology* **77**, 299–315.

The status of the acanthocephalan genus *Arhythmorhynchus*, with particular reference to the validity of *A. brevis*. *Transactions of the American Microscopical Society*, **64**, 133–7.

The Acanthocephalan genus *Corynosoma*. I. The species found in water birds of North America. *Journal of Parasitology*, **31**, 332–40.

1946 Acanthocephala. In *Encyclopedia Britannica*.

Names for the immature stages of the Acanthocephala. (abstract) *Anatomical Record*, **96**, 20.

A review of the influences of bird migration upon avian Acanthocephala. *Anatomical Record*, **96**, 20.

Remarques sur le genre *Moniliformis* (Acanthocephales) et particulierement sur les espèces parasites des rats. *Annales de Parasitologie Humaine et Comparée*, **21**, 142–7.

1947 The Eoacanthocephala of North America, including the description of *Eocollis arcanus*, new genus and new species, superficially resembling the genus *Pomphorhynchus*. *Journal of Parasitology*, **33**, 285–96.

The acanthocephalan genus *Mediorhynchus*, its history and a review of the species occurring in the United States. *Journal of Parasitology*, **33**, 297–315.

An alphabetical index to the generic names of hosts of Acanthocephala of the world included in Anton Meyer's monograph. *American Midland Naturalist*, **38**, 417–26.

Travassosia tumida n. sp., first record of the occurrence of this acanthocephalan genus in North America. *American Midland Naturalist*, **38**, 427–33.

Thorny-headed worms (Acanthocephala) as potential parasites of poultry. *Proceedings of the Helminthological Society of Washington*, **14**, 55–8.

A critical review of terminology for immature stages in acanthocephalan life histories. *Journal of Parasitology*, **33**, 118–25.

On the occurrence of the acanthocephalan genus *Telosentis* in North America. *Journal of Parasitology*, **33**, 126–33.

Analysis of distinction between the acanthocephalan genera *Filicollis* and *Polymorphus*, with description of a new species of *Polymorphus*. *Transactions of the American Microscopical Society*, **66**, 302–13.

1948 Expanding horizons in the recognition of a phylum. *Journal of Parasitology*, **34**, 1–20.

A detailed study of the cement glands in males of the Acanthocephala. (abstract) *Journal of Parasitology*, **34**, 20.

Morphology of *Neoechinorhynchus emydis*, a typical representative of the Eoacanthocephala. I. The praesoma. (abstract) *Anatomical Record*, **101**, 726 (with Wilbur Bullock).

A new species of the acanthocephalan genus *Filisoma* from the dry Tortugas, Florida. *Anatomical Record*, **33**, 487–90 (with Harold W. Manter).

1949 An instance of duplication of the cement glands in an acanthocephalan. *Proceedings of the Helminthological Society of Washington*, **16**, 35–6.

Morphological and phylogenetic interpretation of the cement glands in the Acanthocephala. *Journal of Morphology*, **84**, 427–57.

Pseudoporrorchis teliger, a new species of Acanthocephala from Java. *Parasitology*, **39**, 214–17.

The acanthocephalan genus *Neoechinorhynchus* in the catostomid fishes of North America, with descriptions of two new species. *Journal of Parasitology*, **35**, 500–12.

Review and redescription of the acanthocephalan species *Leptorhynchoides thecatus*. *Transactions of the American Microscopical Society*, **68**, 304–13 (with D.R. Lincicome).

Distribution of *Leptorhynchoides thecatus*, a common acanthocephalan parasitic in fishes. *American Midland Naturalist*, **41**, 421–31 (with D.R. Lincicome).

Preliminary report on the circumpolar distribution of *Neoechinorhynchus rutili* (Acanthocephala) in fresh-water-fishes. *Science*, **109**, 446 (with J.E. Lynch).

1950 Four new species of the acanthocephalan family Neoechinorhynchidae from fresh water fishes of North America, one representing a new genus. *Journal of the Washington Academy of Sciences*, **39**, 398–409 (with R.V. Bangham).

Morphology of *Neoechinorhynchus emydis*, a typical representative of the Eoacanthocephala, I. The praesoma. *Transactions of the American Microscopical Society*, **69**, 288–308 (with W.L. Bullock).

A new species of the acanthocephalan genus *Octospinifer* from California. *Journal of Parasitology*, **36**, 169–73 (with E.C. Haderlie).

The circumpolar distribution of *Neoechinorhynchus rutili*, an acanthocephalan parasite of fresh-water-fishes. *Transactions of the American Microscopical Society*, **69**, 156–71 (with J.E. Lynch).

A new species of the acanthocephalan genus *Arhythmorhynchus* from sandpipers of Alaska. *Journal of Parasitology*, **36**, 278–83 (with R.L. Rausch).

1951 Giant nuclei in the subcuticula of the thorny-headed worm of the hog (*Macracanthorhynchus hirudinaceus*). *Transactions of the American Microscopical Society*, **70**, 37–46.

Speciation and formation of genera in the Acanthocephala. (Program and abstracts.) *Anatomical Record*, **111**, 525–6.

The Acanthocephala of eider ducks. *Proceedings of the Helminthological Society of Washington*, **18**, 81–4 (with R.L. Rausch).

Acanthocephala of passerine birds in Alaska. *Journal of Parasitology*, **37**, 151–9 (with R.B. Williams).

1952 Some host–parasite relationships of the Acanthocephala with special

reference to the organs of attachment. *Experimental Parasitology*, **1**, 305–30.
Acanthocephalan nomenclature introduced by Lauro Travassos. *Proceedings of the Helminthological Society of Washington*, **19**, 1–8.
An additional new species of the acanthocephalan genus *Neoechinorhynchus*. *Journal of Parasitology*, **38**, 53–6 (with H.F. Timmons).
1953 Acanthocephala of North American mammals. *Illinois Biological Monographs*, **23**, 1–179.
A preliminary analysis of the acanthocephalan genus *Corynosoma* in mammals of North America. *Journal of Parasitology*, **39**, 1–13.

4

Classification

Omar M. Amin

The first recognizable account of worms having proboscides armed with hooks was made by Redi (1684) according to Lühe (1904). Later, van Leeuwenhoek (1695) described and illustrated two kinds of acanthocephalans from the intestine of an eel. Leeuwenhoek's first record included a drawing of a proboscis which was regarded by Meyer (1932) as that of *Acanthocephalus anguillae* and by Lühe (1904) as *Acanthocephalus lucii*. Formal binomial names were not used by Leeuwenhoek and earlier observers of acanthocephalans, although binomial names were included in the work of Le Clerc (1715) before the tenth edition of Linnaeus' *Systema Naturae*. The existence of distinct genera and species of Acanthocephala was not recognized until the thirteenth edition of *Systema Naturae*, edited by Gmelin between 1788 and 1793, when the generic name, *Echinorhynchus*, appeared. This name was first proposed in 1776 by Zoega & Müller (in Müller (1776)) for a form (from a fish) similar to the one previously described by Koelreuther (1771) as *Acanthocephalus*. In fact, Koelreuther was probably the first worker to realize that the Acanthocephala were a distinct group of helminths which he named the 'Acanthocephali'. During this period, Pallas, in 1760, 1766 and 1781, described parasitic worms from fish, frogs and mammals, that proved to be acanthocephalans. These included *Macracanthorhynchus hirudinaceus* which he originally named *Taenia hirudinacea* in 1781.

The turn of the century witnessed intensive studies of parasitic worms including the Acanthocephala. Goeze (1782), Zeder (1800), Schrank (1782), Tilesius (1810) and Rudolphi (in publications between 1793 and 1819) described many new species of Acanthocephala and attempted to systematize them. Of the 52 species of Acanthocephala treated by Bremser (1811), 31 were described as 'new'. Bremser classed as a species each worm found in a different host species. All but two turned out later to be synonyms. In 1803, Zeder gave these worms the common name '*Hakenwurmer*' (hooked worms). Rudolphi (1802, 1809) renamed them the 'Acanthocephala' (Greek: *akantho* = spiny; *kephala* = head) and visualized this taxon having ordinal rank with one genus, *Echinorhynchus*. Apparently accepting Linnaeus' allocation of the Vermes as a class, Rudolphi (1808) regarded the 'Vermes acanthocephala' as an order coordinate with 'Vermes trematoda', 'Vermes cestoda' and 'Vermes nematoda', but did not discuss phylogenetic relationships.

In 1821, Westrumb described many species of Acanthocephala, some of which were synonymous. His basic taxonomic characters included the shape and armature of the proboscis, which he considered most important, the length of the neck, and the size, shape and colour of the body. Following Westrumb's system, Diesing (1851) described more species of Acanthocephala and united them and the Gregarinida in the Aprocta which he placed together with the Sipunculoidea.

Most early taxonomic descriptions lacked morphological detail necessary to recognize the species from the original description alone; host and locality information furnished the chief basis for recognition. Lühe's (1904–1905) critical review of early specific descriptions was an important contribution to acanthocephalan taxonomy. Later studies were greatly aided by the foundations laid by Cloquet (1824), Burow (1836), Siebold (1836), Henle (1840), Leuckart (1848, 1862, 1873, 1876), Wagener (1858), Weinland (1856), Pagenstecher (1859, 1863), Greeff (1864), Schneider (1868, 1871), Yarzhinsky (1868–1870), Zalenskii (1870), Linstow (1872), Andres (1878), Baltzer (1880), Saefftigen (1884), Knupffer (1888), Hamann (1889, 1891 *a*, *b*, 1892, 1895), Kaiser (1887, 1891, 1892, 1893, 1913), Bieler (1913), Rauther (1930) and Kilian (1932).

Hamann perceived diversity within the previously accepted uniformity in structure and was the first to propose the fragmentation of the old genus *Echinorhynchus* into three families based on consistent differences in the construction of the proboscis and its receptacle, and the structure of the body wall. His three families (Echinorhynchidae, Gigantorhynchidae, Neorhynchidae) formed the conceptual basis on which the more recent classification of the Acanthocephala was founded. The three major subdivisions, first regarded as orders by Meyer (1931 *b*) and Van Cleave (1936), are essentially Hamann's families elevated.

Uncertainty about the position of the Acanthocephala among the animal phyla was not easily resolved. Leuckart (1848) speculated that common lines of descent ran between the cestodes and Acanthocephala and considered them as the two orders of his class Anenteraeti. Vogt's (1851) Nematelmia, in which he included the gregarines, nematodes, acanthocephalans and gordiacians, was altered to Nemathelminthes by Gegenbauer (1859). In 1891, Hamann drew attention to certain similarities between acanthocephalan and cestode embryos. Cholodkovsky (1897) was first to suggest a close relationship between Acanthocephala and Cestoda and this view was supported by Skrjabin & Shults (1931), Petrochenko (1952) and Van Cleave (1941*b*). Meyer (1932, 1933), however, assembled the Rotifera, Gastrotricha, Kinorhyncha, Priapuloidea, Acanthocephala, Nematomorpha and Nematoda into the Aschelminthes. Van Cleave (1941*b*, 1948) concluded that the Acanthocephala had much more in common with Cestoda than with Nematoda and proposed recognition of the Acanthocephala as an independent phylum.

Systematics of the Acanthocephala had been evolving steadily since Hamann's contributions (1891*a*, 1892, 1895). Lühe (1911) had established new genera and recognized the armature of the body, shape and armature of the proboscis, structure of the proboscis receptacle, shape and size of the eggs, arrangement of the testes, shape and number of the cement glands, position of the cerebral ganglion, size of the lemnisci, and distribution of the lacunar canals and nuclei in the body wall, as taxonomic characteristics. Southwell & MacFie (1925) considered the Acanthocephala to be an order with three suborders in the phylum Nemathelminthes. Each suborder was essentially one of Hamann's (1892) families expanded to accommodate a large number of genera and families. They recognized the importance of the proboscis receptacle, cement glands and subcuticular nuclei, but did not always apply these characters critically at the family and suborder levels. Their three suborders were later elevated by Faust (1929) to ordinal rank with new names proposed for each.

In 1926, Travassos produced a taxonomic scheme which included four families (Neoechinorhynchidae with six genera in two subfamilies, Neoechinorhynchinae and Quadrigyrinae; Echinorhynchidae with 12 genera in two subfamilies, Echinorhynchinae and Centrorhynchinae; Rhadinorhynchidae with six genera; and Gigantorhynchidae with 11 genera in three subfamilies, Gigantorhynchinae, Moniliforminae, and Prosthenorchinae). His system was, however, based on insufficiently constant characters, resulting in the misplacement of certain genera. One year later, Thapar (1927) proposed another scheme which included three orders (Echinorhynchidea with 15 genera in two families, Echinorhynchidae and Oligacanthorhynchidae; Acanthogyridea with 15 genera in two families, Acanthogyridae and Gigantorhynchidae; and Apororhynchidea with one genus in one family, Apororhynchidae). Thapar's scheme was not generally accepted because it misinterpreted trunk spination and was still open to some of the objections to Travassos' (1926) classification. Like Thapar,

Witenberg (1932*a*, *b*) recognized three orders, two of which he accredited to Southwell & MacFie while accepting the third from Thapar. Meyer (1931*b*, 1932, 1933) proposed a scheme in which the Acanthocephala were regarded as a class of the Aschelminthes and were divided into two orders, Palaeacanthocephala and Archiacanthocephala, on the basis of morphology and ontogeny. Meyer attached importance not only to the lacunar system, the presence or absence of protonephridia, the arrangement of proboscis hooks, and the nature of the female ligament sacs, but also to those features which are now described as epizootiology. He considered the position of the main longitudinal vessels of the lacunar system to be of prime importance; the Archiacanthocephala having either dorsal and ventral or dorsal vessels and the Palaeacanthocephala having a pair of lateral vessels. In 1932, Meyer divided the 12 families and 58 genera that had been reported up to that time into his two new orders.

Meyer's system required exceptions to accommodate some genera and families. Van Cleave (1936) removed some of the inconsistencies in Meyer's system by the establishment of a third order, the Eoacanthocephala, on the basis of giant subcuticular nuclei, a single syncytial cement gland and a single-walled proboscis receptacle. Van Cleave (1948) further divided the phylum into two classes and four orders: Class Metacanthocephala with the orders Palaeacanthocephala and Archiacanthocephala, and Class Eoacanthocephala with the orders Gyracanthocephala and Neoacanthocephala.

Petrochenko (1956) devised a system based heavily on acanthor spination in which the Acanthocephala are divided into three subclasses, Neoechinorhynchinea, Echinorhynchinea and Gigantorhynchinea. Like Meyer (1933) and Van Cleave (1941*b*), Petrochenko regarded the Neoechinorhynchinea as the most primitive group; he based this view on the absence of acanthor spines, the small proboscis, with few hooks, the single-walled proboscis receptacle, the syncytial cement gland, giant subcuticular nuclei and more primitive intermediate hosts.

Attention was drawn back to the Meyer–Van Cleave system by Golvan (1959*a*, 1960, 1961, 1962, 1969) who considered the Eoacanthocephala, Palaeacanthocephala and Archiacanthocephala to be classes. Golvan may be criticized for placing too much importance on the number of cement glands and for inconsistent use of trunk spination in his classification. During this period, Yamaguti (1963) arranged the Acanthocephala into four orders. These were the Neoechinorhynchidea, the Echinorhynchidea and the Gigantorhynchidea, which are essentially Petrochenko's subclasses, and the new order Apororhynchidea. The first three of these orders roughly correspond to the Eoacanthocephala, Palaeacanthocephala and Archiacanthocephala of the Meyer–Van Cleave system.

The characterization of the Meyer–Van Cleave concept taxa, Eoacanthocephala, Palaeacanthocephala and Archiacanthocephala, was expanded by Bullock (1969) to include such features as body size, host habitat, distribution of main longitudinal vessels of the lacunar system, number of cement glands and ligament sacs, nature of subcuticular nuclei, proboscis receptacle, embryonic membranes, acanthor spination, presence or absence of protonephridia, trunk spination, and nature of intermediate hosts. This arrangement is now widely accepted and forms the basis for the following classification of the Phylum Acanthocephala.

CLASS Archiacanthocephala Meyer, 1931
Order Apororhynchida Thapar, 1927
Family Apororhynchidae Shipley, 1899
Genus ***Apororhynchus*** Shipley, 1899 (=*Arhynchus* Shipley, 1896; *Neorhynchus* Marval, 1905)
Species ***A. hemignathi*** (Shipley, 1896) Shipley, 1899 (type) [=*Arhynchus hemignathi* Shipley, 1896; *Neorhynchus hemignathi* (Shipley, 1896) Marval, 1905]
A. aculeatum Meyer, 1931
A. amphistomi Byrd & Denton, 1949
A. bivoluerus Das, 1952
A. chauhani Sen, 1975
A. paulonucleatus Hoklova & Cimbaluk, 1971
A. silesiacus Okulewicz & Maruszewski, 1980
Order Gigantorhynchida Southwell & MacFie, 1925
Family Gigantorhynchidae Hamann, 1892 (=Leiperacanthidae Bhalerao, 1937)
Genus ***Gigantorhynchus*** Hamann, 1892
Species ***G. echinodiscus*** (Diesing, 1851) Hamann, 1892 (type) (=*Echinorhynchus echinodiscus* Diesing, 1851)
G. lopezneyrai Diaz-Ungria, 1958
G. lutzi Machado, 1941
G. ortizi Sarmiento, 1954
G. pesteri Tadros, 1966
G. ungriai Antonio, 1958
Genus ***Mediorhynchus*** Van Cleave, 1916 (=*Disteganius* Lehmann, 1953, nom. nud.; *Empodius* Travassos, 1916; *Empodisma* Yamaguti, 1963; *Heteroplus* Kostylev, 1914; *Heteracanthorhynchus* Lundström, 1942; *Leiperacanthus* Bhalerao, 1937; *Micracanthorhynchus* Travassos, 1917)

Species ***M. papillosus*** Van Cleave, 1916 (type) (= *Mediorhynchus colini*, Webster, 1948; *M. bakeri* Byrd & Kellogg, 1971)

M. alecturae (Johnston & Edmonds, 1947) Byrd & Kellogg, 1971 (= *Empodius alecturae* Johnston & Edmonds, 1947)

M. cambellensis Soota, Srivastava & Ghosh, 1969

M. centurorum Nickol, 1969

M. conirostris Ward, 1966

M. corcoracis Johnston & Edmonds, 1950

M. edmondsi Schmidt & Kuntz, 1977

M. emberizae (Rudolphi, 1819) Van Cleave, 1916 (= *Echinorhynchus emberizae* Rudolphi, 1819)

M. empodius (Skrjabin, 1913) Van Cleave, 1924 [= *Gigantorhynchus empodius* Skrjabin, 1913; *Empodius empodius* (Skrjabin, 1913) Travassos, 1916]

M. gallinarum (Bhalerao, 1937) Van Cleave, 1947 (= *Leiperacanthus gallinarum* Bhalerao, 1937; *M. selengensis* Harris, 1973)

M. giganteus Meyer, 1931 [= *Empodius giganteus* (Meyer, 1931) Meyer, 1932; *Empodisma giganteus* (Meyer, 1931) Yamaguti, 1963]

M. grandis Van Cleave, 1916 [= *Heteroplus grandis* (Van Cleave, 1916) Van Cleave, 1918]

M. indicus George, Nadakal, Vijayakumaran & Rajendran, 1981

M. kuntzi Ward, 1960

M. lagodekhiensis Kuraschvili, 1955

M. leptis Ward, 1966

M. lophurae Wang, 1966

M. mattei Marchand & Vassiliades, 1982

M. meiringi Bisseru, 1960

M. micracanthus (Rudolphi, 1819) Meyer, 1932 [= *Echinorhynchus alaudae* Rudolphi, 1819; *E. carrucioi* Condorelli, 1897; *Micracanthorhynchus micracanthus* (Rudolphi, 1819) Travassos, 1917; *M. armenicus* Petrochenko, 1958]

M. mirabilis (Marval, 1905) Travassos, 1924 (= *Gigantorhynchus mirabilis* Marval, 1905)

M. muritensis Lundström, 1942

M. najasthanensis Gupta, 1976

M. numidae (Baer, 1925) Meyer, 1932 [= *Heteroplus numidae* Baer, 1925; *Empodisma numidae* (Baer, 1925) Yamaguti, 1963]

M. orientalis Belopolskaya, 1953 (= *Mediorhynchus bullocki* Gupta & Jain, 1973)

M. oswaldocruzi Travassos, 1923

M. otidis (Miescher, 1841) Van Cleave, 1947 [=*Echinorhynchus otidis* Miescher, 1841; *Heteroplus otidis* (Miescher, 1841) Kostylev, 1914; *Empodius otidis* (Miescher, 1841) Travassos, 1917; *Empodisma otidis* (Miescher, 1841) Yamaguti, 1963]

M. passeris Das, 1951

M. pauciuncinatus Dollfus, 1959

M. petrochenkoi Gvosdev & Soboleva, 1966

M. pintoi Travassos, 1923

M. rajasthanensis Gupta, 1976

M. robustus Van Cleave, 1916 (=*Mediorhynchus garruli* Yamaguti, 1939)

M. rodensis Cosin, 1971

M. sharmai Gupta & Lata, 1967

M. sipocotensis Tubangui, 1935

M. taeniatus (Linstow, 1901) Dollfus, 1936 (=*Echinorhynchus taeniatus* Linstow, 1901; *E. segmentatus* Marval, 1902)

M. tanagrae (Rudolphi, 1819) Travassos, 1921 (=*Echinorhynchus tanagrae* Rudolphi, 1819)

M. tenuis Meyer, 1931

M. textori Barus, Sixl & Majumdar, 1978

M. turnixena (Tubangui, 1931) Webster, 1948 (=*Empodius turnixena* Tubangui, 1931)

M. vaginatus (Diesing, 1851) Meyer, 1932 (=*Echinorhynchus vaginatum* Diesing, 1851)

M. vancleavei (Lundström, 1942) Golvan, 1962 (=*Heteracanthorhynchus vancleavei* Lundström, 1942)

M. wardi Schmidt & Canaris, 1967

M. zosteropis (Porta, 1913) Meyer, 1932 (=*Chentrosoma zosteropis* Porta, 1913)

Order Moniliformida Schmidt, 1972

Family Moniliformidae Van Cleave, 1924

Genus ***Moniliformis*** Travassos, 1915 (=*Echinorhynchus* Zoega in Müller, 1776, part.; *Gigantorhynchus* Hamann, 1892, part.; *Hormorhynchus* Ward, 1917)

Species ***M. moniliformis*** (Bremser, 1811) Travassos, 1915 (type) (=*Echinorhynchus moniliformis* Bremser, 1811; *E. grassi* Railliet, 1893; *E. canis* Porta, 1914; *E. belgicus* Railliet, 1919; *Moniliformis moniliformis siciliensis* Meyer in Petrochenko, 1958; *M. m. agypticus* Meyer in Petrochenko, 1958; *M. dubius* Meyer, 1932)

M. acomysi Ward & Nelson, 1967

M. cestodiformis (Linstow, 1904) Travassos, 1917 (=*Echinorhynchus cestodiformis* Linstow, 1904)

M. clarki (Ward, 1917) Van Cleave, 1924 (= *Hormorhynchus clarki* Ward, 1917)
M. convolutus Meyer, 1932
M. erinacei Southwell & MacFie, 1925
M. gracilis (Rudolphi, 1819) Meyer, 1931 (= *Echinorhynchus gracilis* Rudolphi, 1819)
M. kalahariensis Meyer, 1931
M. monechinus (Linstow, 1902) Petrochenko, 1958 (= *Echinorhynchus monechinus* Linstow, 1902)
M. semoni (Linstow, 1898) Johnston & Edmonds, 1952 (= *Echinorhynchus semoni* Linstow, 1898)
M. spiradentatis McLeod, 1933
M. spiralis Subrahmanian, 1927
M. travassosi Meyer, 1932

Moniliformidae *species inquirenda*
Echinorhynchus appendiculatus Westrumb, 1821 (= *E. soricis* Rudolphi, 1819)
E. myoxi Galli-Valerio, 1929
E. pseudosegmentatus Knupffer, 1888

Genus **Promoniliformis** Dollfus & Golvan, 1963
Species **P. ovocristatus** (Linstow, 1897) Dollfus & Golvan, 1963 (type) [= *Echinorhynchus ovocristatus* Linstow, 1897; *Moniliformis ovocristatus* (Linstow, 1897) Petrochenko, 1958; *Heteracanthorhynchus echinopsi* Horchner, 1962]

Order Oligacanthorhynchida Petrochenko, 1956
Family Oligacanthorhynchidae Southwell & MacFie, 1925 (= Oligacanthorhynchidae Meyer, 1931)
Genus **Macracanthorhynchus** Travassos, 1917 (= *Echinorhynchus* Zoega in Müller, 1776 part.; *Gigantorhynchus* Hamann, 1892, part.)
Species **M. hirudinaceus** (Pallas, 1781) Travassos, 1917 (type) [= *Taenia hirudinacea* Pallas, 1781; *Taenia haeruca* Pallas, 1776; *Echinorhynchus hirudinaceus* (Pallas, 1781); *Gigantorhynchus hirudinaceus* (Pallas, 1781) Hamann, 1892; *Echinorhynchus gigas* (Block, 1782) Meyer 1928; *Hormorhynchus gigas* (Block, 1782) Johnston, 1918]
M. catulinus Kostylev, 1927
M. ingens (Linstow, 1879) Meyer, 1932 [= *Echinorhynchus ingens* Linstow, 1879; *Prosthenorchis ingens* (Linstow, 1879) Travassos, 1917]

Genus **Neoncicola** Schmidt, 1972
Species **N. bursata** (Meyer, 1931) Schmidt, 1972 (type) (= *Oncicola bursata* Meyer, 1931)
N. avicola (Travassos, 1917) Schmidt, 1972 (= *Prosthenorchis avicola* Travassos, 1917)

N. curvata (Linstow, 1897) Schmidt, 1972 [= *Echinorhynchus curvatus* Linstow, 1897; *Prosthenorchis curvatus* (Linstow, 1897) Travassos, 1917]
N. novellae (Parona, 1890) Schmidt, 1972 [= *Echinorhynchus novellae* Parona, 1890; *Prosthenorchis novellae* (Parona, 1890) Travassos, 1917]
N. pintoi (Machado, 1950) Schmidt, 1972 (= *Prosthenorchis pintoi* Machado, 1950)
N. potosi (Machado, 1950) Schmidt, 1972 (= *Prosthenorchis potosi* Machado, 1950)
N. sinensis Schmidt & Dunn, 1974
N. skrjabini (Morosow, 1951) Schmidt, 1972 (= *Oncicola skrjabini* Morosow, 1951)

Genus ***Nephridiorhynchus*** Meyer, 1931 (= *Echinorhynchus* Zoega in Müller, 1776, part.; *Gigantorhynchus* Hamann, 1892 part.)

Species ***N. major*** (Bremser, 1811) Meyer, 1931 (type) [= *Echinorhynchus major* Bremser, 1811; *Gigantorhynchus major* (Bremser, 1811) Porta, 1908]
N. palawanensis Tubangui & Masiluñgan, 1938
N. thapari Sen & Chauhan, 1972

Genus ***Oligacanthorhynchus*** Travassos, 1915 (= *Echinorhynchus* Zoega in Müller, 1776, part.; *Gigantorhynchus* Hamann, 1892, part.; *Echinopardalis* Travassos, 1918; *Hamanniella* Travassos, 1915; *Nephridiacanthus* Meyer, 1931; *Pardalis* Travassos, 1917; *Travassosia*, Meyer, 1931)

Species ***O. spira*** (Diesing, 1851) Travassos, 1915 (type) [= *Echinorhynchus spira* Diesing, 1851; *E. oligacanthoides* Rudolphi, 1819, part.; *E. uromasticus* Fraipoint, 1882; *Gigantorhynchus aurae* (Travassos, 1912) Meyer, 1932]
O. aenigma (Reichensperger, 1922) Meyer, 1932 (= *Acanthocephalus aenigma* Reichensperger, 1922)
O. atrata (Meyer, 1931) Schmidt, 1972 (= *Echinopardalis atrata* Meyer, 1931)
O. bangalorensis (Pujatti, 1951) Schmidt, 1972 (= *Echinopardalis bangalorensis* Pujatti, 1951)
O. carinii (Travassos, 1917) Schmidt, 1972 [= *Hamanniella carinii* Travassos, 1917; *Travassosia carinii* (Travassos, 1917) Meyer, 1932]
O. cati (Gupta & Lata, 1967) Schmidt, 1972 (= *Hamanniella cati* Gupta & Lata, 1967)
O. circumflexus (Molin, 1858) Meyer, 1932 (= *Echinorhynchus circumflexus* Molin, 1858)
O. citilli (Rudolphi, 1806) Kostylev & Zmeev, 1939 (= *Echinorhynchus citilli* Rudolphi, 1806)

O. compressus (Rudolphi, 1802) Meyer, 1932 [= *Echinorhychus compressus* Rudolphi, 1802; *E. cornicis* Rudolphi, 1819; *E. macracanthus* Marval, 1902; *Gigantorhynchus compressus* (Rudolphi, 1802) Marval, 1905]

O. decrescens (Meyer, 1931) Schmidt, 1972 (= *Echinopardalis decrescens* Meyer, 1931)

O. erinacei (Rudolphi, 1793) Meyer, 1932 (= *Echinorhynchus erinacei* Rudolphi, 1793; *E. napaeformis* Rudolphi, 1802; *E. mustelae* Rudolphi, 1819; *E. kerkoides* Westrumb, 1821)

O. gerberi (Baer, 1959) Schmidt, 1972 (= *Nephridiacanthus gerberi* Baer, 1959)

O. hamatus (Linstow, 1897) Schmidt, 1972 [= *Echinorhynchus hamatus* Linstow, 1897; *Gigantorhynchus hamatus* (Linstow, 1897) Porta, 1908; *Nephridiacanthus hamatus* (Linstow, 1897) Meyer, 1932]

O. iheringi Travassos, 1917 (= *Echinorhynchus lagenaeformis* Diesing, 1851, part.)

O. kamerunensis (Meyer, 1931) Schmidt, 1972 (= *Nephridiacanthus kamerunensis* Meyer, 1931)

O. kamtschaticus Hokhlova, 1966

O. lagenaeformis (Westrumb, 1821) Travassos, 1917 (= *Echinorhynchus lagenaeformis* Westrumb, 1821; *E. falconis cyanei* Rudolphi, 1819)

O. lamasi (Freitas & Costa, 1964) Amato, Nickol & Froés, 1979 [= *Echinopardalis lamasi* Freitas & Costa, 1964; *Oncicola lamasi* (Freitas & Costa, 1964) Schmidt, 1977]

O. lerouxi (Bisseru, 1956) Schmidt, 1972 (= *Echinopardalis lerouxi* Bisseru, 1956)

O. longissimus (Golvan, 1962) Schmidt, 1972 (= *Nephridiacanthus longissimus* Golvan, 1962)

O. major (Machado, 1963) Schmidt, 1972 (= *Macracanthorhynchus major* Machado, 1963)

O. manifestus (Leidy, 1851) Van Cleave, 1924 (= *Echinorhynchus manifestus* Leidy, 1851)

O. manisensis (Meyer, 1931) Schmidt, 1972 (= *Nephridiacanthus manisensis* Meyer, 1931)

O. mariemily (Tadros, 1969) comb. n. (= *Echinopardalis mariemily* Tadros, 1969)

O. microcephala (Rudolphi, 1819) Schmidt, 1972 [= *Echinorhynchus microcephala* Rudolphi, 1819; *Hamanniella microcephala* (Rudolphi, 1819) Travassos, 1915]

O. minor Machado, 1964

O. oligacanthus (Rudolphi, 1819) Meyer, 1932 (= *Echinorhynchus oligacanthus* Rudolphi, 1819)

O. oti Machado, 1964

O. pardalis (Westrumb, 1821) Schmidt, 1972 [= *Echinorhynchus pardalis* Westrumb, 1821; *Pardalis pardalis* (Westrumb, 1821) Travassos, 1917; *Echinopardalis pardalis* (Westrumb, 1821) Travassos, 1918]

O. ricinoides (Rudolphi, 1808) Meyer, 1931 (= *Echinorhynchus ricinoides* Rudolphi, 1808; *E. charadriipluvialis* Rudolphi, 1819; *E. coraciae* Rudolphi, in Westrumb, 1821; *E. macracanthus* Bremser in Westrumb, 1821)

O. shillongensis (Sen & Chauhan, 1972) comb. n. (= *Nephridiacanthus shillongensis* Sen & Chauhan, 1972)

O. taenioides (Diesing, 1851) Travassos, 1915 (= *Echinorhynchus oligacanthoides* Rudolphi, 1819, part.; *E. taenioides* Diesing, 1851)

O. thumbi Haffner, 1939

O. tortuosa (Leidy, 1850) Schmidt, 1972 [= *Echinorhynchus tortuosa* Leidy, 1850; *Hamanniella tortuosa* (Leidy, 1850) Van Cleave, 1924]

O. tumida (Van Cleave, 1947) Schmidt, 1972 [= *Travassosia tumida* Van Cleave, 1947; *Hamanniella tumida* (Van Cleave, 1947) Van Cleave, 1953]

Genus ***Oncicola*** Travassos, 1916

Species ***O. oncicola*** (Ihering, 1892) Travassos, 1916 (type)

O. campanulata (Diesing, 1851) Meyer, 1931 (= *Echinorhynchus campanulatus* Diesing, 1851; *E. ovatus* Leidy, 1850)

O. canis (Kaupp, 1909) Hall & Wigdor, 1918 (= *Echinorhynchus canis* Kaupp, 1909)

O. chibigouzouensis Machado, 1963

O. confusus (Machado, 1950) Schmidt, 1972 (= *Prosthenorchis confusus* Machado, 1950)

O. dimorpha Meyer, 1931

O. freitasi (Machado, 1950) Schmidt, 1972 (= *Prosthenorchis freitasi* Machado, 1950)

O. gigas Meyer, 1931

O. juxtatesticularis (Machado, 1950) Schmidt 1972 (= *Prosthenorchis juxtatesticularis* Machado, 1950)

O. luehei (Travassos, 1917) Schmidt, 1972 (= *Prosthenorchis luehei* Travassos, 1917)

O. machadoi Schmidt, 1972 (= *Prosthenorchis travassosi* Machado, 1950)

O. macrurae Meyer, 1931 [= *Echinopardalis macrurae* Meyer, 1931; Witenberg, 1938]

O. magalhaesi Machado, 1962

O. malayanus Toumanoff, 1947

O. martini Schmidt, 1977

O. michaelseni Meyer, 1932

O. micracantha Machado, 1949
O. paracampanulata Machado, 1963
O. pomatostomi (Johnston & Cleland, 1912) Schmidt, 1983 [= *Echinorhynchus pomatostomi* Johnston & Cleland, 1912; *Oligacanthorhynchus pomatostomi* (Johnston & Cleland, 1912) Tubangui, 1933]
O. schacheri Schmidt, 1972
O. sigmoides (Meyer, 1932) Schmidt, 1972 (= *Prosthenorchis sigmoides* Meyer, 1932)
O. spirula (Olfers in Rudolphi, 1819) Schmidt, 1972 [= *Echinorhynchus spirula* Olfers in Rudolphi, 1819; *Prosthenorchis spirula* (Olfers in Rudolphi, 1819) Travassos, 1917]
O. travassosi Witenberg, 1938
O. venezuelensis Marteau, 1977

Genus ***Pachysentis*** Meyer, 1931
Species ***P. canicola*** Meyer, 1931 (type)
P. angolensis (Golvan, 1957) Schmidt, 1972 (= *Oncicola angolensis* Golvan, 1957)
P. dollfusi (Machado, 1950) Schmidt, 1972 (= *Prosthenorchis dollfusi* Machado, 1950)
P. ehrenbergi Meyer, 1931
P. gethi (Machado, 1950) Schmidt, 1972 (= *Prosthenorchis gethi* Machado, 1950)
P. lenti (Machado, 1950) Schmidt, 1972 (= *Prosthenorchis lenti* Machado, 1950)
P. procumbens Meyer, 1931
P. procyonis (Machado, 1950) Schmidt, 1972 (= *Prosthenorchis procyonis* Machado, 1950)
P. rugosus (Machado, 1950) Schmidt, 1972 (= *Prosthenorchis rugosus* Machado, 1950)
P. septemserialis (Machado, 1950) Schmidt, 1972 (= *Prosthenorchis septemserialis* Machado, 1950)

Genus ***Prosthenorchis*** Travassos, 1915
Species ***P. elegans*** (Diesing, 1851) Travassos, 1915 (type) (= *Echinorhynchus elegans* Diesing, 1851)
P. fraterna (Baer, 1959) Schmidt, 1972 (= *Oncicola fraterna* Baer, 1959)
P. lemuri Machado, 1950

Genus ***Tchadorhynchus*** Troncy, 1970
Species ***T. quentini*** Troncy, 1970 (type)
Oligacanthorhynchidae *species inquirenda*
Echinorhynchus amphipacus Westrumb, 1821
E. depressus Nitzsch, 1866
E. hominis Leuckart, 1876
E. magretti Parona, 1885

E. ***pachyacanthus*** Sonsino, 1889
E. ***putorii*** Molin, 1858

CLASS Palaeacanthocephala Meyer, 1931
Order Echinorhynchida Southwell & MacFie, 1925
Family Arhythmacanthidae Yamaguti, 1935
Subfamily Arhythmacanthinae Yamaguti, 1935
Genus ***Heterosentis*** Van Cleave, 1931 (=*Arhythmacanthus* Yamaguti, 1935)
Species ***H. heteracanthus*** (Linstow, 1896) Van Cleave, 1931 (type) [=*Echinorhynchus heteracanthus* Linstow, 1896; *Aspersentis heteracanthus* (Linstow, 1896) Golvan, 1969]
H. fusiformis (Yamaguti, 1935) Tripathi, 1959 (=*Arhythmacanthus fusiformis* Yamaguti, 1935)
H. overstreeti (Schmidt & Paperna, 1978) comb. n. (=*Arhythmacanthus overstreeti* Schmidt & Paperna, 1978)
H. paraplagusiarum (Nickol, 1972) comb. n. (=*Arhythmacanthus paraplagusiarum* Nickol, 1972)
H. plotosi Yamaguti, 1935 [=*Arhythmacanthus plotosi* (Yamaguti, 1935) Schmidt & Paperna, 1978]
H. septacanthus (Sita in Golvan, 1969) comb. n. (=*Arhythmacanthus septacanthus* Sita in Golvan, 1969)
H. thapari (Gupta & Fatma, 1979) comb. n. (=*Arhythmacanthus thapari* Gupta & Fatma, 1979)
Subfamily Neoacanthocephaloidinae Golvan, 1960
Genus ***Acanthocephaloides*** Meyer, 1932
Species ***A. propinquus*** (Dujardin, 1845) Meyer, 1932 (type) (=*Echinorhynchus propinquus* Dujardin, 1845; *E. fabri* Rudolphi, 1819; *E. pumilio* Rudolphi, 1819; *E. kostylevi* Meyer, 1932)
A. distinctus Golvan, 1969
A. incrassatus (Molin, 1858) Meyer, 1932 (=*Echinorhynchus incrassatus* Molin, 1858; *E. devisiana* Molin, 1858; *E. flavus* Molin, 1858)
A. soleae (Porta, 1905) Meyer, 1932 (=*Echinorhynchus soleae* Porta, 1905)
Genus ***Neoacanthocephaloides*** Cable & Quick, 1954
Species ***N. spinicaudatus*** Cable & Quick, 1954 (type)
Subfamily Paracanthocephaloidinae Golvan, 1969
Genus ***Breizacanthus*** Golvan, 1969
Species ***B. chabaudi*** Golvan, 1969 (type)
B. irenae Golvan, 1969
B. ligur Paggi, Orecchia & Della Seta, 1975
Genus ***Euzetacanthus*** Golvan & Houin, 1964
Species ***E. simplex*** (Rudolphi, 1810) Golvan & Houin, 1964 (type)

(= *Echinorhynchus triglae gurnardi* Rathke, 1799; *E. simplex* Rudolphi, 1810)

Genus ***Paracanthocephaloides*** Golvan, 1969

Species ***P. chabanaudi*** (Dollfus, 1951) Golvan, 1969 (type) (= *Acanthocephaloides chabanaudi* Dollfus, 1951)

P. tripathii Golvan, 1969

Family Cavisomidae Meyer, 1932 (= Cavisomatidae Petrochenko, 1956)

Genus ***Cavisoma*** Van Cleave, 1931

Species ***C. magnum*** (Southwell, 1927) Van Cleave, 1931 (type) (= *Oligoterorhynchus magnus* Southwell, 1927)

Genus ***Echinorhynchoides*** Achmerov & Dombrowskaja-Achmerova, 1941

Species ***E. dogieli*** Achmerov & Dombrowskaja-Achmerova, 1941 (type)

Genus ***Caballerorhynchus*** Salgado-Maldonado, 1977

Species ***C. lamothei*** Salgado-Maldonado, 1977 (type)

Genus ***Femogibbosus*** Paruchin, 1973

Species ***F. assi*** Paruchin, 1973 (type)

Genus ***Filisoma*** Van Cleave, 1928

Species ***F. indicum*** Van Cleave, 1928 (type)

F. bucerium Van Cleave, 1940

F. fidum Van Cleave & Manter, 1948

F. hoogliensis Datta & Soota, 1962

F. micracanthi Harada, 1938

F. rizalinum Tubangui & Masilungan, 1946

F. scatophagusi Datta & Soota, 1962

Genus ***Megapriapus*** Golvan, Gracia-Rodrigo & Diaz-Ungria, 1964

Species ***M. ungriai*** (Gracia-Rodrigo, 1960) Golvan, Gracia-Rodrigo & Diaz-Ungria, 1964 (type) (= *Echinorhynchus ungriai* Gracia-Rodrigo, 1960)

Genus ***Neorhadinorhynchus*** Yamaguti, 1939 (= *Neogorgorhynchus* Golvan, 1960)

Species ***N. aspinosus*** (Fukui & Morisita, 1937) Yamaguti, 1939 (type) [= *Rhadinorhynchus aspinosus* Fukui & Morisita, 1937; *Neogorgorhynchus aspinosus* (Fukui & Morisita, 1937) Golvan, 1960; *Pararhadinorhynchus aspinosus* (Fukui & Morisita, 1937) Petrochenko, 1956]

N. atlanticus Gaevskaja & Nigmatullin, 1977

N. madagascariensis Golvan, 1969

N. nudus (Harada, 1938) Yamaguti, 1939 [= *Rhadinorhynchus nudus* Hara, 1938; *Neogorgorhynchus nudus* (Harada, 1938) Golvan, 1960; *Nipporhynchus nudus* (Harada, 1938) Van Cleave & Lincicome, 1940; *Echinorhynchus nudus* (Harada, 1938) Petrochenko, 1956]

Genus ***Paracavisoma*** Kritscher, 1957

Species ***P. impudica*** (Diesing, 1851) Kritscher, 1957 (type) (= *Echinorhynchus impudicus* Diesing, 1851)

Genus ***Pseudocavisoma*** Golvan & Houin, 1964

Species ***P. chromitidis*** (Cable & Quick, 1954) Golvan & Houin, 1964 (type) [= *Cavisoma chromitidis* Cable & Quick, 1954; *Rhadinorhynchoides chromitidis* (Cable & Quick, 1954) Yamaguti, 1963]

Genus ***Rhadinorhynchoides*** Fukui & Morisita, 1937

Species ***R. miyagawai*** Fukui & Morisita, 1937 (type)

Family Diplosentidae Tubangui & Masilungan, 1937

Subfamily Allorhadinorhynchinae Golvan, 1969

Genus ***Allorhadinorhynchus*** Yamaguti, 1959

Species ***A. segmentatus*** Yamaguti, 1959 (type)

Genus ***Golvanorhynchus*** Noronha, Fabio & Pinto, 1978

Species ***G. golvani*** Noronha, Fabio & Pinto, 1978 (type)

Subfamily Diplosentinae Golvan & Houin, 1963

Genus ***Diplosentis*** Tubangui & Masilungan, 1937

Species ***D. amphacanthi*** Tubangui & Masilungan, 1937 (type)

D. manteri Gupta & Fatma, 1979

Genus ***Pararhadinorhynchus*** Johnston & Edmonds, 1947

Species ***P. mugilis*** Johnston & Edmonds, 1947 (type)

P. coorongensis Edmonds, 1973

Family Echinorhynchidae Cobbold, 1876

Subfamily Echinorhynchinae Cobbold, 1876

Genus ***Acanthocephalus*** Koelreuther, 1771 (= *Paracanthocephalus* Achmerov & Dombrowskaja-Achmerova, 1941; *Pseudoechinorhynchus* Petrochenko, 1956)

Species ***A. anguillae*** (Müller, 1780) Lühe, 1911 (type) (= *Echinorhynchus anguillae* Müller, 1780; *E. globulosus* Rudolphi, 1802; *E. linstowi* Hamann, 1891; *E. proteus* Porta, 1905)

A. acutispinus Machado, 1968

A. acutulus Van Cleave, 1931

A. alabamensis Amin & Williams, 1983

A. anthuris (Dujardin, 1845) Lühe, 1911 (= *Echinorhynchus anthuris* Dujardin, 1845)

A. balkanicus Batchvarov, 1974

A. breviprostatus Kennedy, 1982

A. clavula (Dujardin, 1845) Grabda-Kazubska & Chubb, 1968 [= *Echinorhynchus clavula* Dujardin, 1845; *Pseudoechinorhynchus clavula* (Dujardin, 1845) Petrochenko, 1956]

A. correalimai Machado, 1970

A. criniae Snow, 1971

A. curtus (Achmerov & Dombrowskaja-Achmerova, 1941) Yamaguti, 1963 (= *Paracanthocephalus curtus* Achmerov &

Dombrowskaja-Achmerova, 1941; *Acanthocephalus amuriensis* Kostylev, 1941)

A. dirus (Van Cleave, 1931) Van Cleave & Townsend, 1936 (=*Echinorhynchus dirus* Van Cleave, 1931; *A. jacksoni* Bullock, 1962; *A. parksidei* Amin, 1975)

A. domerguei Golvan, Bygoo & Gassmann, 1972

A. echigoensis Fujita, 1920 (=*Acanthocephalus oncorhynchi* Fujita, 1921; *A. aculeatus* Van Cleave, 1931; *A. acerbus* Van Cleave, 1931)

A. elongatus Van Cleave, 1937

A. falcatus (Frölich, 1789) Lühe, 1911 (=*Echinorhynchus falcatus* Frölich, 1789)

A. fluviatilis Paperna, 1964

A. galaxii Hine, 1977

A. goaensis Jain & Gupta, 1981

A. gotoi Van Cleave, 1925

A. graciliacanthus Meyer, 1932 [=*Paracanthocephalus graciliacanthus* (Meyer, 1932) Grabda & Grabda-Kazubska, 1967]

A. haranti Golvan & Oliver in Golvan, 1969

A. hastae Baylis, 1944

A. japonicus (Fukui & Morisita, 1936) Petrochenko, 1956 [=*Filisoma japonicum* Fukui & Morisita, 1936; *Acanthocephaloides japonicus* (Fukui & Morisita, 1936) Yamaguti, 1939]

A. kabulensis Datta & Soota, 1956

A. kashmirensis Datta, 1936

A. lucidus Van Cleave, 1925 (=*Acanthocephalus aratus* Van Cleave, 1925)

A. lucii (Müller, 1776) Lühe, 1911 (=*Echinorhynchus lucii* Müller, 1776; *E. angustus* Rudolphi, 1809)

A. lutzi (Linstow, 1896) Meyer, 1931 (=*Echinorhynchus lutzi* Linstow, 1896)

A. madagascariensis Golvan, 1965

A. minor Yamaguti, 1935

A. nanus Van Cleave, 1925

A. opsariichthydis Yamaguti, 1935

A. parallelotestis Achmerov & Dombrowskaja-Achmerova, 1941

A. paronai (Condorelli, 1897) Meyer, 1932 (=*Echinorhynchus paronai* Condorelli, 1897)

A. pesteri Tadros, 1966

A. ranae (Schrank, 1788) Lühe, 1911 (=*Echinorhynchus ranae* Schrank, 1788; *E. haeruca* Rudolphi, 1809)

A. rauschi (Schmidt, 1969) comb. n. (= *Paracanthocephalus rauschi* Schmidt, 1969; *A. rauschi* Golvan, 1969)
A. serendibensis Crusz & Mills, 1970
A. srilankensis Crusz & Ching, 1976
A. tahlequahensis Oetinger & Buckner, 1976
A. tenuirostris (Achmerov & Dombrowskaja-Achmerova, 1941) Yamaguti, 1963 (= *Paracanthocephalus tenuirostris* Achmerov & Dombrowskaja-Achmerova, 1941)
A. tigrinae (Shipley, 1903) Yamaguti, 1963 (= *Echinorhynchus tigrinae* Shipley, 1903)
A. tumescens (Linstow, 1896) Porta, 1905 (= *Echinorhynchus tumescens* Linstow, 1896)

Genus ***Echinorhynchus*** Zoega in Müller, 1776 (= *Metechinorhynchus* Petrochenko, 1956)

Species ***E. gadi*** Zoega in Müller, 1776 (type) (= *Echinorhynchus candidus* Zoega in Müller, 1776; *E. ekbaumi* Golvan, 1969; *E. lineolatus* Müller, 1777; *E. lophii* Gmelin, 1791; *E. gadicallariae* Viborg, 1795; *E. gadivirentis* Rathke, 1799; *E. acus* Rudolphi, 1802; *E. wachniae* Rudolphi, 1819; *E. socialis* Leidy, 1851; *E. hepaticola* Linstow, 1901; *E. arcticus* Linstow, 1901; *E. vancleavei* Golvan, 1969)
E. abyssicola Dollfus, 1931
E. armoricanus Golvan, 1969
E. attenuatus Linton, 1891
E. baeri Kostylev, 1928 [= *Echinorhynchus sevangi* Dinnik, 1933; *Metechinorhynchus baeri* (Kostylev, 1928) Petrochenko, 1956]
E. briconi Machado, 1959 [= *Metechinorhynchus briconi* (Machado, 1959) Golvan, 1969]
E. calloti Golvan, 1969
E. canyonensis Huffman & Kleiver, 1977
E. cestodicola Linstow, 1905
E. chierchiae Monticelli, 1889
E. cinctulus Porta, 1905 [= *Pseudoechinorhynchus cinctulus* (Porta, 1905) Petrochenko, 1956]
E. cotti Yamaguti, 1939
E. cryophilus (Sokolowskaja, 1962) comb. n. (= *Metechinorhynchus cryophilus* Sokolowskaja, 1962)
E. debenhami Leiper & Atkinson, 1914 [= *Leptorhynchoides debenhami* (Leiper & Atkinson, 1914) Golvan, 1961]
E. dissimilis Yamaguti, 1939
E. gomesi Machado, 1948 [= *Metechinorhynchus gomesi* (Machado, 1948) Golvan, 1969]
E. gracilis Machado, 1948

E. gymnocyprii Liu, Wang & Yang, 1981
E. hexagrammi Baeva, 1965
E. jucundus Travassos, 1923 [= *Metechinorhynchus jucundum* (Travassos, 1923) Petrochenko, 1956]
E. kushiroensis Fujita, 1921 [= *Metechinorhynchus kushiroensis* (Fujita, 1921) Golvan, 1969]
E. lageniformis Ekbaum, 1938 [= *Metechinorhynchus lageniformis* (Ekbaum, 1938) Petrochenko, 1956]
E. lateralis Leidy, 1851 [= *Acanthocephalus lateralis* (Leidy, 1851) Petrochenko, 1956; *Metechinorhynchus lateralis* (Leidy, 1851) Golvan, 1969]
E. laurentianus Ronald, 1957
E. leidyi Van Cleave, 1924 [= *Metechinorhynchus leidyi* (Van Cleave, 1924) Golvan, 1969]
E. lenoki Achmerov & Dombrowskaja-Achmerova, 1941 [= *Pseudoechinorhynchus lenoki* (Achmerov & Dombrowskaja-Achmerova, 1941) Petrochenko, 1956]
E. lotellae Yamaguti, 1939
E. melanoglaeae Dollfus, 1960
E. monticelli Porta, 1904 [= *Pseudoechinorhynchus monticelli* (Porta, 1904) Petrochenko, 1956]
E. oblitus Golvan, 1969
E. orientalis Kaw, 1951
E. paranensis Machado, 1959 [= *Metechinorhynchus paranensis* (Machado, 1959) Golvan, 1969]
E. parasiluri Fukui, 1929 [= *Pseudoechinorhynchus parasiluri* (Fukui, 1929) Petrochenko, 1956]
E. peleci Grimm, 1870
E. rhenanus (Golvan, 1969) comb. n. (= *Metechinorhynchus rhenanus* Golvan, 1969)
E. salmonis Müller, 1784 [= *E. alpinus* Linstow, 1901; *E. pachysomus* Creplin, 1839; *E. phoenix* Schneider, 1903; *E. coregoni* Linkins in Van Cleave, 1919; *E. murenae* Bosc, 1802; *Metechinorhynchus alpinus* (Linstow, 1901) Petrochenko, 1956; *M. salmonis* (Müller, 1784) Petrochenko, 1956]
E. salobrensis Machado, 1948 [= *Metechinorhynchus salobrensis* (Machado, 1948) Golvan, 1969]
E. serpentulus Grimm, 1870
E. truttae Schrank, 1788 [= *Metechinorhynchus truttae* (Schrank, 1788) Petrochenko, 1956]
E. yamagutii Golvan, 1969
E. zanclorhynchi Johnston & Best, 1937

Genus ***Pilum*** Williams, 1976
Species ***P. pilum*** Williams, 1976 (type)

Genus ***Pseudoacanthocephalus*** Petrochenko, 1956

Species ***P. bufonis*** (Shipley, 1903) Petrochenko, 1956 (type) [=*Echinorhynchus bufonis* Shipley, 1903; *Acanthocephalus bufonis* (Shipley, 1903) Southwell & MacFie, 1925; *A. sinensis* Van Cleave, 1937]

P. betsileo Golvan, Houin & Bygoo, 1969

P. bigueti (Houin, Golvan & Bygoo, 1965) Golvan, 1969 (=*Acanthocephalus bigueti* Houin, Golvan & Bygoo, 1965)

P. bufonicola (Kostylev, 1941) Petrochenko, 1956 (=*Acanthocephalus bufonicola* Kostylev, 1941)

P. caucasicus (Petrochenko, 1953) Petrochenko, 1956 (=*Acanthocephalus caucasicus* Petrochenko, 1953)

P. perthensis Edmonds, 1971

P. xenopeltidis (Shipley, 1903) Golvan, 1969 (=*Echinorhynchus xenopeltidis* Shipley, 1903)

Subfamily Yamagutisentinae Golvan, 1969

Genus ***Yamagutisentis*** Golvan, 1969

Species ***Y. rhinoplagusiae*** (Yamaguti, 1935) Golvan, 1969 (type) [=*Acanthocephaloides rhinoplagusiae* Yamaguti, 1935; *Neoacanthocephaloides rhinoplagusiae* (Yamaguti, 1935) Cable & Quick, 1954; *Pseudorhadinorhynchus rhinoplagusiae* (Yamaguti, 1935) Petrochenko, 1956]

Y. neobythitis (Yamaguti, 1939) Golvan, 1969 [=*Acanthocephaloides neobythitis* Yamaguti, 1939; *Neoacanthocephaloides neobythitis* (Yamaguti, 1939) Cable & Quick, 1954; *Pseudorhadinorhynchus neobythitis* (Yamaguti, 1939) Petrochenko, 1956]

Family Fessisentidae Van Cleave, 1931

Genus ***Fessisentis*** Van Cleave, 1931

Species ***F. fessus*** Van Cleave, 1931 (type)

F. friedi Nickol, 1972 (=*Fessisentis vancleavei* Haley & Bullock, 1953)

F. necturorum Nickol, 1967

F. tichiganensis Amin, 1980

F. vancleavei (Hughes & Moore, 1943) Nickol, 1972 (=*Acanthocephalus vancleavei* Hughes & Moore, 1943)

Family Heteracanthocephalidae Petrochenko, 1956

Subfamily Aspersentinae Golvan, 1960

Genus ***Aspersentis*** Van Cleave, 1929

Species ***A. megarhynchus*** (Linstow, 1892) Golvan, 1960 (type) (=*Echinorhynchus megarhynchus* Linstow, 1892; *Aspersentis austrinus* Van Cleave, 1929; *Rhadinorhynchus wheeleri* Baylis, 1929)

A. johni (Baylis, 1929) Chandler, 1934 (=*Rhadinorhynchus johni* Baylis, 1929)

Subfamily Heteracanthocephalinae Petrochenko, 1956
Genus ***Heteracanthocephalus*** Petrochenko, 1956
H. peltorhamphi (Baylis, 1944) Petrochenko, 1956 (type) (= *Rhadinorhynchus peltorhamphi* Baylis, 1944)
H. hureaui Dollfus, 1964
Genus ***Sachalinorhynchus*** Krotov & Petrochenko, 1956
Species ***S. skrjabini*** Krotov & Petrochenko, 1956 (type)
Family Hypoechinorhynchidae Golvan, 1980
Genus ***Bolborhynchoides*** Achmerov, 1959 (= *Bolborhynchus* Achmerov & Dombrowskaja-Achmerova, 1941; *Fresnyarhynchus* Golvan, 1960)
Species ***B. exiguus*** (Achmerov & Dombrowskaja-Achmerova, 1941) Achmerov, 1959 (type) [= *Bolborhynchus exiguus* Achmerov & Dombrowskaja-Achmerova, 1941; *Fresnyarhynchus exiguus* (Achmerov & Dombrowskaja-Achmerova, 1941) Golvan, 1960]
Genus ***Hypoechinorhynchus*** Yamaguti, 1939
Species ***H. alaeopis*** Yamaguti, 1939 (type)
H. magellanicus Szidat, 1950
Family Illiosentidae Golvan, 1960
Genus ***Brentisentis*** Leotta, Schmidt & Kuntz, 1982
Species ***B. uncinus*** Leotta, Schmidt & Kuntz, 1982 (type)
Genus ***Dentitruncus*** Sinzar, 1955
Species ***D. truttae*** Sinzar, 1955 (type)
Genus ***Dollfusentis*** Golvan, 1969
Species ***D. longispinus*** (Cable & Linderoth, 1963) Golvan, 1969 (type) (= *Illiosentis longispinus* Cable & Linderoth, 1963)
D. bravoae Salgado-Maldonado, 1976
D. chandleri Golvan, 1969 [= *Telosentis tenuicornis* (Linton, 1905) Van Cleave, 1947; *Echinorhynchus pristis tenuicornis* Linton, 1905; *Rhadinorhynchus tenuicornis* (Linton, 1905) Van Cleave, 1918]
D. ctenorhynchus (Cable & Linderoth, 1963) Golvan, 1969 (= *Illiosentis ctenorhynchus* Cable & Linderoth, 1963)
D. heteracanthus (Cable & Linderoth, 1963) Golvan, 1969 (= *Illiosentis heteracanthus* Cable & Linderoth, 1963)
Genus ***Goacanthus*** Gupta & Jain, 1980
Species ***G. panajiensis*** Gupta & Jain, 1980 (type)
Genus ***Indorhynchus*** Golvan, 1969
Species ***I. indicus*** (Tripathi, 1959) Golvan, 1969 (type) (= *Rhadinorhynchus indicus* Tripathi, 1959)
Genus ***Metarhadinorhynchus*** Yamaguti, 1959
Species ***M. lateolabracis*** Yamaguti, 1959 (type)
M. thapari Gupta & Gupta, 1975

Genus ***Pseudorhadinorhynchus*** Achmerov & Dombrowskaja-Achmerova, 1941 (= *Hemirhadinorhynchus* Krotov & Petrochenko, 1956)

Species ***P. markewitchi*** Achmerov & Dombrowskaja-Achmerova, 1941 (type)

P. cinereus Gupta & Naqui, 1983

P. cochinensis Gupta & Naqui, 1983

P. dussamicitatum Gupta & Gupta, 1971

P. ernakulensis Gupta & Gupta, 1971

P. leuciscus (Krotov & Petrochenko, 1956) Golvan, 1969 (= *Hemirhadinorhynchus leuciscus* Krotov & Petrochenko, 1956)

P. mujibi Gupta & Naqui, 1983

P. pseudaspii Achmerov & Dombrowskaja-Achmerova, 1941

P. samegaiensis Nakajima, 1975

Genus ***Tegorhynchus*** Van Cleave, 1921 (= *Illiosentis* Van Cleave & Lincicome, 1939)

Species ***T. brevis*** Van Cleave, 1921 (type)

T. africanus (Golvan, 1955) comb. n. (= *Illiosentis furcatus africanus* Golvan, 1955)

T. cetratus (Van Cleave, 1945) Bullock & Mateo, 1970 (= *Illiosentis cetratus* Van Cleave, 1945)

T. edmondsi (Golvan, 1960) comb. n. (= *Illiosentis edmondsi* Golvan, 1960)

T. furcatus (Van Cleave & Lincicome, 1939) Bullock & Mateo, 1970 (= *Illiosentis furcatus* Van Cleave & Lincicome, 1939)

T. pectinarius Van Cleave, 1940

Genus ***Telosentis*** Van Cleave, 1923

Species ***T. molini*** Van Cleave, 1923 (type) (= *Echinorhynchus atherinae* Rudolphi, 1819; *E. acanthosoma* Westrumb, 1821)

T. australiensis Edmonds, 1964

T. exiguus (Linstow, 1901) Van Cleave, 1923 (= *Echinorhynchus exiguus* Linstow, 1901)

T. tenuicornis (Linton, 1892) Van Cleave, 1947 [= *Echinorhynchus tenuicornis* Linton, 1892; *Rhadinorhynchus tenuicornis* (Linton, 1892) Van Cleave, 1918]

Family Polyacanthorhynchidae Golvan, 1956

Genus ***Polyacanthorhynchus*** Travassos, 1920

Species ***P. macrorhynchus*** (Diesing, 1856) Travassos, 1920 (type) (= *Echinorhynchus macrorhynchus* Diesing, 1856)

P. caballeroi Diaz-Ungria & Rodrigo, 1960

P. kenyensis Schmidt & Canaris, 1967

P. rhopalorhynchus (Diesing, 1851) Travassos, 1920 (= *Echinorhynchus rhopalorhynchus* Diesing, 1851)

Family Pomphorhynchidae Yamaguti, 1939

Genus ***Longicollum*** Yamaguti, 1935 (= *Spirorhynchus* Harada, 1935; *Spiracanthorhynchus* Harada, 1938; *Spirorhynchoides* Strand, 1942)

Species ***L. pagrosomi*** Yamaguti, 1935 (type)

L. alemniscus (Harada, 1935) Fukui & Morisita, 1938 [= *Spirorhynchus alemniscus* Harada, 1935; *Longicollum minor* Fukui & Morisita, 1936; *Spiracanthorhynchus alemniscus* (Harada, 1935) Harada, 1938]

L. chabanaudi Dollfus & Golvan, 1963

L. edmondsi Golvan, 1969

L. indicum Gupta & Gupta, 1970

L. lutjani Jain & Gupta, 1980

L. noellae Golvan, 1969

L. riouxi Golvan, 1969

L. sergenti (Choquette & Gayot, 1952) Golvan, 1969 (= *Tenuiproboscis sergenti* Choquette & Gayot, 1952)

Genus ***Pomphorynchus*** Monticelli, 1905

Species ***P. laevis*** (Zoega in Müller, 1776) Van Cleave, 1924 (type) (= *Echinorhynchus laevis* Zoega in Müller, 1776; *E. proteus* Westrumb, 1821, part.)

P. bosniacus Kiskaroly & Cankovic, 1969

P. bufonis Fotedar, Duda & Raina, 1970

P. bulbocolli Linkins in Van Cleave, 1919

P. dubious Kaw, 1941

P. francoisae Golvan, 1969

P. indicus Gupta & Lata, 1967

P. intermedius Engelbrecht, 1957

P. jammuensis Fotedar & Dhar, 1977

P. kashmirensis Kaw, 1941

P. kawi Fotedar, Duda & Raina, 1970

P. kostylewi Petrochenko, 1956

P. lucyi Williams & Rogers, 1984

P. megacanthus Fotedar & Dhar, 1977

P. oreini Fotedar & Dhar, 1977

P. orientalis Fotedar & Dhar, 1977

P. perforator (Linstow, 1908) Meyer, 1932 (= *Echinorhynchus perforator* Linstow, 1908)

P. rocci Cordonnier & Ward, 1967

P. sebastichthydis Yamaguti, 1939

P. tereticollis (Rudolphi, 1809) Meyer, 1932 (= *Echinorhynchus tereticollis* Rudolphi, 1809; *E. dobulae* Schrank, 1790; *E. attenuatus* Müller, 1779, part.; *E. piscinus* Zeder, 1900, part.; *E. longicollis* Pallas in Goeze, 1782, part.)

P. ***tori*** Fotedar & Dhar, 1977
P. ***yamagutii*** Schmidt & Hugghins, 1973
P. ***yunnanensis*** Wang, 1981
Genus ***Tenuiproboscis*** Yamaguti, 1935
T. ***misgurni*** Yamaguti, 1935 (type)
Family Rhadinorhynchidae Travassos, 1923 (=Gorgorhynchidae Van Cleave & Lincicome, 1940; Micracanthorhynchidae Yamaguti, 1963)
Subfamily Golvanacanthinae Paggi & Orecchia, 1972
Genus ***Golvanacanthus*** Paggi & Orecchia, 1972
Species *G.* ***blennii*** Paggi & Orecchia, 1972 (type)
G. ***problematicus*** Mordvinova & Parukhin, 1978
Subfamily Gorgorhynchinae Van Cleave & Lincicome, 1940
Genus ***Australorhynchus*** Lebedev, 1967
Species *A.* ***tetramorphacanthus*** Lebedev, 1967 (type)
Genus ***Cleaveius*** Subramanian, 1927 (=*Mehrarhynchus* Datta, 1940)
Species *C.* ***circumspinifer*** Subramanian, 1927 (type)
C. ***leiognathi*** Jain & Gupta, 1979
C. ***mysti*** (Sahay & Sinha, 1971) comb. n. (=*Mehrarhynchus mysti* Sahay & Sinha, 1971)
C. ***portblairensis*** Jain & Gupta, 1979
C. ***prashadi*** (Datta, 1940) Golvan, 1969 (=*Mehrarhynchus prashadi* Datta, 1940)
C. ***secundus*** (Tripathi, 1959) Golvan, 1969 (=*Mehrarhynchus secundus* Tripathi, 1959)
Genus ***Gorgorhynchoides*** Cable & Linderoth, 1963
Species *G.* ***elongatus*** Cable & Linderoth, 1963 (type)
G. ***bullocki*** Cable & Mafarachisi, 1970
G. ***lintoni*** Cable & Mafarachisi, 1970
Genus ***Gorgorhynchus*** Chandler, 1934 (=*Neoacanthorhynchus* Morisita, 1937)
Species *G.* ***medius*** (Linton, 1908) Chandler, 1934 (type) [=*Echinorhynchus medius* Linton, 1908; *Rhadinorhynchus medius* (Linton, 1908) Van Cleave, 1918; *Gorgorhynchus gibber* Chandler, 1934]
G. ***celebensis*** (Yamaguti, 1954) Golvan, 1969 (=*Rhadinorhynchus celebensis* Yamaguti, 1954)
G. ***clavatus*** Van Cleave, 1940 (=*G. cablei* Golvan, 1969)
G. ***lepidus*** Van Cleave, 1940
G. ***nemipteri*** Parukhin, 1973
G. ***ophiocephali*** Furtado & Lau, 1971
G. ***polymixiae*** Kovalenko, 1981
G. ***robertdollfusi*** Golvan, 1956
G. ***satoi*** (Morisita, 1937) Yamaguti, 1963 (=*Neoacanthorhynchus satoi* Morisita, 1937)
Genus ***Hanumantharaorhynchus*** Chandra, 1983
Species *H.* ***hemirhamphi*** Chandra, 1983 (type)

Genus ***Leptorhynchoides*** Kostylev, 1924 (=*Pleurorhynchus* Nau, 1787)
Species ***L. plagicephalus*** (Westrumb, 1821) Kostylev, 1924 (type) (=*Echinorhynchus plagicephalus* Westrumb, 1821; *E. acipenseris rutheni* Rudolphi, 1819; *E. husonis* Rudolphi, 1819)
L. aphredoderi Buckner & Buckner, 1976
L. thecatus (Linton, 1891) Kostylev, 1924 (=*Echinorhynchus thecatus* Linton, 1891)
Genus ***Metacanthocephaloides*** Yamaguti, 1959
Species ***M. zebrini*** Yamaguti, 1959 (type)
Genus ***Metacanthocephalus*** Yamaguti, 1959
Species ***M. pleuronichthydis*** Yamaguti, 1959 (type)
M. campbelli (Leiper & Atkinson, 1914) Golvan, 1969 [=*Leptorhynchoides campbelli* (Leiper & Atkinson, 1914) Johnston & Best, 1937; *Echinorhynchus campbelli* Leiper & Atkinson, 1914; *E. rennicki* Leiper & Atkinson, 1914; *Metechinorhynchus campbelli* (Leiper & Atkinson, 1914) Petrochenko, 1956]
M. ovicephalus (Zhukov, 1963) Golvan, 1969 (=*Leptorhynchoides ovicephalus* Zhukov, 1963)
Genus ***Micracanthorhynchina*** Strand, 1936 (=*Micracanthorhynchus* Harada, 1935; *Micracanthocephalus* Harada, 1938; *Bolbosentis* Belous, 1952)
Species ***M. motomurai*** (Harada, 1935) Ward, 1951 (type) [=*Micracanthorhynchus motomurai* Harada, 1935; *Micracanthocephalus motomurai* (Harada, 1935) Harada, 1938]
M. cynoglossi Wang, 1980
M. dakusuiensis (Harada, 1938) Ward, 1951 (=*Micracanthocephalus dakusuiensis* Harada, 1938)
M. hemiculturus Demshin, 1965
M. hemirhamphi (Baylis, 1944) Ward, 1951 (=*Micracanthocephalus hemirhamphi* Baylis, 1944)
M. indica Farooqi, 1980
M. lateolabracis Wang, 1980
M. sajori (Belous, 1952) Golvan, 1969 (=*Bolbosentis sajori* Belous, 1952)
Genus ***Paracanthorhynchus*** Edmonds, 1967
Species ***P. galaxiasus*** Edmonds, 1967 (type)
Genus ***Pseudauchen*** Yamaguti, 1963
Species ***P. epinephali*** (Yamaguti, 1939) Yamaguti, 1963 (type) [=*Rhadinorhynchus epinephali* Yamaguti, 1939; *Gorgorhynchus epinephali* (Yamaguti, 1939) Golvan, 1960]
Genus ***Pseudoleptorhynchoides*** Salgado-Maldonado, 1976
Species ***P. lamothei*** Salgado-Maldonado, 1976 (type)

Genus ***Sclerocollum*** Schmidt & Paperna, 1978

Species ***S. rubimaris*** Schmidt & Paperna, 1978 (type)

S. robustum (Edmonds, 1964) Schmidt & Paperna, 1978 [= *Neogorgorhynchus robustus* Edmonds, 1964; *Neorhadinorhynchus robustus* (Edmonds, 1964) Golvan, 1969]

Subfamily Rhadinorhynchinae Lühe, 1912

Genus ***Cathayacanthus*** Golvan, 1969

Species ***C. exilis*** (Van Cleave, 1928) Golvan, 1969 (type) (= *Rhadinorhynchus exilis* Van Cleave, 1928)

Genus ***Megistacantha*** Golvan, 1960

Species ***M. horridum*** (Lühe, 1912) Golvan, 1960 (type) (= *Rhadinorhynchus horridus* Lühe, 1912)

Genus ***Paragorgorhynchus*** Golvan, 1957

Species ***P. albertianum*** Golvan, 1957 (type)

P. chariensis Troncy, 1970

Genus ***Raorhynchus*** Tripathi, 1959

Species ***R. terebra*** (Rudolphi, 1819) Tripathi, 1959 (type) [= *Echinorhynchus terebra* Rudolphi, 1819; *Rhadinorhynchus terebra* (Rudolphi, 1819) Lühe, 1911]

R. inexpectatus Golvan, 1969

R. meyeri (Heinze, 1934) Golvan, 1969 (= *Rhadinorhynchus meyeri* Heinze, 1934)

R. polynemi Tripathi, 1959

R. thapari Gupta & Fatima, 1981

Genus ***Rhadinorhynchus*** Lühe, 1911 (= *Nipporhynchus* Chandler, 1934; *Echinosoma* Porta, 1907; *Protorhadinorhynchus* Petrochenko, 1956)

Species ***R. pristis*** (Rudolphi, 1802) Lühe, 1911 (type) [= *Echinorhynchus pristis* Rudolphi, 1802; *Rhadinorhynchus selkirki* Van Cleave, 1921; *R. alosae* (Hermann, 1782) Meyer, 1932]

R. africanus (Golvan, Houin & Deltour, 1963) Golvan, 1969 (= *Nipporhynchus africanus* Golvan, Houin & Deltour, 1963)

R. asturi Gupta & Lata, 1967

R. atheri (Farooqi, 1981) comb. n. (= *Nipporhynchus atheri* Farooqi, 1981)

R. bicircumspinus Hooper, 1983

R. cadenati (Golvan & Houin, 1964) Golvan, 1969 (= *Nipporhynchus cadenati* Golvan & Houin, 1964)

R. camerounensis Golvan, 1969

R. capensis Bray, 1974

R. carangis Yamaguti, 1939 [= *Nipporhynchus carangis*

(Yamaguti, 1939) Ward, 1951; *Protorhadinorhynchus carangis* (Yamaguti, 1939) Petrochenko, 1956]

R. cololabis Laurs & McCauley, 1964

R. decapteri Parukhin & Kovalenko, 1976

R. ditrematis Yamaguti, 1939 [=*Nipporhynchus ditrematis* (Yamaguti, 1939) Ward, 1951; *Protorhadinorhynchus ditrematis* (Yamaguti, 1939) Ward, 1951]

R. dujardini Golvan, 1969

R. erumeii (Gupta & Fatima, 1981) comb. n. (=*Nipporhynchus erumeii* Gupta & Fatima, 1981)

R. japonicus Fujita, 1920

R. johnstoni Golvan, 1969

R. lintoni Cable & Linderoth, 1963

R. ornatus Van Cleave, 1918 [=*Nipporhynchus ornatus* (Van Cleave, 1918) Chandler, 1934; *N. katsuwonis* (Harada, 1928) Chandler, 1934]

R. plagioscionis Thatcher, 1980

R. polynemi Gupta & Lata, 1967

R. salatrix Troncy & Vassiliades, 1973

R. seriolae (Yamaguti, 1963) Golvan, 1969 (=*Nipporhynchus seriolae* Yamaguti, 1963)

R. trachuri Harada, 1935 [=*Nipporhynchus trachuri* (Harada, 1935) Van Cleave & Lincicome, 1940]

R. vancleavei Golvan, 1969

R. zhukovi Golvan, 1969

Subfamily Serrasentinae Petrochenko, 1956

Genus ***Serrasentis*** Van Cleave, 1923 (=*Echinorhynchus* Müller, 1776, part.; *Echinogaster* Monticelli, 1905; *Echinosoma* Porta, 1907, part.; *Lepidosoma* Porta, 1908)

Species ***S. lamelliger*** (Diesing, 1854) Van Cleave, 1924 (type) [=*Echinorhynchus lamelliger* Diesing, 1854; *Lepidosoma lamelliger* (Diesing, 1854) Porta, 1908]

S. fotedari Gupta & Fatma, 1979

S. nadakali George & Nadakal, 1978

S. mujibi Bilqees, 1972

S. sagittifer (Linton, 1889) Van Cleave, 1923 [=*Echinorhynchus sagittifer* Linton, 1889; *Echinogaster sagittifer* (Linton, 1889) Porta, 1908; *Serrasentis socialis* (Leidy, 1851) Van Cleave, 1924; *S. chauhani* Datta, 1954; *S. longa* Tripathi, 1959; *S. longiformis* Bilqees, 1971; *S. giganticus* Bilqees, 1972; *S. scomberomori* Wang, 1981]

S. sciaenus Bilqees, 1972

S. sidaroszakaio Tadros, Iskandar & Wassef, 1979

Subfamily Serrasentoidinae Parukhin, 1982

Genus ***Serrasentoides*** Parukhin, 1971

Species ***S. fistulariae*** Parukhin, 1971 (type)

Order Polymorphida Petrochenko, 1956

Family Centrorhynchidae Van Cleave, 1916

Genus ***Centrorhynchus*** Lühe, 1911 (= *Echinorhynchus* Zoega in Müller, 1780; *Paradoxites* Lindemann, 1865; *Chentrosoma* Porta, 1906, part.; *Gordiorhynchus* Meyer, 1931; *Travassosina* Witenberg, 1932)

Species *C.* ***aluconis*** (Müller, 1780) Lühe, 1911 (type) [= *Echinorhynchus aluconis* Müller, 1780; *E. otidis* Schrank, 1788; *E. inequalis* Rudolphi, 1808; *Chentrosoma aluconis* (Müller, 1780) Porta, 1909; *Centrorhynchus olssoni* Lundström, 1942; *C. appendiculatum* (Westrumb, 1821) Joyeux & Baer, 1937]

C. ***albensis*** Rengaraju & Das, 1975

C. ***albidus*** Meyer, 1932

C. ***amphibius*** Das, 1950

C. ***asturinus*** (Johnston, 1912) Johnston, 1918 (= *Gigantorhynchus asturinus* Johnston, 1912)

C. ***atheni*** Gupta & Fatma, 1983

C. ***bancrofti*** (Johnston & Best, 1943) Golvan, 1956 (= *Gordiorhynchus bancrofti* Johnston & Best, 1943)

C. ***batrachus*** Das, 1952 (= *C. splendi* Gupta & Gupta, 1970)

C. ***bazaleticus*** Kurachvilli, 1955

C. ***bengalensis*** Datta & Soota, 1954

C. ***bramae*** Rengaraju & Das, 1980

C. ***brevicanthus*** Das, 1949

C. ***brevicaudatus*** Das, 1949

C. ***brumpti*** Golvan, 1965

C. ***brygooi*** Golvan, 1965

C. ***bubonis*** Yamaguti, 1939

C. ***buckleyi*** Gupta & Fatma, 1983

C. ***buteonis*** (Schrank, 1788) Kostylev, 1914 (= *Echinorhynchus buteonis* Schrank, 1788; *E. caudatus* Zeder, 1803; *E. polyacanthoides* Creplin, 1825)

C. ***californicus*** Millzner, 1924

C. ***chabaudi*** Golvan, 1958

C. ***clitorideus*** (Meyer, 1931) Golvan, 1956 (= *Gordiorhynchus clitorideus* Meyer, 1931)

C. ***conspectus*** Van Cleave & Pratt, 1940

C. ***crocidurus*** Das, 1950

C. ***crotophagicola*** Schmidt & Neiland, 1966

C. ***dimorphocephalus*** (Westrumb, 1821) Meyer, 1932 [= *Echinorhynchus dimorphocephalus* Westrumb, 1821; *Prosthorhynchus dimorphocephalus* (Westrumb, 1821) Travassos, 1926]

C. ***dipsadis*** (Linstow, 1888) Golvan, 1956 (= *Echinorhynchus dipsadis* Linstow, 1888)

C. elongatus Yamaguti, 1935
C. falconis (Johnston & Best, 1943) Golvan, 1956 (= *Gordiorhynchus falconis* Johnston & Best, 1943)
C. fasciatus (Westrumb, 1821) Travassos, 1926 (= *Echinorhynchus fasciatus* Westrumb, 1821; *E. motacillae atricapillae* Rudolphi, 1819)
C. freundi (Hartwick, 1953) Golvan, 1956 (= *Gordiorhynchus freundi* Hartwick, 1953)
C. galliardi Golvan, 1956
C. gendrei (Golvan, 1957) Golvan, 1960 (= *Gordiorhynchus gendrei* Golvan, 1957)
C. giganteus Travassos, 1921
C. globocaudatus (Zeder, 1800) Lühe, 1911 (= *Echinorhynchus globocaudatus* Zeder, 1800; *E. tuba* Rudolphi, 1802 in part.)
C. golvani Anantaraman & Anantaraman, 1969
C. grassei Golvan, 1965
C. hagiangensis (Petrochenko & Fan, 1969) comb. n. (= *Gordiorhynchus hagiangensis* Petrochenko & Fan, 1969)
C. hargisi Gupta & Fatma, 1983
C. horridus (Linstow, 1897) Meyer, 1932 [= *Echinorhynchus horridus* Linstow, 1897; *Prosthorhynchus horridus* (Linstow, 1897) Travassos, 1926]
C. indicus Golvan, 1956
C. insularis Tubangui, 1933
C. itatsinsis Fukui, 1929
C. javanicans Rengaraju & Das, 1975
C. knowlesi Datta & Soota, 1955
C. kuntzi Schmidt & Neiland, 1966
C. leptorhynchus Meyer, 1932
C. lesiniformis (Molin, 1859) Bock, 1935 (= *Echinorhynchus lesiniformis* Molin, 1859)
C. longicephalus Das, 1950
C. lucknowensis Gupta & Fatma, 1983
C. mabuiae (Linstow, 1908) Golvan, 1956 (= *Echinorhynchus mabuiae* Linstow, 1908)
C. macrorchis Das, 1949
C. madagascariensis (Golvan, 1957) Golvan, 1960 (= *Gordiorhynchus madagascariensis* Golvan, 1957)
C. magnus Fukui, 1929 (= *Centrorhynchus microrchis* Fukui, 1929)
C. merulae Dollfus & Golvan, 1961
C. microcephalus (Bravo-Hollis, 1947) Golvan, 1956 (*Gordiorhynchus microcephalus* Bravo-Hollis, 1947)
C. microcerviacanthus Das, 1950

C. migrans Zuberi & Farooq, 1974
C. milvus Ward, 1956
C. mysentri Gupta & Fatma, 1983
C. narcissae Florescu, 1942
C. nicaraguensis Schmidt & Neiland, 1966
C. ninni (Stossich, 1891) Meyer, 1932 (= *Echinorhynchus ninni* Stossich, 1891)
C. petrotschenkoi Kuraschvili, 1955
C. polemaeti Troncy, 1970
C. ptyasus Gupta, 1950
C. renardi (Lindemann, 1865) Van Cleave, 1923 (= *Paradoxites renardi* Lindemann, 1865)
C. sholapurensis Rengaraju & Das, 1975
C. simplex Meyer, 1932
C. spilornae Schmidt & Kuntz, 1969 (= *Centrorhynchus andamanensis* Soota & Kansal, 1970)
C. spinosus (Kaiser, 1893) Van Cleave, 1924 (*Echinorhynchus spinosus* Kaiser, 1893)
C. teres (Westrumb, 1821) Travassos, 1926 (= *Echinorhynchus teres* Westrumb, 1821; *E. picae* Rudolphi, 1819 part.)
C. tumidulus (Rudolphi, 1819) Travassos, 1923 (= *Echinorhynchus tumidulus* Rudolphi, 1818; *E. caudatus* Rudolphi, 1819; *E. megacephalus* Westrumb, 1821; *Centrorhynchus tumidulus* Neiva, Cuhna & Travassos, 1914)
C. tyotensis Rengaraju & Das, 1975
C. undulatus Dollfus, 1951
C. wardae Holloway, 1958

Genus ***Sphaerirostris*** Golvan, 1956

Species ***S. picae*** (Rudolphi, 1819) Golvan, 1956 (type) (= *Echinorhynchus picae* Rudolphi, 1819; *E. lobianchii* Monticelli, 1887; *Centrorhynchus picae* Dollfus, 1953)
S. areolatus (Rudolphi, 1819) Golvan, 1956 (= *Echinorhynchus areolatus* Rudolphi, 1819; *E. sigmoides* Westrumb, 1821; *E. orioli* Rudolphi, 1819)
S. erraticus (Chandler, 1925) Golvan, 1956 (= *Centrorhynchus erraticus* Chandler, 1925)
S. lancea (Westrumb, 1821) Golvan, 1956 [= *Echinorhynchus lancea* Westrumb, 1821; *E. vanelli* Goeze, 1782; *Centrorhynchus lancea* (Westrumb, 1821) Skrjabin, 1913; *C. cinctus* (Rudolphi, 1819) Meyer, 1932; *C. embae* Kolodkowski & Kostylev, 1916; *C. scanensis* Lundström, 1942]
S. lanceoides (Petrochenko, 1949) Golvan, 1956 (= *Centrorhynchus lanceoides* Petrochenko, 1949)
S. maryasis (Datta, 1932) Golvan, 1956 (= *Centrorhynchus maryasis* Datta, 1932)

S. opimus (Travassos, 1919) Golvan, 1956 (= *Centrorhynchus opimus* Travassos, 1919)
S. physocoracis (Porta, 1913) Golvan, 1956 (= *Echinorhynchus physocoracis* Porta, 1913)
S. pinguis (Van Cleave, 1918) Golvan, 1956 [= *Centrorhynchus leguminosus* Solovieff, 1912; *C. bipartitus* Solovieff, 1912; *C. pinguis* Van Cleave, 1918; *C. corvi* Fukui, 1929; *Travassosina pinguis* (Fukui, 1929) Witenberg, 1932; *C. skrjarbini* Petrochenko, 1949]
S. reptans (Bhalerao, 1931) Golvan, 1956 (= *Centrorhynchus reptans* Bhalerao, 1931)
S. serpenticola (Linstow, 1908) Golvan, 1956 (= *Echinorhynchus serpenticola* Linstow, 1908)
S. tenuicaudatus (Marotel, 1889) comb. n. [= *Echinorhynchus tenuicaudatus* Marotel, 1889; *Centrorhynchus tenuicaudatus* (Marotel, 1889) Lühe, 1911]
S. turdi (Yamaguti, 1939) Golvan, 1956 [= *Centrorhynchus turdi* Yamaguti, 1939; *Gordiorhynchus turdi* (Yamaguti, 1939) Kamegai, 1963]
S. wertheimae Schmidt, 1975

Family Plagiorhynchidae Golvan, 1960
Subfamily Plagiorhynchinae Meyer, 1931
Genus ***Plagiorhynchus*** Lühe, 1911 (= *Prosthorhynchus* Kostylev, 1915)
Species ***P. crassicollis*** (Villot, 1875) Lühe, 1911 (type)
Subgenus ***Plagiorhynchus*** Lühe, 1911
Species ***P. (P.) charadrii*** (Yamaguti, 1939) Van Cleave, 1951 (= *Prosthorhynchus charadrii* Yamaguti, 1939)
P. (P.) charadriicola (Dollfus, 1953) Golvan, 1956 (= *Prosthorhynchus charadriicola* Dollfus, 1953)
P. (P.) lemnisalis Belopolskaia, 1958
P. (P.) linearis (Westrumb, 1891) Golvan, 1956 (= *Echinorhynchus linearis* Westrumb, 1891)
P. (P.) menurae (Johnston, 1912) Golvan, 1956 (= *Prosthorhynchus menurae* Johnston, 1912)
P. (P.) paulus Van Cleave & Williams, 1950
P. (P.) spiralis (Rudolphi, 1809) Golvan, 1956 [= *Echinorhynchus spiralis* Rudolphi, 1809; *Prosthorhynchus spiralis* (Rudolphi, 1809) Travassos, 1926]
P. (P.) totani (Porta, 1910) Golvan, 1956 [= *Echinorhynchus totani* Porta, 1910; *Prosthorhynchus totani* (Porta, 1910) Meyer, 1932]
Subgenus ***Prosthorhynchus*** Kostylev, 1915
Species ***P. (P.) angrense*** (Travassos, 1926) Schmidt & Kuntz, 1966 (= *Prosthorhynchus angrense* Travassos, 1926)
P. (P.) bullocki Schmidt & Kuntz, 1966

P.* (*P.*) *cylindraceus (Goeze, 1782) Schmidt & Kuntz, 1966 [*Echinorhynchus cylindraceus* Goeze, 1782; *Centrorhynchus cylindraceus* (Goeze, 1782) Kostylev, 1914; *Prosthorhynchus cylindraceus* (Goeze, 1782) Golvan, 1956; *E. transversus* Rudolphi, 1819; *Prosthorhynchus transversus* (Rudolphi, 1819) Travassos, 1926; *P. formosus* Van Cleave, 1918; *P. taiwanensis* Schmidt & Kuntz, 1966]

P.* (*P.*) *gallinagi (Schachtachtinskaia, 1953) Schmidt & Kuntz, 1966 (=*Prosthorhynchus gallinagi* Schachtachtinskaia, 1953)

P.* (*P.*) *genitopapillatus (Lundström, 1942) comb. n. (=*Prosthorhynchus genitopapillatus* Lundström, 1942)

P.* (*P.*) *golvani Schmidt & Kuntz, 1966

P.* (*P.*) *gracilis (Petrochenko, 1958) Schmidt & Kuntz, 1966 (=*Prosthorhynchus gracilis* Petrochenko, 1958)

P.* (*P.*) *limnobaeni (Tubangui, 1933) Golvan, 1956 (=*Prosthorhynchus limnobaeni* Tubangui, 1933)

P.* (*P.*) *longirostris (Travassos, 1926) comb. n. (=*Prosthorhynchus longirostris* Travassos, 1926)

P.* (*P.*) *malayensis (Tubangui, 1935) Schmidt & Kuntz, 1966 (=*Oligoterorhynchus malayensis* Tubangui, 1935)

P.* (*P.*) *nicobarensis (Soota & Kansal, 1970) Zafar & Farooqi, 1981 (=*Prosthorhynchus nicobarensis* Soota & Kansal, 1970)

P.* (*P.*) *ogatai (Fukui & Morisita, 1936) Schmidt & Kuntz, 1966 (=*Porrorchis ogatai* Fukui & Morisita, 1936)

P.* (*P.*) *pittarum (Tubangui, 1935) Schmidt & Kuntz, 1966 (=*Prosthorhynchus pittarum* Tubangui, 1935)

P.* (*P.*) *rectus (Linton, 1892) Van Cleave, 1918 (=*Echinorhynchus rectus* Linton, 1892)

P.* (*P.*) *reticulatus (Westrumb, 1821) Golvan, 1956 (=*Echinorhynchus reticulatus* Westrumb, 1821)

P.* (*P.*) *rheae (Marval, 1902) Schmidt & Kuntz, 1966 (=*Echinorhynchus rheae* Marval, 1902)

P.* (*P.*) *rossicus (Kostylev, 1915) Schmidt & Kuntz, 1966 (=*Prosthorhynchus rossicus* Kostylev, 1915)

P.* (*P.*) *russelli Tadros, 1970

P.* (*P.*) *scolopacidis (Kostylev, 1915) Schmidt & Kuntz, 1966 (=*Prosthorhynchus scolopacidis* Kostylev, 1915)

Plagiorhynchinae *incertae sedis*

P. rostratum (Marval, 1902)

P. urichi Cameron, 1936

Subfamily Porrorchinae Golvan, 1956

Genus ***Lueheia*** Travassos, 1919 (=*Furcata* Werby, 1938)

Species ***L. lueheia*** Travassos, 1919 (type)

L. adlueheia (Werby, 1938) Van Cleave, 1942 (=*Furcata adlueheia* Werby, 1938)
L. cajabambensis Machado & Ibanez, 1967
L. inscripta (Westrumb, 1821) Travassos, 1919 (=*Echinorhynchus inscripta* Westrumb, 1821)

Genus **Oligoterorhynchus** Monticelli, 1914
Species **O. campylurus** (Nitzsch, 1857) Monticelli, 1914 (type) (= *Echinorhynchus campylurus* Nitzsch, 1857)

Genus **Owilfordia** Schmidt & Kuntz, 1967
Species **O. olseni** Schmidt & Kuntz, 1967
O. teliger (Van Cleave, 1949) Schmidt & Kuntz, 1967 (=*Pseudoporrorchis teliger* Van Cleave, 1949)

Genus **Porrorchis** Fukui, 1929 (=*Pseudoporrorchis* Joyeux & Baer, 1935)
Species **P. elongatus** Fukui, 1929 (type)
P. bazae (Southwell & MacFie, 1925) Schmidt & Kuntz, 1967 [=*Echinorhynchus bazae* Southwell & MacFie, 1925; *Prosthorhynchus bazae* (Southwell & MacFie, 1925) Travassos, 1926; *Pseudoporrorchis bazae* (Southwell & MacFie, 1925) Petrochenko, 1958]
P. centropi (Porta, 1910) Schmidt & Kuntz, 1967 [=*Echinorhynchus centropi* Porta, 1910; *Pseudoporrorchis centropi* (Porta, 1910) Joyeux & Baer, 1935]
P. houdemeri (Joyeux & Baer, 1935) Schmidt & Kuntz, 1967 (=*Pseudoporrorchis houdemeri* Joyeux & Baer, 1935)
P. hydromuris (Edmonds, 1957) Schmidt & Kuntz, 1967 (=*Pseudoporrorchis hydromuris* Edmonds, 1957)
P. hylae (Johnston, 1914) Schmidt & Kuntz, 1967 [= *Pseudoporrorchis hylae* (Johnston, 1914) Edmonds, 1956; *Echinorhynchus bulbocaudatus* Southwell & MacFie, 1925; *E. centropusi* Tabungui, 1933; *Gordiorhynchus hylae* (Johnston, 1914) Johnston & Edmonds, 1948; *Pseudoporrorchis bulbocaudatus* (Southwell & MacFie, 1925) Joyeux & Baer, 1935; *P. centropusi* (Tubangui, 1933) Joyeux & Baer, 1935]
P. indicus (Das, 1957) Schmidt & Kuntz, 1967 (=*Pseudoporrorchis indicus* Das, 1957)
P. leibyi Schmidt & Kuntz, 1967
P. maxvachoni (Golvan & Brygoo, 1965) Schmidt & Kuntz, 1967 (=*Pseudoporrorchis maxvachoni* Golvan & Brygoo, 1965)
P. oti Yamaguti, 1939
P. rotundatus (Linstow, 1897) Schmidt & Kuntz, 1967

[=*Echinorhynchus rotundatus* Linstow, 1897; *Pseudoporrorchis rotundatus* (Linstow, 1897) Joyeux & Baer, 1935]

Genus ***Pseudogordiorhynchus*** Golvan, 1957

Species ***P. antonmeyeri*** Golvan, 1957 (type)

Genus ***Pseudolueheia*** Schmidt & Kuntz, 1967

Species ***P. pittae*** Schmidt & Kuntz, 1967 (type)

P. boreotis (Van Cleave & Williams, 1951) Schmidt & Kuntz, 1967 (=*Lueheia boreotis* Van Cleave & Williams, 1951)

Subfamily Sphaerechinorhynchinae Golvan, 1956

Genus ***Sphaerechinorhynchus*** Johnston, 1929

Species ***S. rotundocapitatus*** (Johnston, 1912) Johnston, 1929 (type) (=*Echinorhynchus rotundocapitatus* Johnston, 1912)

S. serpenticola Schmidt & Kuntz, 1966

Family Polymorphidae Meyer, 1931 (=Filicollidae Petrochenko, 1956)

Genus ***Andracantha*** Schmidt, 1975

Species ***A. gravida*** (Alegret, 1941) Schmidt, 1975 (type) (=*Corynosoma gravida* Alegret, 1941)

A. mergi (Lundström, 1941) Schmidt, 1975 [=*Corynosoma mergi* Lundström, 1941; *Hemiechinosoma mergi* (Lundström, 1941) Petrochenko & Smogorjevskaia, 1962]

A. phalacrocoracis (Yamaguti, 1939) Schmidt, 1975 (=*Corynosoma phalacrocoracis* Yamaguti, 1939)

Genus ***Arhythmorhynchus*** Lühe, 1911 (=*Skrjabinorhynchus* Petrochenko, 1956)

Species ***A. frassoni*** (Molin, 1858) Lühe, 1911 (type) (=*Echinorhynchus frassoni* Molin, 1858; *E. roseus* Molin, 1858; *E. rubicundus* Molin, 1859)

A. brevis Van Cleave, 1916 [=*Polymorphus brevis* (Van Cleave, 1916) Travassos, 1926]

A. capellae (Yamaguti, 1935) Schmidt, 1973 [=*Polymorphus capellae* Yamaguti, 1935; *Skrjabinorhynchus capellae* (Yamaguti, 1935) Petrochenko, 1956]

A. comptus Van Cleave & Rausch, 1950

A. distinctus Baer, 1956

A. eroliae (Yamaguti, 1939) Schmidt, 1973 [=*Polymorphus eroliae* Yamaguti, 1939; *Skrjabinorhynchus eroliae* (Yamaguti, 1939) Petrochenko, 1956

A. frontospinosus (Tubangui, 1935) Yamaguti, 1963 (=*Polymorphus frontospinosus* Tubangui, 1935)

A. jeffreyi Schmidt, 1973 (=*Arhythmorhynchus capellae* Schmidt, 1963)

A. johnstoni Golvan, 1960

A. ***limosae*** Edmonds, 1971

A. ***longicollis*** (Villot, 1875) Lühe, 1912 (=*Echinorhynchus longicollis* Villot, 1875; *E. invaginabilis* Linstow, 1902; *E. macrourus* Bremser, 1821; *Arhythmorhynchus anser* Florescu, 1941)

A. ***petrochenkoi*** (Schmidt, 1969) Atrashkevich, 1979 (= *Polymorphus* (*Polymorphus*) *petrochenkoi* Schmidt, 1969)

A. ***plicatus*** (Linstow, 1883) Meyer, 1932 (=*Echinorhynchus plicatus* Linstow, 1883)

A. ***pumiliorostris*** Van Cleave, 1916

A. ***siluricola*** Dollfus, 1929

A. ***teres*** Van Cleave, 1920 (=*Arhythmorhynchus sachalinensis* Krotov & Petrochenko, 1958)

A. ***tigrinus*** Moghe & Das, 1953

A. ***trichocephalus*** (Leuckart, 1876) Lühe, 1912 (=*Echinorhynchus trichocephalus* Leuckart, 1876)

A. ***tringi*** Gubanov, 1952

A. ***uncinatus*** (Kaiser, 1893) Lühe, 1912 (=*Echinorhynchus uncinatus* Kaiser, 1893)

A. ***xeni*** Atrashkevich, 1978

Genus ***Bolbosoma*** Porta, 1908 (=*Echinorhynchus* Zoega in Müller, 1776, part.; *Bolborhynchus* Porta, 1906)

Species ***B. turbinella*** (Diesing, 1851) Porta, 1908 (type) [=*Echinorhynchus balaenocephalus* Owen, 1803; *Bolborhynchus turbinella* (Diesing, 1851) Porta, 1906]

B. balaenae (Gmelin, 1790) Porta, 1908 [=*Echinorhynchus balaenae* Gmelin, 1790; *E. porrigens* Rudolphi, 1814; *Bolbosoma porrigens* (Rudolphi, 1814) Porta, 1908]

B. bobrovoi Krotov & Delamure, 1952

B. brevicolle (Malm, 1867) Porta, 1908 [=*Echinorhynchus brevicollis* Malm, 1867; *Bolborhynchus brevicollis* (Malm, 1867) Porta, 1906]

B. caenoforme (Heitz, 1920) Meyer, 1932

B. capitatum (Linstow, 1880) Porta, 1908 (=*Echinorhynchus capitatum* Linstow, 1880)

B. hamiltoni Baylis, 1929

B. heteracanthis (Heitz, 1917) Meyer, 1932

B. nipponicum Yamaguti, 1939

B. physeteris Gubanov, 1952

B. scromberomori Wang, 1980

B. thunni Harada, 1935

B. tuberculata Skrjabin, 1970

B. turbinella australis Skrjabin, 1972

B. turbinella turbinella (Diesing, 1851) Porta, 1908

B. vasculosum (Rudolphi, 1819) Porta, 1908 (=*Echinorhynchus annulatus* Molin, 1858; *E. aurantiacus* Risso, 1826; *E. serrani* Linton, 1888; *E. vasculosum* Rudolphi, 1819)

Genus ***Corynosoma*** Lühe, 1904 (=*Chentrosoma* Monticelli, 1905; *Echinosoma* Porta, 1907)

Species ***C. strumosum*** (Rudolphi, 1802) Lühe, 1904 (type) (= *Echinorhynchus strumosum* Rudolphi, 1802; *E. gibbosus* Rudolphi, 1809 part.; *Corynosoma osmeri* Fujita, 1921; *C. ambispigerinum* Harada, 1935)

C. alaskensis Golvan, 1958

C. anatarium Van Cleave, 1945

C. australe Johnston, 1937

C. bullosum (Linstow, 1892) Railliet & Henry, 1907 (=*Echinorhynchus bullosum* Linstow, 1892)

C. cameroni Van Cleave, 1953

C. caspicum Golvan & Mokhayer, 1973

C. clavatum Goss, 1941

C. clementi Giovannoni & Fernandes, 1965

C. constrictum Van Cleave, 1918 (=*Echinorhynchus striatus* Goeze in Linton, 1892; *C. bipapillum* Schmidt, 1965)

C. curiliensis Gubanov, 1942

C. enhydri Morozov, 1940 (=*Corynosoma enhydris* Afanasev, 1941)

C. enrietti Molfie & Freitas, 1953

C. falcatum Van Cleave, 1953

C. hadweni Van Cleave, 1953

C. hamanni (Linstow, 1892) Railliet & Henry, 1907 [=*Echinorhynchus hamanni* Linstow, 1892; *E. antarcticum* Rennie, 1907; *Corynosoma antarcticum* (Rennie, 1907) Leiper & Atkinson, 1914; *C. sipho* Railliet & Henry, 1907]

C. kurilensis Gubanov, 1952

C. longilemniscatus Machado, 1961

C. macrosomum Neiland, 1962

C. magdaleni Montreuil, 1958

C. mandarinca Oschmarin, 1963

C. mirabilis Skrjabin, 1966

C. obtuscens Lincicome, 1943

C. otariae Morini & Boero, 1961

C. pacifica Nikolskii, 1974

C. peposacae (Porta, 1914) Travassos, 1926 (=*Echinosoma peposacae* Porta, 1914)

C. pyriforme (Bremser, 1824) Meyer, 1932 (=*Echinorhynchus pyriforme* Bremser, 1824)

C. rauschi Golvan, 1958

C. ***reductum*** (Linstow, 1905) Railliet & Henry, 1907 (= *Echinorhynchus reductus* Linstow, 1905)
C. ***semerme*** (Forssell, 1904) Lühe, 1905 (= *Echinorhynchus semermis* Forssell, 1904)
C. ***septentrionalis*** Treshchev, 1966
C. ***seropedicus*** Machado, 1970
C. ***shackletoni*** Zdzitowiecki, 1978
C. ***similis*** Neiland, 1962
C. ***singularis*** Skrjabin & Nikolskii, 1971
C. ***sudsuche*** Belopolskaia, 1958
C. ***tunitae*** Weiss, 1914
C. ***turbidum*** Van Cleave, 1937
C. ***validum*** Van Cleave, 1953
C. ***ventronudum*** Skrjabin, 1959
C. ***villosum*** Van Cleave, 1953
C. ***wegeneri*** Heinze, 1934

Genus ***Diplospinifer*** Fukui, 1929
Species *D.* ***serpenticola*** Fukui, 1929 (type)

Genus ***Filicollis*** Lühe, 1911
Species *F.* ***anatis*** (Schrank, 1788) Lühe, 1911 (type) (= *Echinorhynchus anatis* Schrank, 1788; *E. filicollis* Rudolphi, 1809; *E. polymorphus* Bremser, 1824; *E. laevis* Linstow, 1905)
F. ***trophimenkoi*** Atrashkevich, 1982

Genus ***Hexaglandula*** Petrochenko, 1950
Species *H.* ***mutabilis*** (Rudolphi, 1819) Petrochenko, 1958 (type) [= *Echinorhynchus mutabilis* Rudolphi, 1819; *Polymorphus mutabilis* (Rudolphi, 1819) Travassos, 1926]
H. ***ariusis*** Bilqees, 1971
H. ***corynosoma*** (Travassos, 1915) Petrochenko, 1958 (= *Polymorphus corynosoma* Travassos, 1915)
H. ***inermis*** (Travassos, 1923) Petrochenko, 1958 (= *Polymorphus inermis* Travassos, 1923)
H. ***karachiensis*** Bilqees, 1971
H. ***paucihamatus*** (Heinze, 1936) Petrochenko, 1958 (= *Polymorphus paucihamatus* Heinze, 1936)

Genus ***Polymorphus*** Lühe, 1911 (= *Parafilicollis* Petrochenko, 1956; *Falsifilicollis* Webster, 1948; *Profilicollis* Meyer, 1931; *Subfilicollis* Hoklova, 1967)
Subgenus ***Polymorphus*** Lühe, 1911
Species *P.* (*P.*) ***minutus*** (Goeze, 1782) Lühe, 1911 (type) [= *Echinorhynchus minutus* Goeze, 1782; *E. boschadis* Schrank, 1788; *E. anatis* Gmelin, 1791; *E. collaris* Schrank, 1792; *Polymorphus boschadis* (Schrank, 1788) Railliet, 1919]
P. (*P.*) ***actuganensis*** Petrochenko, 1949

P.* (*P.*) *acutis Van Cleave & Starrett, 1940
P.* (*P.*) *biziurae Johnston & Edmonds, 1948
P.* (*P.*) *cetaceum (Johnston & Best, 1942) Schmidt & Dailey, 1971 (= *Corynosoma cetaceum* Johnston & Best, 1942)
P.* (*P.*) *chasmagnathi (Holcman-Spector, Mane-Garzon & Dei-Cas, 1977) comb. n. (= *Falsifilicollis chasmagnathi* Holcman-Spector, Mane-Garzon & Dei-Cas, 1977)
P.* (*P.*) *cincli Belopolskaya, 1959
P.* (*P.*) *contortus (Bremser in Westrumb, 1821) Travassos, 1926 (= *Echinorhynchus contortus* Bremser in Westrumb, 1821; *E. collurionis* Rudolphi, 1819)
P.* (*P.*) *corynoides Skrjabin, 1913
P.* (*P.*) *crassus Van Cleave, 1924
P.* (*P.*) *cucullatus Van Cleave & Starrett, 1940
P.* (*P.*) *diploinflatus Lundström, 1942
P.* (*P.*) *gavii Hokhlova, 1965
P.* (*P.*) *kostylewi Petrochenko, 1949
P.* (*P.*) *magnus Skrjabin, 1913
P.* (*P.*) *marchii (Porta, 1910) Meyer, 1932 (= *Corynosoma marchii* Porta, 1910)
P.* (*P.*) *marilis Van Cleave, 1939
P.* (*P.*) *mathevossianae Petrochenko, 1949
P.* (*P.*) *meyeri Lundström, 1942
P.* (*P.*) *miniatus (Linstow, 1896) Travassos, 1926 (= *Echinorhynchus miniatus* Linstow, 1896)
P.* (*P.*) *obtusus Van Cleave, 1918
P.* (*P.*) *paradoxus Connell & Corner, 1957
P.* (*P.*) *phippsi Kostylev, 1922 (= *Echinorhynchus borealis* Gmelin, 1791; *E. anatis mollissimae* Rudolphi, 1809)
P.* (*P.*) *striatus (Goeze, 1782) Lühe, 1911 (= *Echinorhynchus striatus* Goeze, 1782; *E. ardeae* Gmelin, 1789)
P.* (*P.*) *strumosoides Lundström, 1942
P.* (*P.*) *swartzi Schmidt, 1965
P.* (*P.*) *trochus Van Cleave, 1945

Subgenus ***Profilicollis*** Meyer, 1931

Species ***P.* (*P.*) *botulus*** (Van Cleave, 1916) Witenberg, 1932 (type) [= *Filicollis botulus* Van Cleave, 1916; *Profilicollis botulus* (Van Cleave, 1916) Meyer, 1931]
P.* (*P.*) *altmani (Perry, 1942) Van Cleave, 1947 [= *Filicollis altmani* Perry, 1942; *Parafilicollis altmani* (Perry, 1942) Petrochenko, 1956]
P.* (*P.*) *arcticus (Van Cleave, 1920) Van Cleave, 1937 [= *Filicollis arcticus* Van Cleave, 1920; *Profilicollis arcticus* (Van Cleave, 1920) Meyer, 1931]

P. (P.) formosus Schmidt & Kuntz, 1967

P. (P.) kenti Van Cleave, 1947 [= *Parafilicollis kenti* (Van Cleave, 1947) Petrochenko, 1956; *Falsificollis kenti* (Van Cleave, 1947) Yamaguti, 1963]

P. (P.) major Lundström, 1942 [= *Parafilicollis major* (Lundström, 1942) Petrochenko, 1956; *Falsifilicollis major* (Lundström, 1942) Yamaguti, 1963]

P. (P.) sphaerocephalus (Bremser in Rudolphi, 1819) Van Cleave, 1947 [= *Echinorhynchus sphaerocephalus* Bremser, in Rudolphi, 1819; *Filicollis sphaerocephalus* Bremser in Rudolphi, 1819) Travassos, 1928; *Parafilicollis sphaerocephalus* (Bremser in Rudolphi, 1819) Petrochenko, 1956; *Falsifilicollis sphaerocephalus* (Bremser in Rudolphi, 1819) Yamaguti, 1963]

P. (P.) texensis Webster, 1948 [= *Polymorphus* (*Falsifilicollis*) *texensis* Webster, 1948; *Falsifilicollis texensis* (Webster, 1948) Yamaguti, 1963]

Genus ***Southwellina*** Witenberg, 1932 (= *Hemiechinosoma* Petrochenko & Smogorjevskaia, 1962)

Species ***S. hispida*** (Van Cleave, 1925) Witenberg, 1932 (type) [= *Arhythmorhynchus hispidus* Van Cleave, 1925; *A. fuscus* Harada, 1929; *A. duocinctus* Chandler, 1935; *Polymorphus ardeae* Belopolskaya, 1958; *Hemiechinosoma ardeae* (Belopolskaya, 1958) Petrochenko & Smogorjevskaia, 1962; *H. ponticum* Petrochenko & Smogorjevskaia, 1962]

S. dimorpha Schmidt, 1973

S. macracanthus (Ward & Winter, 1952) Schmidt, 1973 (= *Arhythmorhynchus macracanthus* Ward & Winter, 1952)

CLASS Eoacanthocephala Van Cleave, 1936

Order Gyracanthocephala Van Cleave, 1936

Family Quadrigyridae Van Cleave, 1920 (= Acanthogyridae Thapar, 1927; Pallisentidae Van Cleave, 1928)

Subfamily Pallisentinae Van Cleave, 1928

Genus ***Acanthogyrus*** Thapar, 1927 (= *Acanthosentis* Verma & Datta, 1929; *Hemigyrus* Achmerov & Dombrowskaja-Achmerova, 1941)

Subgenus ***Acanthogyrus*** Thapar, 1927

Species ***A. (A.) acanthogyrus*** Thapar, 1927 (type)

A. (A.) guptai Gupta & Verma, 1976

A. (A.) tripathi Rai, 1967

Subgenus ***Acanthosentis*** Verma & Datta, 1929

Species ***A. (A.) antspinis*** (Verma & Datta, 1929) Dollfus & Golvan, 1956 (type) (= *Acanthosentis antspinus* Verma & Datta, 1929; *A. betawi* Tripathi, 1956)

A. (*A.*) ***acanthuri*** (Cable & Quick, 1954) Golvan, 1959 (= *Acanthosentis acanthuri* Cable & Quick, 1954)

A. (*A.*) ***anguillae*** (Wang, 1981) comb. n. (= *Acanthosentis anguillae* Wang, 1981)

A. (*A.*) ***arii*** (Bilqees, 1971) comb. n. (= *Acanthosentis arii* Bilqees, 1971)

A. (*A.*) ***bacailai*** (Verma, 1973) comb. n. (= *Acanthosentis bacailai* Verma, 1973)

A. (*A.*) ***cameroni*** (Gupta & Kajaji, 1969) comb. n. (= *Acanthosentis cameroni* Gupta & Kajaji, 1969)

A. (*A.*) ***dattai*** (Podder, 1938) Dollfus & Golvan, 1956 (= *Acanthosentis dattai* Podder, 1938)

A. (*A.*) ***giuris*** (Soota & Sen, 1956) comb. n. (= *Acanthosentis giuris* Soota & Sen, 1956)

A. (*A.*) ***golvani*** (Gupta & Jain, 1980) comb. n. (= *Acanthosentis golvani* Gupta & Jain, 1980)

A. (*A.*) ***holospinus*** (Sen, 1937) Dollfus & Golvan, 1956 (= *Acanthosentis holospinus* Sen, 1937)

A. (*A.*) ***indica*** (Tripathi, 1959) Chubb, 1982 (= *Acanthosentis hilsai* Pal, 1963; *A. indicus* Tripathi, 1959)

A. (*A.*) ***intermedius*** (Achmerov & Dombrowskaja-Achmerova, 1941) Golvan, 1959 [= *Hemigyrus intermedius* Achmerov & Dombrowskaja-Achmerova, 1941; *Acanthocephalorhynchoides intermedius* (Achmerov & Dombrowskaja-Achmerova, 1941) Petrochenko, 1956]

A. (*A.*) ***maroccanus*** (Dollfus, 1951) Dollfus & Golvan, 1956 (= *Acanthosentis maroccanus* Dollfus, 1951)

A. (*A.*) ***multispinus*** Wang, 1966

A. (*A.*) ***nigeriensis*** Dollfus & Golvan, 1956

A. (*A.*) ***oligospinus*** (Anantaraman, 1969) comb. n. (= *Acanthosentis oligospinus* Anantaraman, 1969)

A. (*A.*) ***papilo*** Troncy & Vassiliades, 1974

A. (*A.*) ***partispinus*** Furtado, 1963

A. (*A.*) ***periophthalmi*** (Wang, 1980) comb. n. (= *Acanthosentis periophthalmi* Wang, 1980)

A. (*A.*) ***scromberomori*** (Wang, 1980) comb. n. (= *Acanthosentis scromberomori* Wang, 1980)

A. (*A.*) ***shashiensis*** (Tso, Chen & Chien, 1974) comb. n. (= *Acanthosentis shashiensis* Tso, Chen & Chien, 1974)

A. (*A.*) ***similis*** (Wang, 1980) comb. n. (= *Acanthosentis similis* Wang, 1980)

A. (*A.*) ***sircari*** (Podder, 1941) Dollfus & Golvan, 1956 (= *Acanthosentis sircari* Podder, 1941)

A. (A.) thapari (Parasad, Sahay & Shambhunath, 1969) comb. n. (= *Acanthosentis thapari* Parasad, Sahay & Shambhunath, 1969)

A. (A.) tilapiae (Baylis, 1948) Dollfus & Golvan, 1956 (= *Acanthosentis tilapiae* Baylis, 1958)

A. (A.) vittatusi (Verma, 1973) comb. n. (= *Acanthosentis vittatusi* Verma, 1973)

Genus ***Devendrosentis*** Sahay, Sinha & Gosh, 1971

Species ***D. garuai*** Sahay, Sinha & Gosh, 1971 (type)

Genus ***Palliolisentis*** Machado, 1960

Species ***P. quinqueungulis*** Machado, 1960 (type)

P. ornatus Machado, 1960

P. polyonca Schmidt & Hugghins, 1973

Genus ***Pallisentis*** Van Cleave, 1928 (= *Neosentis* Van Cleave, 1928; *Farzandia* Thapar, 1931; *Acanthocephalorhynchoides* Kostylev, 1941)

Subgenus ***Farzandia*** Thapar, 1931

Species ***P. (F.) gaboes*** (MacCallum, 1918) Van Cleave, 1928 (type) (= *Echinorhynchus gaboes* MacCallum, 1918)

P. (F.) nagpurensis (Bhalerao, 1931) Baylis, 1933 (= *Farzandia nagpurensis* Bhalerao, 1931)

P. (F.) nandai Sarkar, 1953

Subgenus ***Neosentis*** Van Cleave, 1928

Species ***P. (N.) cleatus*** (Van Cleave, 1928) Harada, 1935 (type) (= *Neosentis cleatus* Van Cleave, 1928)

P. (N.) ophiocephali (Thapar, 1931) Baylis, 1933 (= *Farzandia ophiocephali* Thapar, 1931)

Subgenus ***Pallisentis*** Van Cleave, 1928

Species ***P. (P.) umbellatus*** Van Cleave, 1928 (type)

P. (P.) allahabadii Agarwal, 1958 (= *Pallisentis buckleyi* Tadros, 1966)

P. (P.) basiri Farooqi, 1958

P. (P.) cavasii Gupta & Verma, 1980

P. (P.) cholodkowskyi (Kostylev, 1928) comb. n. [= *Quadrigyrus cholodkowskyi* Kostylev, 1928; *Acanthogyrus cholodkowskyi* (Kostylev, 1928) Golvan, 1959; *Acanthocephalorhynchoides cholodkowskyi* (Kostylev, 1928) Williams, Gibson & Sadighian, 1980]

P. (P.) colisai Sarkar, 1956

P. (P.) fasciata Gupta & Verma, 1980

P. (P.) golvani Troncy & Vassiliades, 1973

P. (P.) gomtii Gupta & Verma, 1980

P. (P.) guntei Sahay, Nath & Sinha, 1967

P. (P.) magnum Saeed & Bilqees, 1971

P.* (*P.*) *panadei Rai, 1967

P.* (*P.*) *ussuriensis (Kostylev, 1941) Golvan, 1959 (= *Acanthocephalorhynchoides ussuriensis* Kostylev, 1941)

Genus ***Raosentis*** Datta, 1947

Species ***R. podderi*** Datta, 1947 (type)

R. thapari Rai, 1967

Genus ***Saccosentis*** Tadros, 1966

Species ***S. pesteri*** Tadros, 1966 (type)

S. ophiocephali Jain & Gupta, 1979

Subfamily Quadrigyrinae Van Cleave, 1920

Genus ***Deltacanthus*** Diaz-Ungria & Gracia-Rodrigo, 1958 (= *Deltania* Diaz-Ungria & Gracia-Rodrigo, 1957; *Acanthodelta* Diaz-Ungria & Gracia-Rodrigo, 1958)

Species ***D. scorzai*** (Diaz-Ungria & Gracia-Rodrigo, 1957) Diaz-Ungria & Gracia-Rodrigo, 1958 (type) [= *Deltania scorzai* Diaz-Ungria & Gracia-Rodrigo, 1957; *Acanthodelta scorzia* (Diaz-Ungria & Gracia-Rodrigo, 1957) Diaz-Ungria & Gracia-Rodrigo, 1958]

Genus ***Quadrigyrus*** Van Cleave, 1920

Species ***Q. torquatus*** Van Cleave, 1920 (type)

Q. brasiliensis Machado, 1941

Q. chinensis Mao, 1979

Q. nickoli Schmidt & Hugghins, 1973

Q. rhodei Wang, 1980

Order Neoechinorhynchida Southwell & MacFie, 1925 (= Neoacanthocephala Van Cleave, 1936)

Family Dendronucleatidae Sokolovskaia, 1962

Genus ***Dendronucleata*** Sokolovskaia, 1962

Species ***D. dogieli*** Sokolovskaia, 1962 (type)

D. petruschewskii Sokolovskaia, 1962

Family Neoechinorhynchidae Ward, 1917 (= Hebesomidae Van Cleave, 1928; Hebesomatidae Yamaguti, 1963)

Subfamily Atactorhynchinae Petrochenko, 1956

Genus ***Atactorhynchus*** Chandler, 1935

Species ***A. verecundus*** Chandler, 1935 (type)

Genus ***Floridosentis*** Ward, 1953

Species ***F. mugilis*** (Machado, 1951) Bullock, 1962 (type) (= *Atactorhynchus mugilis* Machado, 1951; *Floridosentis elongatus* Ward, 1953)

F. pacifica Bravo-Hollis, 1969

Genus ***Tanaorhamphus*** Ward, 1918

Species ***T. longirostris*** (Van Cleave, 1913) Ward, 1918 (type) [= *Neorhynchus longirostris* Van Cleave, 1913; *Neoechinorhynchus longirostris* (Van Cleave, 1913) Van Cleave, 1916]

Subfamily Eocillinae Petrochenko, 1956
  Genus ***Eocollis*** Van Cleave, 1947
    Species ***E. arcanus*** Van Cleave, 1947 (type)
      E. harengulae Wang, 1981
Subfamily Gracilisentinae Petrochenko, 1956
  Genus ***Gracilisentis*** Van Cleave, 1919
    Species ***G. gracilisentis*** (Van Cleave, 1913) Van Cleave, 1919 (type) [= *Neorhynchus gracilisentis* Van Cleave, 1913; *Eorhynchus gracilisentis* (Van Cleave, 1913) Van Cleave, 1914; *Neoechinorhynchus gracilisentis* (Van Cleave, 1913) Van Cleave, 1916]
      G. mugilis sharmai Gupta & Lata, 1967
      G. variabilis (Diesing, 1856) Petrochenko, 1956 (= *Echinorhynchus variabilis* Diesing, 1856)
  Genus ***Wolffhugelia*** Mane-Garzon & Dei-Cas, 1974
    Species ***W. matercula*** Mane-Garzon & Dei-Cas, 1974 (type)
  Genus ***Pandosentis*** Van Cleave, 1920
    Species ***P. iracundus*** Van Cleave, 1920 (type)
Subfamily Neoechinorhynchinae Travassos, 1926
  Genus ***Dispiron*** Bilqees, 1970
    Species ***D. mugili*** Bilqees, 1970 (type)
  Genus ***Gorytocephalus*** Nickol & Thatcher, 1971
    Species ***G. plecostomorum*** Nickol & Thatcher, 1971 (type)
      G. elongorchis Thatcher, 1979
      G. spectabilis (Machado, 1959) Nickol & Thatcher, 1971 (= *Neoechinorhynchus spectabilis* Machado, 1959)
  Genus ***Hebesoma*** Van Cleave, 1928
    Species ***H. violentum*** Van Cleave, 1928 (type)
  Genus ***Hexaspiron*** Dollfus & Golvan, 1956
    Species ***H. nigericum*** Dollfus & Golvan, 1956 (type)
  Genus ***Microsentis*** Martin & Multani, 1966
    Species ***M. wardae*** Martin & Multani, 1966
  Genus ***Neoechinorhynchus*** Stiles & Hassall, 1905 (= *Neorhynchus* Hamann, 1892; *Eorhynchus* Hamann, 1892; *Eosentis* Van Cleave, 1914)
    Species ***N. rutili*** (Müller, 1780) Stiles & Hassall, 1905 (type) (= *Echinorhynchus rutili* Müller, 1780; *E. cobiditis* Gmelin, 1791; *E. clavaeceps* Zeder, 1800)
      N. acanthuri Farooqi, 1980
      N. afghanus Moravec & Amin, 1978
      N. africanus Troncy, 1970
      N. agilis (Rudolphi, 1819) Van Cleave, 1916 (= *Echinorhynchus agilis* Rudolphi, 1819)
      N. aldrichettae Edmonds, 1971

N. ***armenicus*** Mikailov, 1975
N. ***australis*** Van Cleave, 1931
N. ***bangoni*** Tripathi, 1959
N. ***buttnerae*** Golvan, 1956
N. ***carassii*** Roytman, 1961
N. ***carpiodi*** Dechtiar, 1968
N. ***chelonos*** Schmidt, Esch & Gibbons, 1970
N. ***chilkaensis*** Podder, 1937
N. ***chrysemydis*** Cable & Hopp, 1954
N. ***cirrhinae*** Gupta & Jain, 1979
N. ***coiliae*** Yamaguti, 1939
N. ***constrictus*** Little & Hopkins, 1968
N. ***crassus*** Van Cleave, 1919
N. ***cristatus*** Lynch, 1936
N. ***curemai*** Noronha, 1973
N. ***cyanophyctis*** Kaw, 1951
N. ***cylindratus*** (Van Cleave, 1913) Van Cleave, 1919 [= *Norhynchus cylindratus* Van Cleave, 1913; *Eorhynchus cylindratus* (Van Cleave, 1913) Van Cleave, 1914]
N. ***devdevi*** (Datta, 1936) Kaw, 1951 (= *Eosentis devdevi* Datta, 1936)
N. ***distractus*** Van Cleave, 1949
N. ***doryphorus*** Van Cleave & Bangham, 1949
N. ***elongatum*** Tripathi, 1959
N. ***emydis*** (Leidy, 1851) Van Cleave, 1916 (= *Echinorhynchus emydis* Leidy, 1851; *E. hamulatus* Leidy, 1856)
N. ***emyditoides*** Fisher, 1960
N. ***formosanus*** (Harada, 1938) Kaw, 1951 (= *Eosentis formosanus* Harada, 1938)
N. ***golvani*** Salgado-Maldonado, 1978
N. ***hutchinsoni*** Datta, 1936
N. ***ichthyobori*** Saoud, El-Naffar & Abu-Sinna, 1974
N. ***johnii*** Yamaguti, 1939
N. ***karachiensis*** Bilqees, 1972
N. ***limi*** Muzzall & Buckner, 1982
N. ***longilemniscus*** Yamaguti, 1954
N. ***longissimus*** Farooqi, 1980
N. ***macronucleatus*** Machado, 1954
N. ***magnapapillatus*** Johnson, 1969
N. ***magnus*** Southwell & MacFie, 1925
N. ***manasbalensis*** Kaw, 1951
N. ***nematalosi*** Tripathi, 1959
N. ***nigeriensis*** Farooqi, 1981
N. ***notemigoni*** Dechtiar, 1967

N. octonucleatus Tubangui, 1933
N. oreini Fotedar, 1968
N. ovale Tripathi, 1959
N. paraguayensis Machado, 1959
N. paucihamatum (Leidy, 1890) Petrochenko, 1956 (= *Echinorhynchus paucihamatum* Leidy, 1890)
N. prochilodorum Nickol & Thatcher, 1971
N. prolixoides Bullock, 1963
N. prolixus Van Cleave & Timmons, 1952
N. pseudemydis Cable & Hopp, 1954
N. pterodoridis Thatcher, 1981
N. pungitius Dechtiar, 1971
N. quinghaiensis Liu, Wang & Yang, 1981
N. rigidus (Van Cleave, 1928) Kaw, 1951 (= *Eosentis rigidus* Van Cleave, 1928)
N. roonwali Datta & Soota, 1963
N. roseum Salgado & Maldonado, 1978
N. saginatus Van Cleave & Bangham, 1949
N. salmonis Ching, 1984
N. satoi Morisita, 1937
N. simansularis Roytman, 1961
N. strigosus Van Cleave, 1949
N. stunkardi Cable & Fisher, 1961
N. tenellus (Van Cleave, 1913) Van Cleave, 1919 (= *Neorhynchus tenellus* Van Cleave, 1913)
N. topseyi Podder, 1937
N. tsintaoense Morisita, 1937
N. tumidus Van Cleave & Bangham, 1949
N. tylosuri Yamaguti, 1939
N. venustus Lynch, 1936
N. wuyiensis Wang, 1981
N. yalei (Datta, 1936) Kaw, 1951 (= *Eosentis yalei* Datta, 1936)
N. zacconis Yamaguti, 1935

Genus **Octospinifer** Van Cleave, 1919

Species **O. macilentus** Van Cleave, 1919
O. rohitaü Zuberi & Farooq, 1976
O. torosus Van Cleave & Haderlie, 1950
O. variabilis (Diesing, 1851) Kritscher, 1976 (= *Echinorhynchus variabilis* Diesing, 1851)

Genus **Octospiniferoides** Bullock, 1957

Species **O. chandleri** Bullock, 1957 (type)
O. australis Schmidt & Hugghins, 1973
O. incognita Schmidt & Hugghins, 1973

Genus ***Paulisentis*** Van Cleave & Bangham, 1949
Species ***P. fractus*** Van Cleave & Bangham, 1949 (type)
P. missouriensis Keppner, 1974
Genus ***Zeylonechinorhynchus*** Fernando & Furtado, 1963
Species ***Z. longinuchalis*** Fernando & Furtado, 1963 (type)
Family Tenuisentidae Van Cleave, 1936
Genus ***Paratenuisentis*** Bullock & Samuel, 1975
Species ***P. ambiguus*** (Van Cleave, 1921) Bullock & Samuel, 1975 (type) (= *Tanaorhamphus ambiguus* Van Cleave, 1921)
Genus ***Tenuisentis*** Van Cleave, 1936
Species ***T. niloticus*** (Meyer, 1932) Van Cleave, 1936 (type) (= *Rhadinorhynchus niloticus* Meyer, 1932)

APPENDIX

Acanthocephala genera *incertae sedis* assigned to ***Echinorhynchus*** *sensu latu*

E. acanthotrias Linstow, 1883
E. alcedinis Westrumb, 1821
E. astacifluviatilis Diesing, 1851
E. bipennis Kaiser, 1893
E. blenni Rudolphi, 1810
E. corrugatus Sars, 1885
E. dendrocopi Westrumb, 1821
E. diffluens Zenker, 1832
E. eperlani Linstow, 1884
E. galbulae Diesing, 1851
E. garzae Zeder, 1803
E. gazae Gmelin, 1790
E. hexacanthus Dujardin, 1845
E. inflexus Cobbold, 1861
E. labri Rudolphi, 1819
E. lendix (Phipps, 1774) Marvel, 1905 (= *Sipunculus lendix* Phipps, 1774)
E. nardoi Molin, 1859
E. nitzschi Giebel, 1866
E. orestiae Neveu-Lamaire, 1905
E. pardi Huxley in Ihering, 1902
E. pari Rudolphi, 1819
E. platessae Rudolphi, 1809
E. platessoides Gmelin, 1790
E. pleuronectis Gmelin, 1790
E. pleuronectisplatessoides Viborg, 1795
E. praetextus Molin, 1858
E. pupa Linstow, 1905

E. rhytidodes Monticelli, 1905
E. robustus Datta, 1928
E. sciaenae Rudolphi, 1819
E. scopis Gmelin, 1790
E. scorpaenae Rudolphi, 1819
E. sipunculus Schrank, 1788
E. solitarium Molin, 1858
E. stridulae Goeze, 1782
E. strigis Gmelin, 1782
E. taeniaeforme Linstow, 1890
E. tardae Rudolphi, 1809
E. tenuicollis Froelich, 1802
E. urniger Dujardin, 1845

Editors' note

Dr Omar Amin and the editors will be pleased to receive information for addition to the above from users of this book.

5

Functional morphology

Donald M. Miller and T.T. Dunagan

5.1 External features

Acanthocephala are usually white to cream in color but sometimes slightly yellow to orange, with a cylindrical or slightly flattened body. Worms of most species are not more than 10 mm long, but some measure up to 700 mm. Externally, acanthocephalans have several prominent distinguishing characteristics: an invaginable and retractile proboscis with rows of recurved hooks, body-wall folds and, posteriorly, a copulatory bursa in the male and a gonopore in the female. Externally, the proboscis usually has radial symmetry, with its hooks arranged in longitudinal rows or in a spiral pattern (Van Cleave, 1941 *a*). Hooks and spines (smaller and without roots) may be present on any part of the trunk surface. The proboscis is joined to the trunk by the neck which may be emphasized to varying degrees in different species.

Functionally, the body is divided into two major regions, praesoma and trunk or metasoma. The praesoma, which comprises the armed proboscis, the proboscis receptacle, the cerebral ganglion, the lemnisci and various muscles, and the unarmed neck, is responsible for the attachment of the worm to the intestinal mucosa.

The body wall of the trunk encloses the pseudocoel (body cavity) contained within which are ligament sacs, sex organs (see Chapter 7), excretory organs when present (Table 5.1) and genital ganglia in the male. Except for a small area of the neck in some species, the tegument covers the outer surface of the entire worm. Pervading the entire tegument is a set of channels called the lacunar system, which is organized differently in the three classes of Acanthocephala (Table 5.1).

5.1.1 *Hooks and body spines*

Hooks are an important taxonomic feature for these worms, but there is a paucity of data regarding the chemical composition and ultrastructure of the hooks. Hook development has been described by DeGiusti (1949*a*) for *Leptorhynchoides thecatus*, by Harms (1965*a*) for *Octospinifer macilentus* and by Cable & Dill (1967) for *Paulisentis fractus*.

Table 5.1. *Characteristics of acanthocephalan classes (adapted from Bullock, 1969)*

Palaeacanthocephala	Archiacanthocephala	Eoacanthocephala
Body size		
Small to large	Mostly large	Small
Host habitat		
Mostly aquatic	Terrestrial	Aquatic
Lacunar system (main longitudinal vessels)		
Generally lateral	Dorsal and ventral or dorsal only	Dorsal and ventral, anterior end
Cement glands		
Two to eight – multinucleate	Usually eight uninucleate	Usually one, syncytial, giant nuclei, distinct cement reservoir
Trunk spines		
Present or absent	Absent	Present or absent
Subcuticular nuclei		
Numerous amitotic or few highly branched	Few, elongate or branched, fragments stay close together	Very few giant nuclei
Proboscis receptacle		
Closed sac, two muscle layers except in some Polyacanthorhynchidae	Single muscle layer, often modified by ventral cleft or accessory muscles	Closed sac with single muscle layer
Ligament sacs		
Single, ruptured in mature worms; posterior attachment inside uterine bell	Dorsal and ventral, persistent: dorsal sac attaches to uterine bell	Dorsal and ventral in adult; ventral sac attaches to uterine bell
Nephridia		
Absent	Present or absent	Absent
Embryonic membranes		
Usually thin	Usually thick	Thin
Intermediate hosts		
Crustacea	Insects (and millipedes)	Crustacea

The patterns formed by the different arrangements of hooks were reviewed by Van Cleave (1941*a*). Van Cleave & Bullock (1950) noted that in representative species of a number of genera (*Echinorhynchus*, *Leptorhynchoides*, *Acanthocephalus*), the cystacanth had a proboscis with hooks that were fully developed whereas in others, such as *Filicollis* and *Polymorphus*, the proboscis underwent considerable morphological modification, though not in number and arrangement of the hooks. Some differences in hook number may occur between sexes, as in *Echinorhynchus gadi* where the female has slightly more hooks (§7.4). It is also well known that the largest hooks are not always at or near the apex. For example, in many of the Palaeacanthocephala the largest hooks are near the middle of the proboscis. Moreover, in *Telosentis tenuicornis* and *Arhythmorhynchus trichocephalus* the hooks on the ventral surface are larger than those on the dorsal surface. In the Archiacanthocephala and Eoacanthocephala, the largest hooks are at or near the anterior extremity of the proboscis. Recently, meristogram analyses of morphological variants in hook patterns have been utilized to separate species of *Echinorhynchus* (Huffman & Bullock, 1975) and *Pomphorhynchus* (Huffman & Nickol, 1979).

Trunk spines were reviewed by Van Cleave & Bullock (1950). Spines presumably differ from hooks by the absence of a hollow core and by originating from the tegumental 'felt layer' rather than basement membrane. The latter point has been questioned by Hutton & Oetinger (1980). Van Cleave & Bullock (1950) described wide variation in the form, structure, number, location and pattern of the trunk spines which they believed to be secondary holdfast structures. Worms that possess trunk spines acquire them during development in the intermediate host.

Miller & Dunagan (1971), in a brief report on the hooks of *Macracanthorhynchus hirudinaceus*, stated that the greater curvature had a small groove that extended for approximately half of the exposed portion of the blade. The base of this groove led to a tubelike opening which they speculated may have been involved in secretory activity. *Leptorhynchoides thecatus* and a species of *Neoechinorhynchus* did not have a groove or associated tube.

Anantaraman & Ravindranath (1973), using histochemical techniques, reported that hooks in the cystacanth of *Moniliformis moniliformis* were devoid of lipids, acid mucopolysaccharides and proteins, but contained a polysaccharide similar to chitin. These authors indicated that hook rigidity was due to an O-glycoside link between phenolic acid and a polysaccharide which they suggested contained a hexosamine other than glucosamine. These results differed somewhat from those of Crompton (1963), who obtained a positive Millon reaction and a positive ninhydrin–Schiff reaction for protein with *Polymorphus minutus*. Crompton (1963) also noted that the base of hollow hooks had reducing properties and some alkaline phosphatase activity. More recently, Hutton & Oetinger (1980) examined the development of hooks in *M. moniliformis* and determined that hook primordia did not originate from the hypodermal basement membrane but, as Hammond (1967) indicated, may arise from a connective tissue layer between the hypodermis [*sic*] and muscle layers of the body wall. Hutton & Oetinger (1980), while examining larvae at the same stage of development, also observed ultrastructural similarities between the layer of the proboscis wall from which the hooks originated and the circular muscles of the trunk wall.

5.2 Tegumental organization

The body wall of the Acanthocephala is organized into three distinct sections, an outer tegument, a middle group of circular muscles and an inner group of longitudinal muscles. The lacunar channels run throughout these three sections. Barabashova (1971) implicated the tegument of the body wall of acanthocephalans in trophic, metabolic, protective and supporting functions.

The structure of the tegument appeared with light microscopy as a double-layered cuticle, overlying a double-layered hypodermis (Baer, 1961). The outermost layer of cuticle was thin and apparently structureless. Beneath this, the second layer, also thin, had fine striations perpendicular to the surface. The outer part of the hypodermis contained a complex system of fine, interwoven fibers, which appeared felt-like, or as layers of fibers parallel to the surface alternating with felt-like layers. The inner part of the hypodermis contained fibers oriented perpendicular to the surface, its elements inserted into the felt layer distally and into a basal membrane layer proximally (Baer, 1961).

Electron microscopic studies of the body wall have been undertaken by several authors since the preliminary report of Rothman & Rosario (1961) on *Macracanthorhynchus hirudinaceus*. These studies have significantly changed our interpretation of tegumental organization. Consequently, many terms applied earlier to body wall structures are no longer appropriate.

'Tegument' is preferred over such terms as 'cuticle', 'epidermis' and 'dermis' for the outer syncytial layer of the acanthocephalan body wall. The term is advantageous in that it does not bear the connotation of non-living and does not imply a cellular structure. A summary of the various terms utilized by different authors and their correlation with the body wall layers is shown in Table 5.2.

The most external layer of body wall was described as a very thin electron-dense surface coat of a fine filamentous nature. The terms epicuticle (Crompton & Lee, 1965) and Layer I (Nicholas & Mercer, 1965) have been used, but more recently Beermann, Arai & Costerton (1974) and others have justifiably advocated the use of glycocalyx for this layer. Wright & Lumsden (1968) identified acid mucopolysaccharides in the glycocalyx of *Moniliformis moniliformis* and Rothman & Elder (1970) concluded that the PAS-positive reaction observed in its tegument was due primarily to a keratin sulfate-like substance.

Beneath the glycocalyx, the plasma membrane in *Pomphorhynchus laevis* was found to be a typical bilayer, but had numerous infoldings of 15 nm diameter, approximately 200 nm apart (Stranack, Woodhouse & Griffin, 1966). The openings of these infoldings were described as pores by Hammond (1968 *a*) using the scanning electron microscope. These openings and the glycocalyx material which lines them may act as a molecular sieve system to exclude particles of 8.5 nm or larger (Byram & Fisher, 1973).

The tegument has also been referred to as the cuticle and hypodermis (Table 5.2) in early light microscope studies (Meyer, 1933; Baer, 1961); however, electron microscopy has elucidated three layers. Together, they form a thick layer of fibrous syncytial construction that presumably corresponds to an epithelium. The three fibrous strata are an outer striped layer, a middle somewhat thicker feltwork of layers of fibers running in different directions, and an inner layer of radial fibers. The inner layer is much the thickest and is regarded by many as the epidermis proper while the outer striped and the felt layers are assigned to the cuticle. There are no indications of cell divisions and the entire tegument forms a syncytium. Collagen is the main extracellular fibrous protein found in the tegument and Cain (1970) concluded that the function of the tegumental collagen of *Macracanthorhynchus hirudinaceus* is different from the cuticular collagens of other worms.

The striped layer has been interpreted through electron microscopy to be that area of the tegument adjacent to the plasma membrane and pierced by numerous canals or crypts. These crypts have been calculated to give a 20- to 60-fold increase in surface area (Graeber & Storch, 1978). The material contained within them is continuous with material of low relative molecular mass external to the glycocalyx (Wright & Lumsden, 1969). These crypts may serve as pinocytotic invaginations (Edmonds & Dixon, 1966) and acid hydrolase activity has been found in them (Rothman, 1967). In addition to pinocytotic activity, the discovery of acid phosphatase activity (Byram & Fisher, 1974), alkaline phosphatase activity (Bullock, 1958; Crompton, 1963) and leucine aminopeptidase activity (Crompton,

Table 5.2. *Terminology used to describe the tegument*

Species (host)	Glycocalyx (1 μm in adult)	Outer limiting membrane (8.0–12.5 μm)	Striped (5 μm)	Vesicular (5 μm)	Felt (25 μm)
Macracanthorhyncus hirudinaceus (pig) Meyer (1933)	Not named	Cuticle streifenzone	Cuticulare streifenzone	Cuticulare	Fitzfaserschichte
Polymorphus minutus (duck) Crompton & Lee (1965)	Epicuticle	Trilayered membrane	Striped layer	Not named	Felt layer
Moniliformis moniliformis (rat) Nicholas & Mercer (1965)	Layer I. Cuticle	Layer II. Plasma membrane and subplasma membrane	Layer II. Striped layer	Layer III	Layer IV. Felt layer
Pomphorhynchus laevis (fish) Stranack *et al.* (1966)	Cuticle	Cuticle	Striped layer	Felt layer	Felt layer
Acanthocephalus ranae (toad) Hammond (1967)	External material	Plasma membrane	Striped layer	Canal layer	Felt layer
Moniliformis moniliformis (rat) Wright & Lumsden (1968)	Epicuticle	Plasmalemma	Cuticle, striped layer	Cuticle, felt layer	Cuticle, felt layer
Polymorphus minutus (duck) Butterworth (1969*a*)	Not named	Not named	Cuticle, striped layer	Cuticle, felt layer	Cuticle, felt layer
Moniliformis moniliformis (rat) Wright & Lumsden (1969)	Glycocalyx	Plasmalemma	Not studied	Not studied	Not studied
Echinorhynchus gadi (fish) Lange (1970)	Not named	Not named	Epidermis, electron dense layer	Electron dense layer	Fiber layer

	Outer radial (> 60 μm)[a]	Inner radial (> 80 μm)[a]	Inner membrane (8–10 *nm*)	Connective (> 2 μm)
Macracanthorhyncus hirudinaceus (pig) Meyer (1933)	Radiärfibrillenschicht	Radiärfibrillenschicht	Not studied	Not studied
Polymorphus minutus (duck) Crompton & Lee (1965)	Radial layer	Radial layer	Plasma membrane	Connective tissue layer
Moniliformis moniliformis (rat) Nicholas & Mercer (1965)	Layer V	Layer V	Layer VI. Cell membrane	Layer VI. Cell membrane
Pomphorhynchus laevis (fish) Stranack *et al.* (1966)	Radial layer	Vacuolated dermis	Plasma membrane	Folded plasma membrane
Acanthocephalus ranae (toad) Hammond (1967)	Radial fibrillar layer	Radial fibrillar layer	Plasma membrane	Basement membrane
Moniliformis moniliformis (rat) Wright & Lumsden (1968)	Cuticle	Cuticle	Not named	Not named
Polymorphus minutus (duck) Butterworth (1969*a*)	Not observed	Not observed	Not named	Not named
Moniliformis moniliformis (rat) Wright & Lumsden (1969)	Not studied	Not studied	Plasmalemma	Basement lamina
Echinorhynchus gadi (fish) Lange (1970)	Basal zone	Basal zone	Not named	Basal membrane

[a] The depths of these layers are variable and in general are deeper when the animal is larger

The term 'hypodermis' as used by previous authors has always included more than one of the above layers and hence is not listed in this table.

1963) has established the presence of functional lysosomes and mitochondria in this layer (§6.2).

The middle layer of the tegument, the feltwork layer, called the fiber layer by Lange (1970) in *Echinorhynchus gadi*, was found to contain large concentrations of glycogen (von Brand, 1939*b*). In this layer, the endoplasmic reticulum consists of an irregular tubular array of smooth membranes which become progressively more loosely arranged in the deeper layers. Inevitably many of the tubules appear as vacuoles depending upon the plane of section (Stranack *et al.*, 1966) and small roughly spherical mitochondria with few cristae are scattered throughout. Crompton & Lee (1965) attributed a skeletal function to both the striped and feltwork layer and concluded that fibers of the feltwork layer support the trunk spines.

The feltwork layer grades imperceptably into the radial layer in which electron-dense strands may be visible. In this layer, the strands appear to be formed between parallel membranes of endoplasmic reticulum. The radial layer has been implicated as the main metabolic center of the acanthocephalan body (Crompton, 1963) and this conclusion has been supported by the detection of many mitochondria within this layer. In addition, the inner radial layer contains nuclei and lacunar channels.

In Acanthocephala, the number of nuclei has been reported to be approximately constant for each species, at least in the early stages and in many families throughout life (Van Cleave, 1914). The nuclei of the tegument are few in number and more or less fixed in position, being regularly associated with the main channels of the lacunar system. Thus, in the Neoechinorhynchidae, there are nearly always six nuclei in the tegument, five along the main dorsal lacuna and one ventral along the anterior part of the ventral lacuna (Van Cleave, 1919, 1928). In worms of other families that retain the larval condition of the nuclei, there are up to 20 nuclei in the tegument. Their forms range from oval to rosette, or they may be ameboid, highly branched and of relatively large size; some are more than 2 mm in length. In some families of Acanthocephala, such as Polymorphidae and Echinorhynchidae, the larval nuclei fragment during later development so that the tegument is provided with a large number of small nuclei (Van Cleave, 1928).

The plasma membrane on the inner surface of the tegument of *Moniliformis moniliformis* was found to be highly involuted with numerous thin fingerlike elaborations of diameter 45 nm penetrating up to 1.2 μm into the tegument (Byram & Fisher, 1973). Common to the cytoplasm of this region were glycogen particles, vesicles, Golgi complexes and type-five lysosomes. There are strong precedents for expecting ion and water transport to be linked to these membranous elaborations. Internal to the basal membrane is a fine amorphous basal lamina 50 nm thick. It is equally reasonable to assume that the elaborations increase surface area for attachment and provide structural passageways through which lacunar channels course to more medial layers.

Early investigations of the composition of *Macracanthorhynchus hirudinaceus* (von Brand, 1939*a*), *Leptorhynchoides thecatus* and *Echinorhynchus salmonis* (von Brand, 1939*b*, 1940), and *E. gadi*, *E. salmonis*, *Pomphorhynchus bulbocolli*, *Neoechinorhynchus cylindratus* and *N. emydis* (Bullock, 1949*b*) showed that glycogen was found throughout the worms but principally in the body wall. Ward (1952) observed the glycogen content of *M. hirudinaceus* to be 1.86% of the wet mass, while von Brand (1940) found 1.16 to 1.35%. The glycogen in the bursa wall of male *M. hirudinaceus* was found to be highly water-soluble (von Brand, 1939*a*). Crompton (1963) could not detect glycogen in the body wall of *Polymorphus minutus* when standard fixation was used and argued that it may have been lost in the fixation process.

The lipids in male and female *Macracanthorhynchus hirudinaceus* and *Moniliformis moniliformis* were studied by Beames & Fisher (1964), who extracted the whole worm and analyzed the fatty acids by gas liquid chromatography. They found that males had 1.7 times more lipid than females and that both species had C-10 through C-21 fatty acids. Sixteen non-volatile fatty acids were found in *M. hirudinaceus* and 29 in *M. moniliformis*. Dimitrova (1963) reported on the topographic distribution of lipids in *M. hirudinaceus*. Crompton (1965) detected 10 oxidoreductase enzymes in *Polymorphus minutus* and the distribution of these enzymes paralleled the distribution of glycogen in the tissues. Smith, Dunagan & Miller (1979) separated lacunar fluid from the body wall of *M. hirudinaceus* and analyzed both portions. A significant fraction of the lipids occurred in the lacunar fluid.

In toads infected with *Acanthocephalus ranae*, the praesoma wall and lemnisci have been implicated in both the uptake and excretion of lipids (Pflugfelder, 1949; Crompton & Lee, 1965; Hammond, 1968*b*). Hammond (1968*b*) detected red droplets in the lemnisci of the worms 24 h after their hosts had been fed on fat and the dye Scharlack R. By means of electron microscopy, Crompton & Lee (1965) and Hammond (1968*b*) concluded that the lemnisci contained more lipid than the body wall in *Polymorphus minutus* and *A. ranae*, respectively. Hammond argued that lipids were being excreted by the praesoma and he showed autoradiographically that glyceryl trioleate was taken up *in vitro* through the tegument of the trunk of *A. ranae*, whereas Crompton & Lee had suggested that lipids were being absorbed through the praesoma. Hibbard & Cable (1968) also used autoradiography to study the absorption of glucose, tyrosine and thymidine by the body wall of *Paulisentis fractus*. The bursa of the male was found to be especially active in the uptake of materials. Radioactivity associated with thymidine was particularly evident in tissues surrounding the lacunar canals of the tegument. The cuticle has sometimes been presumed to protect acanthocephalans from the digestive mechanisms of the host. The work of Byram & Fisher (1973) challenges this view by suggesting that the glycocalyx serves as a system for binding and inactivating certain host enzymes and deleterious compounds in the immediate environment of the worm.

The organization of the body wall is such that there is a distinct difference between species in its external appearance. Superficially, some appear to be segmented even though they never are. In *Macracanthorhynchus hirudinaceus*, the external appearance is one of alternating interdigitating folds which arise because of the way the musculature attaches internally. In *M. ingens* the pattern is much coarser. The overall pattern in *Moniliformis moniliformis* is one of annulations. The internal pattern of muscular attachment results in a different internal pattern for the lacunar system. In fact, the taxonomic use of lacunar patterns reflects the arrangement of the body wall muscles.

5.3 Lemnisci

Paired lemnisci originate from inner layers of the praesoma and lie in the pseudocoel (Meyer, 1937). There has been a considerable amount of speculation as to their function; the absence of an alimentary tract in acanthocephalans led some early workers to think of them as feeding structures (Weinland, 1856). Only a few studies have actually dealt in any detail with the lemnisci. Graybill (1902) indicated that the lemnisci were attached only at their points of origin and contained no muscles. Each of the lacunae of an individual lemniscus appeared to contain a 'granular coagulum' and to divide into two canals in the proximal region.

Meyer (1937) stated that six giant nuclei are present in the lemnisci of *Macracanthorhynchus hirudinaceus*. Later, von Brand (1939*b*) found the lemnisci of the same species to contain relatively little glycogen and fat. Bullock (1949*b*) found the lemnisci of *Pomphorynchus bulbocolli* to contain large amounts of glycogen and Crompton (1965) detected glycogen deposited in zones of different concentrations in the lemnisci of *Polymorphus minutus*. Bullock also observed lipid (1949*b*) and lipase (1949*a*) in the lemnisci of *Echinorhynchus salmonis*. In some cases there were large fat droplets which did not seem to be homogeneous. Some had the appearance of a fatty shell with a non-fatty core while others appeared to possess several vacuoles or non-fatty droplets in a lipid matrix. Bullock suggested that fat could be directly absorbed from the host tissue by the subcuticula of the praesoma and the lemnisci without the formation of phospholipid (§5.2). Alkaline glycerophosphatase activity could not be detected by Bullock (1949*a*, 1958) in the lemnisci of *Macracanthorhynchus hirudinaceus*. Crompton (1963) confirmed this in *Polymorphus minutus* and also failed to observe muscles in the lemnisci. Hammond (1966*a*) noted that the fluid-filled spaces of the lemnisci of *Acanthocephalus ranae* communicated with those of the proboscis wall. This fluid was forced from the lemnisci when the neck retractor muscles contracted. Later, Hammond (1967) described a connective tissue layer, a granular basement membrane and a highly folded plasma membrane between the neck retractor muscles and the lemnisci in *A. ranae*. Dense patches with indistinct structures were also found to be associated with the plasma membrane. The basement membrane appeared to have many closely packed vacuoles around it together with a skeletal framework of fibers. Ultrastructural examination of the lemnisci of the eoacanthocephalan *Octospinifer macilentus* revealed cytoplasm containing lipid droplets, lysosomes and numerous mitochondria (Beermann *et al.*, 1974). Free ribosomes, Golgi bodies and vesicles of smooth endoplasmic reticulum were also observed throughout. Numerous fibers of varying orientations interlaced the cytoplasm and suggested a structural function. In brief, Beermann *et al.* (1974) described a metabolically active tissue surrounding an extensively infolded plasma membrane whose surface was covered by a loose collagenous connective tissue layer.

Greeff (1864) and later Bullock (1949*b*) suggested that the lemnisci were excretory organs, while Kaiser (1893) and Baer (1961) considered that they had a nutritional function. Hamann (1891*a*) thought that the lemnisci served as fluid reservoirs to aid in the evagination of the proboscis, but Graybill (1902) and Hammond (1966*a*) did not assign this role to them (Meyer (1933), Bullock (1949*a*), Pflugfelder (1949), Crompton & Lee (1965), and Hammond (1967, 1968*b*)) suggested that the lemnisci are involved in metabolism.

In essence, we view the lemnisci, as did Wright (1970), as structures that greatly amplify the free surface area of the body wall exposed to the pseudocoel. The plasma membrane surrounding the lemnisci is an extension of the basement membrane of the inner radial layer of the tegument. Also, the lacunar canals within the lemnisci are continuous with those of the proboscis tegument (Hammond, 1966*a*), but not with the trunk tegument. Smith *et al.* (1979) identified the predominant lipids present in female *Macracanthorhynchus hirudinaceus* and found that the lemnisci had a higher lipid composition than did the trunk body wall or pseudocoel contents (Table 5.3). Essentially, the most abundant non-volatile fatty acids of the phospholipid class were carbon chains with double-bond numbers as follows: C16:1 (10.3%), C18:0 (15.6%), C18:1 (16.7%), C18:2 and C20:2 (15.8%) and C20:2 (10.9%); smaller amounts were found of a large number of other fatty acids. Smith *et al.* (1979) also found that the most abundant non-volatile fatty acids of the neutral lipids were carbon chains with bond numbers as follows: C14:1 (13.1%), C16:0 (12.7%), C18:0 (11.6%) and C18:1 (9.9%). There were smaller amounts of several other fatty acids of the neutral lipid class.

5.4 Lacunar system

The Acanthocephala lack a directional-flow circulatory system but possess instead a loose network of tubelike cavities without linings or pumps; it is probable that the trunk wall musculature acts as the motive force for fluid flow. These cavities are the channels of the lacunar system which interconnect, forming a definite pattern in some species. Even in those genera such as *Centrorhynchus* where the lacunar system has been described as netlike and without a regular pattern, there are usually a pair of main channels that traverse the length of the body with numerous connections to the remainder of the system. In archiacanthocephalans and eoacanthocephalans, the large channels are located dorsally and ventrally in the tegument, although the ventral channel is occasionally missing (Meyer, 1933). In palaeacanthocephalans, the main channels are placed laterally (Table 5.1).

It is now generally accepted that the lacunar system of the trunk wall is distinct from that of the praesoma, the latter presumably having no connection with the channels of the former. However, this conclusion has yet to be confirmed by dye-injection techniques (Miller & Dunagan, 1976). Workers that originally described the lacunar system and those who have written general accounts (Rauther, 1930; Baer, 1961) have all indicated that it is restricted to the inner radial layer of the tegument and does not open to the outside or communicate with any body structure interior to

Table 5.3. *Lipid composition of female* Macracanthorhynchus hirudinaceus (*adapted from Smith* et al., *1979*)

Tissue	Wet mass (mg)	Lipid mass (mg)	Percentage of lipid
Body wall	616	39	4.7
Pseudocoel	982	23	2.3
Lemnisci	308	32	10.4

the muscles. However, it is now clear that the worm surface has numerous infoldings with direct openings to the 'hypodermal' interior which, in turn, may provide access to the lacunar system (Rothman & Rosario, 1961; Crompton & Lee, 1963; Nicholas & Hynes, 1963*a*; Nicholas & Mercer, 1965; Edmonds & Dixon, 1966). Recent studies (Miller & Dunagan, 1976, 1978) using a dye-injection technique have extended the understanding of the components involved in connecting the various anastomosing channels of the system (Fig. 5.1; Table 5.4). In addition, these authors described the rete system, consisting of a series of connecting thin-walled tubes covering the entire medial surface of the longitudinal muscles of *Macracanthorhynchus hirudinaceus*. The rete system is located between the longitudinal and circular muscles in *Oligacanthorhynchus tortuosa*. No connection was observed between the rete system and the lacunar system in either species.

Description of the medial longitudinal channel (MLC) has been omitted from most accounts of lacunar systems. Rauther (1930) illustrated the dorsal and ventral lacunar channels under the term 'median Hauptkanale'. He also depicted the medial longitudinal lacunar channels but did not label these structures in his illustration of a cross-section of *Macracanthorhynchus hirudinaceus*. Kilian (1932) studied the general morphology of the body wall of *Oligacanthorhynchus microcephala*, a large acanthocephalan from South American opossums. In his discussion of the lacunar system and body wall, he did not mention a MLC along the medial surface of the longitudinal muscles, but did show a structure referred to as a 'Ringmuskelmarkbeutel' in the proper position for the MLC.

The MLC(s) is (are) a prominent feature of the inside surface, being observed approximately one third of the distance from the dorsal lacunar channel. Because the channels are closer to the dorsal surface and are conspicuous in histological sections, Dunagan & Miller (1978*b*) suggested that they may be utilized for the determination of body planes in histological sections. During development, each MLC is apparently formed by numerous cells that have a complicated pattern of interconnection. Each of the interconnecting cells is attached to longitudinal muscles to surround the point where they pass between the muscle bundles via the radial canal to join the tegumental part of the lacunar system. The path of the channel so formed is somewhat irregular as the earlier connections between cells were not in a straight line. This discontinuous attachment permits other systems to pass under the MLC (Dunagan & Miller, 1978*b*).

A pair of MLCs is located adjacent to the circular muscles and connected to circular lacunar channels called primary ring canals (PRCs) (Miller & Dunagan, 1976). The PRCs are connected to the MLCs by regularly spaced transverse mediolateral connections termed radial canals

Table 5.4. *Lacunar components (adapted from Miller & Dunagan, 1976)*

Abbreviation	Structure
MLC	Medial longitudinal channel
RC	Radial canal
PRC	Primary ring canal
SRC	Secondary ring canal
DLC	Dorsal lacunar channel
VLC	Ventral lacunar channel
SLC	Secondary longitudinal canal
CMC	Circular muscle canal
LMC	Longitudinal muscle canal

Fig. 5.1. Schematic diagram of the lacunar system in *Macracanthorhynchus hirudinaceus*. DLC, dorsal lacunar channel; HC, hypodermal canal; MLC, medial longitudinal channel; PRC, primary ring canal; SRC, secondary ring canal; VLC, ventral lacunar channel. (From Miller & Dunagan, 1976.)

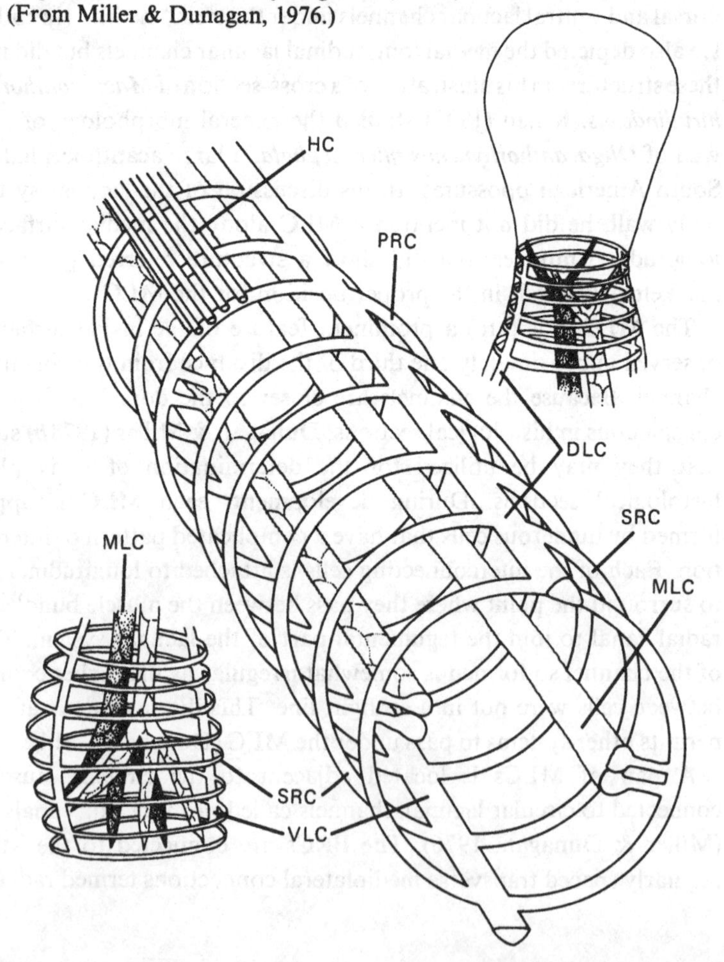

(RCs). These small tubular connections pass between the circular and longitudinal muscles and are easily overlooked in fixed tissue. However, when the dorsal or ventral lacunar channels of living worms are injected with dye or Indian ink, the substance moves progressively through the channels as follows; either dorsal or ventral lacunar channel (DLC, VLC), to PRCs, to RCs and finally into the MLC. Meyer (1931 *a*) described the MLC under the name '*linke* (*rechte*) *Muskelbeutel*', but gave little detail of its morphology, function or possible link with the '*ventrales* (*dorsales*) *hypodermis langefass*'. Miller & Dunagan (1976) demonstrated that the secondary ring canals were actually the dye-filled lumens of large circular muscles. Thus, the hollow muscles provide an extension of the lacunar ducts and both are integrated into one network enclosed on the pseudocoel surface by the MLC and on the outside by the DLC and VLC. Whether or not lumen contents actually move in a prescribed direction has not been determined.

5.5 Muscles

Few detailed accounts of proboscis receptacle morphology have been published. The terminology used by Kaiser (1893) or Kilian (1932) for the musculature associated with the praesoma has not found uniform translation. Basically, the proboscis receptacle consists of one or two muscle layers enclosed in a thin sheath. A stylized diagram of a section through the area of the cerebral ganglion, Fig. 5.2, shows the major muscle groups observed in the proboscis receptacle at this level, and Table 5.5 gives some of the terms used by different authors. Species belonging to the Palaeacanthocephala are described as having a double-walled receptacle, whereas those belonging to the Eoacanthocephala and Archiacanthocephala have a single-walled receptacle. Species of *Moniliformis* were thought to be an exception, but Wanson & Nickol (1975) concluded that in *M. moniliformis* the receptacle was not double-walled. They also indicated that *M. moniliformis* differs from other Archiacanthocephala in that the muscular layer of the receptacle wall does not form the ventral cleft observed in other archiacanthocephalans. Perhaps a more obvious difference in *Moniliformis* is the spiral arrangement of the protrusor muscles located outside the proboscis sheath. This design may give the proboscis a twisting motion during eversion.

Nickol & Holloway (1968) found that the receptacle wall in *Corynosoma hamanni* possesses two layers of radially arranged muscle fibers. Both muscle layers extended anteriorly to the junction with the proboscis wall, but the outer layer is very thin at the closed end of the receptacle. Each muscle layer of the receptacle wall has an associated layer of thin reticular tissue resembling connective tissue. Wanson & Nickol (1975) made similar observations on *Plagiorhynchus cylindraceus* and also described the proboscis sheath, a thick non-muscular acellular covering of the outer surface of the receptacle.

The thick receptacle wall of *Macracanthorhynchus hirudinaceus* appears to be perforated by numerous tubelike channels that allow the fluid from the receptacle lumen to enter the receptacle wall muscles. The fluid apparently runs throughout the length of the proboscis receptacle and fills most of the core of the proboscis from its base to the apical organ. There is no information on the composition of the fluid, but it stains a deep magenta color with the PAS reaction suggesting that carbohydrates with a 1,2-glycol structure are present.

Fig. 5.2. Diagrammatic cross-sections of the muscle groups surrounding the cerebral ganglion. (*a*) *Macracanthorhynchus hirudinaceus* (Redrawn from Rauther, 1930). (*b*) *Moniliformis moniliformis* (Redrawn from Kaiser, 1893). The proboscis sheath surrounds the ganglion and the receptacle wall. The spiral arrangement of the external muscles of the proboscis in *M. moniliformis* is strikingly different from the arrangement in *M. hirudinaceus*. CG, cerebral ganglion; CM, circular muscles; DACI, dorsal apical cone invertor muscle; DLC, dorsal lacunar channel; DPR, dorsal proboscis retractor muscle; DRP, dorsal receptacle protrusor muscle; LEM, lemniscus; MVLR, midventral longitudinal receptacle muscle; PR, proboscis retractor muscle; PS, proboscis sheath; RW, receptacle wall muscle; SSC, sensory support cell; VLC, ventral lacunar channel; VPR, ventral proboscis retractor muscle; VRP, ventral receptacle protrusor muscle.

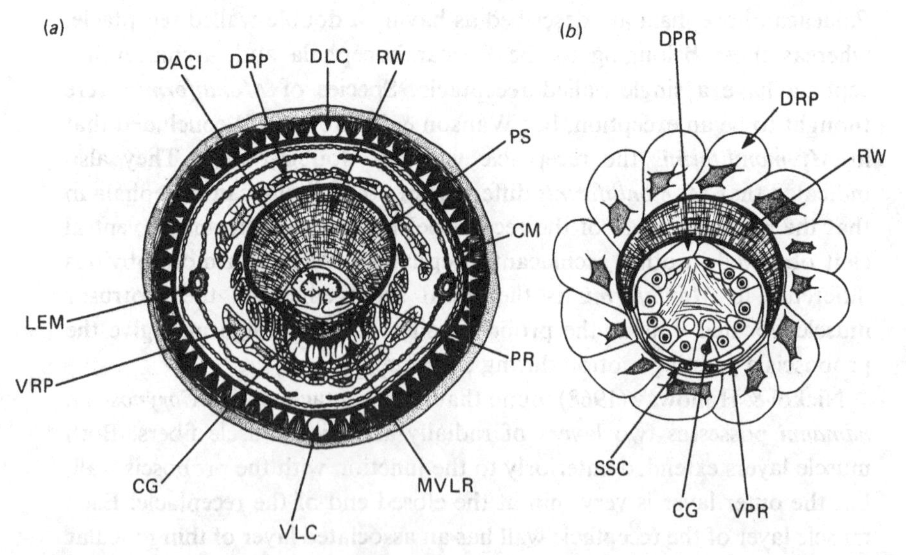

Table 5.5. *Muscle nomenclature*

Macracanthorhynchus hirudinaceus Kaiser (1893)	*Oligacanthorhynchus microcephala* Kilian (1932)	*Oligacanthorhynchus microcephala* Hyman (1951)	*Oncicola* sp. Schmidt (1972*b*)	*Macracanthorhynchus hirudinaceus* Dunagan & Miller (1974)	*Prosthenorchis elegans* Wanson & Nickol (1975)
Protrusor receptaculi ventralis	Protrusor receptaculi ventrales	Ventral receptacle protrusors	—	Ventral receptacle protrusors	—
Protrusor receptaculi lateralis	Protrusor receptaculi lateralis	—	—	Lateral receptacle protrusors	Receptacle protrusors
Sarkolemmahüll-membrane	Sarcolemma	Ventral wall of receptacle	Non-muscular layer	Proboscis sheath	Non-muscular sheath
Retractor proboscidis	Retractor proboscidis	Proboscis retractors	Primary retractors	Proboscis retractors	Proboscis invertors
Fibrillenplatten des Receptaculum	Receptaculum proboscidis	Dorsal receptacle wall	Muscular wall	Receptacle wall	Receptacle wall
Protrusor receptaculi dorsalis	Protrusor receptaculi dorsalis	Dorsal receptacle protrusor	Primary dorsal protrusors	Dorsal receptacle protrusor	—
Inner Deckmuskel des Receptaculum	Ventraler Belagmuskel	Midventral receptacle muscle	Longitudinal band	Midventral longitudinal muscle	Ventral receptacle muscle
Inner Deckmuskel des Receptaculum	Ventraler Belagmuskel	Midventral receptacle muscle	Primary ventral protrusors	Midventral longitudinal muscles	Ventral receptacle muscle
Ganglion cephalicum	Ganglion cephalicum	Cerebral ganglion	Brain	Cerebral ganglion	Ganglion
—	Retractor proboscidis	—	—	Dorsal apical cone invertors	—
Markraum des Rüsselscheide	Markbeutel des Receptaculi proboscidis	—	—	Medullary fluid	—

The importance of the proboscis receptacle in the taxonomy of the Oligacanthorhynchidae was discussed by Schmidt (1972*b*). He pointed out that the muscular wall did not insert on the inside of the proboscis but was attached near the inner apex. Schmidt suggested that the muscles associated with the outside surface of the receptacle in most archiacanthocephalans may be homologous with the outer muscle layer of the

Fig. 5.3. Diagrammatic cross-sections through the proboscis of *Macracanthorhynchus hirudinaceus* (*a* through *f*, anterior to posterior). CPM, circular muscle of proboscis; DACI, dorsal apical cone invertor muscle; DRF, dorsal receptacle flexor muscle; DRP, dorsal receptacle protrusor muscle; DRR, dorsal receptacle retractor muscle; HR, hook retractor muscle; LPN, lateral posterior nerve; LRP, lateral receptacle protrusor muscle; M, medullary fluid; PR, proboscis retractor muscle; PS, proboscis sheath; RM, retinacular muscle; RW, receptacle wall muscle; VACI, ventral apical cone invertor muscle; VLR, ventral longitudinal retractor muscle; VRP, ventral receptacle protrusor muscle; VRR, ventral receptacle retractor muscle. (Redrawn from Dunagan & Miller, 1974.)

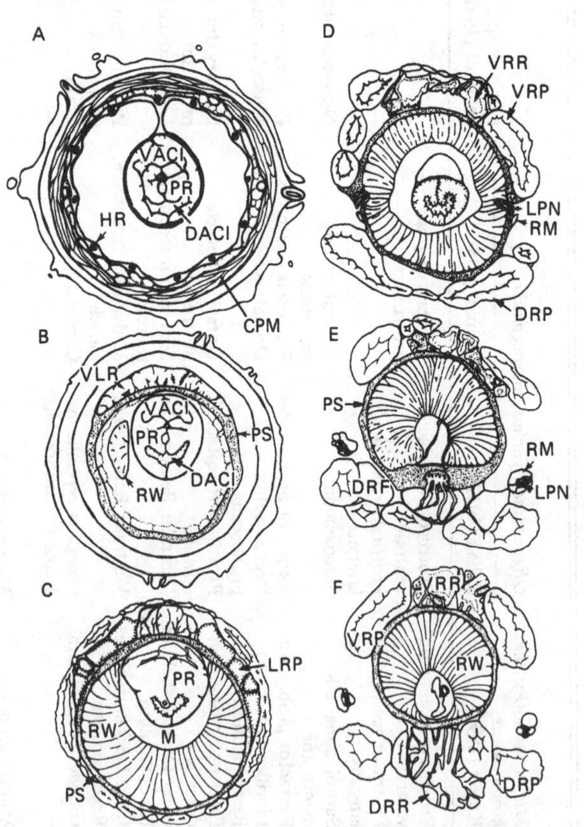

receptacle in palaeacanthocephalans. This observation was supported by Wanson & Nickol (1975).

Van Cleave & Bullock (1950) examined the praesoma of *Neoechinorhynchus emydis* from turtles and used the name 'invertors of the proboscis' for the muscles originating at its tip and extending posteriorly through its length into the receptacle. These muscles were found to continue through the muscular wall of the receptacle to become the receptacle retractors which are inserted into the trunk body wall. They also described a

Fig. 5.4. A diagrammatic representation of the body wall of *Macracanthorhynchus hirudinaceus*. The diagram shows the relationship of the lacunar and rete systems to the muscles and tegument. Note that the anastomosing channels making up this vascular system allow a fluid connection between the lumen of the longitudinal canals in the tegument and the medial longitudinal channel in the pseudocoel. CM, circular muscles; DLC, dorsal lacunar channel; HC, hypodermal canal; HD, hypodermal duct; LM, longitudinal muscle; MLC, medial longitudinal channel; N, nucleus; PRC, primary ring canal; RC, radial canal; R, rete muscles. (From Miller & Dunagan, 1976.)

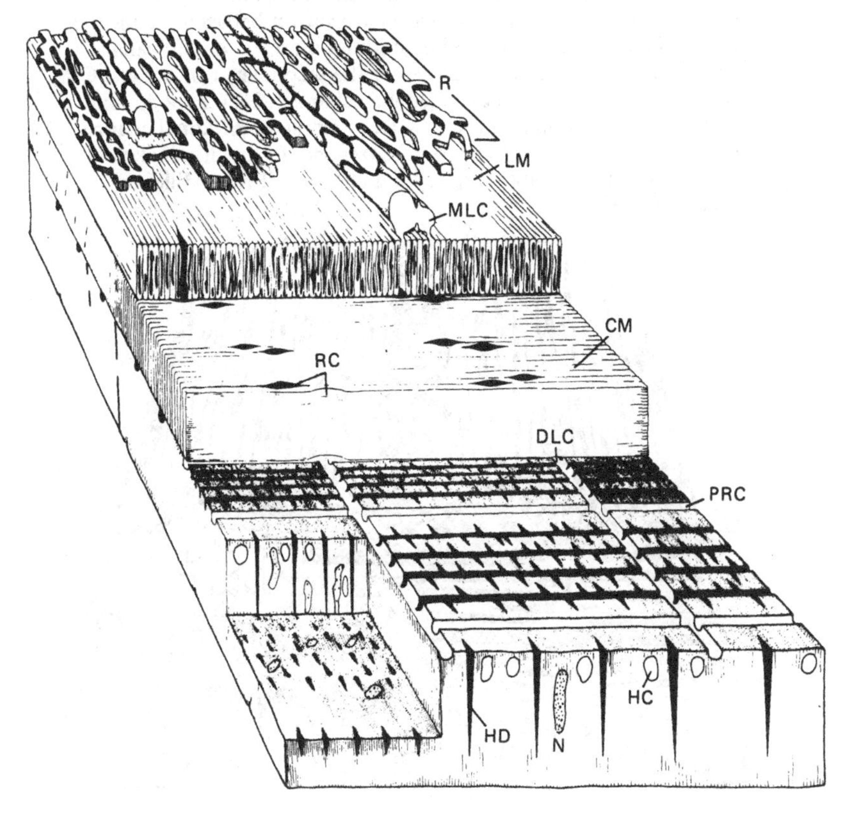

muscular sling which was seen as a small sacculate structure outside the posterior ventral margin of the receptacle with a thick sheet of fibers extending anterodorsally from it over each lateral surface of the receptacle wall.

Fourteen muscles were illustrated by Dunagan & Miller (1974) in the praesoma of *Macracanthorhynchus hirudinaceus* (Fig. 5.3). The role of these and other muscles in the movement of the proboscis was discussed.

5.5.1 *Body wall muscles*

The body wall organization of muscular layers is similar in most acanthocephalan species and consists of an inner longitudinal and an outer circular layer enclosed by the tegument. Recent studies (Miller & Dunagan, 1976, 1977, 1978) have shown considerable variation in their attachment

Fig. 5.5. A diagrammatic representation of the body wall of *Oligacanthorhynchus tortuosa*. The rete network is located between circular and longitudinal muscles. All muscles are loosely packed. The rete system with its medial longitudinal channels is located on the medial surface of the longitudinal muscle and adjacent to the pseudocoel. Lateral posterior nerves are shown between the medial longitudinal channel and the longitudinal muscles. Radial canals penetrate the muscle layers and enter the secondary ring canals. CM, circular muscles; DLC, dorsal lacunar channel; HD, hypodermal duct; LM, longitudinal muscle; MLC, medial longitudinal channel; R, rete muscles; RC, radial canal; SLC, secondary longitudinal canal; T, tegument; VLC, ventral longitudinal channel. (From Miller & Dunagan, 1978.)

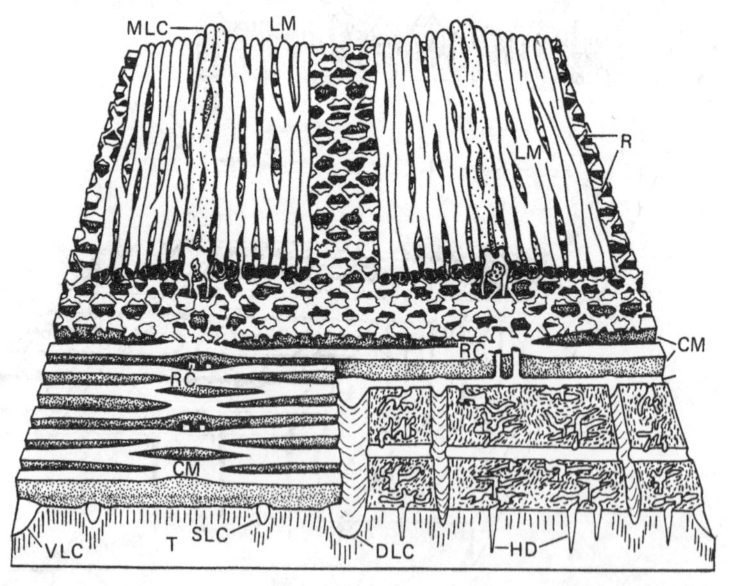

and internal distribution. Two species that have great differences in the size and distribution of the inner longitudinal muscles and the kind of pattern that they present are *Macracanthorhynchus hirudinaceus* (Fig. 5.4) and *Oligacanthorhynchus tortuosa* (Fig. 5.5). In addition, there is a difference between these two species in the position of the rete system (Miller & Dunagan, 1976). Work on other species indicates that there are differences in the overall pattern of the longitudinal musculature as well as other muscle systems.

An interesting aspect of acanthocephalan musculature is their tubular nature (Miller & Dunagan, 1977) and intimate involvement in the circulation of lacunar fluid (Fig. 5.6). Because the muscles are arranged in an

Fig. 5.6. Scanning electron micrographs of razor sections of *Oligacanthorhynchus tortuosa* at the level of the testis. (*a*) Cross-section through the entire body. Longitudinal muscles are shown in cross-section as are the lacunar canals in the tegument. Testis is shown in the middle of the pseudocoel. (*b*) A razor cut through the body shows the circular muscles in cross-section (see box on (*a*)). (*c*) Longitudinal section which cuts through the medial longitudinal channel. (*d*) A cross-section showing longitudinal and circular muscle. (From Miller & Dunagan, 1977.)

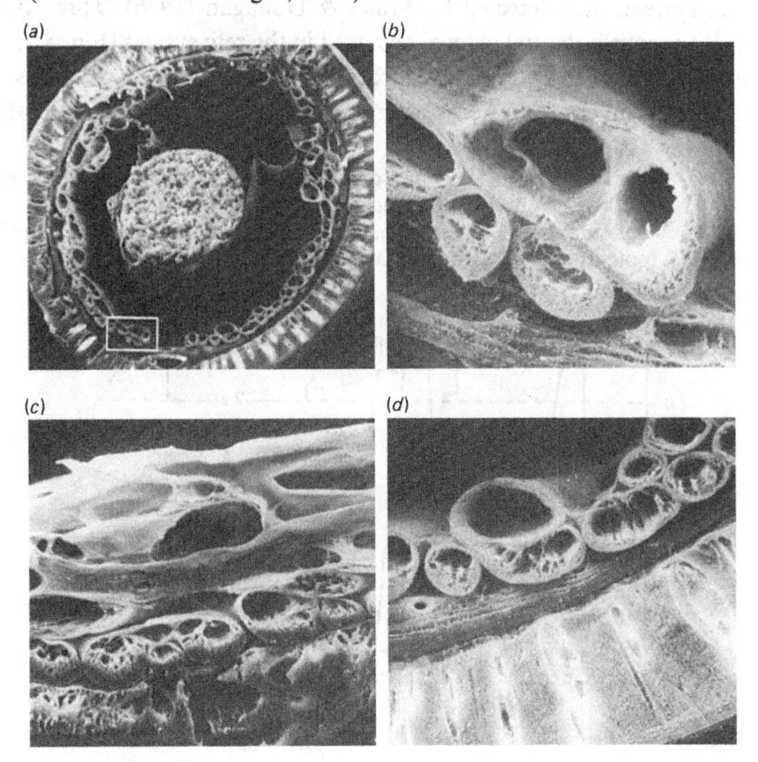

anastomosing network of tubes, they effectively provide for their own nourishment and waste removal. Presumably, contraction of circular muscles forces lacunar fluid into the longitudinal muscle system and contraction of the longitudinal muscles tends to force lacunar fluid into the circular muscle system.

In studies by Hightower, Miller & Dunagan (1975), both the circular and longitudinal muscles of *Macracanthorhynchus hirudinaceus* were found to be electrically inexcitable by direct methods, but overall body stimulation could be utilized to excite either set. In addition to electrical stimulus, acetylcholine caused contraction while atropine blocked it. Evidence was also obtained for a through-conduction system located near the medial longitudinal channel. On the basis of the difference in response to stimuli, a separate activation system for each set of muscles was postulated. The tegumental layer of the body wall was not necessary for contraction; however, its presence did alter the tension-time and relaxation-time parameters.

Subsequently, Wong, Miller & Dunagan (1979) demonstrated that contractions of the body were associated with electrical potential patterns that could be recorded from an extensive thin-walled tubular network, the rete system, first detected by Miller & Dunagan (1976). Three types of spontaneous potential change occurred in the rete system (Fig. 5.7; Wong *et al.*, 1979). The potentials recorded depended upon the positions of the electrode in the anterior–posterior axis and the medial–lateral axis.

Fig. 5.7. Terminology and diagrammatic representation of potentials recorded from *Macracanthorhynchus hirudinaceus* body wall musculature. A, aberrant potential; L, large potential; S, synaptic-like potential. (Redrawn from Miller, Wong & Dunagan, 1981.)

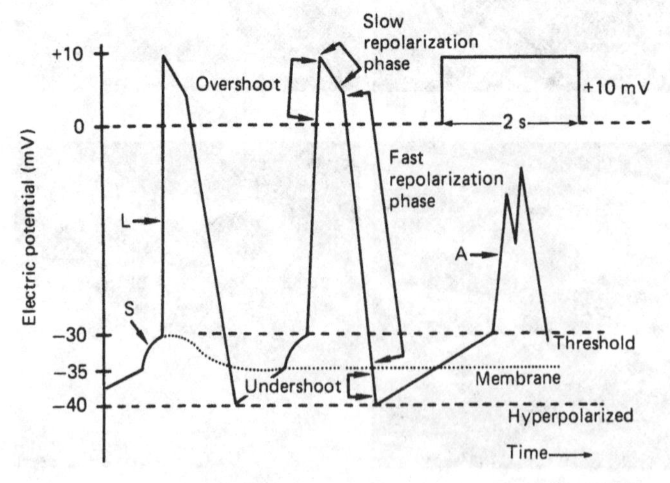

Tetrodotoxin eliminated the large potentials but not the smaller ones, while ouabain lengthened the time for depolarization of the large potentials and depolarized the membrane potential. It was suggested that the rete system activated the body wall muscles in the Acanthocephala (Wong *et al.*, 1979).

More recently, Miller, Wong & Dunagan (1981) determined that spontaneous spike potentials were dependent upon calcium flux. The membrane potential was depolarized by acetylcholine, potassium-free medium, calcium ions and chloride-free medium. Spontaneous potentials increased in number with acetylcholine and calcium at concentrations above 3mM, but were decreased in number in chloride-free and calcium-free media.

5.6 Nervous system

Fourteen general descriptions of acanthocephalan ganglia occur in the literature (Table 5.6) but only a few of these deal with work done in the last 50 years.

5.6.1 *Cerebral ganglion*

Topographic views of the dorsal and ventral surfaces of the cerebral ganglion of *Macracanthorhynchus hirudinaceus*, *Moniliformis moniliformis* and *Oligacanthorhynchus tortuosa* have been prepared by Miller, Dunagan & Richardson (1973) and by Dunagan & Miller (1975, 1981). Most of the cell bodies of this ganglion are on the ventral surface (Fig. 5.8) and a large and extensive neuropile separates these cells from those on the dorsal surface. Decussation, while common, has not been observed in all cells. Ipsilateral and contralateral neurites from the same cell body are common in the larger cells. Indeed, some cells such as the very large posteriormost cell have processes that extend anteriorly as well as posteriorly. Cell bodies are paired on both surfaces between each half of the ganglion with a few located centrally between the halves. The junctions formed between neurons and intracellular material have been described by Golubev & Sal'nikov (1979) who found six types in the cerebral ganglion of *Echinorhynchus gadi*.

The cytoarchitecture of the individual cells differs considerably. Since most cells are matched between halves of the ganglion, however, characteristics observed in one cell are also found in its counterpart. Nuclei are usually large and round with distinct nucleoli; several centrally located cells have two nuclei. Perinuclear rings are common. Most of the differences between ganglion cells occur in the cell granulation and reflect different neurotransmitters. Crompton (1963) found cholinesterase activity in cerebral ganglion cells of *Polymorphus minutus*.

Table 5.6. *Ganglia from adult acanthocephalans*

Species	Cerebral ganglion: Cell number (reference)	Cerebral ganglion: Size (μm)	Genital ganglion: Cell number	Host
Acanthocephalus anguillae	50 (Hamann, 1891*a*)	400 wide	—	Fish
Acanthocephalus lucii	— (Kaiser, 1893)	120–130 long 90–95 wide 65–70 thick	—	Fish
Acanthocephalus ranae	54 (Kaiser, 1893)	145–155 long 100–106 wide (Hamann, 1891*a*)	15	Frog
Arhythmorhynchus uncinatus	— (Kaiser, 1893)	240–250 long 110–115 wide 50–60 thick	—	Bird
Bolbosoma balaenae	— (Kaiser, 1893)	185–195 long 98–105 wide 46–50 thick	—	Marine mammal
Bolbosoma turbinella	73 (Harada, 1931)	225 long 160 wide	14–15	Marine mammal
Corynosoma hamanni	— (Nickol & Holloway, 1968)	250 long 90 wide	—	Weddel seal
Gracilisentis gracilisentis	108–109 (Van Cleave, 1914)	100 long	18	Fish
Macracanthorhynchus hirudinaceus	86 (Kaiser, 1893)	320–350 long 330–340 wide 140–150 thick	19–20	Swine
Moniliformis moniliformis	88 (Dunagan & Miller, 1975)	150–170 long 130–135 wide 100–105 thick (Kaiser, 1893)	19	Rat
Neoechinorhynchus emydis	94 (Van Cleave, 1914)	—	—	—
Oligacanthorhynchus microcephala	— (Hamann, 1891*a*)	—	30–40	Anteater
Oligacanthorhynchus microcephala	80 (Kilian, 1932)	160–220 long 130–220 wide 49–90 thick	30	Opossum
Oligacanthorhynchus tortuosa	83 (Dunagan & Miller, 1981)	400 long 190 wide 90 thick	—	Opossum

The shape and size of the cerebral ganglion varies among species and is affected by fixation procedures. In *Macracanthorhynchus hirudinaceus* the dorsal surface may be concave or convex depending upon the preparation technique. Nevertheless, certain cells can easily be identified as common among species. The general design of this ganglion is the same in male and female worms; whether it is identical in both sexes is unknown.

Harada (1931) described an isolated tripolar nerve cell in *Bolbosoma turbinella* located behind the cerebral ganglion but within the proboscis receptacle. He stated that two nerve fibers originated from the sides of this cell and eventually reached the proboscis tip as the anterior median nerves. Connections between this cell and the cerebral ganglion were not described but his figure 'A' suggests that they occur. We have not yet observed such a cell.

5.6.2 *Peripheral nervous system from cerebral ganglion*

Most work on the peripheral nervous system of acanthocephalans has been confined to those of the larger species (see Bullock & Horridge,

Fig. 5.8. Map of the cerebral ganglion of *Oligacanthorhynchus tortuosa* with a dorsal and ventral view. Numbers in the cells are arranged to have even numbers on the right and numbers ending in zero along the midline. Cell limits determined from reconstruction of serial sections. (Redrawn from Dunagan & Miller, 1981.)

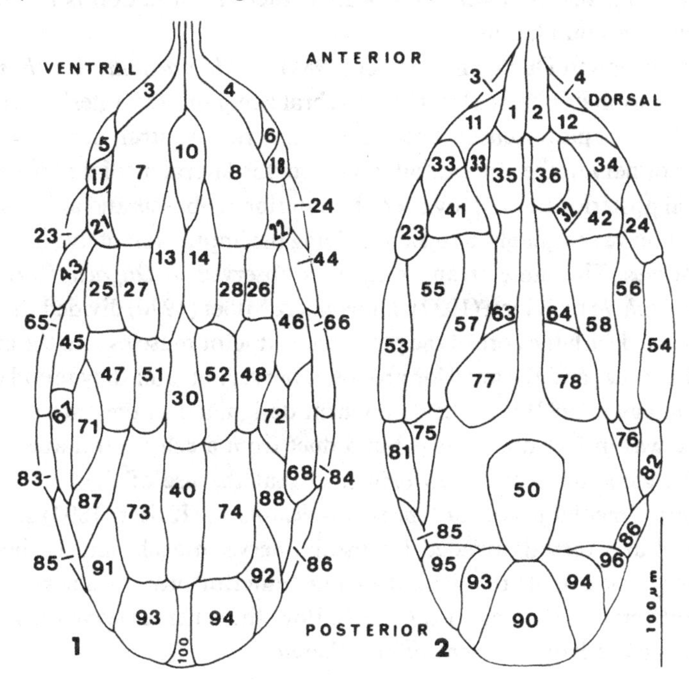

1965; Nicholas, 1967). This dealt with nerves associated with the cerebral ganglion and largely ignored nerves originating from the male genital ganglion (Kaiser, 1893; Brandes, 1899). The following descriptions on parts of the peripheral nervous system have appeared: Kilian (1932) for *Oligacanthorhynchus microcephala*; Harada (1931) for *Bolbosoma turbinella*; Dunagan & Miller (1970, 1979) for *Macracanthorhynchus hirudinaceus*; Dunagan & Miller (1976, 1977, 1978*b*) for *Moniliformis moniliformis*; and Makhanbetov (1972, 1974, 1975*a*, *b*) and Ivanova & Makhanbetov (1975) for *Polymorphus phippsi*. In *P. phippsi* the general plan for the nervous system is similar to that of other species, but there are four commissures in the anterior part of the praesoma.

As understood at present, the nervous system is more complex in male worms; for example in male *Moniliformis moniliformis* three ganglia-associated nerve tracts are known. Of these, the bursal ganglion remained undiscovered until found by Dunagan & Miller (1977).

There is still some confusion about the acanthocephalan nervous system, even for *Macracanthorhynchus hirudinaceus*. For example, when considering the cerebral ganglion nerves, there are said to be five peripheral nerves, yet Rauther (1930) stated, '*Bei M. hirudinaceus...gehm vom Gehirn – ganglion nach vorn, 1 Mediannerv, 1 Ventralnerv and 2 Switennerven aus*...'. Further complications arise from the lack of agreement on the nomenclature, but this is typical of any area where information is in the early stages of accumulation.

According to Dunagan & Miller (1984), in *Macracanthorhynchus hirudinaceus* six major nerves leave the cerebral ganglion: the anterior proboscis nerve, the apical sensory nerve, the anterior ventral nerve (=lateral anterior nerve), the anterior lateral nerve, the lateral medial nerve and the lateral posterior nerve. However, the anterior proboscis and apical sensory nerves leave the ganglion together, giving the impression of five rather than six nerves. The same arrangement was observed in *Oligacanthorhynchus microcephala* by Kilian (1932). Dunagan & Miller (1970) divided the nerves associated with the core of the proboscis retractor muscles into the anterior medial nerve and the anterior proboscis nerve, but later they found (Miller & Dunagan, 1983) that the component designated as the anterior medial nerve was in fact not a nerve but a duct from a cell ventral and anterior to the ganglion. Thus, they explained that the use of the original term 'anterior medial nerve' or '*vorder Median-nerv*' (Kaiser, 1893) should be revised and they designated the medial nerve in each pair as the apical sensory nerve while retaining the name 'anterior proboscis nerve' for the lateral nerve of each pair (Table 5.7). Both these nerves leave the ganglion together but have different origins therein.

The anterior lateral nerve consists of two pairs of neurites that together leave the anterior midventral surface of the cerebral ganglion. The paired anterior ventral nerves, each of which consists of a bundle of seven processes, leave the ganglion anterior to the anterior lateral nerves and extend to the lateral sensory organs. The paired lateral medial nerves leave the ganglion along its midlateral margin and extend posteriorly before doubling back to run anteriorly. Each lateral medial nerve consists of two processes. Each of the paired lateral posterior nerves consists of approximately 22 processes. This pair of nerves leaves the ganglion from the dorsal surface along the posterior lateral margin. Each bundle of processes is enclosed by the retinacular muscle while passing through the pseudocoel.

Little is known about the nerve endings, but in *Polymorphus phippsi* they are seen as circular plates, 7–18 μm in diameter, and are particularly clear in the retractor muscles of the proboscis and the longitudinal muscles of the body wall (Makhanbetov, 1975*a*, *b*).

5.6.3 *Genital ganglia*

The posteriorly located genital ganglia are apparently found in all male worms and consist, in *Bolbosoma turbinella* and *Oligacanthorhynchus microcephala*, of two lateral cell accumulations connected by a small dorsal and a large ventral commissure (Harada, 1931; Kilian, 1932). Dunagan & Miller (1978*a*, 1979) were unable to confirm the presence of a ventral commissure in *Macracanthorhynchus hirudinaceus*. The genital ganglia of *Polymorphus phippsi* are linked by a ventral commissure containing five or six fibers and a dorsal commissure with three or four (Ivanova & Makhanbetov, 1975). The two commissures in *P. phippsi* originate from the symmetrical ends of each horseshoe-shaped ganglion. According to Kilian (1932), the genital ganglia in *O. microcephala* are connected by a dorsal commissure containing eight fibers and a ventral commissure with 15 fibers. Seven large fibers originate from each ganglion (Kilian, 1932)

Table 5.7. *Nerves originating from the cerebral ganglion of* Macracanthorhynchus hirudinaceus (*adapted from Miller & Dunagan, 1983*)

Number	Name	Abbreviation
1	Anterior proboscis nerve	APN
	Apical sensory nerve	ASN
	Anterior proboscis nerve	APN
2	Anterior ventral (=lateral anterior) nerve	AVN
3	Anterior lateral nerve	ALN
4	Lateral medial nerve	LMN
5	Lateral posterior nerve	LPN

but that appears to include the two commissures, leaving five (unnamed by Kilian) to innervate the male reproductive system. The number of cells in each pair of ganglia (Table 5.6) varies among species.

These ganglia are located, in *Macracanthorhynchus hirudinaceus*, below

Fig. 5.9. Scanning electron montage of the posterior of *Macracanthorhynchus hirudinaceus*, illustrating the genital ganglion with respect to the bursa. Textural differences in the musculature of the genital sheath are apparent as are the presence of protrusor, bursal depressor and longitudinal muscles. B, bursa; BD, bursal depressor muscle; BS, bursal sheath; BUP, bursal protrusor; GES, genital sheath; GG, genital ganglion; LM, longitudinal muscle; P, protrusor muscle; T, tegument.

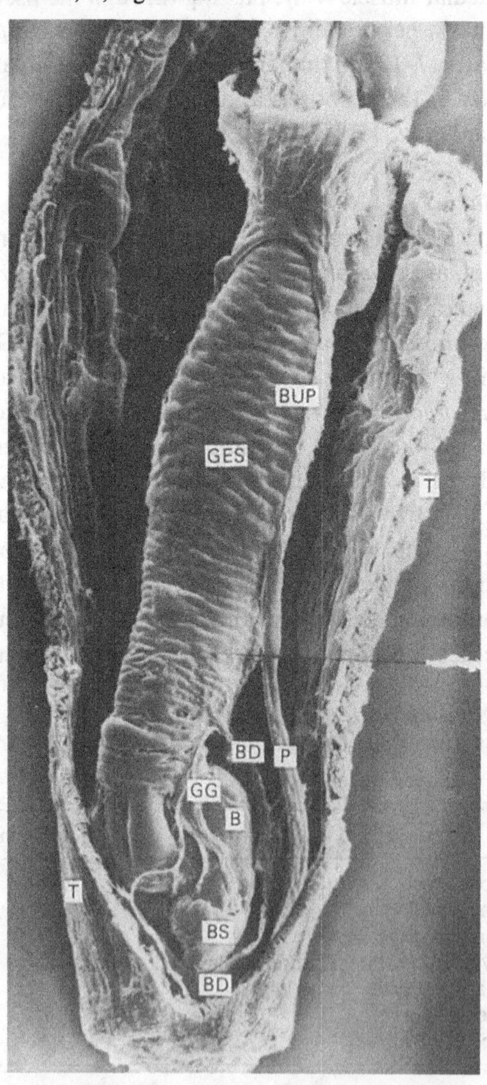

the insertion of the bursal depressor muscle and near the ejaculatory duct musculature and Saefftigen's pouch (Fig. 5.9). This position enables the dorsal commissure to pass medially to the ejaculatory duct musculature and Saefftigen's pouch and externally to thick bursal musculature. In a species of *Neoechinorhynchus*, the dorsal and ventral commissures are external to the muscles and ducts associated with the reproductive apparatus. Makhanbetov (1974) described one branch of the lateral posterior nerve as being connected to the genital ganglion in *Polymorphus phippsi*; it is generally believed that fibers of the lateral posterior nerve connect with the genital ganglia.

According to Dunagan & Miller (1979), the nerves originating from each genital ganglion of male *Macracanthorhynchus hirudinaceus* consist of (1) the anterior depressor nerve which originates from the ventral lateral margin and proceeds anteriorly to the origin of the bursal depressor muscle; (2) the penis nerve which originates from the ventral lateral margins and innervates receptors in the penis tip; (3) the inner and outer bursal nerves from the posterior margin which extend to receptors in the ventral surface of the bursal muscle; (4) the protrusor nerve originating from the posterior margin and extending to the protrusor and posterior margins of the bursal depressor muscles; (5) the lateral bursal nerve originating from the posterior lateral margin and extending to receptors on the outer surface of the bursal muscle; (6) the core nerve from the ventral lateral surface and extending medially to the ejaculatory duct and servicing muscles in this region; and (7) the posterior depressor nerve running to the protrusor muscle. To our knowledge no neurophysiological measurement has been made on any component in this system.

5.6.4 *Bursal ganglion*

Moniliformis moniliformis has a small ganglion consisting of four large club-shaped cells arranged in two pairs along the medial surface of the dorsal longitudinal muscles and near to the pseudocoel (Dunagan & Miller, 1977). Thus the bursal ganglion is located medially to the dorsal lacunar channel (Fig. 5.10). In the anterior–posterior axis the ganglion is located about 1 mm from the posterior terminus of a worm with an invaginated bursa. The cytoplasm of the ganglion cells is finely granulated and each cell contains a single nucleus with distinct nucleolus. Each cell is bipolar with ipsilateral and contralateral neurites. The ganglion may provide neurites to the bursal muscles and the longitudinal body wall muscles.

According to Kaiser (1893), Leuckart discovered at the posterior end of female *Macracanthorhynchus hirudinaceus* two ganglia which were smaller but otherwise similar to the male genital ganglia. This observation remains unverified.

5.7 Sensory receptor system

The organization and functioning of the sensory receptors in the Acanthocephala is poorly known. It has been suggested that the sense receptors are either tactile, chemical, glandular or vestiges of a digestive tract. Not all species possess an apical or lateral sensory organ if, indeed, these structures are sensory. However, since Meyer's monograph (1933), these structures have been considered to be sensory without supporting evidence. *Gracilisentis gracilisentis* appears to lack both of these organs (Jilek & Crites, 1979).

The apical organ varies in form from a papilla to a depression surrounded by a small elevated rim devoid of surface extensions. The core of this organ in *Macracanthorhynchus hirudinaceus* is formed from a combination of apical sensory nerves (Dunagan & Miller, 1984) and a support cell duct (Miller & Dunagan, 1983). All these structures are

Fig. 5.10. A cross-section of *Moniliformis moniliformis* through the bursal ganglion. The four ganglion cells are in two pairs with fibers extending in ipsilateral and contralateral directions. Notice the ganglion position in relation to the attachment of the bursal depressor muscle to the body wall. BD, bursal depressor muscle; BG, bursal ganglion; CM, circular muscles; DLC, dorsal lacunar channel; LM, longitudinal muscles of body wall; P, pseudocoelom; T, tegument. (Redrawn from Dunagan & Miller, 1977.)

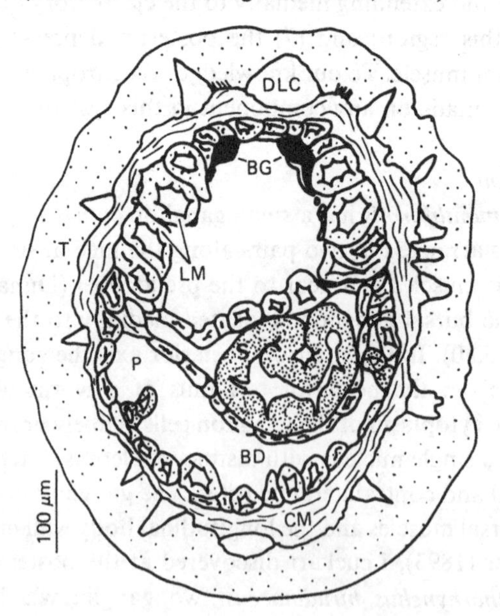

enclosed by a large binucleate cone-shaped cell (Fig. 5.11). Van Cleave & Bullock (1950) observed that *Neoechinorhynchus emydis* has an apical organ divided into two regions; the anterior part possesses three nuclei and is formed from the hypodermis. Nickol & Holloway (1968) described the apical organ of *Corynosoma hamanni* as pyriform with two large nuclei. von Haffner (1943) found the apical organs in *Oligacanthorhynchus thumbi* and *Gigantorhynchus echinodiscus* to be different. In *G. echinodiscus* two apical papilla are present with separate openings to the outside and in *O. thumbi* there is a single papilla. The posterior extensions of the sensory support cell apparently do not fuse in *G. echinodiscus* but extend to each of the apical papillae as separate ducts. We believe that von Haffner (1943)

Fig. 5.11. A schematic representation of the cone-shaped organ in *Macracanthorhynchus hirudinaceus*. APN, anterior proboscis nerve; ASO, apical sensory organ; N, nucleus; SSCD, sensory support cell duct; SN, sensory nerve; T, tegument. (Redrawn from Dunagan & Miller, 1983.)

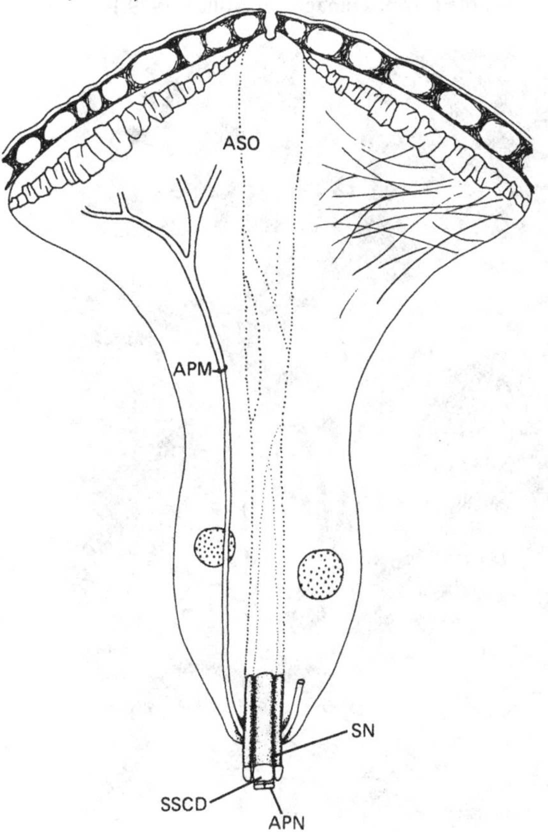

mistook these ducts for median nerves, but his descriptions of the sensory support cell are generally similar to ours on *M. hirudinaceus* and *Moniliformis moniliformis*. The apical organ is penetrated by nerves which coil or twist into a clump near the external surface. However, Harada (1931) stated that nerve fibers do not penetrate the sensory papillae in *Bolbosoma*

Fig. 5.12. Receptors in the caudal area of *Macracanthorhynchus hirudinaceus*. (*a*) Cross-section of lateral margin of bursal muscle at point of entry of lateral posterior nerve. Note that the large axons (white markers) are highly vesiculate at this point. (*b*) Cross-section through the bursal muscle showing the large vesiculate receptors (arrowed) of the inner and outer bursal nerve. (*c*) Cross-section of the penis tip showing three receptors (white markers). (*d*) Enlargement of lower receptor (arrowed) shown in (*b*). (*e*) Cross-section of lateral margin of bursal muscle showing the two receptors (shafted arrows) of the lateral bursal nerve and axons (arrow heads) of the inner and outer bursal nerves. The latter are on the dorsal surface of the bursal muscle. (*f*) Enlargement of the upper left receptor shown in (*e*). B, bursa. (Adapted from Dunagan & Miller, 1979.)

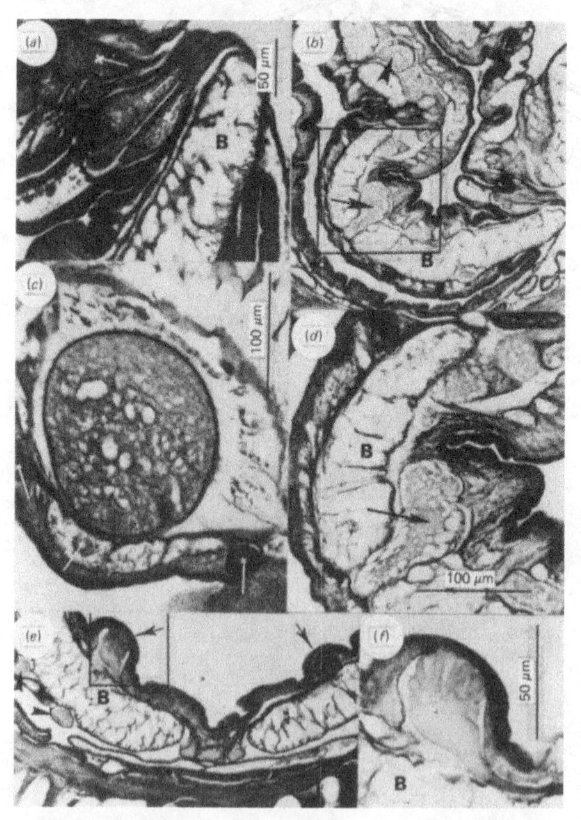

turbinella, which he described as being supplied only by the '*Stutzzelle*' (support cell). von Haffner (1943) observed each of the two papillae of the apical organ of *G. echinodiscus* to contain a series of parallel branched endings of the median nerve. This same design occurs in *M. moniliformis* and probably in other species.

Until recently, the large ductlike process that penetrates the posterior margin of the apical organ was identified as the axon of the anterior medial nerve from the cerebral ganglion. In *Macracanthorhynchus hirudinaceus* this duct is not nervous tissue but is a process from a multinucleate cell located ventrally and slightly anterior to the cerebral ganglion (Miller & Dunagan, 1983). Thus, there is no anterior medial nerve in *M. hirudinaceus*.

Lateral organs, which may be sensory receptors, occur as paired structures opposite each other on the neck of the proboscis of some species. These organs are small depressions surrounded by a slightly elevated ring of tegument. At the center of the resulting crater lie the termini of neurites from separate cerebral ganglion cells. The detailed organization of these structures has not been determined and we do not understand their presence in some species and not in others.

Receptors associated with the reproductive system are well known in male worms. In *Macracanthorhynchus hirudinaceus*, the lateral bursal nerves from each of the paired genital ganglia extend over the outer surface of the bursa musculature to the anterior margin of the bursal sac whereupon each of the nerves enters the musculature and terminates in a pair of uniquely designed (Fig. 5.12) fingerlike receptors located on the inside of the posterior rim of the bursa (Dunagan & Miller, 1979). The inner and outer bursal nerves are also associated with receptors located on the ventral surface of the bursal muscle near its posterior margin. These pear-shaped receptors (Fig. 5.12) are highly vesiculated and individually innervated. Neither type is directly exposed to the outside, but each is covered by a layer of tegument from the lining of the bursal lumen.

5.7.1 *Sensory support cell*

The multinucleate sensory support cell (Fig. 5.13) was most recently described (Miller & Dunagan, 1983) from *Macracanthorhynchus hirudinaceus*. A similar cell ('*Stutzzelle*') had previously been described from *Bolbosoma turbinella* by Harada (1931) who found two cells joined along their posterior margins and located between the dorsal proboscis retractors and the proboscis receptacle. A cell extension from each anterior margin went to the lateral sensory organ on the same side. In contrast, the support cell in *M. hirudinaceus* is observed (Fig. 5.13) as a cap on the midventral longitudinal receptacle muscle which is ventral and anterior to the cerebral ganglion. This cell has four large extensions which originate one from each corner. The two posterior processes follow the ventral receptacle retractor muscles through the proboscis receptacle. Once inside the receptacle wall, they turn anteriorly and fuse into a tubelike duct that extends to the apical organ (Fig. 5.13). Proboscis retractor muscles enclose this duct throughout the proboscis. In previous descriptions (Brandes, 1899; Dunagan & Miller, 1970) this duct had been erroneously identified as the anterior medial nerve. The two anterior sensory support cell processes move laterally over the outside surface of the proboscis receptacle until they turn anteriorly and terminate in the lateral sensory organs. Harada (1931) called this a glial cell in *B. turbinella*, but no information is available about its function. As there is no apical organ in *B. turbinella* the presence of a support cell is interesting.

Fig. 5.13. Sensory support cell in *Macracanthorhynchus hirudinaceus*. Diagram illustrating the relationship of the sensory support cell and the anterior lateral nerve to the midventral longitudinal receptacle muscle and the proboscis sheath. ALN, anterior lateral nerve; MVLR, midventral longitudinal receptacle muscle; PS, proboscis sheath; SSC, sensory support cell. (Redrawn from Miller & Dunagan, 1983.)

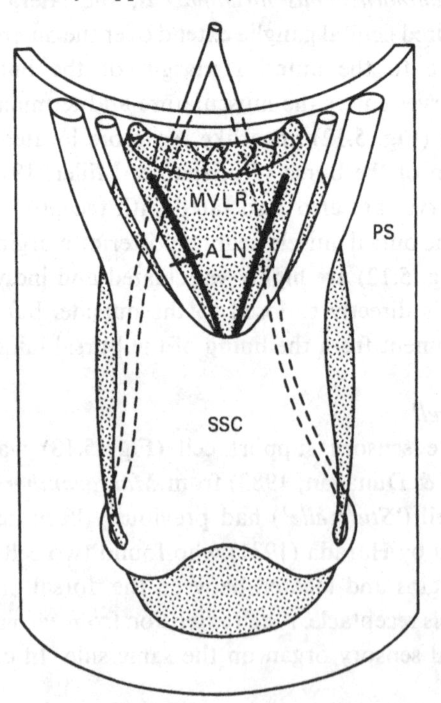

5.8 Excretory system

Types of excretory system in non-arthropod invertebrates can be classified into two categories of nephridia: (1) metanephridia, which possess openings into the body cavity and are confined to coelomate animals, and (2) protonephridia, which are found in both coelomate and pseudocoelomate animals (see Hyman, 1951). The protonephridia may be further divided into flame cells, containing a group of cilia, and solenocytes containing a single flagellum.

Only flame cells have been described for the Acanthocephala and Golvan (1959*b*) divided the arrangements of these into two types: '*type ramifie*' (dendritic) based on Kilian's (1932) description of *Oligacanthorhynchus microcephala*, and '*type capsulaire*' (sac) based on Meyer's (1931*c*) description of *O. taenioides*. There is no difficulty in separating the two types of flame cell arrangement. The dendritic type (Fig. 5.14) consists of a central canal from which numerous smaller canals arise and terminate in

Fig. 5.14. Diagrammatic representation of the dendritic type of excretory system in a male *Oligacanthorhynchus microcephala*. CED, cement duct; EXB, excretory bladder; EXC, excretory canal; FB, flame bulb; GP, gonopore; PE, penis; PRN, protonephridia; T, tegument; UGC, urogenital canal; VD, vas deferens; VE, vas efferens.

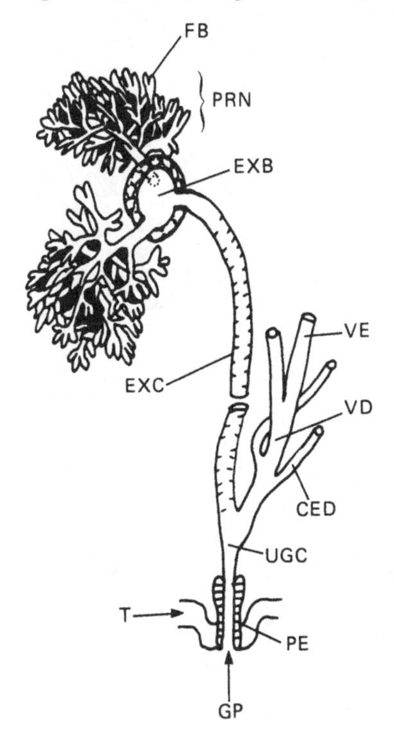

blind pouches containing cilia. The sac type (Fig. 5.15) consists of a bladderlike receptacle into which the flame cells open directly. In either design the excretory canals eventually join the vas deferens or the uterus.

The dendritic type of excretory system is seen in *Oligacanthorhynchus microcephala* where in male worms the cauliflower-like protonephridia are symmetrically located on the dorsal median line of the anterior rim of the ejaculatory duct constrictor muscle (Kilian, 1932). About 500 or 600 flame cells, each measuring 20–45 μm by 15–20 μm and containing 10–20 cilia, are found in *O. microcephala*. The dendritic arrangement of the flame cells involves three orders of branching (Fig. 5.14). Together the protonephridia empty through the urogenital canal (Fig. 5.14), the muscular terminus of which was named the 'ductus ejaculatorius' by early authors. The excretory canal is lined with cilia throughout its length except for its anterior terminus.

The only difference between the male and female systems occurs in the

Fig. 5.15. An outline of the sac (capsulaire) type of excretory system from a female *Oligacanthorhynchus taenioides*. If the excretory system is present, both sexes in any given species will possess only one type of excretory system. CO, common oviduct; EXC, excretory canal; FB, flame bulb; NC, nephridial canal; OVD, oviduct; U, uterus; UGC, urogenital canal.

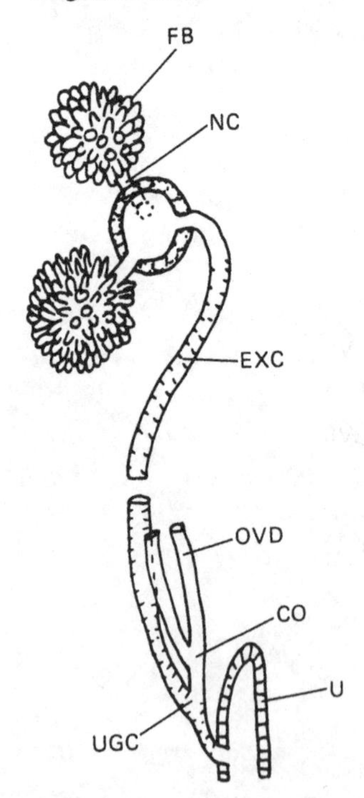

number and type of ducts associated with the excretory canal. In males, the excretory canal joins the vas deferens, forming the urogenital canal. In females the ciliated excretory canal joins the oviduct to form a short urogenital canal which discharges into the uterus either terminally or subterminally, depending on the species.

An example of the sac type of excretory system was described by Meyer (1931*c*) for *Oligacanthorhynchus taenioides* and by von Haffner (1942*a*, *c*) for *O. thumbi*. The major difference between the sac and dendritic types is in the organization. In the sac type, the flame cells empty into a common vesicle rather than a series of branching tubes (Fig. 5.15). It should be noted, however, that von Haffner (1942*a*) suggested a new term, '*Leitungs-schlauch*', for the terminal part of the system, to include components previously designated as genital sheath, ejaculatory duct, vas efferens, vas deferens, cement gland ducts, excretory canal and various muscles. He considered that this term more accurately reflected function of the terminal part of the urogenital system because it contains ducts that transport cement gland excretions, excretory products and sex secretions.

von Haffner (1942*b*) felt that Kilian (1932) was in error in stating that an excretory system does not occur in *Gigantorhynchus echinodiscus*. According to von Haffner, one is present, but is highly modified and flame cells are absent. Instead a specialized cell containing a ciliated vesicle performs the excretory function. This rudimentary excretory organ was believed to be homologous with a part of the excretory canal of other archiacanthocephalans.

Most descriptions of Acanthocephala do not include comments on the presence or absence of a protonephridial system. Indeed, Van Cleave (1953) in his introduction to the acanthocephalans of North American mammals had but a single sentence on this system. In a letter to Rolf Kilian, Anton Meyer wrote he had found protonephridia in worms of the following genera: *Pachysentis*, *Oncicola*, *Prosthenorchis*, *Oligacanthorhynchus*, *Nephridiorhynchus* and *Macracanthorhynchus* (Kilian, 1932). Schmidt (1972*b*) rearranged the taxonomy of the class Archiacanthocephala and partly based his classification on the presence or absence of protonephridia. According to him, this class has three orders, Oligacanthorhynchida, Giganthorhynchida and Moniliformida, and only species of Oligacanthorhynchida possess protonephridia (see Chapter 4). However, it is evident from von Haffner's (1942*b*) study that some of the Gigantorhynchida should be restudied for the presence of modified protonephridia.

The functions of the protonephridia, although not investigated, might include osmoregulation, detoxification and excretion. It is generally recognized that aquatic invertebrates eliminate nitrogen as ammonia, and acanthocephalans might do likewise. Because most acanthocephalans apparently lack a protonephridial system, it is usually assumed that the tegument is involved in the exchange of ions, water and metabolic wastes. Parshad & Guraya (1977*b*) concluded that, in the absence of excretory organs, *Sphaerirostris pinguis* eliminated waste products through the body wall and did not regulate its osmotic pressure.

Branch (1970*a*) found that the body wall fluid of male and female *Moniliformis moniliformis* was hyperosmotic to rat intestinal fluid, but if the exchangeable cations only were considered, the worms were approximately isotonic with their environment. The distributions of potassium, sodium and chloride ions were determined histochemically. Defined bands were found in the tegument with potassium ions concentrated beneath the outer surface and sodium and chloride ions near the circular muscle. Branch (1970*b*) postulated the existence of sodium and potassium transport pumps in the tegument. Tanaka & MacInnis (1980), who found the osmotic pressure of the pseudocoelomic fluid of *M. moniliformis* to be 291 mosmol/kg water, suggested that high concentrations of glutamate in the pseudocoelomic fluid may contribute to osmoregulation. Denbo (1971) found that 30% sea water caused the smallest disturbance in sodium concentration or volume in *Macracanthorhynchus hirudinaceus* and observed that the pseudocoelomic fluid of *M. hirudinaceus* was maintained hyperosmotic to the environment between external osmotic concentrations of 0 to 500 mosmol. He also considered that *M. hirudinaceus* was capable of some degree of volume, osmotic and ionic regulation. In contrast, Crompton & Edmonds (1969) measured the osmotic pressure of the pseudocoelomic fluid of *Polymorphus minutus* and found it similar to that of a simultaneous sample of host intestinal contents. They concluded that *P. minutus* was probably an osmoconformer and Crompton (1970) considered that this view could be applied to all acanthocephalans. Moeller (1978) observed that the osmotic pressure of the intestinal contents of fish remained constant even when they moved from brackish to fresh water. He suggested that acanthocephalans of fish were not normally subjected to significant changes in osmotic pressure. However, when female *Echinorhynchus gadi* were placed in media of different salinities, the worms were not able to stabilize their water balance, presumably through lack of osmoregulation. Mettrick, Budziakowski & Podesta (1979) found that the osmotic pressure of the gut luminal contents of rats infected with *Moniliformis moniliformis* increased by 20–40 mosmol/l and that movement of water and various ions was disturbed in the infected intestine. These findings suggest that the pathological response of a host to an acanthocephalan parasite may provide an osmotically suitable environment.

5.9 Reproductive system

5.9.1 *Female reproductive system*

The female reproductive system consists of: (1) gonads from which ovarian balls develop to produce oocytes and eventually eggs (=shelled acanthors) (Chapter 7); (2) ligament sac(s) which, if they persist, contain the developing eggs; and (3) an efferent duct system which comprises a uterine bell, uterus and vagina (Figs 5.16 and 5.17) (Parshad & Crompton, 1981).

Ligament sacs, hollow tubes associated with the reproductive system, in the pseudocoel extend from the proboscis receptacle to the uterine bell. In Archiacanthocephala and Eoacanthocephala, these tubes are paired, one dorsal and one ventral with each in contact with the other (Fig. 5.16*a*). In Palaeacanthocephala, the single ligament sac ruptures early in sexual development (Bullock, 1969). The ligament sacs fail to persist in those members of the Eoacanthocephala that have been examined. Anteriorly in *Moniliformis moniliformis* and probably other archiacanthocephalans, the two sacs fuse to form a single sac; posteriorly, the dorsal ligament sac terminates in the uterine bell (Fig. 5.16*a*). The ventral ligament sac ends

Fig. 5.16. (*a*) Efferent duct system of *Moniliformis moniliformis*. (After Asaolu, 1980.) (*b*) Cross-section of wall of anterior bell chamber of *Polymorphus minutus*. (After Whitfield, 1968.) BW, bell wall; CM, circular muscles; DLS, dorsal ligament sac; GP, gonopore; LAP, lappet cells (uterine bell); LM, longitudinal muscle of body wall; LP, lateral pocket; M, medullary fluid; U, uterus; V, vagina; VL, vascular layer; VLS, ventral ligament sac.

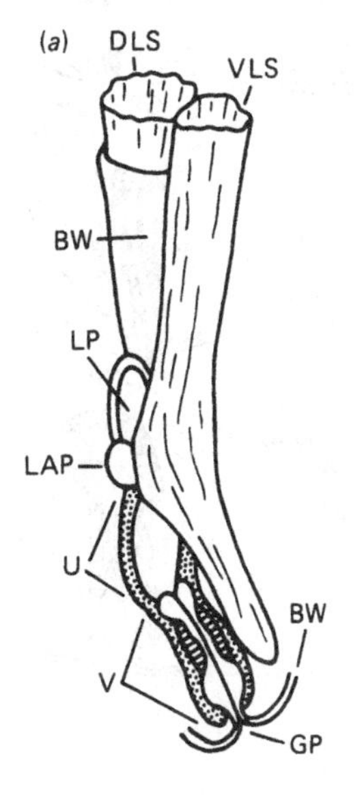

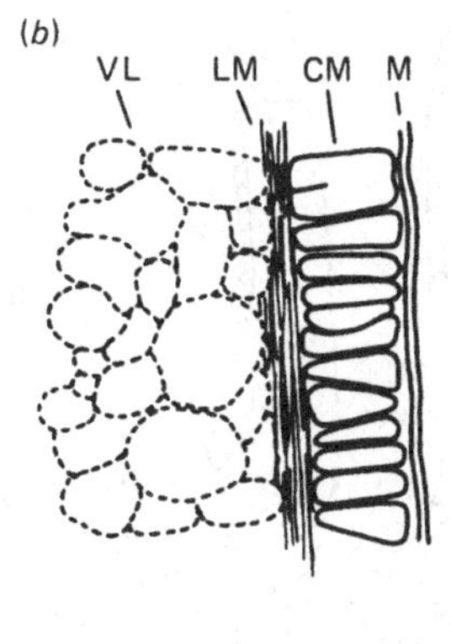

blindly near the genital sphincter and is attached to the uterine bell through the ventral opening of the bell complex (Asaolu, 1980). In *M. moniliformis*, the dorsal part of the dorsal ligament sac and the ventral part of the ventral ligament sac are attached to the dorsal and ventral body walls, respectively, but the lateral parts of the sacs are free (Asaolu, 1980). An early illustration of the relationship of the ligament sacs to the female efferent duct system is Meyer's (1931*c*, Fig. 2) diagram of *Oligacanthorhynchus taenioides*. Asaolu's work (1980) confirms this general design. In eoacanthocephalans, the uterine bell is associated with the ventral ligament sac (Bullock, 1969).

Kaiser (1893), Rauther (1930) and von Haffner (1942*a*) described

Fig. 5.17. (*a*) Lateral view of bell complex in *Polymorphus minutus*. (After Whitfield, 1968.) (*b*) Ventral view of bell complex of *Moniliformis moniliformis*. (After Asaolu, 1980.) (*c*)–(*d*) Vaginal glands associated with gonopore. (*c*) *Acanthogyrus antspinus*. (After Verma & Datta, 1929.) (*d*) *Acanthogyrus dattai*. (After Podder, 1938.) (*e*) Vagina of *Moniliformis moniliformis*. (After Asaolu, 1980.) ABC, anterior bell chamber; AVMC, anterior ventral median cell; BW, bell wall; DLAC, dorsal ligament attachment cell; DLS, dorsal ligament sac; DMC, dorsal medial cell; GP, gonopore; GSM, genital sphincter muscle; ILV, inner layer of vagina; LP, lateral pocket; MWC, median wall cell; PVMC, posterior ventral medial cell; T, tegument; U, uterus; UD, uterine duct; UDC, urogenital canal; V, vagina; VAC, ventral accessory cell; VG, vaginal glands; VLAC, ventral ligament attachment cell; VO, ventral opening; VSM, vaginal sheath muscles.

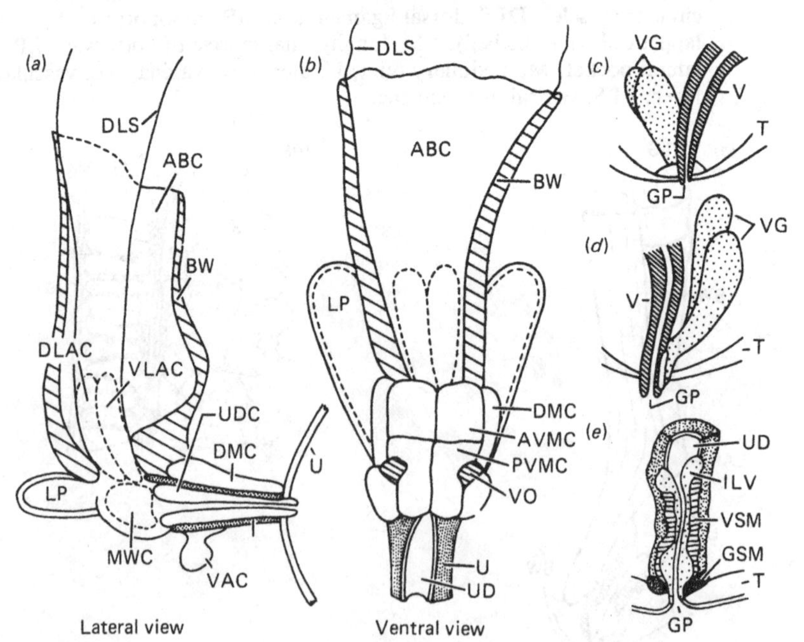

openings between the two ligament sacs. These permit the circulation of eggs from the ventral ligament sac into the larger dorsal ligament sac and through the uterine bell, a process which facilitates the retention of immature eggs. von Haffner (1942*a*) stated that the sacs were constructed of numerous collagen fibers and were rich in elastic components in certain areas.

The ligament strand lies between the two ligament sacs or along the ventral face of the single ligament sac and is in the same position in either sex. It contains numerous nuclei and is considered to be syncytial in nature. von Haffner (1942*a*) described the strand as having primary and secondary branching in the anterior end of *O. thumbi*, a feature common to many acanthocephalans. von Haffner (1942*c*) believed this strand represented a vestigial intestine, a view previously expressed by Schneider (1871) but opposed by Kaiser (1893) who considered the strand to be germinal tissue. von Haffner (1942*c*) further believed that the genital systems of both sexes had originated from a cloaca and inferred support for this view from embryological observations by Meyer (1936, 1937).

The efferent duct system begins with the anterior opening of the bell wall and terminates in the gonopore. The system between these two points is complex but has been conveniently divided into three regions: (1) uterine bell; (2) uterus; and (3) vagina. Certain differences exist in the overall organization of the bell structure as described by Yamaguti & Miyata (1942) and Machado (1950) on the one hand and Whitfield (1968) and Asaolu (1980) on the other. Only a few species have been examined in detail and it may be unwise to generalize.

Whitfield (1968) interpreted the bell wall of the anterior chamber in *Polymorphus minutus* as an outer layer of circular muscle, a middle layer of longitudinal muscle and an inner thick vascular layer (Fig. 5.16*b*). In *Moniliformis moniliformis*. the egg-containing dorsal ligament sac merges with the anterior end of the bell wall in a specialized area called the ligament attachment syncytium (Asaolu, 1980). However, in *P. minutus* the dorsal ligament sac, while attached to the dorsal outer rim of the anterior chamber, continues to the posterior opening of this chamber (Fig. 5.17*a*). In *Fessisentis fessus* and *P. minutus*, strong muscular activity in the anterior chamber moves eggs from the anterior to the posterior part of the chamber by peristalsis (Dunagan & Miller, 1973; Whitfield, 1970).

Eggs move from the anterior bell chamber into a selector apparatus which in *Moniliformis moniliformis* (Asaolu, 1980) consists of two lateral pockets, two dorsal median cells, two anterior ventral median cells, two posterior ventral median cells and two lappet cells (Asaolu, 1980). Yamaguti & Miyata (1942) found slightly different numbers of certain of these cells; they used only whole mounts. Whitfield (1968, 1970) gave a detailed histological description of the uterine bell of *Polymorphus minutus*, but suggested that some of his conclusions regarding cellular organization were tentative because he used only light microscopy. Nevertheless, his work provides a clear understanding of the bell cytology as it relates to egg movement (Fig. 5.18*a–f*). In *P. minutus* there are two lateral pockets, two median wall cells, two ventral accessory cells, one median dorsal cell and a sheathing syncytium around the tube of the uterine duct. A duct from the anterior bell chamber leads into an egg-sorting or selector apparatus.

Fig. 5.18. Operation of bell complex as an egg-sorting apparatus. Immature eggs follow sequence (*a,b,c*) into ventral ligament sac or body cavity. Mature eggs (shaded) follow (*a, b, e, f*) sequence into the uterus. Mature or immature (unshaded) eggs may enter lateral pockets (*d*) and follow (*d, b, c*) sequence if immature and (*d, b, e, f*) sequence if mature. Eggs cannot move from (*b*) to (*a*). (Modified from Whitfield, 1970.) ABC, anterior bell chamber; BW, bell wall; CO, common oviduct; E, egg; LP, lateral pocket; MWC, median wall cell U, uterus; UD, uterine duct; VO, ventral opening.

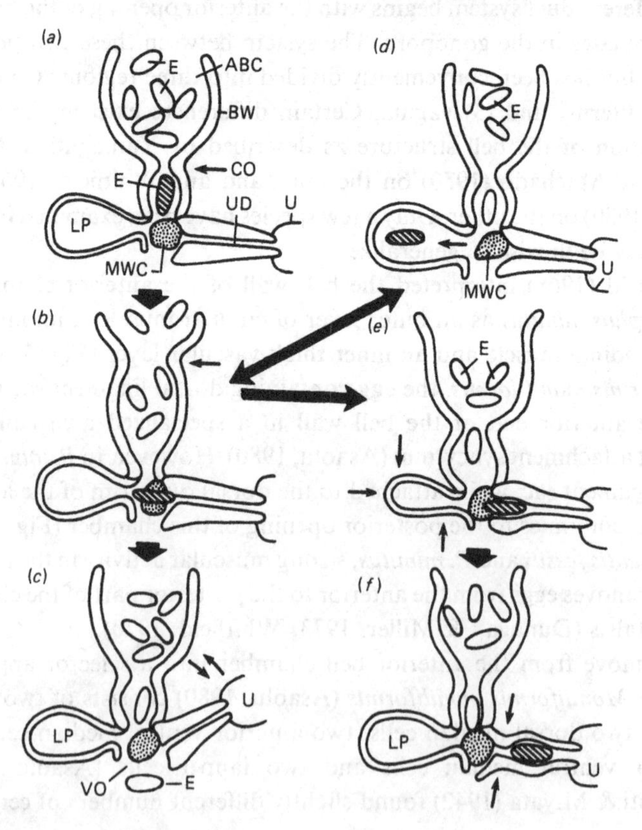

The median wall cells are laterally compressed and attached to the posterior part of two pairs of ligament attachment cells that form the plug in the posterior opening of the anterior bell chamber. The median wall cells also form a partition in the sagittal plane of the selector apparatus over the lateral surfaces of which eggs may move. These cells, therefore, form the inner wall of the 'Y' junction observed by Whitfield (1970) and Asaolu (1980). In *P. minutus*, the outer wall is formed by the lappets of the uterine duct cells which enclose the median wall cells. The ventral free edges of the lappets form the bell opening into the body cavity or, in certain species, into the ventral ligament sac. The dorsal surface is covered by the sheathing syncytium and median dorsal cell and prevents egg movement in that direction. The lateral pockets are attached anteriorly to the ventral base of the bell wall and laterally to the lappets. The lumen of each pocket is thus open to one of the egg channels formed by the median wall cells. The significance of these separate pathways leading either to the ventral bell opening or the uterus is not fully understood. Eggs are free to move into and out of each pocket whose walls consist mostly of circular muscle (Whitfield, 1968). Each of the channels formed by the median wall cells leads posteriorly to a duct formed by the uterine duct cells. These two paired uterine ducts empty into the uterus. von Haffner (1942*b*) wrote of these as two funnel-shaped depressions which were transformed into two oviducts entering a ciliated urogenital canal. Asaolu (1980) concluded that for *M. moniliformis* there is only one duct leading into the uterus.

The movement of eggs through this maze with its several options may appear complicated but there are actually only three possibilities: (1) into the uterus; (2) return through the ventral bell opening; or (3) enter lateral pocket and then either route (1) or (2). Whitfield's (1970) account of this process showed that in *Polymorphus minutus* four muscular events were responsible for moving eggs from the anterior chamber of the bell through the selector apparatus. These events were aided by fluid flow accompanying muscular activity. The sequence of events was bell wall peristalsis, bell flexion and extension, lateral pocket contraction and uterine duct peristalsis. Muscular activity of the bell would force eggs into either of the two channels formed between the lateral surface of each median wall cell and the adjacent lappets of the uterine duct cells. Somehow when an egg was in this position its length and thus its maturity could be assessed. If too short and thus immature, the egg would fail to enter the uterine duct and would pass through an opening adjacent to the lappets of the uterine duct cells into the body cavity or ligament sac to retrace its route. The ejection of immature eggs would occur during bell flexion. If the egg were too long to be discharged by this process, it would be forced by the contraction of the ipsilateral pocket musculature into the uterine duct where peristalsis would force it through the duct into the uterus.

Meyer (1933) and Yamaguti (1963) doubted that egg-sorting occurred in the uterine bell because of the appearance of immature eggs in the uterus. Kaiser (1893), Van Cleave (1953), Bullock (1969), Whitfield (1970) and Asaolu (1980) adopted the egg-sorting view and Whitfield (1970) suggested that the immature eggs were misplaced because of either early activity of the selector apparatus before becoming fully mature or events associated with the death of the worm during its preparation for study. Clearly, the general finding of mature eggs in host feces only implies some sort of selection mechanism.

Entrance to the uterus is guarded by the selector apparatus anteriorly and by a vaginal sphincter posteriorly. The time for which eggs remain in the uterus has not been determined and it is not known whether this has any functional significance. The figures given by Petrochenko (1956, 1958) and Yamaguti (1963) indicate that the uterus and vagina have not generally been illustrated in detail. Exceptions include Meyer's (1931*c*) drawing of *Oligacanthorhynchus taenioides*, which shows the uterus to be uniformly thick and muscular, and Yamaguti's (1939) figure of *Bolbosoma nipponicum*, where the uterus is thin anteriorly and thick posteriorly. There seems little question that the uterus is muscular throughout. Asaolu (1980) described an outer circular muscle layer, a middle non-muscular cytoplasmic layer and a thin inner layer of fibrous material in the uterus of *Moniliformis moniliformis*. The uterus of *Echinorhynchus gadi* has similar structure (Khatkevich, 1975). However, the organization of these layers may differ from species to species. Verma & Datta (1929) concluded that *Acanthogyrus antspinus* has a uterus with nucleated flask-shaped glands, but they may have been referring to the lateral pockets of the uterine bell.

The vagina of *Moniliformis moniliformis* is joined to the uterus at an angle of 45 degrees ventrally (Asaolu, 1980). The vagina of this species consists of a narrow lumen enclosed by an inner dumbell-shaped syncytium, a middle smooth muscle layer organized as a sphincter, and an outer layer which is an extension of the muscle layer of the uterine wall (Fig. 5.17*e*).

Interesting exceptions to this pattern are the vagina of *Neoechinorhynchus buttnerae* which is peculiarly coiled (Golvan, 1956*c*; Schmidt & Huggins, 1973) and the vagina of *N. johnii* which is unusually modified (Yamaguti, 1939). Specialized cells, presumed to be glandular (Fig. 5.17*c*, *d*), occur in association with the vagina (Verma & Datta, 1929; Podder, 1938). Verma & Datta (1929) stated that two nucleated, club-shaped glands lie outside the vaginal wall and communicate with the vagina by a common pore close to its external aperture. They thought that these glands probably secreted a fluid to lubricate the genital opening.

In most species the gonopore is subterminal and eggs reach its opening by moving in single file through the vagina. However, the position of the gonopore varies, as has been found in *Fessisentis fessus* by Nickol (1972). In *Moniliformis moniliformis* the vaginal sphincter regulates the opening and closing of the gonopore (Asaolu, 1980). For varying periods after insemination, the gonopore is usually blocked by a copulatory cap which must be lost before egg release can begin. Whitfield (1970) concluded that in *Polymorphus minutus*, egg release did not begin until the uterus was filled with eggs.

5.9.2 *Male reproductive system*

The male reproductive system includes two testes, vasa efferentia, vas deferens, seminal vesicle, cement glands, cement reservoir, Saefftigen's pouch, pyriform glands, copulatory bursa and penis; some of these features are illustrated in Fig. 5.19 and are described further by Parshad & Crompton (1981). Variations in these components may occur according to the species.

The spherical or ovoid testes, usually arranged in tandem, are suspended in the anterior, middle or posterior portion of the body cavity by the ligament strand, enclosed by the ligament sac and attached to its ventral wall. Both the dorsal and ventral ligament sacs arise early in development from the base of the proboscis receptacle and extend posteriorly to merge with the genital sheath musculature. Because the ligament strand and ligament sacs are believed to be endodermal in origin, the space between the ligament complex and the body wall is considered a pseudocoel (von Haffner, 1942*c*). In each order of Acanthocephala, the dorsal ligament sac generally persists to enclose the testes; a few exceptions where rudimentary ventral sacs persist have been cited by Kilian (1932).

Monorchic males are known from several species (Table 5.8). Fourteen of 208 male *Acanthocephalus dirus* were monorchic (Bullock, 1962), and male *Fessisentis vancleavei* usually have only one testis (Nickol, 1972; Buckner & Nickol, 1978). The testis of a monorchic male is larger than a testis from a diorchic male and Bullock (1962) assumed that monorchidism arose from fusion. He had observed that there are usually two vasa efferentia in monorchic male *A. dirus*. Fotedar (1968) observed in *Neoechinorhynchus oreini* one large testis which was slightly demarcated in the middle and suggesting that fusion had occurred. Crompton (1972*b*), using surgical methods, established that one monorchic specimen of

Moniliformis moniliformis was capable of successful insemination of females.

The vasa efferentia extend posteriorly for a short distance from the testes before fusing to form the vas deferens; for the most part the vasa efferentia are located outside the genital sheath and are enclosed within the ligament sac as they proceed posteriorly. Asaolu (1981) concluded that in *Monili-*

Fig. 5.19. Diagrammatic representation of the male reproductive system. As drawn, the diagram is not representative for any particular species but incorporates many different structures that may occur. BLC, bursal lacunar canal; CEM, cement gland; GES, genital sheath; GG, genital ganglion; GP, gonopore; LMSP, longitudinal muscle of Saefftigen's pouch; PE, penis; PRN, protonephridia; SCM, Saefftigen's pouch circular muscle; SP, Saefftigen's pouch; TES, testis; VD, vas deferens; VE, vas efferens.

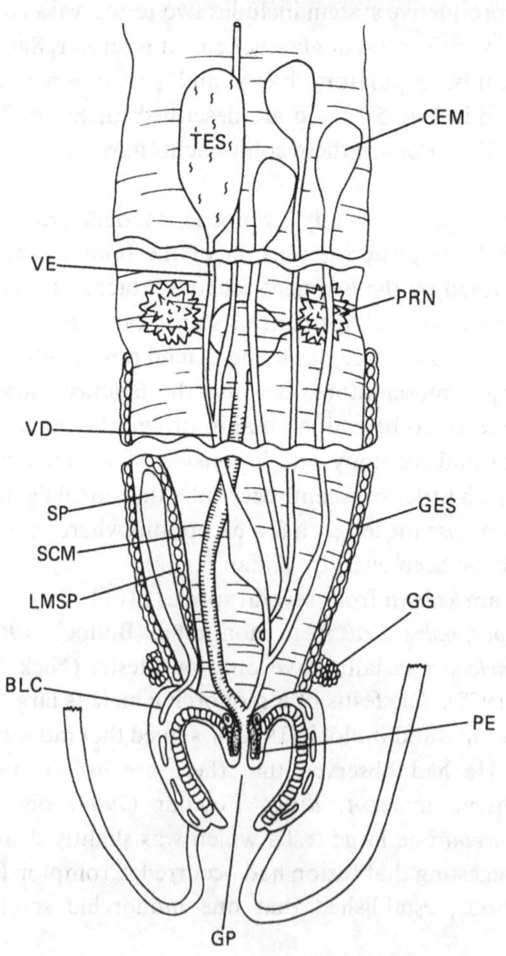

formis moniliformis the vasa efferentia are formed from outgrowths of the ventral wall of the ligament sac and communicate with the testes by means of several openings. The vas deferens (also called the common sperm duct) is mostly contained within the genital sheath. There is no agreement about the appropriate term for the structure formed by the entry of the vas deferens into the genital sheath. Terms used have included sperm duct, spermatic duct, ejaculatory duct, cirrus canal, cirrus duct and common sperm duct.

In some species a vas efferens or the vas deferens is enlarged to form a seminal vesicle for sperm storage. Posteriorly within the genital sheath, the vas deferens merges with either a single duct or multiple ducts from the cement glands to form the common genital canal. In *Moniliformis moniliformis* each cement duct enters the vas deferens separately (Asaolu, 1981), although most authors indicate the fusion of cement ducts before entry to the vas deferens in species with multiple cement glands. Slightly posteriad, the common genital canal may fuse with the ciliated excretory canal, if this is present, to form the urogenital canal, Chandler (1946*b*) observed, in addition to the previously described components of the male system of *Moniliformis moniliformis*, two blind sacs arising from the vas deferens. These blind sacs are also present and are ciliated in *Oligacanthorhynchus tortuosa* and *Macracanthorhynchus hirudinaceus* according to Chandler (1946*b*).

Van Cleave (1949) investigated the morphology and phylogenetic significance of the cement glands and recognized three general types (Fig.

Table 5.8. *Occurrences of monorchidism in acanthocephalans*

Species	Reference
Acanthocephalus anguillae	Petrochenko (1956)
Acanthocephalus dirus	Bullock (1962); Amin (1975*b*)
Acanthocephalus lucii	Petrochenko (1956)
Acanthocephalus ranae	Petrochenko (1956)
Acanthocephalus tenuirostris	Petrochenko (1956)
Centrorhynchus elongatus	Kobayashi (1959)
Corynosoma constrictum	Schmidt (1965)
Fessisentis fessus	Nickol (1972)
Fessisentis friedi[a]	Buckner & Nickol (1978)
Fessisentis necturorum[a]	Buckner & Nickol (1978)
Fessisentis vancleavei[a]	Buckner & Nickol (1978)
Moniliformis moniliformis	Crompton (1972*b*)
Neoechinorhynchus oreini	Fotedar (1968)

[a] Monorchidism is usual.

5.20). The cement gland complex, attached to the ligament strand, consists of a single syncytial gland in the Eoacanthocephala or of as many as eight elongate, tubular or flask-shaped glands in the Archiacanthocephala and Palaeacanthocephala. Cement gland ducts arise from the glands and run posteriorly towards the penis. Sometimes these ducts fuse and in some species they become enlarged and serve as cement reservoirs. A separate cement reservoir occurs in the Eoacanthocephala.

The fused ducts, which are usually paired, open into the common genital canal from the mid-dorsal side. Van Cleave (1949) believed that the syncytial cement glands of the Eoacanthocephala are the most primitive. The cement glands of the Echinorhynchidae pass through a stage of development characteristic of the syncytial cement glands of Neoechinorhynchidae. As Parshad & Crompton (1981) pointed out, there are exceptions which Van Cleave's descriptions do not cover. For example, there is structural continuity between the four cement glands of *Polymorphus*

Fig. 5.20. Cement gland types. (Redrawn from Van Cleave, 1949.) Type A (eoacanthocephalan) has a single syncytial cement gland with giant nuclei. Type B (archiacanthocephalan) has up to eight cement glands, each with a single giant nucleus. Type C (palaeacanthocephalan) may have several glands, each of which is syncytial with several nuclear fragments. CED, cement duct; CER, cement reservoir; N, nucleus.

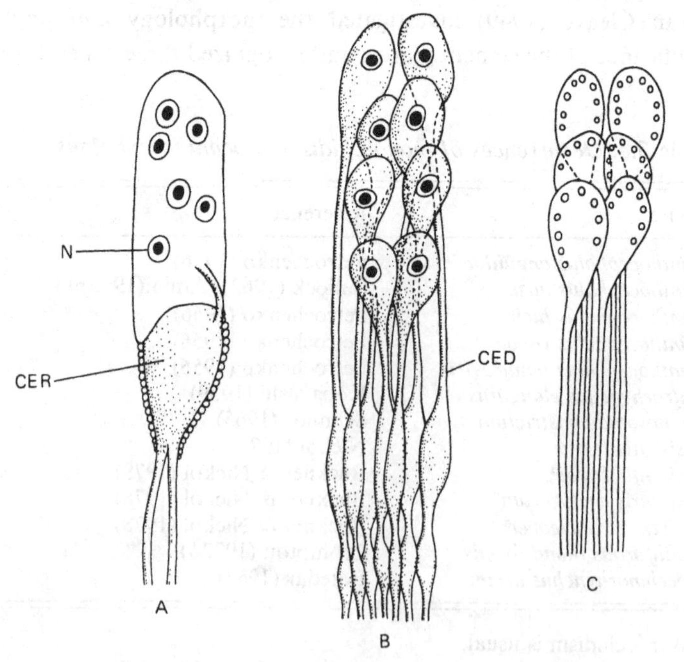

minutus. Also, cement glands frequently vary in number as seen in *Acanthocephalus dirus* where the range is from none to 12 (Amin, 1975*b*).

The cement gland secretions in *Moniliformis moniliformis* are stained intensely by Heidenhain's iron hematoxylin and by mercuric bromophenol blue in *Sphaerirostris pinguis* (Parshad & Crompton, 1981). Because of the staining properties and the presence of cytoplasm rich in rough endoplasmic reticulum, protein is assumed (Haley & Bullock, 1952; Parshad & Crompton, 1981) to be a component of cement gland secretions of *S. pinguis*, *Neoechinorhynchus emydis* and *Echinorhynchus gadi*. With the exception of an unidentified species of *Polymorphus* and *Corynosoma wegeneri*, alkaline phosphatase activity seems to be absent from the cement glands (Bullock, 1958).

The secretions of the cement glands are assumed to contribute to the formation of the copulatory caps which are often observed around the gonopore of the females and sometimes on males (Abele & Gilchrist, 1977; Parshad & Crompton, 1981). The caps are quite hard and the material from which they are formed has presumably become polymerized following secretion. Whitfield (see Parshad & Crompton, 1981) described the pyriform gland in *Polymorphus minutus* and Asaolu (1981) identified it in *Moniliformis moniliformis*. Dunagan & Miller (1973) discovered two types of gland cells at the margin of the bursa of *Fessisentis fessus*. One cell is

Fig. 5.21. Schematic diagram of genitourinary canal and penis structure. (Redrawn from Petrochenko, 1956.) BM, muscular cap of bursa; GP, gonopore; SE, septum.

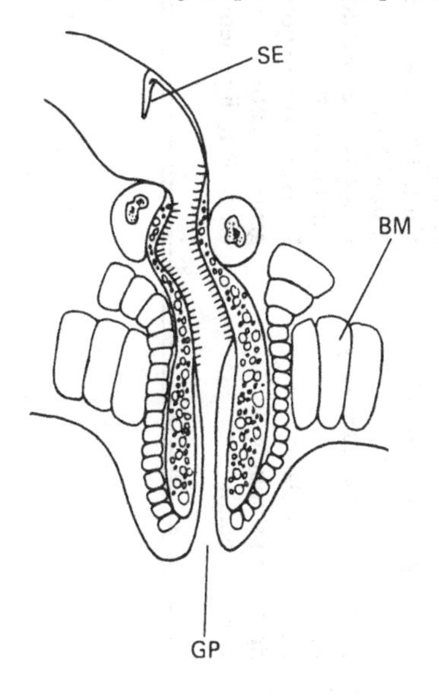

Table 5.9. *Muscles of the reproductive tract of male* Moniliformis moniliformis (*adapted from Dunagan & Miller, 1978*c)

Name	Origin	Insertion	Probable function
Anterior bursal accessory muscle	Inside dorsal surface of genital sheath	Medial ventral surface of bursal depressor	Unknown
Bursal cap muscles	Bursal cap	Body wall	Eversion of bursa
Bursal depressor muscles	Body wall adjacent to genital opening	Lateral surface of body wall and genital sheath	Eversion of bursa
Ejaculatory duct muscle	Inside ventral surface of genital sheath	Anterior ventral surface of bursal cap	Evacuation of Saefftigen's pouch
Bursal protrusor muscle	Ventral body wall adjacent to genital pore	Lateral outside surface of genital sheath	Fine control of everted bursa
Posterior bursal accessory muscle	Inside ventral surface of genital sheath	Medial ventral surface of bursal depressor	Unknown
Circular muscle of Saefftigen's pouch	Extends from region of genital sheath	Extends to posterior end of genital ganglion	Empty fluid from Saefftigen's pouch
Genital sheath	Extends to ligament sac	Extends to surface of bursa	Contains genitourinary ducts and perhaps assists in ejaculation

reniform with homogenous contents and the other is spatulate with granular contents. These cells are located in a convenient position for the release of some catalyst that might be involved in cap formation (Dunagan & Miller, 1973).

The penis is a sharp-cone-shaped structure which projects into the lumen of the bursa copulatrix (Fig. 5.21). The lumen of the penis becomes continuous with the ciliated urogenital canal which merges with the descending unciliated seminal duct. Surrounding the penis canal is a urogenital syncytium and exterior to this a layer of muscles.

The copulatory structures consist of the muscular Saefftigen's pouch (*Markbeutal of Saefftigen*), the eversible campanulate bursa and the penis. The bursa is equipped for its role with complex musculature which has been described for *Moniliformis moniliformis* by Dunagan & Miller (1978*c*) (Table 5.9). Under the harmonious mechanism of these muscles, the bursa spreads over the posterior extremity of the female, followed by attachment as the edge of the bursal musculature contracts. Then, the gonopore region of the female is pulled into the body of the male, constricted and in effect, separated from the rest of the trunk. Golvan (1958) wrote, '*Habituellement invaginée dans la partie terminale du tronc, cette bourse se deploie et exterieur lors de la copulation et enserre l'extremité posterieur du tronc de la femelle*'. Yamaguti (1963), however, believed evagination of the bursa to be a postmortem phenomenon.

Saefftigen's pouch is enclosed in a sheath of circular muscle within the genital sheath. The pouch is compressed against the wall of the sheath by additional diagonal muscles. The spongy medulla of the pouch contains fluid which moves back and forth through the stalk of the pouch into the apical caplike portion of the bursa.

The genital sheath (Fig. 5.19) is a complex arrangement of circular muscles enclosing the cement gland ducts, sperm ducts, excretory duct and urogenital duct. Kaiser (1893) termed it the '*Genitalscheide*' as did Kilian (1932). Meyer (1933) called it the 'ductus ejaculatorius' but von Haffner (1942*a*) made a specific point of changing the name to '*Leitungsschlauch*'. Posteriorly, the genital sheath terminates at the muscular cap of the bursa while anteriorly it joins the ligament sac. The thickness of the sheath varies considerably. At the end near the ligament sac it is very thick, but it becomes progressively thinner posteriorly.

6

Feeding, nutrition and metabolism

Jane A. Starling

6.1 Introduction

Most studies of feeding, nutrition, metabolism and related phenomena in acanthocephalans have been limited to three species: *Moniliformis moniliformis*, *Macracanthorhynchus hirudinaceus* and *Polymorphus minutus*. The dearth of rigorously controlled studies on even these three species leaves enormous gaps in our understanding of acanthocephalan biology. It is becoming evident, however, that most biochemical and physiological phenomena in helminths are not unique to organisms adapted to a parasitic lifestyle. Differences between helminths and other organisms, and differences among helminths, are often a matter of degree or of subtle variations upon a common theme rather than manifestations of strikingly distinct mechanisms. This premise underlies much of the discussion in this review. Where details from specific studies on acanthocephalans are lacking, the discussion draws heavily upon our knowledge of other helminths or of free-living organisms.

This review attempts to integrate the information available on the Acanthocephala into the context of these more thoroughly understood systems. I have endeavored to identify specific questions whose pursuit is likely to prove fruitful for our efforts to understand acanthocephalan physiology and to broaden our understanding of other helminths as well. A number of highly speculative suggestions are introduced; it is my fervent hope that these speculations will be attacked vigorously in the laboratory, so that more firmly based explanations may evolve.

6.2 The acanthocephalan digestive–absorptive surface

It was assumed that nutrient assimilation was a major function of the acanthocephalan body surface long before any systematic studies of acanthocephalan feeding mechanisms were performed. Because of the intimate association between the adult praesoma and the host intestinal mucosa, there has been occasional speculation that the praesoma might take up nutrients directly from host tissues (Byrd & Denton, 1949; Pflugfelder, 1949; Van Cleave, 1952). However, Edmonds (1965) demonstrated that adult *Moniliformis moniliformis* obtain their nutrients from the contents of the host intestine rather than directly from host tissues. Furthermore, autoradiographic studies of the uptake of glucose, amino acid, nucleoside and triacylglycerol into *Acanthocephalus ranae* (Hammond, 1968*b*) and *Paulisentis fractus* (Hibbard & Cable, 1968) clearly identified the metasoma tegument, whose surface lies exposed within the host intestinal lumen, as the site of nutrient assimilation.

6.2.1 *Structure of the feeding surface*

The general morphology of the acanthocephalan tegument is described elsewhere in this volume (§5.3). The following overview presents a brief perspective of the development of our concept of the syncytial epithelium that forms the acanthocephalan tegument, and highlights the features which are critical to a discussion of nutrient absorption and feeding mechanisms.

Crompton (1963) performed a detailed light microscopic study of the body wall of adult *Polymorphus minutus*. He described an acidic, carbohydrate-rich 'layer' at the helminth surface which he named the 'epicuticle' because it appeared to be external to the 'cuticle'. This misleading terminology has persisted in the literature (e.g. Nicholas, 1973; Graeber & Storch, 1978) even though we now realize through electron microscopy that the 'epicuticle' is a typical glycocalyx like those demarcating the external functional boundaries of all eukaryotic cells (Wright & Lumsden, 1968, 1969; Lee, 1972; Byram & Fisher, 1973, 1974; Lumsden, 1975). In order to understand the acanthocephalan digestive–absorptive surface, we must recognize that much of the tegument glycocalyx is a manifestation of the oligosaccharide-bearing portions of integral membrane glycoproteins, and hence a critical morphological and functional component of the plasma membrane itself (see Houslay & Stanley, 1982).

Crompton (1963) suggested that the striped layer of the acanthocephalan tegument might represent some sort of permanent structure, oriented perpendicular to the helminth surface, which provided conduits for nutrient absorption through the 'cuticle'. Crompton & Lee (1965) identified these putative conduits in transmission electron microscope (TEM) images of *Polymorphus minutus* as 'canals' whose lumina communicated directly with the environment through 'pores' of reduced bore. Crompton & Lee recognized that the 'pores' and 'canals' in the striped layer were lined by a trilaminate plasma membrane continuous with that at the body surface; however, they construed the surface membrane and the electron-lucid layer beneath it to be a 'cuticle', and they interpreted the underlying non-canalicular portion of the striped layer as a 'homogeneous material' which Lee (1966) concluded was also a metabolically inactive component of the 'cuticle'. A contemporaneous TEM study of the adult *Moniliformis moniliformis* tegument (Nicholas & Mercer, 1965) introduced similar misinterpretations of the acanthocephalan surface, and ultrastructural examinations of other acanthocephalans which appeared shortly after the reports by Crompton & Lee and Nicholas & Mercer consistently referred to the cuticular nature of the tegument surface (Stranack, Woodhouse & Griffin, 1966; Lee, 1966; Hammond, 1967; Nicholas, 1967; Bird & Bird, 1969).

Wright & Lumsden (1969) described the cytoplasmic character of the non-canalicular component of the striped layer in the *Moniliformis moniliformis* tegument, and concluded that the features interpreted as 'cuticle' in earlier TEM studies were typical of the apical cytoplasm of many columnar epithelial cells. A more extensive study by Byram & Fisher (1973) provided more details of the typical cytoplasmic features of the *M. moniliformis* tegument and underscored Wright & Lumsden's demonstration of its epithelial nature.

Most authors have retained some variant of the 'pore and canal' terminology introduced by Crompton & Lee (1965). Byram & Fisher (1973) preferred to describe the regularly arrayed bag-like invaginations of the adult *Moniliformis moniliformis* surface as crypts, whose apices narrow into slender necks through which the crypt lumina communicate with the external environment. The term crypt is appropriate as a general term for these invaginations. Not only does it emphasise their bag-like (as opposed to tubular) features in many species, but it is also more descriptive of the surface invaginations of non-adult forms. The crypts occur as labyrinthine cavities whose membranous lining is extensively evaginated to form microvillar projections in acanthors (Wright & Lumsden, 1970), and these cavities may give an impression of vesicular structures in developing acanthellae and cystacanths (Butterworth, 1969*a*). More importantly, in contrast to the connotation of a conduit having a specific physical terminus implied by the term 'canal', the term 'crypt' suggests a cavity lacking specific polarity except at its entrance; this image is more consistent with the probable functional organization of the membranes lining the tegument crypts.

Although early interpretations of acanthocephalan ultrastructure singled out the crypts as the only possible sites of nutrient uptake, Wright &

Lumsden (1969) emphasized that the surface plasma membrane outside the crypts might also be involved in nutrient assimilation, but they cautioned against assuming that the plasma membrane at the tegument surface would be functionally identical to that lining the crypts. Cytological observations indicate several structural differences between the surface and crypt regions of the *Moniliformis moniliformis* tegument membrane, including the staining intensity of the glycocalyx (Wright & Lumsden, 1969; Byram & Fisher, 1973, 1974). These differences suggest regional heterogeneity in biochemical composition of the apical plasma membrane of the tegument, and it is probable that there is a significant dichotomy between the portions of the plasma membrane at the free surface and those portions lining the crypts, in terms of the specific membrane proteins which they contain.

6.2.2 *Digestive properties of the acanthocephalan surface*

The possibility that the acanthocephalan body surface has digestive capabilities has received substantial attention. Both histochemical and physiological investigations have detected hydrolytic enzyme activities associated with the tegument surface and there is good evidence that some of those activities are of intrinsic origin.

Surface hydrolase activities. Histochemical studies have revealed phosphatase activities associated with the apical portion of the adult tegument of several species of Archiacanthocephala and Palaeacanthocephala, but alkaline phosphatase activity appears lacking in the striped layer of some palaeacanthocephalans and of all eoacanthocephalans which have been examined (Table 6.1). It is now presumed that the phosphohydrolase activities associated with the striped layer of the adult tegument represent surface enzymes accessible to substrates in the host intestinal lumen. However, this presumption has been verified only for *Moniliformis moniliformis* (Rothman, 1967; Byram & Fisher, 1974). Starling & Fisher (1975) observed that the glucose liberated when glucose 6-phosphate (G6P) is hydrolyzed by the surface phosphatase of adult *M. moniliformis* can be absorbed by the helminth, and it is reasonable to suggest that the surface phosphatase may convert phosphate esters naturally present in the intestinal lumen (e.g. nucleotides, see Pappas & Read, 1974) into molecular species which can be absorbed into the tegument.

Crompton (1963) demonstrated leucine aminopeptidase activity associated with the apical region of the *Polymorphus minutus* tegument. It is probable that Crompton identified a surface hydrolase, but experiments designed to demonstrate peptidase activity specifically exposed at the surface of *P. minutus* have not been performed. Uglem & Beck (1972) found aminopeptidase activities against a variety of artificial substrates in homogenates of *Neoechinorhynchus cristatus*, *N. crassus* and *N. emydis*. Aminopeptidase activity was detected in the apical region of the *N. cristatus* tegument histochemically, and hydrolysis of some peptides by intact *N. cristatus in vitro* suggested that the aminopeptidase activity is exposed at the tegument surface (Uglem, 1972*a*).

Uglem, Pappas & Read (1973) demonstrated convincingly that adult *Moniliformis moniliformis* has a surface aminopeptidase which hydrolyzes alanyl and leucyl peptides. In the presence of methionine, which competes with leucine and alanine for absorption by *M. moniliformis* (Rothman & Fisher, 1964), worms incubated *in vitro* with leucylleucine or alanyl peptides released significant amounts of leucine or alanine into the incubation medium. When methionine was omitted from incubations, 93% of the leucine liberated by hydrolysis of leucylleucine was absorbed by the acanthocephalans. Furthermore, the presence of alanyl or leucyl peptides significantly inhibited the absorption of [^{3}H]-leucine from the incubation medium. The surface aminopeptidase activity of *M. moniliformis* appears to be specific for relatively small peptides; Ruff, Uglem & Read (1973) found no evidence for surface proteolytic activity when they incubated *M. moniliformis* with the artificial protease substrate azoalbumin.

Table 6.1. *Histochemical and cytochemical evidence for alkaline phosphatase activity at the tegument surface of adult Acanthocephala*[a]

ARCHIACANTHOCEPHALA
Macracanthorhynchus hirudinaceus (+)[b]; *Moniliformis moniliformis* (+)[b,c,d]; *Oncicola canis* (encysted immature forms) (+/−)[b]
PALAEACANTHOCEPHALA
Acanthocephalus sp. (−)[b]; *Corynosoma wegeneri* (+)[b]; *Echinorhynchus salmonis* (+)[f]; *E. gadi* (+)[b,f]; *Fessisentis vancleavei* (+)[b]; *Leptorhynchoides thecatus* (−)[b]; *Polymorphus minutus* (+)[g]; *Polymorphus* sp. (+)[b]; *Pomphorhynchus bulbocolli* (+)[b,e]; *Tegorhynchus furcatus* (+)[b]; *Telosentis tenuicornis* (+)[b]
EOACANTHOCEPHALA
Atactorhynchus verecundus (−)[b]; *Gracilisentis gracilisentis* (−)[b]; *Neoechinorhynchus cristatus* (−)[b]; *N. cylindratus* (−)[b,e]; *N. emydis* (−)[e]; *N. pseudemydis* (−)[b]; *N. saginatus* (−)[b]; *Octospinifer macilentus* (−)[b]; *Tanaorhamphus longirostris* (−)[b]

[a] (+) = positive reaction for alkaline phosphatase; (−) = no reaction
[b] Bullock (1958)
[c] Rothman (1967) (electron microscopic study)
[d] Byram & Fisher (1974) (electron microscopic study)
[e] Bullock (1949*a*)
[f] Lange (1970)
[g] Crompton (1963)

Ruff *et al.* (1973) found that *Moniliformis moniliformis* displayed significant surface amylolytic activity, hydrolyzing as much as 1.3 mg starch/g worm dry mass per min. These authors emphasized that the presence of amylase activity at the tegument surface would be advantageous because maltose, the product of starch hydrolysis by α-amylase, will support fermentation (Laurie, 1957) and glycogenesis (Laurie, 1959) in *M. moniliformis in vitro*. Studies of the effects of dietary carbohydrate on *M. moniliformis* growth *in vivo* (Nesheim *et al.*, 1977, 1978; Parshad, Crompton & Nesheim, 1980) also suggested that surface amylolytic activity might be significant to the nutrition of this acanthocephalan worm (§6.4).

Laurie's observation that *Moniliformis moniliformis* can metabolize maltose *in vitro* (Laurie, 1957, 1959) provided indirect evidence for surface maltase activity in this acanthocephalan. Similarly, Dunagan (1962) obtained tenuous evidence suggesting disaccharidase activities hydrolyzing maltose and trehalose at the surface of a species of *Neoechinorhynchus* from turtles; survival *in vitro* was significantly increased by addition of glucose, maltose, galactose or trehalose (in decreasing order of effectiveness) to culture media.

Starling (1972) and Starling & Fisher (1975) obtained additional evidence for maltase activity at the surface of adult *Moniliformis moniliformis*. Starling (1972) found that radioactivity is absorbed by *M. moniliformis* incubated in [^{14}C]-maltose, but concluded that the maltose was hydrolyzed to glucose before absorption occurred. Starling & Fisher (1975) observed that maltose, but not trehalose, sucrose, or lactose, inhibited the uptake of [^{14}C]-glucose by intact *M. moniliformis*. They noted that the relation between maltose concentration and inhibition of [^{14}C]-glucose absorption indicated that maltose did not inhibit [^{14}C]-glucose uptake directly (see below), and they concluded that the apparent inhibition of [^{14}C]-glucose uptake by maltose (and by G6P) was secondary to the liberation of free glucose as a result of hydrolytic activity at the worm's surface.

The rate at which *Moniliformis moniliformis* absorbed glucose moieties from maltose and the degree to which maltose inhibited [^{14}C]-glucose absorption in the experiments conducted by Starling & Fisher were fairly low. However, their medium (KRT; Read, Rothman & Simmons, 1963) contained tris-(hydroxymethyl)aminomethane (Tris^{+}), which is a potent inhibitor of maltase and other α-glucosidases (see Semenza, 1968). Hence, the studies by Starling & Fisher probably underestimated the capacity of *M. moniliformis* to hydrolyze maltose at its surface. Since maltose is nearly as effective as mannose in supporting glycogenesis and metabolism in

M. moniliformis in the absence of $Tris^{+}$ ions (Laurie, 1957, 1959), it would appear that the surface maltase of adult *M. moniliformis* is fairly active. Although the maltase at the surface of *M. moniliformis* does not appear to compete effectively with that of the host intestinal mucosa for free maltose (Read & Rothman, 1958; Parshad, Crompton & Nesheim, 1980; §6.4.1), it may play a crucial role in the utilization of the products released as a result of amylolytic activity at the helminth surface.

Bullock (1949*a*) detected lipase activity histochemically in the apical region of the teguments of *Echinorhynchus salmonis* and *Pomphorhynchus bulbocolli*, but could not demonstrate lipase activity in *Neoechinorhynchus emydis* or *N. cylindratus*. Ruff *et al.* (1973) could find no biochemical evidence for surface lipase activity in *Moniliformis moniliformis* using *N*-methylindoxylmyristate as a substrate. These results may reflect systematic differences between acanthocephalan classes in terms of the occurrence of putative surface lipase activities, or they may be fortuitous. It should be noted that the depositions of reaction product in the apical regions of the *E. salmonis* and *P. bulbocolli* teguments, which Bullock interpreted as positive evidence for lipase activity, were modest, and may have been artifacts of his long incubations. In any case, these palaeacanthocephalans should be re-examined biochemically for potential lipid hydrolysis before the presence of surface lipase can be accepted.

Location of the surface hydrolases. Rothman (1967) and Byram & Fisher (1974) examined the distribution of the alkaline phosphatase and acid phosphatase activities at the surface of *Moniliformis moniliformis* using EM cytochemistry. Deposits of the electron-dense reaction products diagnostic of phosphatase activities were observed within the crypts of the tegument, but not in association with the glycocalyx and plasma membrane at the free surface outside the crypts. Significantly, the phosphatase reaction product was not dispersed throughout the crypt lumina; rather, it was restricted to regions lying close to the luminal surface of the plasma membrane. The membrane-associated phosphatase activity appeared to be distributed along the entire expanse of the plasma membrane lining the crypts, but diminished abruptly at the neck demarcating the crypt entrance (Byram & Fisher, 1974). This distribution suggests that the phosphatase activity either is a manifestation of enzymes which are integral components of the crypt plasma membrane or resides with proteins which are tightly and specifically adsorbed to it.

The surface aminopeptidase activities of acanthocephalans have not been localized cytologically. Crompton's (1963) histochemical study of leucine aminopeptidase activity in the *Polymorphus minutus* tegument revealed intense activity throughout the striped layer; if this activity is indeed attributable to a surface (as opposed to cytoplasmic) hydrolase, then one would expect that an appropriate EM cytochemical examination would confirm that most, and possibly all, of the surface aminopeptidase is present within the crypts. The interpretation of physiological studies of the surface aminopeptidase activity of *Moniliformis moniliformis* (Uglem *et al.*, 1973) is consistent with this expectation.

Uglem *et al.* found that leucylleucine inhibited the uptake of [^{14}C]-leucine (0.1 mM) by adult *Moniliformis moniliformis* by as much as 50% at peptide concentrations presumed to be saturating for the surface aminopeptidase. They argued that the amino acids derived from peptide hydrolysis could compete successfully with [^{14}C]-leucine absorption by the worm only if the leucine uptake sites and the aminopeptidase activity were located in close proximity and/or if peptide hydrolysis and leucine absorption occurred in unstirred regions, where diffusion of peptide hydrolysis products away from the sites of leucine absorption was impeded. They nominated the crypts as the logical location for these unstirred regions, and found support for this proposition in their observations on leucine aminopeptidase activity of cystacanth larvae.

Uglem *et al.* (1973) demonstrated that mechanically exsheathed *Moniliformis moniliformis* cystacanths, activated by incubation in sodium taurocholate, had significant surface aminopeptidase activity. The activated cystacanths absorbed [^{14}C]-leucine, and [^{14}C]-leucine absorption was significantly inhibited by methionine and leucine. However, [^{14}C]-leucine uptake was unaffected by the presence of leucylleucine. Noting Byram's (1971) observation that the surface crypts of *M. moniliformis* cystacanths are both fewer in number and much shallower than those of the adult tegument, Uglem *et al.* concluded that the failure of leucylleucine hydrolysis to inhibit [^{14}C]-leucine uptake by activated cystacanths was a consequence of the architecture of the poorly developed crypts, which were too shallow and open to impede diffusion of peptide hydrolysis products released within them away from the sites of leucine absorption.

The results presented by Starling & Fisher (1975) suggest that at least some of the supposed maltase activity at the surface of *Moniliformis moniliformis* is also present within the tegument crypts. The basis for this assertion is the similarity between maltose and G6P as inhibitors of [^{14}C]-glucose absorption. The relation between the inhibition of glucose uptake by *M. moniliformis* and increasing concentrations of G6P or maltose in the incubation medium cannot be expressed as the rectangular hyperbola observed for 2-deoxyglucose, which competes directly with glucose for absorption (Fig. 6.1). The upwardly concave curves observed for the inhibition of [^{14}C]-glucose absorption by maltose and G6P indicate that there was an intervening kinetic process (e.g. hydrolysis) involved in the effects of these compounds on radioactive glucose uptake. In the present context, the important feature of the maltose and G6P inhibition curves in Fig. 6.1 is their shape at low inhibitor concentrations.

Starling & Fisher (1975) noted that when they attempted to extrapolate the kinetic curve for inhibition of [^{14}C]-glucose absorption by G6P to lower inhibitor concentrations, they obtained the anomalous result that a 12% residual inhibition persisted in the absence of any inhibitor. They surmised that this nonsensical result represented a separation of the inhibition of [^{14}C]-glucose uptake into two components. Starling & Fisher reasoned that the '12% residual inhibition' indicated a highly efficient inhibition of a part of the total [^{14}C]-glucose uptake, and that this inhibition component was essentially saturated at the lowest G6P concentration tested. They proposed that the '12% residual inhibition' represented inhibition of [^{14}C]-glucose uptake across the membranes lining the crypts, where the surface phosphohydrolase activity is localized (Rothman, 1967; Byram & Fisher, 1974)

Fig. 6.1. The competitive inhibition of 2mM [^{14}C]-glucose uptake into 32-day-old female *Moniliformis moniliformis* by 2-deoxyglucose (open circles), and the indirect inhibition of [^{14}C]-glucose uptake by maltose (closed circles) and glucose 6-phosphate (closed squares). (Modified from Starling & Fisher, 1975.)

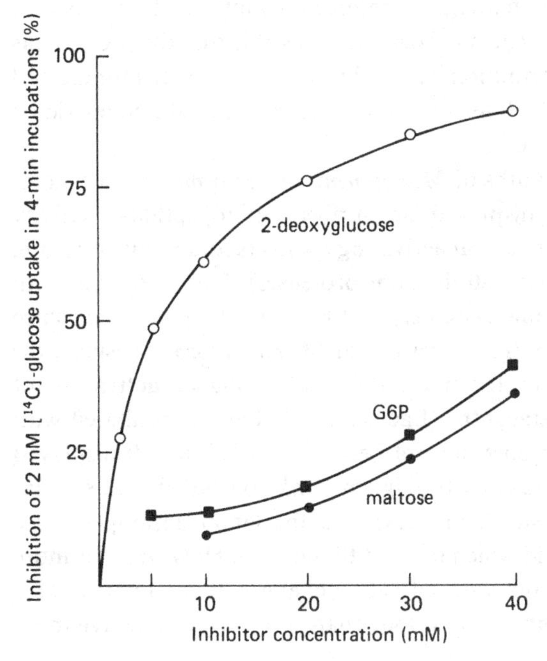

and where the products of G6P hydrolysis might build up to significant levels. They suggested that the second component represented inhibition of [^{14}C]-glucose uptake across the plasma membrane at the free surface of the tegument, which depended on the diffusion of free glucose out of the crypts following G6P hydrolysis by the crypt phosphatase. The kinetic curve for inhibition of [^{14}C]-glucose uptake in the presence of maltose (Fig. 6.1) resembles that for G6P inhibition of [^{14}C]-glucose absorption in that extrapolation to lower maltose concentrations suggests inhibition in the absence of an inhibitor. If this interpretation of the kinetics for G6P inhibition of [^{14}C]-glucose uptake is correct, then it is possible that maltose hydrolysis occurs within the crypts of the *M. moniliformis* tegument.

Origin of the surface hydrolases. Uglem *et al.* (1973) established that the aminopeptidase activity at the surface of *Moniliformis moniliformis* is intrinsic to the helminth and not an adsorbed host product. Repeated washings of worms in incubation saline had no effect on the surface peptidase activity, nor could the aminopeptidase activity be removed from the acanthocephalan's surface by incubation in divalent cation-free medium containing a chelating agent. Furthermore, surface aminopeptidase activity could be demonstrated for artificially activated cystacanth larvae which had never been exposed to the alimentary canal of the definitive host. The evidence suggests that the *M. moniliformis* surface aminopeptidase is a specific membrane protein integrally intercalated into the lipid phase of the plasma membrane. If that is the case, it is possible that the protein has a non-random lateral distribution within the apical plasma membrane and that patches of aminopeptidase-containing membrane might be restricted to the crypts of the tegument.

Freshly isolated cystacanths of *Moniliformis moniliformis* and mechanically exsheathed larvae displayed no surface aminopeptidase activity unless treated with any of several activating agents (sodium taurocholate, lipase, Triton X-100, Tween 80, but not proteases). Larvae treated with sodium taurocholate became extremely motile, as would be expected since bile salts are necessary to the activation of *M. moniliformis* cystacanths during infection (Graff & Kitzman, 1965) and enhance activation of *Polymorphus minutus* cystacanths (Lackie, 1974). Larvae incubated with lipase or non-ionic detergents did not become motile but still displayed significant aminopeptidase activity. Uglem *et al.* concluded that sodium taurocholate and other surfactants exposed the larval aminopeptidase activity by removing a lipid which masked the enzyme. However, it is more likely that these surfactants induced subtle changes in the hydrophobic phase of the surface membrane and that changes in membrane dynamics resulted in the expression of aminopeptidase activity (see Houslay & Stanley, 1982).

Lumsden (1975) suggested that the acid phosphatase and alkaline phosphatase activities associated with the crypt plasma membranes of the *Moniliformis moniliformis* tegument are also enzymes of intrinsic origin. There have been no studies bearing specifically on this hypothesis, but the circumstantial evidence is in its favor.

Bullock (1958) found no evidence for alkaline phosphatase activity in larval and encysted juvenile forms of several acanthocephalan species whose adults show intense histochemical reactions for alkaline phosphatase activity in the apical portion of their teguments. The absence of alkaline phosphatase activity in these immature forms suggests that the activity does not arise before establishment in the alimentary canal of the definitive host; this point does not distinguish between adsorption of a host protein and developmental activation of an enzyme intrinsic to the helminth, as occurs with the surface aminopeptidase of *Moniliformis moniliformis* (Uglem *et al.*, 1973). Bullock (1958) identified several palaeacanthocephalan species whose adults invariably lacked surface alkaline phosphatase activity; in some cases, these phosphatase-negative acanthocephalans were recovered from host species which also harbored Acanthocephala showing intense histochemical reactions for surface alkaline phosphatase. The simplest conclusion to be drawn from these observations is that the presence of alkaline phosphatase in the apical tegument is determined by the acanthocephalan and not the host.

An intrinsic origin of the surface phosphatase activities would be confirmed if it were possible to identify the enzymes in secretory vesicles or as a part of vesicular plasma membrane precursors within the tegumentary cytoplasm. Byram & Fisher (1973, 1974) described as lysosomes five distinct types of vesicular elements in the *Moniliformis moniliformis* tegument on the basis of their unmistakable acid phosphatase activity. Among this diversity of morphologically distinct acid phosphatase-containing vesicles, 'Type 1 lysosomes' were identified as virgin lysosomes because, unlike the other designated lysosomal types, they were never observed to contain material either undergoing digestion or taken up into the tegument by pinocytosis, and because they occurred in close proximity to the crypts, assumed to be involved in pinocytosis. Although the identification of these vesicles as virgin lysosomes may be correct, it seems only to depend on the possession of acid phosphatase activity. Since 'Type 1 lysosomes' are in the same size class (0.1–0.15 μm, incorrectly cited as 1.0–1.5 μm in Bryam & Fisher, 1973) as the Thorotrast-staining ovoid vesicles presumed to be carrying new membrane (or secretory material) to the tegument free surface, it is tempting to speculate that the 'Type 1 lysosomes' in the *M. moniliformis* tegument may be vesicles of newly formed membrane components in transit to their destination as part of the plasma membrane of the crypts.

Unlike the surface aminopeptidase, and most probably the surface phosphohydrolase(s) as well, the surface amylase activity of adult *Moniliformis moniliformis* is clearly of host origin. Ruff *et al.* (1973) found that the amylase activity leached gradually from the helminth surface into the medium during successive washings. Amylase activity could not be restored to the worm's surface by incubation in porcine pancreatic extract, but incubation in rat gut contents increased the surface amylolytic activity of intact helminths. The host origin of the surface amylase activity was confirmed immunologically; amylase collected from the surface of *M. moniliformis* was insensitive to crude antiserum against porcine pancreatic amylase, but was significantly inhibited by antiserum to crude rat pancreatic extract.

Because α-amylase of bacterial and porcine origin had been shown to undergo enhancement of activity when associated with the surface of cestodes (Taylor & Thomas, 1968; Read, 1973), Ruff *et al.* (1973) looked for possible increases in amylase activity upon binding to the surface of *Moniliformis moniliformis*, but no such enhancement was found. Ruff *et al.* also examined the effects of *M. moniliformis* on the activities of trypsin, chymotrypsin and lipase, each of whose activity is reduced upon incubation with the cestode *Hymenolepis diminuta* (Pappas & Read, 1972*a*, *b*; Ruff & Read, 1973). Again, they found that *M. moniliformis* had no interaction with the activities of any of these hydrolytic enzymes.

6.2.3 *Functional organization of the digestive–absorptive surface: a hypothesis*

The results of ultrastructural studies of the acanthocephalan tegument have emphasized that the crypts provide an extensive amplification of the surface area available for nutrient uptake (Hammond, 1967; Bird & Bird, 1969; Wright & Lumsden, 1969; Byram & Fisher, 1973; Beermann, Arai & Costerton, 1974; Graeber & Storch, 1978). Morphometric calculations indicate that the degree of surface amplification ranges from 20- to 62-fold for different species (Byram, 1971; Byram & Fisher, 1973; Graeber & Storch, 1978), but these estimates may severely overstate the functional surface area involved in the uptake of solutes of low relative molecular mass present in the host intestinal fluids.

The microvillar surfaces of the cestode tegument and vertebrate intestinal mucosa represent obvious amplifications of the total surface area of the apical plasma membrane. However, it does not appear that the microvilli constitute a significant functional amplification of the surface area involved in the uptake of small solutes *in vivo*. This apparent anomaly is a consequence of the unstirred water organized within the interstices of the microvilli; the hydrodynamic unstirred layer forms a substantial diffusion barrier which may be rate-limiting for the absorption of many solutes (Wilson & Dietschy, 1974; Befus & Podesta, 1976; Podesta, 1977). Podesta (1977) estimated that the functional surface area of *Hymenolepis diminuta* available for absorption is only slightly greater than the surface area would be without microvilli. Wilson & Dietschy (1974) determined that the effective surface area of the rat ileal mucosa available for transport is less than 1% of the anatomical surface area; this represents an area approximately equal to the surface area of the tips of the intestinal villi if there were no microvilli present.

The lumina of the crypts in the acanthocephalan tegument communicate with the external environment through necks of very small bore. The outer openings of the pores in the tegument of *Acanthocephalus ranae* appear to be about 100 nm in diameter in scanning electron microscopic (SEM) images (Hammond, 1968*a*), while their inner diameters in TEM images are approximately 50 nm (Hammond, 1967). In *Moniliformis moniliformis*, the neck openings measure 12–40 nm in diameter in TEM images (Byram & Fisher, 1973) and appear to average about 20 nm (Wright & Lumsden, 1969). Since the surface glycocalyx of some acanthocephalan species is so dense that it partially or completely masks the pores in SEM images (Hammond, 1968*a*; Graeber & Storch, 1978), and since the plasma membrane lining the necks of the crypts bears an elaborate glycocalyx (Wright & Lumsden, 1969; Byram & Fisher, 1973), the effective openings of the crypts for unimpeded exchange of fluids are probably much smaller than the membrane-to-membrane dimensions measured from transmission electron micrographs.

We must conclude that the surface crypts of the acanthocephalan tegument are unstirred regions comparable to the intervillar spaces of the vertebrate small intestine. As a consequence, the plasma membrane lining the tegument crypts may make a relatively minor contribution to the total uptake of solutes such as glucose and free amino acids *in vivo*, despite the fact that the crypt membranes represent 80–98% of the anatomical surface area of the acanthocephalan tegument. Starling & Fisher (1975) estimated that only about 12% of the glucose entering *M. moniliformis* incubated in 2mM [^{14}C]-glucose was absorbed across the membranes lining the crypts. Uglem *et al.* (1973) found that about 50% of the uptake of 0.1mM [^{3}H]-leucine by *M. moniliformis* was not inhibited by the products of leucylleucine hydrolysis (§6.2.2); this suggests that significantly less than 50% of the [^{14}C]-leucine absorption observed in their experiments occurred across the crypt plasma membranes where aminopeptidase activity is believed to be localized.

Although unstirred layer effects may limit the access of small solutes in the host intestine to the crypt plasma membranes, they should greatly increase the facility with which the products of hydrolytic enzymes localized within the crypts are absorbed across the crypt membranes. Products liberated by hydrolytic activities within the unstirred crypts should not be diluted significantly by exchange with extracrypt fluids and should therefore enjoy a kinetic advantage for absorption across the crypt membranes. The unstirred layer associated with the brush border of the vertebrate intestinal mucosa may provide a similar kinetic advantage to the absorption of products liberated by some hydrolases of the microvillar membrane (Warden *et al.*, 1980). Thus, the surface crypts of the acanthocephalan tegument may be morphological adaptations for maximizing absorption of the products of surface hydrolytic enzymes. As such, the crypts might best be described as extracytoplasmic digestive organelles.

6.3 Feeding mechanisms and related phenomena

6.3.1 *Transport of small organic solutes*

Small solutes generally enter or leave cells via mediated transport events effected by specific protein components of the plasma membrane. These transport proteins bind one or more molecules of an appropriate solute and subsequently release the solute at the trans (opposite) side of the membrane as a consequence of poorly understood conformational changes in the transporter protein.

General characteristics of membrane transport. In many cases, the simplest criterion for identifying absorption mediated by specific transporters is the demonstration that solute uptake is a non-linear function of its concentration. If appropriate corrections are made for the effects of ordered water and solvent flow (see Wilson & Dietschy, 1974; Podesta, 1977; Podesta *et al.*, 1977; Cornford & Oldendorf, 1979), the *unidirectional* transport of a solute can be described by an expression similar to the Michaelis–Menten equation for enzyme kinetics. The equation for unidirectional inward flux of a solute is:

$$J_{in} = (J_{max})([S]_o)/([S]_o + K_t)$$

where J_{in} is the rate of (radioactive) solute uptake (=influx), J_{max} is the maximal influx expected at infinite solute concentration, and $[S]_o$ is the extracellular concentration of the transported solute. K_t is the transport version of the Michaelis constant, K_m, and identifies the solute concentration at which the influx is half-maximal. K_t is a complex combination of kinetic constants, and it will approximate the solute–transporter dissociation constant, K_S, only if the conformational step is rate-limiting for the entire process (see Stein, 1967; Christensen, 1975; Segel, 1975).

The Michaelis–Menten expression for unidirectional uptake of a solute does not define situations in which the transporter has more than one solute binding site, nor does it describe fully the kinetics for transporters which may be dependent upon sodium ions. However, studies of transport in acanthocephalans have not reached a level of sophistication which warrants discussion here of multiple substrate kinetics. Sodium ion-dependent processes can generally be treated by analogy to allosteric enzymes for which sodium ions are activators and potassium ions frequently inhibitors. The specific kinetic effects of sodium and other effector ions can usually be ignored as long as the ionic conditions are held constant in individual experiments, but results obtained under different ionic conditions may be incompatible. The interested reader should consult Stein (1967), Schultz & Curran (1970), Christensen (1975), Segel (1975), Wilson (1978), Levitt (1980) or Carafoli & Scarpa (1982) for details of more kinetically complex systems and treatment of net flux kinetics.

Two solutes transported by the same transporter will compete with each other for binding to the transport locus. Kinetically, this phenomenon can be treated in the same manner as competitive inhibition of enzymes which obey Michaelis–Menten kinetics. The competitive inhibition of the uptake of a radioactive solute can be investigated by increasing concentrations of non-radioactive solute of the same molecular species. This approach has been applied in several studies of membrane transport in parasites to determine the 'uninhibitable component' of solute absorption. The uninhibitable component has generally been treated as a diffusional component of solute uptake (Uglem & Read, 1973; Pappas & Read, 1975). However, except in the case of lipophilic molecules, rates of simple diffusion across membranes may be too low to be of kinetic significance because of the low solubilities of polar molecules in the hydrophobic phase of the membrane (Stein, 1967; Houslay & Stanley, 1982). It is likely that most of the diffusion measured in transport studies on helminths is a consequence of two other phenomena: (1) entrapment of solute in unstirred layers at the absorptive surface (Wilson & Dietschy, 1974; Podesta, 1977; Podesta *et al.*, 1977; Cornford & Oldendorf, 1979); and (2) solvent drag associated with the osmotic flow of water (Stein, 1967), most of which appears to occur non-specifically through the polar regions created by transporter proteins or ion-specific channels (Houslay & Stanley, 1982).

Some transport systems can mediate the net movement of solute from a compartment of low to one of higher electrochemical potential (active transport) while others cannot. The latter facilitated diffusion systems confer selectivity upon solute uptake and provide a mechanism for rapid transfer of polar solutes across cell membranes; the net movement of solutes via facilitated diffusion systems is always downhill. Facilitated diffusion will mediate uptake of a radioactive solute from a medium having a low solute concentration into tissue where the concentration of that solute is relatively high. However, this is not active transport: the uptake of radioactive tracer is compensated by a simultaneous efflux of non-radioactive solute such that the net movement of solute is outward in the direction required by the electrochemical gradient. Some studies of membrane transport in helminths have incorrectly interpreted the uptake of radioactive solute from solutions of low concentration as evidence for active transport because they failed to recognize that transport systems mediate bidirectional movement.

Active transport phenomena are divided into two categories based upon their energy-coupling mechanisms. In primary active transport systems, the energy source is a chemical reaction which is tightly coupled to solute translocation. Primary active transport systems include: ion-transporting ATPases, in which discrete steps in the ATP hydrolysis reaction induce conformational changes in the transporter (Glynn & Carlish, 1975; Levitt, 1980; Carafoli & Scarpa, 1982); and the electron transport chains of the inner mitochondrial membrane, the chloroplast thylakoid membrane and the bacterial plasma membrane, which couple oxidation–reduction reactions to the specific transmembrane movement of protons against an electrochemical gradient (Christensen, 1975; Wilson, 1978; Houslay & Stanley, 1982).

Secondary active transport systems utilize the ion gradients created by primary active transport systems. The transporter mediates the simultaneous uphill movement of the actively transported solute and downhill movement of the driving ion, either in the same direction (symport or co-transport) or in the opposite direction (antiport or counter-transport). Proton gradients drive secondary active transport across bacterial, mitochondrial and chloroplast membranes, and proton symports are also present in the plasma membranes of eukaryotic cells having proton-pumping ATPases (e.g. yeast) (Wilson, 1978; LaNoue & Schoolwerth, 1979; Gunn, 1980). In metazoans, secondary active transport usually involves sodium ion symports for accumulation of organic nutrients (Crane, 1965, 1968; Schultz & Curran, 1970; Christensen, 1975; Pappas & Read, 1975; Podesta, 1982) or sodium ion antiports for active extrusion of ions (Na^+–Ca^{2+} antiports: Gamj, Murer & Kinne, 1979; Gunn, 1980; DiPolo & Beaugé, 1983).

Some assumptions about the energetics of secondary active transport driven by sodium ion coupling have been questioned, especially in the case of amino acids (Potashner & Johnstone, 1971; Johnstone, 1972, 1974; Podesta, 1980). However, it is clear that sodium ion symport is involved in most cases of active transport of glucose into metazoan cells (Schultz & Curran, 1970; Pappas & Read, 1975; Starling, 1975; Wilson, 1978; Podesta, 1982), and sodium ion (or proton) symports are linked to the concentrative uptake of some non-cationic amino acids as well (Schultz & Curran, 1970; Christensen *et al.*, 1973; Christensen, 1975; Wilson, 1978; Johnstone & Laris, 1980; Ahearn, 1982; Podesta, 1982).

Carbohydrate absorption. Crompton & Lockwood (1968) measured the apparent uptake of glucose into *Polymorphus minutus in vitro* by monitoring the disappearance of [^{14}C]-glucose from the medium during a 60-min incubation. They reported that glucose uptake was a saturable function of external concentration, with half-saturation occurring at about 1.4mM glucose. However, it should be noted that the transport characteristics reported by Crompton & Lockwood probably reflect characteristics of steady state metabolism as well as transport because of the lengthy incubations in their experiments.

Crompton & Lockwood reported that *Polymorphus minutus* accumulated glucose against a concentration gradient, but this conclusion is questionable. Tissue glucose was estimated on the basis of radioactivity, rather than chemical determinations. Such estimates can be confounded by intermixing of absorbed [^{14}C]-glucose with non-radioactive glucose from internal stores or by exchange of tissue glucose for radioactive glucose in the incubation medium. Crompton & Lockwood indicated that worms incubated for 60 min in 8.6mM [^{14}C]-glucose had an apparent tissue glucose concentration of 0.92mM, and that this value rose to 1.33mM during a subsequent 30-min incubation in 0.7mM [^{14}C]-glucose. These results seem inconsistent with a transport system having an apparent K_t of approximately 1.4mM. We must conclude that Crompton & Lockwood's radioactive estimates of internal glucose concentration probably did not provide a reliable index of true glucose content.

Starling & Fisher (1975) examined the kinetics of monosaccharide uptake by female *Moniliformis moniliformis in vitro* and found that the glucose transporter also accepts 2-deoxyglucose, mannose, *N*-acetylglucosamine, 3-O-methylglucose, fructose and galactose (in decreasing order of efficacy) as substrates (Table 6.2). Glucose, mannose, fructose and galactose were phosphorylated rapidly after their absorption (see below and §6.5.2), but *N*-acetylglucosamine, 2-deoxyglucose and 3-O-methylglucose, which are not metabolized by *M. moniliformis* (Laurie, 1957), were phosphorylated only sparingly (Starling, 1972). Since there is good agreement between the K_t and K_i values for efficiently transported sugars (Table 6.2) whether or not they are metabolized, these parameters probably reflect properties of the transport system and not of enzymes involved in postabsorptive metabolism. However, the lower apparent maximal rates for uptake of non-metabolized sugars (Table 6.2) suggest that there may have been significant efflux of absorbed sugars during the 4-min incubations used by Starling & Fisher, especially in the case of poorly phosphorylated hexoses (see below).

Inhibitor studies indicated that the glucose transporter did not interact with L-arabinose, D-lyxose, D-ribose, D-xylose, L-rhamnose, L-sorbose, L-sorbitol, D-mannitol, glucose 6-phosphate (G6P), lactose, maltose, sucrose, trehalose, gluconic acid, glucuronic acid or glucuronolactone, nor was glucose absorption inhibited by arsenate, molybdate, fluoride, or ouabain. Neither phlorizin nor iodoacetate (1mM) inhibited glucose transport, as would be expected since these reagents do not affect metabolism of exogenous sugars by intact *Moniliformis moniliformis* (Laurie, 1957). Glucose uptake was inhibited by *p*-chloromercuribenzoate, whose effect on glucose absorption probably accounts for part of its inhibition of glucose metabolism by *M. moniliformis* (Laurie, 1957).

Starling & Fisher (1975) interpreted their kinetic data as evidence that

Table 6.2. *Kinetic parameters for the absorption of hexoses by 32-day-old female* Moniliformis moniliformis in vitro (*after Starling & Fisher, 1975*)

Hexose	K_t (mM)	K_i[a] (mM)	J_{max}[b]
Glucose	2.5	2.5	20.2
2-Deoxyglucose	2.5	2.7	11.0
Mannose	4.1	3.9	24.7
N-Acetylglucosamine	3.9	8.1	10.5
3-O-Methylglucose	18.6	14.7	16.4[c]
Fructose	27[c]; (7.6)[d]	26.3	35[c]
Galactose	80[c]	(58)[e]	29[c]

[a] As inhibitors of glucose or 2-deoxyglucose uptake
[b] In μmol radioactive substrate absorbed/g ethanol-extracted dry mass per 4 min
[c] Values must be considered inaccurate because of failure to correct for unstirred layer effects (see text)
[d] Estimated K_t for putative fructose-preferring transporter (see text)
[e] Single determination

the *Moniliformis moniliformis* tegument contains a second, fructose-preferring transporter which may also interact with *N*-acetylglucosamine, 3-O-methylglucose and galactose. Podesta (1982) suggested that the kinetic anomalies which led Starling & Fisher to postulate a second, fructose-preferring, transport system may have been an artifact of unstirred layer effects. Thus, the conclusion that *M. moniliformis* has a fructose-preferring transporter warrants further investigation.

Starling & Fisher found that total replacement of sodium ions in the incubation media by potassium, lithium, choline or Tris ions had no effect on the rate of glucose absorption in 4-min incubations. The apparent sodium ion-independence of glucose uptake by *Moniliformis moniliformis* was correlated with the absence of microvilli at the acanthocephalan surface, noting that sodium ion-dependent glucose transport mechanisms in metazoans appear restricted to brush border epithelia (Starling, 1975). Podesta (1982) has argued that such ion replacement studies may be ineffective in completely removing sodium ions from the vicinity of surface transporters. The argument suggests that sodium ions diffusing out of tissues may become trapped within unstirred layers at the surface of the tissue, where they can interact with sodium ion-dependent transporters despite the absence of sodium ions in the medium. Indeed, there is evidence that sodium ions trapped in surface unstirred layers may have confounded attempts to study sodium ion-sensitivity of amino acid uptake by cestodes. Nevertheless, the types of ion replacement studies which suggested sodium ion independence of glucose absorption by *M. moniliformis* have been effective in identifying the sodium ion dependence of hexose uptake by *Hymenolepis diminuta* (Starling, 1975) and by other cestodes (Pappas & Read, 1975).

Podesta (1982) cited unpublished studies which indicated an increase in the transmural sodium ion current during glucose absorption by *Moniliformis moniliformis*. Such an increase is consistent with a direct coupling between glucose uptake and sodium ion flux, but could be accounted for by concurrent (but not coupled) fluxes secondary to glucose absorption. It is not clear from the brief description in Podesta (1982) whether the possibility that the observed increases in sodium ion flux were only indirectly linked to glucose absorption has been excluded.

Even if the circumstantial evidence suggesting that sodium ion uptake accompanies glucose absorption by *Moniliformis moniliformis* is verified, it is clear that the glucose transporters of the *M. moniliformis* tegument are different from the sodium ion-coupled glucose transport systems described for cestodes. The cestode glucose transporters resemble those of the luminal brush border membranes of the vertebrate intestine and the renal proximal tubule: they can mediate glucose accumulation against a concentration gradient; they are highly sensitive to sodium ions, the influx of which is stoichiometrically coupled to glucose movement; they are fairly exacting in their steric requirements for an equatorial hydroxyl group at C-2 of the pyranose ring, and interact poorly, if at all, with mannose, 2-deoxyglucose and fructose (Crane, 1960, 1968; Schultz & Curran, 1970; Honegger & Semenza, 1973; Pappas & Read, 1975; Starling, 1975; Kinne, 1976). Kinetically, the *M. moniliformis* glucose transporters resemble the facilitated diffusion systems found in mammalian erythrocytes, hepatocytes, myocytes, adipocytes and basolateral membranes from enterocytes and renal tubule cells; these facilitated diffusion systems are much less sensitive to phlorizin than are sodium ion-coupled transporters, and they invariably accept mannose and 2-deoxyglucose as substrates (Stein, 1967; Clausen, 1975; Kinne, 1976; Plagemann, Wohlhueter, Graff, Erbe & Wilkie, 1981).

The evidence indicates that glucose transport into *Moniliformis moniliformis* occurs by facilitated diffusion systems which do not mediate glucose accumulation against a concentration gradient. Radioactive glucose, mannose, fructose and galactose were phosphorylated rapidly after their absorption by *M. moniliformis* incubated in media buffered with either Tris-maleate (Starling & Fisher, 1978, 1979) or phosphate plus bicarbonate (Starling, unpublished; §6.5.2). Some of the hexose phosphate formed from glucose and other hexoses was utilized for general metabolism, glycogenesis and other synthetic processes, but much was converted into trehalose. Preincubating worms in unlabeled glucose prior to incubation in [^{14}C]-glucose lowered the [^{14}C]-glucose uptake rate (Starling & Fisher, 1975) but did not reduce significantly the degree to which absorbed glucose was converted to other metabolites. These observations indicate that there is no accumulation of free glucose within the tegumentary compartment where glucose is absorbed. Hence, it appears that *M. moniliformis*, and probably other acanthocephalans, are dependent upon a metabolic sink phenomenon rather than active transport to insure that absorbed hexoses are retained within helminth tissues.

The effects of dietary sugars on the growth of *Moniliformis moniliformis in vivo* (Parshad, Crompton & Nesheim, 1980; §6.4.1) are consistent with the properties of the surface glucose transporter as reported by Starling & Fisher (1975). Parshad *et al.* found that diets containing 3% (by mass) mannose produced slightly better growth (dry mass) than did 3% fructose diets, but that dietary glucose supported growth relatively poorly: glucose levels of 24 g/100 g diet were required to sustain the same amount of worm growth that was observed in rats fed the diet containing 3% fructose. These results suggest that *in vivo* the rat intestinal mucosa effectively depletes glucose, but not mannose or fructose, from the intestinal lumen before significant quantities of sugar can be absorbed by *M. moniliformis*.

Glucose uptake by the well-stirred mammalian intestine is considered to have an apparent K_t comparable to that determined for glucose uptake by *Moniliformis moniliformis* (Starling, 1975), but the mammalian intestine may have a second glucose transporter with a significantly higher affinity for glucose (Honegger & Semenza, 1973). Furthermore, the glucose transporters of the intestinal brush border are sodium ion coupled; they can accumulate glucose, even when luminal glucose is low, with little glucose efflux from the mucosa into the intestinal lumen. In contrast, the mammalian intestine absorbs mannose and fructose via transporters which are not sodium ion coupled and cannot mediate accumulation against a concentration gradient (Crane, 1960, 1968; Honegger & Semenza, 1973). The fructose transporter of the hamster small intestine has a significantly higher K_t than the apparent K_t for fructose uptake by *M. moniliformis* (Honegger & Semenza, 1973; Starling & Fisher, 1975), and mannose is very poorly absorbed by the rat intestine (Kohn, Dawes & Duke, 1965). That mannose was slightly more effective than fructose in supporting *M. moniliformis* growth *in vivo* (Parshad, Crompton & Nesheim, 1980) is consistent with the lower apparent K_t for mannose absorption by *M. moniliformis in vitro* (Table 6.2; Starling & Fisher, 1975).

Parshad, Crompton & Nesheim (1980) found galactose to be even less effective than glucose in supporting *Moniliformis moniliformis* growth *in vivo*. Although the glucose transporters of the mammalian intestine have lower affinities for galactose than for glucose, galactose is actively transported and effectively metabolized (Crane, 1968; Honegger & Semenza, 1973; Berman *et al.*, 1976). In contrast, galactose is a poor substrate for the hexose transporters at the *M. moniliformis* surface (Table 6.2; Starling & Fisher, 1975). Since galactose is a moderately good substrate for the glucose transporters of the mammalian intestine, it would be of interest to determine whether the addition of significant quantities of galactose to glucose-containing experimental diets produced positive effects on *M. moniliformis* growth *in vivo*.

Maltose and G6P inhibited the absorption of [^{14}C]-glucose by *Moniliformis moniliformis* (Starling & Fisher, 1975), and *M. moniliformis* incubated in either [^{14}C]-G6P or [^{14}C]-maltose for 1–4 min absorbed radioactivity at significant rates (Starling, 1972). Examination of the radioactive species present in worms incubated in [^{14}C]-maltose suggested that the radioactivity was absorbed as free glucose rather than as the disaccharide (Starling, 1972). Furthermore, kinetic analysis indicated that the effects of G6P and maltose on [^{14}C]-glucose uptake were caused by free glucose liberated when these compounds were hydrolyzed at the worm's surface and not by the compounds themselves (Starling & Fisher, 1975; §6.2.2). Hence, *M. moniliformis* does not absorb maltose, as suggested by Crompton (1970), but can recover glucose moieties liberated from maltose and from G6P by the action of hydrolases associated with its surface.

Dunagan (1962) provided circumstantial evidence that the tegumentary glucose transporters of species of *Neoechinorhynchus* mediate galactose absorption; addition of galactose to culture media (balanced salts solutions) increased the survival of helminths significantly, but galactose was less effective than an equivalent amount of glucose in prolonging survival *in vitro*. Dunagan reported that phlorizin inexplicably increased the survival of worms in the absence of added glucose or galactose, but that the effects of glucose and galactose in extending worm survival were abolished in the presence of phlorizin. The latter observation at first suggests that phlorizin inhibits glucose and galactose uptake by *Neoechinorhynchus*. However, the fact that phlorizin prolonged worm survival *in vitro* suggests another interpretation. Phlorizin is a 2′β-glucoside of phloretin. If *Neoechinorhynchus* has a surface β-glucosidase similar to the phlorizin hydrolase found in the vertebrate intestinal brush border (Malathi & Crane, 1969; Warden *et al.*, 1980), release of glucose due to phlorizin hydrolysis could account for the survival-prolonging effects of phlorizin. In that event, it is not clear whether the abolition of the survival-promoting effects of glucose and galactose in the presence of phlorizin was caused by the glycoside itself or by phloretin, which is an inhibitor of many facilitated diffusion systems for glucose transport (Stein, 1967; Kinne, 1976).

Dunagan (1962) found that maltose and trehalose were as effective as comparable amounts of galactose in prolonging the survival of species of *Neoechinorhynchus in vitro*. Crompton (1970) interpreted this observation as evidence that maltose and trehalose are absorbed by *Neoechinorhynchus*. However, it is more likely that the survival-prolonging effects of maltose and trehalose were a consequence of the hydrolysis of these disaccharides, either by antibiotic-resistant microbes in the culture system or by hydrolases present at the surface of the helminths.

Amino acid absorption. Amino acid transport into metazoan cells involves as many as seven separate transport systems and a given amino acid may be transported by several such systems simultaneously (Schultz & Curran, 1970; Christensen *et al.*, 1973; Christensen, 1975; Pappas & Read, 1975). Multiple transport systems for neutral amino acids with distinct but overlapping affinities are legion. However, the specificities and relative abundances of these systems may differ between tissues in the same animal, and they show considerable variation across phylogenetic lines (Schultz & Curran, 1970; Christensen, 1975; Pappas & Read, 1975; Ahearn, 1982; Podesta, 1982).

Rothman & Fisher (1964) examined the uptake of several neutral amino acids by *Moniliformis moniliformis* and *Macracanthorhynchus hirudinaceus* in 3-min incubations (Table 6.3). They observed that [^{14}C]-methionine uptake exhibited saturation kinetics and was inhibited by leucine, isoleucine, alanine and serine. They also examined the kinetics for [^{14}C]-leucine, [^{14}C]-isoleucine, [^{14}C]-alanine and [^{14}C]-serine absorption and determined the K_i s for methionine inhibition of the uptakes of these amino acids (Table 6.3). Although Rothman & Fisher treated their data as though they were dealing with a single amino acid transport system, the kinetic

Table 6.3. *Some kinetic parameters for* L-*amino acid uptake by* Moniliformis moniliformis *and* Macracanthorhynchus hirudinaceus (*adapted from Rothman & Fisher, 1964*)

		Moniliformis		*Macracanthorhynchus*	
Parameter	Substrate	Females	Males	Females	Males
J_{max} [a]	[^{14}C]-methionine	1.17	1.43	0.5	0.67
	[^{14}C]-leucine	0.87	1.9	0.67	— [b]
	[^{14}C]-isoleucine	0.5	2.87	0.3	—
	[^{14}C]-alanine	5.8	5.4	0.23	0.47
	[^{14}C]-serine	2.6	4.53	—	11.77
K_t (mM)	[^{14}C]-methionine	0.7	0.8	1.0	0.9
	[^{14}C]-leucine	0.3	0.5	0.9	—
	[^{14}C]-isoleucine	0.3	0.2	—	—
	[^{14}C]-alanine	1.8	1.3	0.4	0.3
	[^{14}C]-serine	0.9	1.3	—	10.0
K_i (mM)	[^{14}C]-methionine				
	I[c] = leucine	0.6	0.4	1.0	0.5
	isoleucine	0.7	0.9	5.5	1.0
	alanine	0.3	0.5	30.0	1.9
	serine	0.6	0.3	9.0	1.0
	I = methionine				
	S[c] = [^{14}C]-leucine	0.8	1.0	19.0	—
	[^{14}C]-isoleucine	0.5	1.3	0.6	—
	[^{14}C]-alanine	5.2	11.0	N.E.[d]	1.3
	[^{14}C]-serine	5.3	2.4	N.E.[d]	3.5

[a] In μmol [^{14}C]-substrate absorbed/min per g ethanol-extracted dry mass for 3-min incubations
[b] Not determined
[c] I = inhibitor tested; S = substrate
[d] No apparent inhibitory effect

parameters in Table 6.3 suggest that they measured transport mediated by at least two systems for each of the amino acids studied. Furthermore, Rothman & Fisher's results suggest that the specificities and/or relative abundances of these transport systems differed between sexes of the same species and between species; for example, methionine appeared to be a good inhibitor of [^{14}C]-leucine uptake by male and female *M. moniliformis*, but a poor inhibitor of [^{14}C]-leucine absorption by *M. hirudinaceus* females.

Branch (1970*c*) measured the uptake of radioactivity by *Moniliformis moniliformis* incubated for 1 min in 0.1–10mM [^{14}C]-leucine. In contrast to the results of Rothman & Fisher (1964) and of Edmonds (1965), who also reported evidence for inhibition of leucine absorption by serine and valine, Branch found no significant inhibition of [^{14}C]-leucine uptake by methionine and no sign that leucine absorption by *M. moniliformis* reached a plateau at high leucine concentrations. Branch interpreted the apparently non-saturable component of leucine uptake at high substrate concentrations as evidence for a second, low-affinity, leucine transport system, and suggested that Rothman & Fisher did not observe this component either because of the low leucine concentrations used or because of the presence of Tris$^+$ ions in their incubation media. The basis for the latter suggestion was the premise that Tris$^+$ ions might substitute for sodium ions in binding to a sodium ion-dependent transporter, thereby inhibiting its activity.

Uglem & Read (1973) re-examined the kinetics of leucine and alanine absorption by *Moniliformis moniliformis* and concluded that the uptakes of leucine and alanine involve both mediated transport components, which are completely saturated at amino acid concentrations of about 4mM, and non-saturable diffusion components, which contribute significantly to amino acid uptake only at concentrations exceeding 2mM. Although Uglem & Read suggested that Branch's results similarly represented the combined effects of mediated absorption and free diffusion, the apparently non-saturable component of leucine uptake reported by Branch was significantly greater than that observed by Uglem & Read.

It is possible that the larger non-saturable component observed in Branch's experiments was a mechanical artifact or a factor relating to the age or strain of the helminths examined. However, the incubation medium used by Rothman & Fisher and Uglem & Read (KRT; Read *et al.*, 1963) contained Tris$^+$ and maleate and lacked the phosphate and bicarbonate ions (and presumably glucose) present in the Hanks' saline used in Branch's studies. If Branch's saline contained glucose, it is possible that there was a higher solvent drag component as a result of concomitant glucose and leucine absorption. Uglem & Read observed that substitution of phosphate-buffered saline for the KRT of their incubation media did not affect leucine uptake. However, it is not clear whether Uglem & Read's studies of potential $Tris^+$ ion effects on leucine uptake were performed at a single amino acid concentration or over a broad range. For this reason, the possibility that $Tris^+$ ions may alter the characteristics of amino acid transport into *M. moniliformis* at high amino acid concentrations does not appear to be resolved.

Moniliformis moniliformis accumulates methionine (Rothman & Fisher, 1964) and leucine (Uglem & Read, 1973) against marked (10–25-fold) concentration gradients *in vitro*. Diffusive loss of accumulated amino acids beyond the boundaries of the surface unstirred layers appears to be minimal; Uglem & Read observed no efflux of [^{14}C]-leucine from preloaded worms during postincubations in 1–20mM unlabeled leucine. Branch (1970*c*) reported the accumulation of radioactivity by *M. moniliformis* incubated for 2 h in [^{14}C]-leucine, [^{14}C]-alanine, or [^{14}C]-serine, but not [^{14}C]-glutamate, as tentative evidence for the accumulation of the neutral amino acids against a concentration gradient. Although Uglem & Read verified the accumulation of leucine against a concentration gradient, alanine was found to be extensively metabolized during prolonged incubations and did not rise in concentration in the worms. Hence, the apparent accumulation of alanine suggested by Branch seems to have been an artifact of metabolism, and perhaps of exchange between tissue alanine pools and [^{14}C]-alanine in the medium as well. Branch's results for serine accumulation may have been similarly complicated by metabolism and exchange for endogenous amino acids (Crompton & Ward, 1984).

Uglem & Read (1973) found that substitution of $Tris^+$ ions for the sodium ions in their incubation media had no detectable effect on the uptake of [^{14}C]-leucine by *Moniliformis moniliformis* in either 2-min or 90-min incubations, nor did the presence of leucine (0–20mM) in the incubation medium significantly change the net influx of ^{22}sodium ions. They were unable to demonstrate sodium ion sensitivity for the uptakes of alanine, methionine, lysine or glutamate by *M. moniliformis*. In order to reconcile the accumulation of leucine by *M. moniliformis* incubated in sodium-free medium with the concept that organic solute accumulation in animal tissues is energetically coupled through sodium ion symports, Uglem & Read suggested that the concentrative leucine uptake which they observed involved sodium ion-coupled active transport across some internal membrane.

Among the well-defined systems for neutral amino acid transport in the Ehrlich ascites cell, System L (leucine-preferring) is notable for its independence of sodium ions (Schultz & Curran, 1970; Christensen *et al.*,

1973; Christensen, 1975; Podesta, 1982). Despite its sodium ion independence, System L can mediate amino acid accumulation against a considerable concentration gradient (Christensen *et al.*, 1973). Hence, the absence of a sodium ion effect upon leucine accumulation by *Moniliformis moniliformis* is consistent with many characteristics of leucine transport in other systems, as is the sodium ion-independent uptake of the cationic amino acid lysine (Christensen *et al.*, 1973; Christensen, 1975). There is evidence that ATP hydrolysis is closely linked to concentrative uptake of amino acids transported by System L and by other systems (Johnstone, 1972, 1974; Christensen *et al.*, 1973), and it has been proposed that the linkage of ATP hydrolysis to concentrative amino acid uptake involves creation of a proton gradient within the membrane (Christensen *et al.*, 1973; Christensen, 1975). Thus, active transport of some amino acids into metazoan cells may be driven by proton gradients, as is the case in bacteria and yeast, rather than by sodium ion gradients.

In contrast to leucine and cationic amino acids, the absorption of alanine, methionine or glutamate by vertebrate membrane systems involves a sodium ion-dependent component (Schultz & Curran, 1970; Christensen, 1975). Podesta (1982) suggested that the failure of Uglem & Read to demonstrate sodium ion dependence for the uptake of these amino acids by *Moniliformis moniliformis* may have been caused by sodium ions trapped within unstirred layers during supposedly sodium-free incubations. Similar unstirred-layer effects (Podesta, 1982) may have resulted in the incorrect conclusion that methionine uptake by the cestode *Hymenolepis diminuta* is sodium ion insensitive (Read *et al.*, 1963; Pappas, Uglem & Read, 1974). Indeed, the data presented by Read *et al.* (1963) demonstrate a significant inhibition of methionine uptake by *H. diminuta* when sodium ions in the incubation medium were replaced by potassium ions, but only at methionine concentrations approaching or exceeding the apparent K_t for its transport. Since it is not clear whether Uglem & Read (1973) performed their sodium ion replacement studies at more than one concentration of the amino acids examined, their conclusion that amino acid absorption by *M. moniliformis* is entirely sodium ion independent should be reinvestigated.

Lipid absorption. Nothing is known about the mechanisms by which Acanthocephala absorb lipid. The accumulation of lipid within developing larval forms has been described in morphological studies (Butterworth, 1969*a*), but has not been examined physiologically. The size of *Moniliformis moniliformis* and its dispersion in the host intestine can be affected by the nature of lipids used in experimental diets (Crompton *et al.*, 1983; §6.4.1), but it is not clear that this effect is related to lipid absorption by the helminths.

Hammond (1968*b*) used TEM autoradiography to examine the uptake of radioactivity by *Acanthocephalus ranae* incubated in [^{3}H]-trioleylglycerol emulsified in host bile. The autoradiograms revealed uptake of trioleylglycerol (or its hydrolysis products) only by the metasoma tegument, thereby refuting suggestions that lipid absorption occurs in the praesoma (Bullock, 1949*b*; Pflugfelder, 1949; Lange, 1970). Hammond implied that the lipid absorption had occurred by pinocytosis, but apparently there was no microscopic evidence of pinosome formation.

Hammond (1968*b*) assumed that the [^{3}H]-trioleylglycerol in which he incubated *Acanthocephalus ranae* was absorbed as the intact triglyceride. However, Hammond did not indicate whether the host bile used to emulsify the trioleylglycerol was free of lipase activity; furthermore some palaeacanthocephalans may possess surface lipase activity (Bullock, 1949*a*; §6.2.2). In any case, it is likely that most of the saponifiable lipid reaching the absorptive surfaces of adult acanthocephalans *in vivo* is in the form of emulsified free fatty acids and monoacylglycerols formed by the action of host intestinal lipases (Mettrick, 1980; Shiau, 1981; Carey, Small & Bliss, 1983). Similarly, the sterols and carotenoids absorbed from the host intestine probably encounter the acanthocephalan surface as emulsified free alcohols rather than as acyl esters.

As a hypothesis, absorption of lipid into the acanthocephalan tegument may be mechanistically similar to lipid absorption into the mucosal epithelium of the vertebrate intestine. Unfortunately, the mechanisms by which lipids enter intestinal absorptive cells are poorly understood (Carey *et al.*, 1983). It is generally accepted that fatty acids, monoacylglycerols and sterols cross the apical plasma membrane of intestinal absorptive cells by free diffusion (Johnston, 1968; Sallee, Wilson & Dietschy, 1972; Shiau, 1981). However, the predominance of diffusional lipid uptake in some portions of the mammalian intestine does not preclude the existence of mediated transport systems for lipid uptake elsewhere, although bile acid absorption in the jejunum is diffusional, mediated transport systems for bile acid uptake operate in the ileum (Wilson & Treanor, 1975).

Barrett (1981) suggested that the accumulations of neutral lipids found in adult helminths, including acanthocephalans (§6.5.3), may be a consequence of a need to absorb large amounts of lipid in order to satisfy a requirement for a particular fatty acid or lipophilic vitamin. This suggestion assumes that helminths somehow control the rate of lipid uptake from the host intestinal lumen but that they have no mechanisms for selective absorption of particular lipid components. However, there is circumstantial evidence that Acanthocephala do exert selectivity in the uptake of at least some classes of lipids: carotenoids and steroids provide examples.

The hemolymph of *Gammarus pulex*, the intermediate host of *Polymorphus minutus*, contains at least seven distinct α- and β-carotenoids, of which astaxanthin and lutein are the predominant constituents (Barrett & Butterworth, 1968; §6.5.3). It is reasonable to assume that the diet of ducks contains a variety of carotenoids as well. Astaxanthin is the only carotenoid found in either cystacanths or adults of *P. minutus* (Barrett & Butterworth, 1968), while adult *Filicollis anatis* has β-carotene as its only carotenoid (Barrett & Butterworth, 1973). Similarly, a comparison between the sterol compositions of *Moniliformis moniliformis* and *Macracanthorhynchus hirudinaceus*, and the sterol contents of the intestinal fluids of their respective hosts, suggests that steroid uptake by these archiacanthocephalans may not reflect the simple molar ratios of steroids present in the host intestinal lumen (Barrett, Cain & Fairbairn, 1970; Table 6.8).

Nucleoside absorption. Hibbard & Cable (1968) demonstrated that [^{3}H]-thymidine was rapidly absorbed into the metasoma tegument of *Paulisentis fractus*. The absorbed thymidine was incorporated into nuclear DNA in ovarian masses, acanthors and testes and into extranuclear (mitochondrial?) DNA, but not nuclear DNA, in body wall tissues. The incorporation of absorbed [^{3}H]-thymidine into DNA in *P. fractus* suggests that this acanthocephalan contains thymidine kinase, and possibly other enzymes of salvage pathways for nucleoside metabolism. Interestingly, thymidine kinase has been demonstrated in *Moniliformis moniliformis* (Farland & MacInnis, 1978). Page & MacInnis (1971) indicated that thymidine and uridine are absorbed by *M. moniliformis* and *Macracanthorhynchus hirudinaceus* by mediated transport mechanisms, but technical details have not been published.

6.3.2 *Ion transport and osmotic phenomena*

Adults. The components of the body walls of *Moniliformis moniliformis* and *Macracanthorhynchus hirudinaceus* are typical in having low, non-equilibrium concentrations of cytoplasmic sodium ions and elevated levels of cellular potassium ions (Table 6.4). The potassium ion levels reported by Branch (1970*a*; Table 6.4) are extremely high compared to values from other sources cited in Table 6.4 and expectations for cellular potassium ion levels outside marine environments. Branch's reported potassium levels may possibly be technical artifacts, as may be the very high osmotic pressures for *M. moniliformis* tissues and rat intestinal fluids reported by Branch (1970*b*) (see below); perhaps there was significant fluid loss from the samples which Branch examined. Because of the uncertainty in the values reported for tissue potassium ion levels in *M. moniliformis* and *M. hirudinaceus*, and because the cellular membrane potentials are not known, there is no direct proof at present that potassium ions are accumulated against an electrochemical gradient in the body wall tissues of these archiacanthocephalans. It is clear that the cellular components of the archiacanthocephalan body wall have non-equilibrium distributions of sodium and probably of potassium with respect to both the host intestinal fluids and the pseudocoelomic fluid (Table 6.4).

Branch (1970*a*) postulated sodium ion and potassium ion pumps to account for the apparent non-equilibrium distributions of sodium and potassium in *Moniliformis moniliformis* and *Macracanthorhynchus hirudinaceus* tissues. Exchange studies using ^{24}sodium ions and ^{42}potassium ions suggested active sodium ion and potassium ion transport in *M. moniliformis*, but the evidence is questionable because the calculations depended upon accurate measurements of tissue and extracorporeal sodium and potassium levels. Furthermore, the data give little information about the

Table 6.4. *The major cations (mEq/l ± 1 standard deviation) of archiacanthocephalan body wall and pseudocoelomic fluids and of rat intestinal fluids*

	Na^+	K^+	Ca^{2+}
Macracanthorhynchus hirudinaceus			
Female body wall (anterior)[a]	30.5 ± 13.1	257.2 ± 37.1	—[b]
Female body wall (posterior)[a]	29.2 ± 5.7	285.5 ± 31.0	—
Female pseudocoel (anterior)[a]	71.5 ± 10.5	91.5 ± 7.4	—
Female pseudocoel (posterior)[a]	66.7 ± 5.7	105.9 ± 21.4	—
Whole pseudocoel (pooled sexes)[c]	86.0	10.0	2.5
Moniliformis moniliformis			
Female body wall[d]	27.9	159.1	9.6
Female pseudocoel[d]	73.7	43.6	17.3
Whole female worms[d]	36.9	105.7	12.8
Whole male worms[d]	64.4	225.1	—
Rat intestinal fluids			
Whole small intestine[d]	111.0	23.2	—
Jejunal fluids[e]	76	4.6	0.43
Ileal fluids[e]	55	6.0	—

[a] Calculated from Branch (1970*a*)
[b] Not reported
[c] From Denbo (1971)
[d] From Branch (1970*a*)
[e] From Mettrick & Podesta (1974)

quantitative capacity of the ion transporting mechanisms. At present the best evidence for the existence of active cation transport by acanthocephalans seems to be their capacity to withstand the ionic atrocities perpetrated on them by early investigators. Species of *Neoechinorhynchus* maintained in 85–145mM sodium chloride solutions containing small amounts of calcium chloride managed to retain normal tissue hydration levels for 2–4 days and to survive for as long as 2 weeks *in vitro* (Gettier, 1942; Van Cleave & Ross, 1944); apparently *Neoechinorhynchus* has a creditable ability to retain tissue potassium ions and exclude sodium ions under unfavorable ionic conditions.

The ultrastructure of the acanthocephalan tegument suggests an epithelium adapted for water transport (Crompton & Lee, 1965; Stranack *et al.*, 1966). The basal membrane of the metasoma tegument is highly infolded (Crompton & Lee, 1965; Nicholas & Mercer, 1965; Stranack *et al.*, 1966; Hammond, 1967; Wright & Lumsden, 1969; Lange, 1970; Wright, 1970; Byram & Fisher, 1973; Beermann *et al.*, 1974), as is the pseudocoel-facing membrane of the lemnisci (Hammond, 1967; Wright, 1970; Lange, 1970; Beermann *et al.*, 1974). The swollen appearance of the extracytoplasmic spaces between the tegument basal infoldings of acanthocephalans fixed slowly in hypo-osmotic fixatives (Nicholas & Mercer, 1965) is consistent with the morphology associated with water inflow according to Diamond's standing osmotic gradient model (Diamond, 1971; Podesta, 1982).

If the digitiform basal infoldings of the metasoma tegument and the pseudocoelomic aspect of the lemnisci are associated with osmotically driven water movement, ion-coupled ATPases probably exist, responsible for formation of standing osmotic gradients associated with the tegument basal membrane. Histochemical studies (Branch, 1970*a*) suggest high levels of sodium ions and chloride ions associated with the basal tegument membrane of *Moniliformis moniliformis*, as might be expected if sodium ion pumps or Na^+–Cl^--ATPases are present in the membranes of the basal infoldings. In any event, there is little doubt that all acanthocephalan syncytia have ion-specific ATPases responsible for maintaining normal cytoplasmic ion concentrations, and it is likely that these include ion pumps associated with the basal membrane of the tegument.

Adult acanthocephalans swell in dilute fluids or in moderately hyperosmotic solutions of permeant salts, and they rapidly lose body water in hyperosmotic solutions of non-permeant solutes (Van Cleave & Ross, 1944; Branch, 1970*b*; Denbo, 1971). Branch (1970*b*) used these properties to measure the apparent permeabilities of *M. moniliformis* to several cations and anions in unbuffered hyperosmotic solutions of single salts.

As might be anticipated, *Moniliformis moniliformis* was found to be permeable to potassium, sodium, lithium, chloride and bicarbonate ions (Branch, 1970*b*). Hyperosmotic potassium chloride solutions (190mM) induced immediate osmotic swelling, whereas sodium chloride or lithium chloride solutions (190mM) induced shrinkage followed by swelling. The initial shrinkage in the presence of sodium chloride suggests active sodium ion extrusion, which would retard net sodium ion influx and associated osmotic swelling. The comparable response to lithium chloride suggests that lithium ions can compete for the sodium ion transporters, as is generally true in metazoan cells (Glynn & Karlish, 1975). Osmotic responses to sodium bicarbonate resembled those to sodium chloride, except that water regain after the initial shrinkage was slower. It is not clear whether this observation indicates a limited permeability to bicarbonate ions or evidence for a bicarbonate extrusion mechanism, but the former is unlikely.

The chloride salts of calcium and magnesium induced rapid osmotic water loss and allowed little or no water regain, as did the sodium salts of sulfate and nitrate. Such behavior indicates that magnesium and calcium ions and divalent anions did not enter *Moniliformis moniliformis* under the ionic conditions of Branch's experiments. Whether significant fluxes or exchanges of calcium ions and/or magnesium ions might be observed in the presence of more balanced external salt solutions was not examined, but some exchange of these ions would be expected.

Denbo (1971) measured the steady state osmotic responses of *Macracanthorhynchus hirudinaceus* in several dilutions of artificial sea water (0–600 mosmol/l). Inorganic ions accounted for approximately 60% of the osmolytes in pseudocoelomic fluid from worms incubated in 30% sea water, and trehalose or other non-diffusible organic solutes probably formed the remainder of the osmolytes. Incidentally, Tanaka & MacInnis (1980) found that *Moniliformis moniliformis* pseudocoelomic fluids contained high concentrations of amino acids which might be involved in osmoregulation.

The concentration of sodium ions in *Macracanthorhynchus hirudinaceus* pseudocoelomic fluid remained fairly stable in worms incubated in 10–30% artificial sea water, even though the pseudocoelomic fluid osmotic pressure for worms incubated in 30% sea water was about 100 mosmol/l higher than that for worms incubated in 10% sea water; in contrast, pseudocoelomic chloride ion concentration decreased dramatically upon incubation in hypotonic medium (Denbo, 1971; Table 6.5). Denbo concluded that the pseudocoelomic sodium ion concentration was highly regulated within the ion concentration range expected for the porcine intestine, but that chloride ions were distributed passively. However, he was unable to provide a satisfactory explanation for the enormous anion gap observed for the pseudocoelomic fluid from *M. hirudinaceus* incubated in 10% sea water. It is probable that the anion gap was filled by diffusible organic anions, as occurs in mammalian intestinal fluids (Mettrick & Podesta, 1974). Since carbon dioxide is a metabolic product of acanthocephalans (§6.6.2), it would be of interest to determine whether bicarbonate ion levels rise significantly in the pseudocoels of acanthocephalans incubated in ionically balanced but hypo-osmotic media.

The steady state osmotic pressures of pseudocoelomic fluid from *Macracanthorhynchus hirudinaceus* incubated in sea water dilutions of 500 mosmol/l or less were higher than the osmotic pressure of the incubation medium, and the osmotic pressure differential decreased with increasing osmolality of the medium (Denbo, 1971). Worms incubated in dilute sea water containing a non-permeant osmolyte (sucrose) resisted loss of tissue water when metabolically active, but dehydrated rapidly when poisoned with sodium fluoride. Denbo concluded that *M. hirudinaceus* (and possibly other archiacanthocephalans) are weak hyperosmotic regulators within the osmotic pressure range normally encountered in the host intestine.

Branch (1970*a*) presented data which seemed to indicate that the pseudocoelomic fluid of *Moniliformis moniliformis* was mildly hyperosmotic to rat intestinal fluid and that the body wall fluid was extremely hyperosmotic to that of the pseudocoel. Although Branch's results appear consistent with the conclusion that archiacanthocephalans are hyperosmotic regulators, his measurements of osmolality gave unusual results. For example, the value given for rat intestinal fluid is 150 mosmol/l hypertonic to that for serum, yet these fluids normally have similar steady state osmolalities (Mettrick & Podesta, 1974), and the value for female

Table 6.5. *The ion concentrations (mEq/l) of the pseudocoelomic fluid from* Macracanthorhynchus hirudinaceus *following incubation (3 h, 39 °C) in 10% or 30% artificial sea water (adapted from Denbo, 1971)*

Fluid compartment	Na^+	K^+	Ca^{2+}	Cl^-
10% Sea water incubation				
Incubation medium	44.0	0.92	0.46	53.0
Pseudocoelomic fluid	84.2	15.0	1.30	15.0
30% Sea water incubation				
Incubation medium	132.0	2.76	1.35	159.0
Pseudocoelomic fluid	86.0	10.0	2.5	101.0

pseudocoelomic fluid is almost twice that measured by Tanaka & MacInnis (1980).

Crompton (1970) suggested that acanthocephalans may be osmoconformers rather than osmoregulators, an assumption based on two observations. First, *Moniliformis moniliformis* from starved hosts (Read & Rothman, 1958) lost much more fresh tissue mass than could be accounted for by glycogen depletion alone. This excess mass loss might be attributed to osmotic flux into a dehydrated host gut, but the argument ignores the effects that sudden carbohydrate deprivation might have on helminth energy metabolism and the consequences for ion regulatory activities. Secondly, Crompton noted that the osmotic pressure of the pseudocoelomic fluid from 1-day-old *Polymorphus minutus* adults tended to be similar to (actually slightly higher than) that of the luminal contents from uninfected duck intestine in the region normally occupied by this palaeacanthocephalan (Crompton & Edmonds, 1969). *Echinorhynchus gadi* appears to adjust to new steady state osmotic conditions when incubated in hypo-osmotic balanced salt solutions (Khlebovich & Mikhailova, 1976). This response is consistent with a lack of osmoregulatory capacity, but the extent to which this is related to ionic and osmotic conditions in the host intestine is unclear.

In the vertebrate intestine, dietary water causes transient dilution of gastric and upper duodenal fluids while food ingestion causes extensive mucosal, pancreatic and biliary secretions. However, the luminal osmotic pressure in most of the small intestine remains fairly stable as a result of concurrent water and solute absorption (Read, 1950*a*; Mettrick & Podesta, 1974; Mettrick, 1980). Some authors consider that infection with *Moniliformis moniliformis* (Mettrick, Budziakowski & Podesta, 1979) results in elevation of the osmotic pressure of the rat intestinal fluids by 20–40 mosmol/l, presumably because of the effects of helminth metabolic products on intestinal fluid and solute fluxes. This apparent shift in steady state intestinal osmolality does not alter the fact that the osmotic pressure of the luminal fluids in the intestine fluctuates within fairly narrow limits around a homeostatic set-point (Mettrick & Podesta, 1974). Hence, evolutionary pressures would not be expected to select for marked osmoregulatory capabilities in adult acanthocephalans. However, some acanthocephalans do exhibit tissue ion and volume regulation, and it may have been the consequences of these phenomena which Denbo (1971) interpreted as hyperosmotic osmoregulation.

Larvae. Edmonds (1966) found that acanthors of *Moniliformis moniliformis* hatched in simple saline solutions, but that activation and hatching required salt concentrations above 100mM. Maximal hatching occurred in 300mM solutions of sodium bicarbonate or in 300mM sodium chloride supplemented with 20mM sodium bicarbonate; solutions of sodium chloride supplemented with sodium bicarbonate appeared to be slightly more effective hatching agents than osmotically comparable solutions of sodium bicarbonate alone. Neither 20mM sodium bicarbonate nor 0.1–0.8M sodium chloride alone induced hatching, but 33mM phosphate buffer appeared to substitute fairly effectively for sodium bicarbonate as a supplement to the sodium chloride hatching solution. Several potassium or sodium salts could be substituted for sodium chloride as the activating salt, but only so long as 20mM sodium bicarbonate was present (Edmonds, 1966).

The apparent requirements for either bicarbonate or phosphate ions as a supplement to sodium chloride in the hatching medium suggest that multiple or complementary ionic effects are involved. Since carbon dioxide (or bicarbonate) is involved in activation of *Moniliformis moniliformis* cystacanths (Graff & Kitzman, 1965), it may be that sodium bicarbonate has a comparable role in acanthor hatching. In any case, bicarbonate or phosphate may play a specific induction role in hatching which is separate from the requirement for a relatively high salt concentration. Neither sucrose alone nor sucrose supplemented with sodium bicarbonate induced hatching of *M. moniliformis* acanthors. Clearly, the effects of high external salt solutions on hatching are not merely manifestations of elevated external osmolality, although osmotic factors are likely to be involved.

Eggs collected directly from macerated female *Moniliformis moniliformis* showed much higher release of acanthors *in vitro* if they were stored in 60% sucrose for 1 h before the activation treatment (Lackie, 1972*a*). Lackie suggested that exposure to sucrose mimicked the desiccation normally experienced by eggs in the fecal pellet of the host. *Moniliformis moniliformis* eggs freshly collected from the female pseudocoel, in saline osmotically equivalent to about 150mM sodium chloride (KRT; Read *et al.*, 1963), hatched within 5 min of stimulation by the Edmonds procedure (Wright, 1971); in contrast, eggs stored 3 days in cold (5 °C) 85mM sodium chloride did not commence hatching until about 30 min after stimulation, and the process continued for about 3 h (Edmonds, 1966). These observations suggest that the rate and success of *M. moniliformis* hatching increase with factors which favor a high internal osmotic pressure in the shelled acanthor.

Wright (1971) implied that the ionic requirement for hatching of *Moniliformis moniliformis* acanthors might be associated with the polyionic properties of the acanthor shell (envelope II). It is also possible that hatching events may require flux of permeant ion pairs and associated water flow into the stimulated acanthor. Shortly after the induction of hatching, intact envelopes undergo distortion (Wright, 1971) which could be either a response of the polyanionic component of envelope II to the ions of the hatching fluid, or a result of osmotic changes caused by fluid flux.

Fairbairn (1970) suggested that the high concentration of trehalose (0.4M) in the perivitelline fluids of *Ascaris* eggs might be responsible for the osmotic water flow into activated eggs necessary for hatching. The trehalose content of freshly isolated shelled acanthors of *Moniliformis moniliformis* (16 μmol/g fresh mass; McAlister & Fisher, 1972) does not appear high enough to contribute substantially to the total osmotic pressure of acanthors, but if the trehalose were localized external to the acanthor, its osmotic effect could be significant.

The rate of activation of *Polymorphus minutus* cystacanths following brief treatment in sodium taurocholate was inversely related to the osmotic pressure of the incubation medium within the osmolality range equivalent to 100–220mM sodium chloride (Lackie, 1974). This observation indicates that the osmotic pressure difference across the cyst wall or the body wall is critical to *P. minutus* cystacanth activation. Indeed, it appears that osmotic water flow into the pseudocoelomic fluid through the body wall provides fluid pressure necessary for proboscis evagination (A.M. Lackie, 1975).

6.3.3 *Endocytosis in Acanthocephala*

Electron micrographs of the acanthocephalan body wall reveal numerous vesicles in all cytoplasmic regions of the tegument. Most early investigators interpreted these vesicular structures as evidence of pinocytosis (Crompton & Lee, 1965; Nicholas & Mercer, 1965; Hammond, 1967), though some mistakenly envisioned the vesicular elements as evidence for membranous channels spanning the breadth of the tegument (Stranack *et al.*, 1966).

Edmonds & Dixon (1966) attempted to demonstrate pinocytotic activity in adult *Moniliformis moniliformis* incubated in a mixture of carbon particles and thorium dioxide particles. Thin sections of the experimental worms examined by TEM revealed abundant thorium dioxide in the lumina of the surface crypts, and these results were taken as evidence of uptake 'in a manner that resembles pinocytosis'. However, the micrographs did not show that particles had actually entered the tegument within vesicular pinosomes.

Byram & Fisher (1974) obtained direct EM evidence for pinocytosis in acanthocephalans. *Moniliformis moniliformis* incubated with horseradish peroxidase (HRP) had dense accumulations of peroxidase reaction product in the crypt lumina, and peroxidase activity was also abundant in lysosomes within the apical regions of the tegument. Significantly, HRP was rarely observed to be bound to the glycocalyx at the helminth's free surface (outside the crypts). Byram & Fisher concluded that the HRP accumulated within the surface crypts and was then taken up by pinocytosis at the crypt plasma membrane, and that the pinosomes formed during this process subsequently fused with cytoplasmic lysosomes.

It is doubtful that Byram & Fisher (1974) intended to imply a concentrative receptor-mediated binding phenomenon (Silverstein, Steinman & Cohn, 1977; Goldstein, Anderson & Brown, 1979) when they described the accumulation of peroxidase within the surface crypts of *M. moniliformis*. It is likely, however, that HRP bound non-specifically to the glycocalyx of the crypt plasma membrane prior to its uptake by endocytosis. We do not know whether acanthocephalans possess specific surface receptors involved in receptor-mediated endocytosis of some nutrients. The presence of coated vesicles in the apical tegument of *Echinorhynchus gadi* (Lange, 1970) is not proof of receptor-mediated pinocytosis, since clathrin-coated vesicles are involved in many forms of intracytoplasmic vesicle movement (Pearse & Bretscher, 1981).

Byram & Fisher (1974) also investigated the uptake of ferritin and thorium dioxide by *Moniliformis moniliformis*. They found that ferritin and thorium dioxide micelles bound heavily to mucus-like material found in patches along the outer aspects of the glycocalyx at the tegument surface, but they observed only occasional binding of single ferritin molecules or thorium dioxide micelles to what appeared to be individual filaments of the surface glycocalyx. Furthermore, individual thorium dioxide micelles were seen within the crypts very infrequently, and ferritin molecules were not seen within the crypt lumina or associated with the surface of the crypt plasma membranes.

Byram & Fisher (1974) suggested that the filamentous material in the neck at the entrance to each crypt (Byram & Fisher, 1973) functioned as a molecular sieve that prohibited entry of the large ferritin molecules and thorium dioxide micelles, both of which had estimated diameters in excess of 8.5 nm. In support of this hypothesis, neither Byram & Fisher nor Wright & Lumsden (1968) observed penetration of thorium dioxide into crypts of fixed *Moniliformis moniliformis* stained *en bloc* even though thorium dioxide bound readily to the crypt glycocalyx in thin sections.

Byram & Fisher (1974) described 'electron-opaque concentrations of material', up to 100 nm in diameter, in the crypt lumina and within apical cytoplasmic lysosomes of *Moniliformis moniliformis* incubated in ferritin or thorium dioxide. The densities of the accumulations precluded identification of individual ferritin particles, but unstained sections of worms incubated in thorium dioxide appeared to have small thorium dioxide particles within the accumulations, both in the crypts and in cytoplasmic lysosomes. Incubation in ferritin appeared to increase substantially the number of lysosomes present in the cytoplasm adjacent to the crypts (Byram & Fisher, 1974).

It is possible that the 'mucus-like material' associated with ferritin and thorium dioxide particles may have been aggregated filaments of the surface glycocalyx and that the electron-dense aggregates observed in the crypt lumina and in cytoplasmic lysosomes of worms incubated in ferritin or thorium dioxide were derived from portions of the glycocalyx with which these tracers bound outside the crypts (the 'mucus-like material'). This view does not necessarily contradict Byram & Fisher's suggestion that free ferritin molecules and thorium dioxide particles were excluded from the crypts by some mechanism. Rather, it suggests a lateral migration of (cross-linked?) membrane glycoproteins bearing ferritin or thorium dioxide ligands from the plasma membrane of the free surface to regions of the plasma membrane lining the crypts, followed by endocytosis of the ligand-complexed glycoproteins at the crypt plasma membrane. If pinocytosis occurs only at the crypt plasma membranes, and not at the free surface, lateral migration of membrane components into the plasma membrane of the crypts would be expected to occur in the normal course of membrane cycling and renewal (Silverstein *et al.*, 1977; Pearse & Bretscher, 1981).

The differences between the observations of Edmonds & Dixon (1966) and Byram & Fisher (1974) with respect to *Moniliformis moniliformis* incubated with thorium dioxide deserve some comment. Formation of endocytic vesicles requires active membrane movements caused by ATP-dependent aggregations of actin and associated contractile proteins at the cytoplasmic surface of the membrane (Silverstein *et al.*, 1977). In contrast, lateral migrations of membrane proteins within the membrane are generally ATP-independent phenomena (Houslay & Stanley, 1982). The incubation medium used by Byram & Fisher (KRT supplemented with sodium bicarbonate) contained calcium ions, whereas that used by Edmonds & Dixon (0.8% sodium chloride) did not. It is possible that the absence of calcium ions from the incubation medium used by Edmonds & Dixon had no effect on lateral migrations of glycoproteins bearing thorium particles into the crypt plasma membranes, but prohibited the initiation of the ATP-dependent protein aggregations necessary for endocytosis.

6.4 Nutrition

6.4.1 *Nutritional studies* in vivo

Acanthocephalan nutritional requirements are ultimately satisfied by the presence of essential substances in the host diet but, with the exception of carbohydrates, nutritional requirements of helminths are difficult to identify *in vivo* because of exchange between the host intestinal lumen and host tissues via both mucosal transfer mechanisms and exocrinoenteric circulation (Read, 1950*a*, 1961; Read *et al.*, 1963; Mettrick & Podesta, 1974; Mettrick, 1980; Roberts, 1980).

Carbohydrates. Crompton and his colleagues found that *Moniliformis moniliformis* becomes established in rats fed on diets containing cellulose as the sole carbohydrate ('carbohydrate-free diets') with the same efficiency as in rats on isoenergetic high-starch diets (Nesheim *et al.*, 1977, 1978; Crompton *et al.*, 1983). The worms grew little in hosts fed on 'carbohydrate-free diets' and the rate at which established worms were lost from the small intestine was greater than in rats fed on diets containing adequate carbohydrate.

The size (dry mass) attained by *Moniliformis moniliformis* in hosts on 'carbohydrate-free diets' was correlated with the rats' ability to maintain adequate hepatic and serum carbohydrate levels through gluconeogenesis. *Moniliformis moniliformis* grown for 42 days in mature male rats fed 'carbohydrate-free diets' were significantly larger than comparably aged worms whose hosts were weanling male rats at the time of infection (Nesheim *et al.*, 1978); the mature hosts on 'carbohydrate-free diets' had liver glycogen levels similar to those in rats fed low-starch diets (3.6 g starch/100 g diet) (Nesheim *et al.*, 1977), whereas the actively growing young rats were hypoglycemic on diets containing less than 3% (by mass) starch and had virtually no hepatic glycogen stores on 'carbohydrate-free diets' (Nesheim *et al.*, 1978). These results suggest that *M. moniliformis* growth may have been sustained by small amounts of glucose which leaked from host tissues into the intestinal lumen, as appears to occur in parenterally fed rats infected with *Hymenolepis diminuta* (Castro *et al.*, 1976). Indeed, Nesheim *et al.* (1977) found that the intestinal fluids from mature rats fed 'carbohydrate-free diets' contained significant levels of glucose.

The growth of both *Moniliformis moniliformis* and its host (immature male rat) were directly related to the starch content (0–4 g/100 g diet) of experimental diets, but the presence of the parasite caused no discernible effect on host growth (Nesheim *et al.*, 1978). *Moniliformis moniliformis* was considered not to have affected host growth on starch-limited diets because its mass was too slight to remove significant quantities of glucose from the intestine during the period when rat growth was maximal.

Moniliformis moniliformis grew poorly in rats fed diets containing 3% glucose, galactose, maltose, lactose or trehalose as the carbohydrate source, but dietary fructose, mannose and sucrose supported growth well (Parshad, Crompton & Nesheim, 1980). Indeed, 35-day-old worms from rats fed 3% mannose or fructose diets were larger than 42-day-old worms from rats fed 4% dietary starch in earlier studies (Nesheim *et al.*, 1978). Mannose diets supported *M. moniliformis* growth most effectively, but diets containing 3% fructose supported worm growth as well as or better than 24% glucose diets. *Moniliformis moniliformis* growth was directly related to dietary glucose content over the range 0–36% glucose (Parshad, Crompton & Nesheim, 1980); in contrast, dietary fructose levels greater than 3 g/100 g diet produced little additional worm growth (Crompton *et al.*, 1983). Sucrose supported *M. moniliformis* growth about half as effectively as fructose on a per gram basis (Parshad, Crompton & Nesheim, 1980), which suggests that only the fructose moiety of the disaccharide, released into the intestinal lumen as a result of sucrase activity at the intestinal brush border, was available to the helminths.

It is clear that *Moniliformis moniliformis* grew well in rats on mannose- and fructose-containing diets, and actually stunted the growth of rats fed on low-fructose diets (Crompton *et al.*, 1981), probably because the helminth tegument is better equipped to absorb these hexoses than is the rat intestinal mucosa (§6.3.1). Since diets containing 3% glucose or maltose supported *M. moniliformis* growth *in vivo* only slightly better than 'carbohydrate-free diets', it is also clear that *M. moniliformis* competes poorly with the rat intestinal mucosa for absorption of these starch hydrolysis products if they are free in the intestinal lumen. As Parshad, Crompton & Nesheim (1980) noted, the fact that *M. moniliformis* matures in rats on natural (starch-containing) diets suggests that the amylolytic and maltolytic activities occurring at the worm's surface (§6.2.2, §6.2.3) are important to its normal nutrition.

The severe stunting of *Moniliformis moniliformis* introduced into rats fed carbohydrate-deficient diets appears to be a reversible delay of normal development rather than a permanent impairment. When infected rats maintained on a 3% glucose diet were transferred to a 3% fructose diet, the helminths showed significant increases in mass and accelerated sexual development (Crompton, Singhvi & Keymer, 1982). Crompton *et al.* (1982) commented on the obvious significance of the ability of *M. moniliformis* to bide its time 'in a state of temporary suspended animation' until its host finds a diet containing adequate carbohydrate.

They also suggested that this ability might be of significance to the survival of established acanthocephalan infections in fasting or hibernating hosts.

Starvation of hosts for 48 h causes significant shedding of *Moniliformis moniliformis* (Burlingame & Chandler, 1941) or *Polymorphus minutus* (Nicholas & Hynes, 1958; Hynes & Nicholas, 1963) present as established infections, but does not necessarily result in complete elimination of the helminths. Read & Rothman (1958) found that the glycogen stores of 4-week-old *M. moniliformis* were severely depleted when hosts were fasted for 48 h. Rats placed on starch-free diets on day 28 postinfection contained few female *M. moniliformis* when autopsied 12 days later, and male worms appeared to have undergone no growth while the hosts were maintained on carbohydrate-deficient diets (Read & Rothman, 1958). Among acanthocephalans from fish, host starvation adversely affects *Pomphorhynchus laevis* in goldfish (Kennedy, 1972), but apparently not *Echinorhynchus truttae* in *Salmo trutta* (Awachie, 1972*b*). Species of *Neoechinorhynchus* and *Acanthocephalus ranae* have been found to survive for months in unfed turtles and frogs (Van Cleave & Ross, 1944; Pflugfelder, 1949), but the possibility that the worm burden of reptilian and amphibian hosts declines during host fasting has not been examined systematically.

The deleterious effect of host starvation on established infections of acanthocephalans from mammals and birds suggests that mature worms are less able to withstand the exigencies of carbohydrate-deficient diets than are newly introduced infections. The fact that acanthocephalans in reptilian and amphibian hosts maintained at low temperatures appear to survive long periods of host fasting may simply reflect lower metabolic rates in helminths from poikilothermic hosts. *Neoechinorhynchus* from turtles maintained at 25 °C in the absence of exogenous carbohydrate retained at least marginal glycogen reserves for longer than 16 days (Dunagan, 1964). However, it is also possible that acanthocephalans whose hosts normally undergo long periods of fasting may be adapted to enter a state of temporary suspended animation even as adults, at least to the extent that their basal metabolism can be maintained by glucose which diffuses from host tissues into the intestinal lumen.

Lipid characteristics. Nesheim *et al.* (1977) observed that *Moniliformis moniliformis* gained mass more rapidly during the first few weeks of development in mature male rats fed high levels of dietary starch (59%, by mass) than in rats on 3.6% starch diets, but that worms from rats fed low-starch diets attained higher maximal mass and persisted longer. These differences, however, may reflect more than dietary starch content. The high-starch diet contained maize oil as its lipid component, while fatty acids derived from maize oil comprised the lipid constituent of the low-starch diet. Crompton *et al.* (1983) observed that substitution of maize oil for fatty acids in experimental diets resulted in smaller, significantly overdispersed worms in mature female rats autopsied 35 days postinfection. Since host sex significantly influences the size and longevity of *M. moniliformis* (Graff & Allen, 1963; Crompton & Walters, 1972; Crompton *et al.*, 1981), care should be taken extrapolating observations made using mature female hosts to situations involving male hosts. Nonetheless, it is possible that the adverse effects on *M. moniliformis* infections attributed to a high-starch diet by Nesheim *et al.* (1977) may have been related to the lipid component of the diet.

Amino acids. Transfer of rats harboring *Moniliformis moniliformis* to diets lacking added leucine, lysine, or methionine and cysteine for 14 days had no effect on helminth growth, but caused significant losses in mass in the hosts (Nesheim *et al.*, 1977). As the results of this study suggest, it is unlikely that dietary amino acid deficiencies will affect the growth of intestinal helminths until the host suffers severe amino acid deficiency.

The vertebrate intestinal lumen contains a large pool of free amino acids derived from hydrolysis of secretory digestive proteins, degradation of moribund mucosal cells and probably from exchange between the mucosal epithelium and the intestinal lumen, as well as directly from dietary protein (Read, 1971; Mettrick & Podesta, 1974; Mettrick, 1980). Although the composition of the luminal amino acid pool may change transiently after ingestion of protein having an unusual amino acid composition or experimental diets lacking some amino acids, the molar ratios of amino acids in the intestinal pool oscillate about remarkably stable homeostatic set-points (see Read *et al.*, 1963; Read, 1971; Mettrick & Podesta, 1974; Mettrick, 1980). Indeed, Read *et al.* (1963) concluded that *Hymenolepis diminuta* depends upon the relative constancy of the amino acid ratios in the host intestinal lumen to regulate the balance of its own amino acid uptake, and studies of diurnal changes in amino acid pools in this cestode appear to support that conclusion (see Mettrick, 1980).

That amino acids from host tissues contribute significantly to the intestinal pool available to acanthocephalans is evident from results obtained by Edmonds (1965). Having demonstrated that orally administered [^{14}C]-leucine and [^{32}P]-inorganic phosphate ($^{32}P_i$) were absorbed by *Moniliformis moniliformis* and the host intestinal mucosa with comparable time courses, Edmonds examined the fates of intraperitoneally injected tracers. Intraperitoneally injected $^{32}P_i$ and [^{14}C]-leucine entered the host intestinal mucosa fairly rapidly: maximal mucosal labeling with both isotopes peaked between 60 and 120 min postinoculation and then declined to lower levels. No significant transfer of $^{32}P_i$ from host tissues to *M. moniliformis* during the 4-h period of the experiments was observed. In contrast, some [^{14}C]-leucine was evident in the helminths within 1 h postinoculation, and the [^{14}C]-leucine content of the worms increased steadily over the remaining 3 h during which samples were examined.

6.4.2 *Site selection and nutrition*

Burlingame & Chandler (1941) and Holmes (1961) noted an anteriad emigration of *Moniliformis moniliformis* within the rat intestine during the first 3 weeks postinfection. Similar developmental emigrations have been reported for several acanthocephalan species, as well as a variety of other helminths (Crompton, 1973, 1975; Holmes, 1973; Mettrick & Podesta, 1974). Developmental emigrations have generally been interpreted as site-finding behavior which optimizes the positioning of helminth feeding surfaces with respect to nutrient gradients or other physiological parameters in the gut (Burlingame & Chandler, 1941; Holmes, 1961, 1962*a*; Crompton & Whitfield, 1968*b*; Mettrick & Podesta, 1974; Mettrick, 1980). The concept that developmental emigration is a response to intestinal gradients is reinforced by the diel migratory activities in hymenolepid tapeworms and other platyhelminths (Crompton, 1973; Holmes, 1973; Mettrick & Podesta, 1974; Arai, 1980) and by evidence which suggests that diel migration by *Hymenolepis diminuta* is a positive response to physiological gradients correlated with glucose concentration or other signals associated with host feeding (Mettrick & Podesta, 1974; Arai, 1980).

Crompton *et al.* (1983) found that 5-week-old *Moniliformis moniliformis* in rats fed low amounts of fructose had more anterior mean attachment positions than worms in rats given high levels of dietary fructose. In rats fed diets containing 3% fructose or mannose, the anteriad extent of the mean attachment position of *M. moniliformis* was inversely related to the efficacy of the sugar in promoting worm growth, the mean attachment position in rats given mannose diets being posterior to that in rats fed fructose diets (Parshad, Crompton & Nesheim, 1980). These observations implicate intestinal sugar gradients in determining the optimal attachment site for *M. moniliformis* within the rat intestine.

Increased population densities extend the limits of attachment sites for *Moniliformis moniliformis* (Burlingame & Chandler, 1941; Holmes, 1961), *Polymorphus minutus* (Crompton & Harrison, 1965) and *Echinorhynchus truttae* (Awachie, 1966), as well as other helminths (Crompton, 1973; Holmes, 1973; Mettrick & Podesta, 1974). Burlingame & Chandler attributed the reduced size of *Moniliformis moniliformis* in large infections to competition for nutrients, and Holmes (1961, 1962*a*) specified carbohydrate as the most likely limiting factor. Holmes (1961, 1962*a*) found that concurrent infections of *M. moniliformis* and *Hymenolepis diminuta* in rats caused a marked posteriad displacement of the cestodes once the acanthocephalans had emigrated to their anteriad positions. Since Holmes was unaware of the diel migratory patterns of *H. diminuta* in rats, but not in hamsters (see Arai, 1980), he did not examine effects of concurrent infection on this parameter, nor did he state the time of day at which hosts were sacrificed for examination.

Concurrent infections in rats caused significant (>50%) reductions in the sizes of *Hymenolepis diminuta* recovered 8 weeks postinfection and moderate reductions in the sizes of the *Moniliformis moniliformis* (Holmes, 1961, 1962*a*). Concurrent infections of *M. moniliformis* and *H. diminuta* in hamsters had no effect on the size or distribution of either helminth (Holmes, 1962*b*). Holmes concluded that the effects of concurrent infection of rats with *M. moniliformis* and *H. diminuta* resembled the crowding effect in monospecific infections, which at that time was considered to be due solely to competition for carbohydrate. The absence of effect in hamsters was explained by suggesting that competition for carbohydrate within the hamster intestine was less critical because the host absorbed a smaller proportion of the available carbohydrate (Holmes, 1961, 1962*a*, *b*).

There is now evidence that the crowding effect in cestodes involves more than competition for nutrients, although carbohydrate availability is important (Roberts, 1980). Products secreted by *Hymenolepis diminuta* may affect the rate of proglottid production in crowded infections (Insler & Roberts, 1980), and indirect effects resulting from changes in host intestinal physiology (Mettrick & Podesta, 1974; Mettrick, 1980) may also play a role. It is not known whether *Moniliformis moniliformis* and *H. diminuta* excretory products are involved in interspecific stunting in concurrent infections in rats, although Holmes (1961) indicated that the interspecific effects appeared to be greater than predicted from monospecific crowding effects. If products released by *M. moniliformis* and by *H. diminuta* did affect the growth of the other species in rats, it is possible that acanthocephalan products may have intraspecific growth regulating effects as well. However, the absence of effects in concurrent infections in hamsters (Holmes, 1962*b*) underscores the importance of host physiology to interactions between species of helminth.

Crompton (1970) made the amusing suggestion that the posteriad displacement of *Hymenolepis diminuta* in concurrent infections with *Moniliformis moniliformis* in rats might be caused by the effects of the ethanol excreted by *M. moniliformis* (§6.6.2) on the coordination of the cestodes. However, alterations in the site specificity of one or both species in interspecific infections occur among other acanthocephalan–cestode pairs and in other forms of interspecific infections (Crompton, 1973; Holmes, 1973). It is unlikely that all of these phenomena can be attributed to inebriation of one member of the pair.

6.4.3 *Attempts to culture acanthocephalans* in vitro

Adult *Moniliformis moniliformis*, adult *Octospinifer macilentus*, adult *Polymorphus minutus*, and adult stages of species of *Neoechinorhynchus* have been maintained *in vitro* for days, weeks and months, respectively, by Nicholas & Grigg (1965), Harms (1965*b*), Crompton & Ward (1967*a*) and Dunagan (1962). Although little growth occurred in mature adults maintained *in vitro*, mature acanthor production (Dunagan, 1962) or release (Nicholas & Grigg, 1965) and copulatory activity (Dunagan, 1962) persisted in some instances. Some development of *M. moniliformis* acanthellae *in vitro* has been achieved starting with acanthellae dissected from cockroaches (J. M. Lackie, 1975) but not from artificially hatched acanthors (Lackie & Rotheram, 1972). Acanthellae did not complete development to form cystacanths *in vitro*.

Jensen (1952) was able to cultivate infective cystacanths of *Pomphorhynchus bulbocolli* to nearly normal preadult or adult forms in a medium containing peptone broth, horse serum, yeast extract and glucose. Females eventually developed ovarian balls, males appeared morphologically mature and frequently everted their copulatory bursae, but copulation did not occur (Lackie, 1978). Some growth of juveniles from activated cystacanths has been reported by Tobias & Schmidt (1977) for *Moniliformis moniliformis* and by Young & Lewis (1977) for *Corynosoma constrictum*, but the system producing growth of *C. constrictum* juveniles (chick chorioallantoic membrane) was inappropriate for growth of activated *Polymorphus paradoxus* or *P. marilis*.

Attempts to cultivate acanthocephalans *in vitro* have provided little insight into their nutritional requirements. All media producing significant growth have contained undefined components (yeast extract, peptone, chick chorioallantoic membrane, egg yolk or blood serum), often supplemented with complete vitamin and amino acid mixtures (Lackie, 1978).

As Lackie (1978) indicated, insufficient attention has been paid to the choice of balanced salts solutions for culture media. It is possible that inadequate osmotic or ionic balance was a factor contributing to mortalities of *Neoechinorhynchus* observed by Dunagan (1962) and *Moniliformis moniliformis* by Nicholas & Grigg (1965) *in vitro*, since gradual swelling preceded loss of mobility in declining cultures. It is not clear whether the moderately successful cultivation of *Pomphorhynchus bulbocolli* juveniles (Jensen, 1952), in what amounts to serum-supplemented bacterial culture media, resulted from serendipity or from broad osmotic and ionic tolerances by this cosmopolitan parasite of fish. Harms (1965*b*) suggested that the inadequacy of Jensen's media for cultivation of *Octospinifer macilentus*, which occupies the same region of the host intestine as *P. bulbocolli* in shared host species, might be related to stricter physiological requirements, and thence the narrower host range, of the former acanthocephalan.

Nicholas & Grigg (1965) reported that *Moniliformis moniliformis* adults appeared to survive better under 95% nitrogen–5% carbon dioxide than under 75% nitrogen–5% carbon dioxide–20% oxygen, and they concluded that 'a reduction in oxygen tension below that of air is beneficial'. Van Cleave & Ross (1944) also noted that species of *Neoechinorhynchus* from turtles appeared to survive longer in simple saline solutions 'from which contact with free atmospheric air is excluded'. However, the more successful attempts to culture acanthocephalans *in vitro* (Jensen, 1952; Dunagan, 1962; J. M. Lackie, 1975) appear to have been conducted under aerobic conditions. Furthermore, neither the vertebrate intestinal lumen nor the arthropod hemocoel is anoxic (Crompton, 1970; Mettrick & Podesta, 1974; Befus & Podesta, 1976; Mettrick, 1980).

Although acanthocephalan energy metabolism appears to be primarily anaerobic (§6.6), we must assume that some oxygen is required for anabolic hydroxylation reactions essential for normal growth and development (Saz, 1981). For example, the connective tissue underlying the acanthocephalan tegument contains numerous collagen-like fibrils (Crompton & Lee, 1965; Nicholas & Mercer, 1965; Hammond, 1967; Byram & Fisher, 1973; Beermann *et al.*, 1974). Collagen isolated from *Macracanthorhynchus hirudinaceus* appears to have a fairly typical hydroxyproline and hydroxylysine content (Cain, 1970), and the formation of these residues presumably involves post-translational hydroxylations by oxygen-dependent mixed-function oxidases (von Brand, 1973; Metzler, 1977; Barrett, 1981).

Tobias & Schmidt (1977) obtained some growth of juvenile *Moniliformis moniliformis in vitro* under 90% nitrogen–10% carbon dioxide, but they observed no tissue differentiation. It is likely that the oxygen-free atmosphere which they employed was one of the factors limiting the growth of their cultures.

6.5 Tissue carbohydrates and lipids and their metabolism

6.5.1 *Glycogen*

Glycogen abounds in virtually all tissues of adult acanthocephalans, in shelled acanthors and in cystacanths (von Brand, 1939*a*, *b*, 1940; Miller, 1943; Bullock, 1949*b*; Crompton, 1965, 1970; Crompton & Lee, 1965; Nicholas & Mercer, 1965; Hammond, 1967; Wright & Lumsden, 1969, 1970; Lange, 1970; Wright, 1970, 1971; Whitfield, 1969; Byram & Fisher, 1973; Beermann *et al.*, 1974). von Brand (1939*a*, 1940), Bullock (1949*b*) and Crompton (1965) reported histochemical evidence for glycogen within the lacunar canals apparently permeating the tegument of several acanthocephalan species, but these observations may have been fixation artifacts (Butterworth, 1969*a*). The glycogen stores in adult acanthocephalans turn over dramatically between host feeding periods (Read & Rothman, 1958). Those in infective acanthors and cystacanths appear to supply energy reserves essential to the events of activation and infection (Crompton, 1970; Horvath, 1971).

von Brand (1939*a*) confirmed that glycogen is the major storage polysaccharide in *M. hirudinaceus* by demonstrating its susceptibility to amylase digestion. von Brand (1939*a*) and von Brand & Saurwein (1942) also reported the presence of small amounts of what they believed to be a galactose-containing polysaccharide in *M. hirudinaceus* tissues. The primary basis for their presumption was that the material was resistant to amylase digestion. Laurie (1959) also observed small amounts of polysaccharide resistant to hydrolysis by α- and β-amylases in extracts of *Moniliformis moniliformis*, but paper chromatography revealed that glucose was the only product of acid hydrolysis. Although the presumptive identification of 'galactogen' in *M. hirudinaceus* has been cited in numerous review articles, there is no evidence that the material was not merely a limit dextrin remaining after amylolytic digestion of glycogen.

Glycogen in adults. Many of the quantitative values reported for the glycogen content of acanthocephalans are questionable, either because the nutritional history of the host was not known or because the values were calculated on the basis of tissue wet mass, which can change dramatically either with host feeding behavior (Crompton, 1970, 1972*a*) or in response to inappropriate osmolality of collection salines (§6.3.2). Patent female *Moniliformis moniliformis* recovered from unstarved hosts at 09.00 h have glycogen levels equal to 22–24% of the total worm dry mass (Körting & Fairbairn, 1972; Starling & Fisher, 1978). Glycogen levels approaching 20% of the tissue dry mass have also been reported for species of *Neoechinorhynchus* (Dunagan, 1964). The glycogen content of male

M. moniliformis from unstarved hosts appears to be comparable to (Read & Rothman, 1958) or slightly lower than (Körting & Fairbairn, 1972) that of females of the same age. Significantly, Graff & Allen (1963) found that *M. moniliformis* glycogen levels were strongly affected by host sex.

Glycogen levels ranging from 0.6 to 2.4 g/100 g wet mass have been reported for 'freshly isolated' *Macracanthorhynchus hirudinaceus* females (von Brand, 1939*a*; Ward, 1952; Donahue *et al.*, 1981), and von Brand (1940) found that glycogen accounted for 7.4% and 12.7% of the dry masses of female and male *M. hirudinaceus*, respectively. Although *M. hirudinaceus* may have a lower maximal glycogen content than its smaller relatives, the values measured in this species were probably confounded by host fasting before slaughter or by endogenous metabolism prior to analysis.

Unstarved female *Moniliformis moniliformis* (68–118 days old) depleted 18% (range: 13–29%) of their glycogen stores during a 4-h incubation in the absence of exogenous carbohydrate, while males had significantly lower glycogen consumption rates (4–5% depletion in 4 h) (Körting & Fairbairn, 1972). Attempts by von Brand (1940) and Ward (1952) to measure glycogen consumption by *M. hirudinaceus in vitro* (20- and 43-h incubations, respectively) gave much lower glycogen depletion rates than those observed for *M. moniliformis*. However, the *M. hirudinaceus* studied by von Brand and by Ward had low initial glycogen levels, and inspection of Ward's data indicates that the glycogen consumption rate observed in her experiments was directly correlated with the initial levels of glycogen in experimental worms. Glycogen consumption by species of *Neoechinorhynchus* maintained *in vitro* also decreased as worm glycogen stores fell below 8% of the tissue solids (Dunagan, 1964).

Glycogen depletion and synthesis rates *in vivo* are also dependent on tissue glycogen levels. The depleted glycogen stores of *Moniliformis moniliformis* from starved hosts were significantly replenished *in vivo* when hosts were given a single starch meal (Read & Rothman, 1958) and *M. moniliformis* from fasted hosts incubated *in vitro* in glucose, fructose, mannose or maltose restored their glycogen stores almost to 'unstarved' levels within 1–2 h (Laurie, 1959; Kilejian, 1963). Read & Rothman (1958) observed significant diel changes in the tissue glycogen levels of *M. moniliformis* in normally feeding hosts: the glycogen levels of 4-week-old worms increased almost two-fold between 14.00 h and 23.00 h as a consequence of the nocturnal feeding activity of the host; however the glycogenesis rate in these 'normally cycling' *M. moniliformis* was less than that seen in worms from fasted hosts.

Laurie (1959) found that the glycogen content of 6-week-old *Moniliformis*

moniliformis was 6.7% of the tissue dry mass after the hosts were fasted overnight, and Kilejian (1963) detected comparable glycogen levels in 4-week-old *M. moniliformis* from rats starved for 24 h (6.6% and 5.3% of the tissue dry masses for male and female worms, respectively). Comparison of these values with those recorded by Starling & Fisher (1978) for worms from normally fed hosts suggests that 4- to 6-week-old *M. moniliformis* exhaust about 75% of their glycogen reserves during a 24-h period of host starvation. Read & Rothman (1958) found that starvation of hosts for 48 h caused a precipitous fall in the glycogen stores of 4-week-old *M. moniliformis* (to approximately 5% of the maximum tissue levels measured in their experiments), but continued host starvation (up to 80 h) did not cause further decreases in helminth glycogen. Presumably, helminth basal metabolism was maintained by glucose entering the intestinal lumen from host tissues during prolonged host starvation (§6.4.1).

Enzymic basis for regulation of glycogen metabolism. Glycogen metabolism in acanthocephalans is presumably regulated primarily through the effects of glycogen levels and cellular messengers (e.g. calcium ions, cyclic nucleotides) on the states of phosphorylation of glycogen synthase and glycogen phosphorylase, and by the effects of metabolic effectors (G6P and AMP) on the activities of these enzymes (see Stalmans & Hers, 1973; Hers, 1976; Roach, 1981; Li, 1982). Preliminary studies have shown that glycogen synthase in *Moniliformis moniliformis* exists in both dephosphorylated and G6P-dependent phosphorylated forms, and that these forms can be interconverted by incubating worm homogenates under appropriate conditions (Starling & Hollenberg, unpublished; §6.5.2). Donahue *et al.* (1981) published a brief report on glycogen synthase and glycogen phosphorylase from *Macracanthorhynchus hirudinaceus*. They demonstrated activation of glycogen phosphorylase by AMP (K_a = 3.2mM), from which we can infer that the enzyme exists in both the phosphorylated *a* form and the AMP-dependent (dephospho) *b* form. Donahue *et al.* suggested that *M. hirudinaceus* glycogen synthase had a K_m (uridine diphosphoglucose) of 0.22mM and a K_a of 0.25mM for activation by G6P; the significance of these kinetic parameters is not clear since they apparently did not measure G6P stimulation of the dephosphorylated (I) form independently of the requisite activation of the G6P-dependent phosphorylated forms.

In mammalian systems, the dephosphorylations of G6P-dependent glycogen synthase *b* and AMP-independent glycogen phosphorylase *a* are inhibited by high glycogen levels and stimulated by high levels of glucose and G6P (see Stalmans & Hers, 1973; Hers, 1976). Glycogen appears to inhibit glycogen synthase phosphatase(s) directly; its effects on phosphorylase phosphatase(s) may be more complex (Hers, 1976; Li, 1982). Glycogen inhibits dephosphorylation of glycogen synthase *b* isolated from *Hymenolepis diminuta*, and the proportion of total glycogen synthase activity in the I form is significantly lower in preparations from unstarved tapeworms than in preparations from hosts fasted for 24 h and then fed 1 h before worm isolation (Dendinger & Roberts, 1977). Hence, there is evidence that tissue glycogen levels regulate the rate of glycogen synthesis in *H. diminuta* through their effects on glycogen synthase *b* dephosphorylation. Thus, glycogen is likely to exert a similar effect on the phosphorylation states of acanthocephalan glycogen synthases and glycogen phosphorylases.

Glycogen during development. Ovarian balls and oocytes contain little glycogen, but large glycogen stores accumulate during development of the shelled acanthors (von Brand, 1939*a*, 1940; Crompton, 1965). The glycogen in *Macracanthorhynchus hirudinaceus* and *Polymorphus minutus* acanthors disappears rapidly after acanthor activation (Miller, 1943; Crompton, 1970). Miller (1943) suggested that glycogen is depleted during acanthor hatching and the vigorous motions of the *Moniliformis moniliformis* acanthor during hatching (Wright, 1971) support this view.

Glycogen does not appear to accumulate in acanthocephalan larvae until the development of early acanthellae has progressed significantly. During the 4 months required for larval *Macracanthorhynchus hirudinaceus* to develop, glycogen was not detectable histochemically until 22 days postinfection, and accumulations of tegumental glycogen stores were evident only after 54 days (Miller, 1943). Glycogen accumulation in developing *Polymorphus minutus* larvae follows a similar temporal pattern (Crompton, 1970).

Cystacanths of *Macracanthorhynchus hirudinaceus* (Miller, 1943) and *Polymorphus minutus* (Crompton, 1970) show strong histochemical reactions for glycogen within the basal regions of the tegument, but the quantities of glycogen accumulated within cystacanths may vary considerably between species or orders. Horvath (1971) found that glycogen accounted for 12% of the dry mass of *Moniliformis moniliformis* cystacanths, whereas glycogen comprised only 1.5% of the dry mass of *P. minutus* cystacanths (Butterworth, 1969*a*). It is not possible to determine from the data available whether the higher glycogen levels in *M. moniliformis* cystacanths, as compared to *P. minutus* cystacanths, are representative of general differences between Archiacanthocephala and Palaeacanthocephala or whether they reflect higher hemolymph carbohydrate levels in

Periplaneta americana than in the gammarid hosts of *P. minutus* (see Crompton, 1970, for host carbohydrate levels).

Glycogen metabolism in activated cystacanths. Crompton (1970) felt that the glycogen content of *Polymorphus minutus* cystacanths might be inadequate for the metabolic needs of the larva, even though one would expect the metabolic demands of this 'resting stage' to be relatively small. However, there is little reason to believe that cystacanth metabolic needs cannot be met by exogenous carbohydrate within the hemocoel of the intermediate host. According to Crompton (1970), Butterworth (1969*b*) found that *P. minutus* cystacanths incubated in [^{14}C]-glucose in 'saline adjusted to comply with some of the parasite's environmental conditions' used 11.6 nmol glucose/mg dry mass per h, and data cited by Crompton (1970) imply that glucose uptake exceeded the metabolic rate considerably. It would appear that the glycogen content of *P. minutus* cystacanths is sufficient to sustain metabolism of the *unactivated* cystacanth for several hours after ingestion by the definitive host, but it would probably be depleted rapidly after cystacanth activation unless spared by exogenous carbohydrate in the host intestine.

Cystacanths of *Moniliformis moniliformis* activated *in vitro* consumed approximately 30% of their glycogen reserves (35 μg/mg cystacanth dry mass) during a 2-h aerobic incubation in glucose-free medium, whereas unactivated cystacanths showed no detectable glycogen consumption (Horvath, 1971). Freshly activated cystacanths incubated in [^{14}C]-glucose incorporated little radioactivity into glycogen during a 5-min pulse ($<$0.2 nmol glucose/50 cystacanths). Similar incubations 60 and 120 min after activation resulted in incorporation into glycogen of 4.8 and 15.1 nmol [^{14}C]-glucose/50 cystacanths, respectively (Horvath, 1971); these values translate into apparent glycogenesis rates of 9.2 and 28.6 μg/mg cystacanth dry mass per h. Since exogenous glucose would be expected to spare glycogen, the rate and extent at which endogenous glycogen is depleted from activated cystacanths in a glucose-containing environment would probably be significantly lower than the rates Horvath measured *in vitro* in the absence of exogenous glucose.

Horvath (1971) interpreted the increased glycogenesis rates in *Moniliformis moniliformis* cystacanths at 60 and 120 min postactivation as evidence that glycogenesis is regulated by endogenous glycogen levels. This is compatible with the effects of glycogen levels on glycogen synthesis in mammalian tissues, cestodes and *M. moniliformis* adults (see above). However, the design of Horvath's experiments means that these increases may also reflect changes either in levels of other allosteric effectors of glycogen synthase or in permeability of the activated cystacanths to glucose.

6.5.2 *Trehalose*

Trehalose (1-[α-D-glucopyranosyl]-α-D-glucopyranoside) comprises 2–4% of the tissue solids in adult *Moniliformis moniliformis* (Fairbairn, 1958; Laurie, 1959; McAlister & Fisher, 1972; Starling, 1975; Starling & Fisher, 1978). The non-reducing disaccharide is present in the pseudocoelomic fluid, ovarian balls and acanthors taken from female worms. Its levels in acanthors are significantly greater than in isolated ovarian balls or in pseudocoelomic fluid; the highest trehalose levels in adult worms occur in the body wall (McAlister & Fisher, 1972).

It appears that trehalose is also present in *Macracanthorhynchus hirudinaceus*. Tissue homogenates fortified with ATP and uridine diphosphoglucose (UDPG) can incorporate [^{14}C]-glucose into [^{14}C]-trehalose (Fisher, 1964), and both trehalose 6-phosphate synthase (Donahue *et al.*, 1981) and trehalase (Dunagan & Yau, 1968; Donahue *et al.*, 1981) have been demonstrated in crude preparations. However, the actual trehalose content of *M. hirudinaceus* tissues has still to be measured.

Crompton & Lockwood (1968) were unable to find trehalose in extracts from 7-day-old *Polymorphus minutus* adults. Paper chromatography of extracts from *P. minutus* which had been incubated in [^{14}C]-glucose did not reveal any radioactive trehalose, nor could trehalose be demonstrated chemically when paper chromatograms were developed with reagents which clearly identified authentic trehalose standards and trehalose present in extracts from *Moniliformis moniliformis*. Thus, it would appear that trehalose is not a major sugar in *P. minutus*. Crompton & Lockwood (1968) suggested that the presence of trehalose in *M. moniliformis* and *M. hirudinaceus* might reflect evolutionary adaptations on the part of archiacanthocephalans to the presence of trehalose in the hemocoels of their intermediate hosts, and that the lack of the disaccharide in *P. minutus* could be correlated with the presence of little (Fairbairn, 1958) or no (Butterworth, cited in Crompton & Lockwood, 1968) trehalose in the gammarid hemocoel.

The role of trehalose in Moniliformis moniliformis. Trehalose has various functions in the organisms in which it occurs. In insects, it is the blood sugar by which glycosyl moieties absorbed from the gut or stored as fat body glycogen are transferred to the muscles (Wyatt, 1967; Friedman, 1978). Its synthesis and hydrolysis are coordinated with spore formation and germination in some fungi (Van Assche, Van Laere & Carlier, 1978) and cellular slime molds (Killick & Wright, 1975; Jackson, Chan & Cotter, 1982). Trehalose accumulation in *Artemia salina* (Clegg, 1965) and in soil nematodes (Loomis, Madin & Crowe, 1980; Womersley & Smith, 1981) as they enter anhydrobiosis has been interpreted in terms of the osmotic pressure exerted by trehalose and other solutes, but recent studies have implicated trehalose as a critical membrane-stabilizing agent under anhydrobiotic or cryobiotic conditions (Crowe, Crowe & Mouradian, 1983). Trehalose has been identified as a reserve carbohydrate in non-embryonic tissues of intestinal nematodes because its levels decline in parallel with glycogen during incubation *in vitro* (Harpur, 1963; Roberts & Fairbairn, 1965). Fairbairn (1970) has also suggested that trehalose present in the perivitelline fluid of ascarid eggs may be involved in hatching.

The substantial levels of trehalose in *Moniliformis moniliformis* undoubtedly contribute to the osmolality of the body tissues and extracellular fluids. Denbo (1971) suggested that trehalose may maintain the osmolality of *Macracanthorhynchus hirudinaceus* pseudocoelomic fluids. Eggs of *M. hirudinaceus* are resistant to desiccation and moderate freezing (Kates, 1942), and the membrane-stabilizing properties of trehalose (Crowe *et al.*, 1983) may be involved in this resistance. There is also some evidence that elevated internal osmotic pressure, to which trehalose may contribute, is important in the activation and hatching of *M. moniliformis* eggs (§6.3.2).

Laurie (1959) found significant levels of trehalose in *Moniliformis moniliformis* from fasted hosts. When the fasted acanthocephalans were incubated for 2–4 h in glucose or other glycogenic sugars, tissue glycogen levels increased two- to three-fold, but there was no discernible change in worm trehalose content. Although the trehalose may be used as an emergency storage disaccharide when tissue glycogen levels are severely depleted, Laurie's results suggest that trehalose does not serve as a primary storage carbohydrate for energy metabolism in *M. moniliformis.*

Trehalose as a glucose shuttle. In contrast to Laurie's observation that net trehalose synthesis does not occur when fasted *Moniliformis moniliformis* are incubated in glycogenic sugars, Starling & Fisher (1978, 1979) found significant formation of [^{14}C]-trehalose by female *M. moniliformis* incubated in [^{14}C]-glucose, [^{14}C]-mannose or [^{14}C]-fructose for brief periods. More than 50% of the [^{14}C]-glucose absorbed by *M. moniliformis* from unstarved hosts during 1–4-min incubations was incorporated into trehalose; only a small amount of radioactivity was present in glycogen, even after a 40-min postincubation in unlabeled glucose. Starling & Fisher noted that their results were consistent with the concept that trehalose might serve as a metabolic sink for trapping glucose within acanthocephalan tissues after its absorption (Crompton, 1970; McAlister & Fisher, 1972; §6.3.1). Moreover, they extended the model to suggest that trehalose synthesized in the tegument during glucose uptake served as a shuttle to carry glucose moieties to non-tegumental tissue compartments.

One might question this model on the grounds that transient incorporation of glucose into trehalose is energetically expensive. Trehalose synthesis is a two-step process (Cabib & Leloir, 1958; McAlister & Fisher, 1972; Elbein, 1974). The first step, catalyzed by trehalose phosphate synthase (TPS), is the transfer of the glucosyl moiety of uridine diphosphoglucose (UDPG) to the anomeric carbon of G6P to form trehalose 6-phosphate (T6P); T6P is then cleaved to trehalose by a specific trehalose 6-phosphate phosphatase (TPP):

$$\mathrm{UDPG + G6P \xrightarrow{TPS} UDP + T6P}$$

$$\mathrm{T6P + H_2O \xrightarrow{TPP} trehalose + P_i}$$

Since the syntheses of G6P and UDPG from free glucose require one and two ATP-equivalents, respectively, the cost of trehalose synthesis is 1.5 ATP/glucosyl residue. However, in an organism like *Moniliformis moniliformis*, which does not accumulate glucose by sodium ion-coupled active transport (§6.3.1), the transient storage of glucose in trehalose before its transfer (as glucose) to other body compartments is less expensive than the other alternative: transient storage in tegumentary glycogen. Although 50% of the ATP expenditure required for glycogen synthesis (2 ATP/glucosyl residue) is recovered as glucose 1-phosphate when glycogen is mobilized for energy metabolism, this energy recovery is lost if the glucose 1-phosphate is hydrolyzed to glucose for transfer to other body compartments.

The glucose shuttle model resolves the apparent contradiction between the observation that trehalose levels in *Moniliformis moniliformis* are fairly stable (Laurie, 1959) and results indicating high rates of trehalose synthesis during glucose absorption (Starling & Fisher, 1978, 1979). Some of the components of the original model, however, require modification. Starling & Fisher based the model on the observation that ethanolic extracts of *M. moniliformis* contain substantial quantities of free glucose (Fairbairn, 1958; Starling & Fisher, 1978). They reasoned that the large glucose pools believed to be present in the acanthocephalans must be physically separated from the tegument cytoplasm; otherwise, they should not have observed rapid conversion of [^{14}C]-sugars absorbed *in vitro* into other biochemical species (trehalose and biochemical intermediates). However, the glucose levels reported by Fairbairn and by Starling & Fisher appear to have been artifacts resulting from hydrolysis of glycogen and/or trehalose during ethanol extraction. Cornish, Wilkes & Mettrick (1981*b*) found the glucose content of samples prepared from freeze-clamped 9-week-old *M. moniliformis* females to be only 0.5 μmol/g fresh tissue, and Starling (unpublished) has confirmed this observation. Hence, the large extrategumental glucose pools presumed by Starling & Fisher do not appear to exist, although the apparent glucose concentration in the female *M. moniliformis* pseudocoel (1.67mM; Tanaka & MacInnis, 1980) suggests that there is some glucose compartmentation within the acanthocephalan body.

The glucose shuttle model also suggested that the conversion of trehalose formed in the tegument into glucose in non-tegumentary compartments might involve a membrane-associated trehalase resembling the hydrolase-related transport systems described for the vertebrate intestine (Malathi *et al.*, 1973; Ramaswamy *et al.*, 1974; Crane, 1975). The model implied that coupling of trehalose hydrolysis to the transfer of glucose moieties across the tegument basal membrane might provide energy for the accumulation of extrategumental glucose pools. Efforts to demonstrate a membrane-bound trehalase have, however, been unsuccessful (§6.6.7).

Starling (unpublished) has re-examined the initial metabolism of [^{14}C]-glucose absorbed by *Moniliformis moniliformis* using procedures designed to minimize artifacts of tissue processing. Preliminary results (Table 6.6) suggest that the observations made by Starling & Fisher (1978, 1979) may contain quantitative errors introduced by their tissue processing procedure but they confirm that much of the [^{14}C]-glucose absorbed by unstarved *M. moniliformis* during brief incubations is incorporated into trehalose rather than glycogen.

The major differences between *Moniliformis moniliformis* from fasted and non-fasted rats (Table 6.6) were in trichloroacetic acid (TCA)-insoluble material and in the fraction designated 'X'. The apparent incorporation of glucose into glycoprotein probably stems from completion of post-translational glycosylations suspended in the fasting state. The radioactive component of fraction 'X' is believed to be an acid-labile ligand in a protein or lipoprotein intermediate, but the significance of its apparent rapid synthesis and turnover in worms from normally fed hosts, but not in worms from fasted hosts, is not clear.

A large proportion of the absorbed [^{14}C]-glucose pulse was incorporated rapidly into trehalose by *Moniliformis moniliformis* from both fasted and normally fed hosts, although trehalose synthesis appears to have been slightly more rapid in fasted worms (Table 6.6). Worms from fasted hosts incorporated more [^{14}C]-glucose into glycogen than did worms from

Table 6.6. *The incorporation of radioactivity into trehalose, glycogen, and acid-insoluble material by 51-day-old female* Moniliformis moniliformis *incubated for 60 s in 10 mM [^{14}C]-glucose*[a]

Preincubation[b] (min)	Postincubation[b] (min)	% Total radioactivity in homogenates[c]				
		Glucose[d]	Trehalose[e]	Glycogen	TCA-insoluble[f]	'X'[g]
Host starved 24 h						
0	0	9.5±0.2	40.6±3.7	1.7±0.1	6.8±0.6	1.8±0.1
0	5	6.5±2.9	55.0±2.3	2.7±0.2	10.3±0.2	1.4±0.4
20	0	12.5±5.1	39.4±5.2	3.2±1.8	14.0±3.7	1.9±0.4
20	5	8.6±4.3	44.8±0.9	6.9±0.8	18.5±4.1	1.3±0.1
Host not starved						
0	0	9.5±0.9	12.4±0.2	1.0±0.1	0.7±0.2	15.0±0.4
0	5	3.0±0.1	44.1±4.3	2.5±0.4	0.8±0.2	2.6±0.1
20	0	13.5±3.7	18.6±0.7	1.1±0.5	0.3±0.1	10.2±0.9
20	5	8.6±3.5	54.6±1.1	1.4±0.6	0.7±0.2	3.2±2.1

[a] Starling (unpublished). Worms collected at 10.00 h. Media containing 112 mM NaCl, 1.2 mM $MgSO_4$, 2.4 mM $CaCl_2$, 4.8 mM KCl, 2 mM NaH_2PO_4, 10 mM $NaHCO_3$ and glucose + mannitol to 40 mM were gassed 30 min with 95% N_2–5% CO_2 prior to use; the pH of gassed media was 7.1. Incubations (37 °C) were in beakers open to the atmosphere. Samples were rinsed (5 s) in saline after final incubation, freeze-clamped (−70 °C) and homogenized in cold (−20 °C) 80% (v/v) ethanol. The ethanol-soluble fraction was partially evaporated under vacuum, and the remaining aqueous fraction analyzed by paper chromatography after extraction with $CHCl_3$ (Starling & Fisher, 1978). Protein and nucleic acids were precipitated from the ethanol-insoluble fraction with 5% (w/v) trichloroacetic acid (TCA) and glycogen was precipitated from the TCA-soluble fraction with ethanol

[b] Samples held in glucose-free medium prior to incubation in 10 mM [^{14}C]-glucose (1 mCi/mmol) or to preincubation in 2 mM unlabeled glucose. Postincubation media contained 2 mM glucose

[c] Mean ± standard deviation for 3 samples of four worms each

[d] May contain some contamination from edge of lactate peak

[e] May contain some contamination from edge of fumarate peak

[f] Precipitated by 5% (mass/v) TCA

[g] 'X' = unidentified ethanol-insoluble radioactivity rendered ethanol-soluble by TCA treatment

normally fed hosts, and glycogen labeling was intensified in fasted worms by preincubation in glucose. The latter observation is consistent with a conversion of glycogen synthase into the more active dephospho form as a consequence of glucose absorption in worms containing low, but not high, glycogen levels (§6.5.1). Synthesis of trehalose (Table 6.6) from the absorbed [^{14}C]-glucose pulse greatly exceeded glycogenesis from newly absorbed [^{14}C]-glucose, both for fasted and non-fasted worms, even though glycogenesis would be expected to have occurred in fasted worms under the conditions of the experiments (Laurie, 1959; Kilejian, 1963).

Trehalose phosphate synthase and glycogen synthase in Moniliformis moniliformis. Starling & Hitt performed preliminary kinetic studies on partially purified trehalose phosphate synthase (TPS) from *Moniliformis moniliformis.* At pH 8.0, where enzyme activity is maximal, the apparent K_m for G6P was 2.75mM. In the presence of 2mM G6P, the UDPG concentration giving half-maximal velocity was about 80μM; at 10mM G6P the UDPG saturation curve was moderately sigmoid, and the $S_{0.5}$ for UDPG increased to about 130μM. UDPG saturation curves at pH 7.0 resembled those obtained at pH 8.0. These kinetic values are strikingly different from those reported by McAlister & Fisher (1972), which might have been artifacts of the extremely low TPS activities in their crude preparations.

Much of the glycogen synthase isolated at 10.00 h from unfasted *Moniliformis moniliformis* is in the G6P-dependent form. Starling & Hollenberg performed preliminary studies on the kinetics of glycogen synthase preparations whose activity in the absence of G6P was only 15% of the activity in the presence of 5mM G6P ('activity ratio' = 0.15). In the presence of 5mM G6P, the approximate UDPG concentration giving half-maximal glycogen synthase activity was 0.5mM; this value agrees with those reported for glycogen synthases from mammalian tissues and from cestodes (Stalmans & Hers, 1973; Mied & Bueding, 1979; Roach, 1981). Unlike low-activity-ratio enzymes from mammalian tissues and from the cestode *Hymenolepis diminuta*, however, highly phosphorylated glycogen synthase from *M. moniliformis* was not fully activated in the presence of 5mM G6P. Its activity in the presence of 15mM G6P was twice that seen in the presence of 5mM G6P, and the apparent K_a was 11.4mM G6P.

These results suggest that, in the absence of other regulatory factors, the TPS in *Moniliformis moniliformis* body walls should out-compete highly phosphorylated forms of glycogen synthase for their common substrate, UDPG. The low $S_{0.5}$ of TPS for UDPG suggests that TPS can probably compete effectively with the G6P-independent form of glycogen synthase under some conditions. A variety of metabolites may, however, alter the relative activities of glycogen synthase and TPS *in vivo*.

TPS from *Moniliformis moniliformis* is inhibited by trehalose (McAlister & Fisher, 1972) and its relative molecular mass (220000 daltons) suggests that it may be subject to several other forms of allosteric modulation. Starling & Abel have obtained circumstantial evidence that TPS and trehalase from *M. moniliformis* may be regulated by protein phosphorylation/dephosphorylation in the same manner as the enzymes of glycogen metabolism. For example, incubation of homogenates with Mg^{2+}–ATP + cyclic AMP results in an increase in trehalase activity and a decrease in TPS activity.

Glycogen synthase activity can be modulated by a variety of other phosphorylated compounds in addition to G6P (Mied & Bueding, 1979). *Moniliformis moniliformis* glycogen synthase *b* was not inhibited by 10mM inorganic phosphate (P_i) in the presence of 10mM G6P, nor did P_i stimulate G6P-independent glycogen synthase activity in high-activity-ratio preparations. These properties are different from the responses of the glycogen synthases of *Hymenolepis diminuta* and mammalian muscle to P_i (Stalmans & Hers, 1973; Mied & Bueding, 1979).

6.5.3 *Lipids*

Adult acanthocephalans contain lipid deposits which histochemical and cytological studies show as large and small droplets in virtually all tissues (von Brand, 1939; Bullock, 1949*b*; Crompton, 1963; Crompton & Lee, 1965; Nicholas & Mercer, 1965; Hammond, 1967; Wright & Lumsden, 1968; Lange, 1970; Byram & Fisher, 1973; Beermann *et al.*, 1974; Parshad & Guraya, 1977*c*). Acanthors seem to lack large lipid deposits (Bullock, 1949*b*). Acanthellae of *Polymorphus minutus* accumulate increasing amounts of lipid during development to the cystacanth, in which lipids account for 28% of the dry mass (Butterworth, 1969*a*; Barrett & Butterworth, 1971). Comparable lipid accumulations probably occur during development in other species and Butterworth (1969*a*) suggested that the vacuolated structures identified as the lacunar canals of the cystacanth tegument of many species may actually be areas vacated by lipid droplets during inadequate fixation.

Lipid composition. Chemical analyses of total body lipids have been performed on adult *Macracanthorhynchus hirudinaceus* by von Brand (1939*a*, 1940) and Beames & Fisher (1964), on adult *Moniliformis moniliformis* by Beames & Fisher (1964), on adult *Echinorhynchus salmonis* by

Table 6.7. *The major lipid compositions of* Macracanthorhynchus hirudinaceus, Moniliformis moniliformis *and* Sphaerirostris pinguis

	Macracanthorhynchus[a]		*Moniliformis*[a]		*Sphaerirostris*[b]	
	Females	Males	Females	Males	Females	Males
Total lipid (% wet mass)	0.9	1.7	4.2	7.2	4.3	5.1
% Total lipid						
Phospholipids	46.7	45.9	18.6	9.0	11.1	13.6
Total neutral lipids	54.3	51.8	78.8	88.3	—	—
Unsaponifiable	21.4	37.8	9.7	7.9	14.1	19.2
Triacylglycerols	—	—	—	—	52.3	53.4
Fatty acyl esters	30.2	15.5	62.5	63.5	—	—
Free fatty acids	—	—	—	—	16.1	14.3

[a] Beames & Fisher (1964)
[b] Parshad & Guraya (1977*c*)

Vysotskaya & Sidorov (1973), on adult *Sphaerirostris pinguis* by Parshad & Guraya (1977*c*) and on both cystacanths and adults of *Polymorphus minutus* by Barrett & Butterworth (1971). The major lipid compositions of *M. hirundinaceus*, *M. moniliformis* and *S. pinguis* are summarized in Table 6.7. Smith, Dunagan & Miller (1979) determined the total lipid levels of the body wall (4.7 g/100 g wet mass), pseudocoel contents (2.3 g/100 g wet mass) and lemnisci (10.4 g/100 g wet mass) of female *M. hirudinaceus* and found significantly higher total body lipid levels than did von Brand (1939*a*, 1940) and Beames & Fisher (1964). The reason for the disparities in the values reported for *M. hirudinaceus* lipids is unclear, but they may reflect differences in the hydration states and ages of the worms examined.

Two features which appear representative of all acanthocephalans are revealed by the data in Table 6.7. Males have more total lipid than females, and all species appear to contain considerable amounts of neutral lipids. The fatty acids of the neutral lipids and phospholipids of adult *Macracanthorhynchus hirudinaceus* and *Moniliformis moniliformis* (Beames & Fisher, 1964; Smith *et al.*, 1979) and of cystacanths of *Polymorphus minutus* (Barrett & Butterworth, 1971) share characteristics common to the lipids of many helminths (Barrett, 1981). A high proportion of the fatty acyl chains are unsaturated (80% of the total saponifiable neutral lipids in the studies reported by Beames & Fisher), and there appears to be a predominance of long chain (C_{18}, C_{20}, C_{22}) fatty acids (Table 6.10).

Interestingly, neutral lipid accounts for 93% of the total lipid in *Polymorphus minutus* cystacanths, yet very little of it is in the form of triacylglycerols (Barrett & Butterworth, 1971). Wax esters comprise 86% of the cystacanth neutral lipid, and cholesterol esters an additional 9%. In adults, wax and sterol esters account for approximately 25% of the neutral lipid fraction.

Phosphatidyl choline and phosphatidyl ethanolamine are the major phospholipids of *Macracanthorhynchus hirudinaceus* (Beames & Fisher, 1964), although small amounts of sphingomyelin are also present (Smith *et al.*, 1979). Beames & Fisher identified small amounts of plasmalogen in both *M. hirudinaceus* and *Moniliformis moniliformis* (which contains slightly more phosphatidyl choline than *M. hirudinaceus*), but relatively small amounts of phosphatidyl serine and phosphatidyl ethanolamine were found in *M. moniliformis* (Beames & Fisher, 1964).

Cholesterol was the only sterol detected in *Polymorphus minutus* (Barrett & Butterworth, 1971). In contrast, Beames & Fisher (1964) found that cholesterol accounted for only 50–80% of the unsaponifiable lipids in *Macracanthorhynchus hirudinaceus* and *Moniliformis moniliformis*. Barrett, Cain & Fairbairn (1970) analyzed the sterols of these archiacanthocephalans and of the intestinal contents of their hosts (Table 6.8). They found that *M. hirudinaceus* contained high proportions of campesterol and β-sitosterol; the reduced derivatives of these phytosterols (campestanol and stigmastanol) and of cholesterol (cholestanol), which were not detected in the host intestinal contents, were present in both *M. hirudinaceus* and *M. moniliformis*. *Hymenolepis diminuta* does not contain the phytosterols found in *M. moniliformis* (Ginger & Fairbairn, 1966). Hence, it would appear that the sterol contents of intestinal helminths are determined by more factors than the lipid content of the host diet.

Several acanthocephalans are known to contain carotenoid pigments (Van Cleave & Rausch, 1950; Barrett & Butterworth, 1968; Smith, 1969; Barrett, 1981) dissolved in tissue lipid droplets (Barrett & Butterworth, 1968). The archiacanthocephalan *Macracanthorhynchus hirudinaceus* and the eoacanthocephalan *Neoechinorhynchus pseudemydis* contain lutein, a xanthophyll derivative of α-carotene, as their only carotenoid (Barrett & Butterworth, 1973). *Pallisentis nagpurensis* has been found to contain xanthophyll derivatives of both α- and β-carotene (Ravindranathan & Nadakal, 1971). Only β-carotene and astaxanthin, a β-carotene derivative, have been identified to date in the Palaeacanthocephala: *Pomphorhynchus laevis* and *Filicollis anatis* adults contain unesterified β-carotene, while *Rhadinorhynchus ornatus* adults and *Polymorphus minutus* cystacanths and adults contain only esterified astaxanthin (Barrett & Butterworth, 1968, 1973). The function of carotenoid pigments in adult acanthocephalans is unknown. The exclusivity with which particular carotenoids are absorbed

Table 6.8. *The sterols of* Macracanthorhynchus hirudinaceus, Moniliformis moniliformis *and the intestinal contents of their hosts (from Barrett, Cain & Fairbairn, 1970)*

	Percentage of total lipids			
Sterol	*Macracanthorhynchus hirudinaceus*	Pig gut[a]	*Moniliformis moniliformis*	Rat gut
Cholesterol	57.5	99	84	90
Campesterol	16	0.5	1.4	1.8
β-Sitosterol	21	0.5	10	6.8
Brassicosterol?	0[b]	0	0.3	1.4
Cholestanol	1.2	0	1.0	0
Campestanol	1.9	0	1.0	0
Stigmastanol	2.4	0	2.3	0

[a] Fasted before slaughter; sterol content may not be representative
[b] Zero values indicate below limits of detection

from among the spectrum present in host intestinal or hemocoel fluids (Barrett & Butterworth, 1968) suggests some form of selective pressure for uptake of specific carotenoids. The possibility that cystacanth pigmentation may increase the probability of predation by the definitive host (see Chapter 9) might explain such selective pressure.

Lipid metabolism. Lipid oxidation is essential to energy metabolism in eggs and free-living larval stages of some nematodes, and there is evidence that lipid oxidation may be important to larval stages of some trematodes and in coracidia, but not plerocercoids, of pseudophyllidean cestodes (von Brand, 1973, 1979; Barrett, 1976*b*, 1981; Barrett & Körting, 1977). The possibility that acanthellae developing within the intermediate host are able to catabolize lipids has not been investigated, but it does not appear that cystacanths of *Polymorphus minutus* can do so (Butterworth, 1967; Crompton, 1970). There is no evidence that lipid oxidation is of metabolic significance in adult helminths (see Fairbairn, 1970; von Brand, 1973, 1979; Barrett & Körting, 1976; Barrett, 1981; Saz, 1969, 1981). Barrett (1981) mentioned lipid oxidation as a possible source of reducing equivalents in acanthocephalans which produce succinate (§6.6.2). Körting & Fairbairn (1972) did not detect lipid fermentation in *Moniliformis moniliformis* incubated anaerobically, but these results should not yet be extended to all acanthocephalan species.

Barrett *et al.* (1970) found that *Moniliformis moniliformis* cannot synthesize sterols from [^{14}C]-acetate. When they injected [^{14}C]-mevalonate into *Macracanthorhynchus hirudinaceus*, radioactivity was incorporated into the non-saponifiable lipid fraction, but none was present in sterols. These observations suggest that archiacanthocephalans resemble other helminths in their lack of capacity to synthesize steroids *de novo* (von Brand, 1973, 1979; Barrett, 1981).

Barrett *et al.* also examined the ability of *Macracanthorhynchus hirudinaceus* to dealkylate or otherwise metabolize β-sitosterol. It should be noted that *M. hirudinaceus* and *Moniliformis moniliformis* contain significant amounts of stigmastanol, which is a reduction product of β-sitosterol, yet Barrett *et al.* could not detect stigmastanol in their analysis of host intestinal fluids. Since Barrett *et al.* could not demonstrate formation of stigmastanol from β-sitosterol by *M. hirudinaceus*, it would appear that stigmastanol and other reduced sterols present in *M. hirudinaceus* accumulate from traces appearing periodically in the host intestinal lumen.

Moniliformis moniliformis incubated in [^{14}C]-acetate *in vitro* incorporated radioactivity into fatty acids and their esters (Barrett *et al.*, 1970). Although *M. moniliformis* has not been examined for its ability to synthesize fatty acids *de novo*, the possibility is remote since *de novo* synthesis of long-chain fatty acids occurs in few species of helminths (von Brand, 1973, 1979; Barrett, 1981; Ward, 1982). The incorporation of [^{14}C]-acetate into fatty acids by *M. moniliformis* probably involved elongation of pre-existing fatty acid chains, as has been demonstrated in several species of helminths (von Brand, 1973, 1979; Barrett, 1981). Fatty acid elongation may also occur in *Macracanthorhynchus hirudinaceus* (Smith *et al.*, 1979; see below).

Körting & Fairbairn (1972) demonstrated some, but not all, enzymes of a putative β-oxidation cycle and of a glyoxylate cycle in preparations from *Moniliformis moniliformis* cystacanths and adults (Table 6.9). They failed to detect enoyl-CoA hydratase in their survey of adults and found only low apparent activities for the dehydrogenases of the β-oxidation cycle. However, since Körting & Fairbairn found it necessary to centrifuge adult *M. moniliformis* homogenates, a significant proportion of the

Table 6.9. *Enzymes of lipid metabolism in* Moniliformis moniliformis[a,b]

Enzyme	Adults	Cystacanths
Palmityl-CoA synthetase		
$C_{12:0}$	0.8	—
$C_{14:0}$	10	—
$C_{16:0}$	3.3	1.4
$C_{18:0}$	0.4	—
Acyl-CoA dehydrogenase	1.8	4.0
Enoyl-CoA hydratase	0	50
β-Hydroxyacyl-CoA dehydrogenase	1.0	0
Acetyl-CoA acyltransferase	23	13
Citrate synthase	0	7.1
Aconitate hydratase		
citrate to *cis*-aconitate	0	2.3
isocitrate to *cis*-aconitate	0	6.8
Isocitrate lyase	3.6	—
Malate synthase	0	1.2
Succinate dehydrogenase	49	—
Fumarate hydratase	0; 160[c]	0(?)
Malate dehydrogenase	1008	1333

[a] From Körting & Fairbairn (1972) unless otherwise indicated. Cystacanths = 700-*g* supernates; adults = 5000-*g* supernates from sucrose homogenates; — = not measured

[b] Activity in nmol product/min per mg protein

[c] Wilkes, Cornish & Mettrick (1982*a*). 100000-*g* supernate from preparation lacking mitochondrial integrity but containing reducing agents

mitochondrial fraction might have been lost from the supernatant fractions which they assayed. Furthermore, several of the enzyme activities measured in their survey were significantly lower than those observed by other investigators (see Table 6.11), and it is possible that the conditions used were unsatisfactory for preserving the activity of some enzymes (Wilkes, Cornish & Mettrick, 1982*a*; §6.6.2). Ward & Crompton (1969) observed small amounts of *n*-butyrate and *n*-valerate among the products excreted by *M. moniliformis* incubated anaerobically *in vitro* (Table 6.15); a reasonable pathway for the synthesis of these volatile acids would be reverse β-oxidation using acetyl-CoA and propionyl-CoA as substrates.

It is appropriate to ask whether the apparent fatty acid chain elongation demonstrated in *Moniliformis moniliformis* (Barrett *et al.*, 1970) is of biological significance. Its rate is about 10^4 times lower than the rate of anaerobic glycogen catabolism in the same species (Körting & Fairbairn, 1972), so its role in energy metabolism seems doubtful. However, fatty acid chain elongation may contribute to the formation of specific membrane lipids or reproductive processes.

The fatty acid composition of neutral lipids and phospholipids from tissues of female *Macracanthorhynchus hirudinaceus* (Smith *et al.*, 1979; Table 6.10) suggests that the products of fatty acid elongation have a non-random distribution. The fatty acids of the neutral lipids from the lemnisci and body wall appear to differ slightly, while all C_{14} lipids and saturated C_{16} lipids are significantly under-represented in the pseudocoel neutral lipids and in tissue phospholipids. Comparison of the phospholipid composition with the neutral lipid composition of the body wall and lemnisci (Table 6.10) reveals an interesting, though perhaps fortuitous, pattern. The phospholipid fatty acid composition resembles the pattern of body wall and lemniscal neutral lipid fatty acids elongated by two carbon atoms.

We know almost nothing about the biosynthesis and turnover of phospholipids in acanthocephalans. The occurrence of longer-chain fatty acids in membrane phospholipids of *Macracanthorhynchus hirudinaceus* can only be a matter of speculation. Perhaps the enzymes involved in the addition of the nitrogenous component to phosphatidic acid or diacylglycerol require substrates bearing longer fatty acyl chains. Phospholipid fatty acyl group turnover appears to occur at relatively high rates in helminths (Barrett, 1981), but whether this is related to specific membrane lipid restructuring (Barrett, 1981), or results from lysophospholipid formation during endocytosis and transfer of vesicles among membranous structures (Silverstein *et al.*, 1977) is not known.

Reproductive tissues from acanthocephalans have not been examined for their capacity to incorporate [^{14}C]-acetate into fatty acids. The possibility that the ovarian balls are responsible for the very-long-chain, unsaturated acids which Smith *et al.* (1979) identified in the pseudocoel lipids of *Macracanthorhynchus hirudinaceus* should be explored. Mitochondria, in which fatty acid elongation is postulated to occur in helminths (Barrett & Körting, 1977; Barrett, 1981), are abundant in both the germinative and non-germinative components of maturing and patent ovarian balls (Parshad & Crompton, 1981; Chapter 7). The supporting syncytium surrounding developing oocytes of *Moniliformis moniliformis* contains smooth endoplasmic reticulum (Atkinson & Byram, 1976), with which fatty acid desaturation systems might be associated. Lipid droplets accumulate within the supporting syncytium and within oocytes during development, but diminish as the ovarian balls become senescent (Atkinson & Byram, 1976).

Table 6.10. *The partial fatty acid composition of the neutral lipids and phospholipids from some tissues of female* Macracanthorhynchus hirudinaceus (*after Smith, Dunagan & Miller, 1979*)[a]

	Neutral lipids			
Fatty acid	Body wall	Lemnisci	Pseudocoel	Body wall phospholipids[b]
$C_{14:0}$	4.8	4.0	1.8	1.5
$C_{14:1}$	10.1	11.3	3.5	3.9
$C_{16:0}$	17.6	15.0	5.3	4.7
$C_{16:1}$	—	—	—	13.0
$C_{18:0}$	13.4	13.9	8.2	14.3
$C_{18:1}$	13.8	10.5	9.4	15.9
$C_{18:2}$; $C_{20:0}$	10.4	8.5	15.0	13.3
$C_{18:3}$; $C_{20:1}$	6.3	6.6	11.7	—
$C_{20:2}$	—	—	—	12.0
$C_{20:3}$	5.9	5.6	11.1	6.8
$C_{22:1}$	3.3	5.0	7.0	5.1
unknown	—	—	—	3.9
$C_{24:1}$	2.9	4.9	4.3	1.9
unknown	3.4	5.0	6.0	—
$C_{22:6}$	2.7	6.2	2.9	0.9
unknown	2.6	0.2	9.2	0.2
unknown	trace[c]	0.2	trace[c]	4.6

[a] Methyl esters identified by gas chromatography as % of total lipids. Lipids of fewer than 14 carbons omitted for clarity; (—) = none detected
[b] Phospholipids of other fractions had similar compositions
[c] Less than 0.1% of total lipid

6.6 Energy metabolism

Carbohydrate is the main substrate for acanthocephalan energy metabolism and it is reasonably clear that acanthocephalans metabolize glucose and glycogen via a conventional Embden–Meyerhof pathway as far as phosphoenolpyruvate (PEP). The evidence suggests that the further metabolism of PEP involves anaerobic reactions common to other helminths, but the sequence of steps leading to the formation of some end-products, the ATP yield, and, indeed, the full complement of metabolic end-products excreted by acanthocephalans, are still uncertain.

6.6.1 *Glycolytic reactions*

The enzymes of glycolysis and related pathways in acanthocephalans are shown in Table 6.11 (§6.5.1, §6.5.2). Cornish, Wilkes & Mettrick (1981*b*) determined the levels of all Embden–Meyerhof pathway intermediates except 3-phosphoglycerol phosphate in extracts from freeze-clamped *Moniliformis moniliformis* adults (Table 6.12) and their results are consistent with the operation of a conventional glycolytic pathway.

Dunagan & Scheifinger (1966*b*) observed that crude soluble preparations from *Macracanthorhynchus hirudinaceus* phosphorylated glucose and fructose at comparable rates, that galactose phosphorylation was significantly slower and that mannose was not phosphorylated. Their data suggest that *M. hirudinaceus* has a fructose-phosphorylating enzyme which does not interact with glucose or mannose. Starling (1972) found a similar fructose-specific kinase whose activity was not inhibited by glucose or 2-deoxyglucose in soluble preparations from *Moniliformis moniliformis*. Superficially, the results suggest that the hexokinase from *M. hirudinaceus* does not react with mannose, in contrast to the hexokinases of mammalian tissues (Colowick, 1973; Purich, Fromm & Rudolph, 1973) and probably that from *M. moniliformis* (Starling, 1972) but it is not clear that Dunagan & Scheifinger's preparations retained any true hexokinase activity. Their assays had high endogenous rates of G6P formation (measured using G6P dehydrogenase) and addition of glucose effected no increase in G6P production. Dunagan & Scheifinger attributed the endogenous G6P production to phosphorylation of glucose present in their preparations, but they may have been observing glycogenolysis coupled to phosphoglucose isomerase activity.

G6P dehydrogenase and 6-phosphogluconate dehydrogenase have moderate to substantial activities in acanthocephalan tissues (Table 6.11) and the complement of pentose phosphate pathway enzymes has been identified, at lower activity levels, in *Macracanthorhynchus hirudinaceus* and in species of *Neoechinorhynchus* by Saxon & Dunagan (1975, 1976).

Table 6.11. *Glycolytic and related enzymes in acanthocephalans*

Enzyme	Activity (nmol product/min per mg protein)		
	Moniliformis moniliformis		*Macracanthorhynchus hirudinaceus*
	Adults	Cystacanths	
Hexokinase	present[a]	50[b]	37[c]; (0.16)[d]
Phosphoglucose isomerase	—[e]	480[b]	42[c]
Phosphoglucomutase	—	—	64[c]
Phosphofructokinase	—	480[b]	87[c]; (1.25)[d]
Fructose 1,6-bisphosphatase	1.3[f]	2.4[f]	—
Aldolase[g]	present[h]	1250[b]	—
Triosephosphate isomerase	present[h]	—	—
Enolase	—	320[b]	—
Pyruvate kinase	31[f]	0[b]; 8[f]	—
Phosphoenolpyruvate carboxykinase	90[f]; 300[i]; 374[j]	68[f]; 1100[j]	—
Lactate dehydrogenase[g]	557[f]	5900[b]; 590[f]	present[k]
Alcohol dehydrogenase (NADH)	0[f]	0[f]	0[c]
(NADPH)	45[f]	55[f]	—
Pyruvate decarboxylase	2.8[f]	—	—
Malate dehydrogenase	1008[f]	1333[f]; 4500[b]	1039/441[k,l]
Malic enzyme (NAD^+)	57–70[j]	2[j]	0[d]
($NADP^+$)	12–38[j]; 0[f]	1.2[f]; 2[j]	—
Glycerol phosphate dehydrogenase	—	470[b]	—
Glucose 6-phosphate dehydrogenase	53[f]	1370[b]; 42[f]	70[c]; 10.6[m]
6-Phosphogluconate dehydrogenase	92[f]	1040[b]; 88[f]	20.4[n,o]

[a] Starling (1972). Inhibited by mannose; a fructose phosphorylating kinase not inhibited by glucose is also present
[b] Horvath (1972). 12000-*g* supernate; preparation lacked mitochondrial integrity
[c] Dunagan & Scheifinger (1966*b*). 34000-*g* supernate
[d] Donahue *et al.* (1981). 10000-*g* supernate. Activity in μmol/min per g fresh tissue
[e] Not reported
[f] Körting & Fairbairn (1972). Homogenized in 0.25 M sucrose. Adults, 5000-*g* supernates (may have low mitochondrial enzymes). Cystacanths, 700-*g* supernates
[g] Reported present in species of *Neoechinorhynchus* by Dunagan (1964)
[h] Read (unpublished) cited by Read (1961)
[i] Cornish, Wilkes & Mettrick (1981*a*)
[j] Horvath & Fisher (1971). 12000-*g* supernate; preparation lacked mitochondrial integrity
[k] Dunagan & de Luque (1966). Starch gel electrophoresis; one lactate dehydrogenase isozyme; three malate dehydrogenase isozymes
[l] Dunagan & Scheifinger (1966*a*). Activity in soluble/particulate fractions from 34800-*g* centrifugation
[m] Stein (1971). $NADP^+$-dependent glucose 6-phosphate dehydrogenase purified and partially characterized
[n] Saxon & Dunagan (1975). Low activities for transaldolase, transketolase, phosphoriboseisomerase and xylulose 5-phosphate-3-epimerase also reported
[o] Saxon & Dunagan (1976). *M. hirudinaceus* and species of *Neoechinorhynchus* compared at different assay temperatures. Transaldolase, transketolase, phosphoriboseisomerase, and xylulose 5-phosphate-3-epimerase also reported

Stein (1971) partially purified G6P dehydrogenase from *M. hirudinaceus* and examined its saturation characteristics with respect to G6P and magnesium ions, but did not study its possible regulatory properties. Phosphogluconate metabolism serves primarily as a source of NADPH and pentose phosphates for anabolic metabolism and probably makes only minor contributions of substrates to pathways of energy metabolism.

The mass action ratios calculated from metabolite levels measured in freeze-clamped *Moniliformis moniliformis* and *Macracanthorhynchus hirudinaceus* indicate that the reaction catalyzed by phosphofructokinase is far away from equilibrium (Cornish *et al.*, 1981*b*; Donahue *et al.*, 1981). This non-equilibrium state identifies phosphofructokinase as a regulatory enzyme (Rolleston, 1972), and phosphofructokinase may be the rate-

Table 6.12. *Metabolite levels determined for freeze-clamped* Moniliformis moniliformis *and* Macracanthorhynchus hirudinaceus *adults*[a]

	Moniliformis moniliformis[b]		*Macracanthorhynchus hirudinaceus*[c]	
Metabolite	Males	Females	Freshly isolated	Starved 48 h
ATP	1822	1593	3476	2119
ADP	789	800	1708	2089
AMP	134	76	222	585
Inorganic phosphate	—[d]	—	5244	9241
Glucose	1467	553	—	—
Glucose 1-phosphate	—	—	40	0.1
Glucose 6-phosphate	64	64	436	186
Fructose 6-phosphate	10	19	7	28
Fructose 1,6-bisphosphate	50	50	30	5
Dihydroxyacetone phosphate	48	39	—	—
Glyceraldehyde 3-phosphate	13	11	—	—
3-Phosphoglycerate	855	557	—	—
2-Phosphoglycerate	107	55	—	—
Phosphoenolpyruvate	417	246	—	—
Pyruvate	36	24	—	—
Succinate	4436	3071	—	—
Lactate	2683	1155	—	—
Malate	1119	964	—	—
Adenylate charge	0.81[e]	0.81[e]	0.80	0.66

[a] In nmol/g wet tissue
[b] Data from Cornish, Wilkes & Mettrick (1981*b*) for 9-week-old worms collected from Wistar rats at 10.00 h
[c] From Donahue *et al.* (1981); all specimens appear to have been females
[d] Not reported
[e] Calculated from the adenylate levels reported by Cornish *et al.* (1981*b*)

controlling enzyme of the Embden–Meyerhof pathway in acanthocephalans (Bryant, 1975, 1978; Barrett, 1981). Cornish *et al.* (1981*b*) also identified hexokinase and pyruvate kinase in *M. moniliformis* as having non-equilibrium mass action ratios characteristic of regulatory enzymes. Thus, *M. moniliformis* appears unremarkable with respect to the Embden–Meyerhof pathway steps catalyzed by regulatory enzymes (Barrett, 1981).

The hexokinase mass action ratio determined by Cornish *et al.* (1981*b*) for whole female *Moniliformis moniliformis* (0.06) is five to six orders of magnitude smaller than the apparent equilibrium constant for the hexokinase-catalyzed reaction. Cornish *et al.* suggested that their results may have overestimated the true mass action ratio for the hexokinase reaction because of G6P produced as a result of glycogenolysis. Unless the glucose levels which Cornish *et al.* measured for whole *M. moniliformis* are unrepresentative of glucose levels in the *M. moniliformis* tegument (for example, because of compartmentation of glucose within non-tegumentary tissues), the extent to which *M. moniliformis* hexokinase appears to be down-regulated seems incompatible with the importance of hexose phosphorylation and trehalose synthesis to net carbohydrate absorption in *M. moniliformis* (Starling & Fisher, 1978, 1979; §6.3.1). It is unlikely that *M. moniliformis* hexokinase is as sensitive to inhibition by G6P as are the non-hepatic isoenzymes from mammalian tissues (Colowick, 1973; Purich *et al.* 1973) as helminth hexokinases generally have higher K_i s for G6P inhibition than their mammalian counterparts (Supowit & Harris, 1976; Barrett, 1981).

6.6.2 *Terminal pathways in metabolism*

Most of the metabolic products formed by acanthocephalans (Table 6.13) are similar to those of other helminths (von Brand, 1973, 1979; Barrett, 1981). Ethanol, a major metabolic product of *Moniliformis moniliformis* (Crompton & Ward, 1967*b*; Ward & Crompton, 1969; Körting & Fairbairn, 1972) has been identified from relatively few helminths, but may be more widespread than the literature suggests. Ethanol does not appear to be produced by *Polymorphus minutus* (Crompton & Ward, 1967*a*) although Crompton (1965) detected histochemical localization of alcohol dehydrogenase activity in or near mitochondria in *P. minutus* tissues. Whether ethanol is excreted by other acanthocephalans is not known. Beitinger & Hammen (1971) did not examine possible volatile products formed by *Echinorhynchus gadi*, and Dunagan (1963, 1964) examined only lactate excretion by species of *Neoechinorhynchus*.

Crompton & Ward (1967*a*) determined the radioactive products excreted

Table 6.13. *End-products of metabolism in Acanthocephala*

Species	Gas phase	Products	Volatiles	Reference
Moniliformis moniliformis[a]	Air	Acetate > formate > lactate	Not examined	Laurie (1959)
Moniliformis moniliformis[b]	100% N_2	Ethanol ≫ formate > lactate > acetate > propionate, *n*-butyrate, *n*-valerate	Accounted for	Crompton & Ward (1967*b*); Ward & Crompton (1969)
Moniliformis moniliformis	95% N_2–5% CO_2	Ethanol ≫ lactate > acetate, propionate > succinate	'No higher volatile acids'	Körting & Fairbairn (1972)
Neoechinorhynchus spp.[c]	Air?	Lactate (+other products, presumably)	Not examined	Dunagan (1963, 1964)
Polymorphus minutus[d]	100% N_2	Succinate ≧ lactate	<2% volatile	Crompton & Ward (1967*a*)
Echinorhynchus gadi[e]	100% N_2	Succinate ≦ lactate	Not examined	Beitinger & Hammen (1971)
	Air	Lactate ≫ succinate		

[a] Lactate excretion greater with exogenous substrates. Males excreted more lactate than females

[b] Ethanol also excreted aerobically, but other aerobic products not measured

[c] Lactate excretion declined as proportion of total (undetermined) products with prolonged (several days) incubation

[d] Based on radioactivity from exogenous [^{14}C]-glucose only

[e] Based on radioactivity from exogenous [^{14}C]-glucose only. Excretion products collected in postincubation media after 3 h incubation in [^{14}C]-glucose. Less than 8% of radioactivity in worm acids accounted for in excreted products

by *Polymorphus minutus* incubated for 4 h in [^{14}C]-glucose, but their results may not reflect the true stoichiometry of end-product formation because of concurrent endogenous glycogen metabolism. Succinate accounted for 48% of the radioactivity in the organic acid fraction, lactate 34–41% and the remaining radioactivity did not appear to be associated with any specific component, but 12–16% of the radioactivity was not accounted for. Excretion of succinate and lactate alone cannot account for metabolic redox balance (see below). Similar caveats apply to the results obtained for *Echinorhynchus gadi* by Beitinger & Hammen (1971) (Table 6.13). Furthermore, it is not possible to interpret Beitinger & Hammen's results quantitatively because they determined the products of metabolism by placing *E. gadi*, which had been incubated 3 h (25 °C) in [^{14}C]-glucose, into cold (2–5 °C) substrate-free saline for 12 h and then examined the non-volatile radioactive components of the saline chromatographically. Beitinger & Hammen wrote that the radioactivity in succinate and lactate in the media represented '7.9% of that found in acids retained by the worms', but they did not state whether the rates of excretion were similar under aerobic and anaerobic conditions, nor did they measure the amount of radioactivity in volatile products.

The pathways involved in metabolic end-product formation in Acanthocephala are probably similar to those in other helminths. Lactate production is believed to occur via the conventional glycolytic pathway. Formation of other products may diverge from the glycolytic pathway at the level of PEP. The pathways by which PEP is converted into succinate and other products in helminths are depicted in Fig. 6.2. The major components of this scheme appear common to a large number of helminths (Saz, 1971, 1981; von Brand, 1973, 1979; Fioravanti & Saz, 1980; Barrett, 1981; Komuniecki, Komuniecki & Saz, 1981*a*) as well as to some free-living facultatively anaerobic invertebrates (Holwerda & de Zwaan, 1980).

In many helminths, some or all of the PEP formed in the Embden–Meyerhof pathway is carboxylated to oxaloacetate (OAA) by phosphoenolpyruvate carboxykinase (PEPCK) (Fig. 6.2, [3]):

$$\mathrm{PEP + CO_2 + GDP(IDP) \xrightarrow{Mn^{2+}} OAA + GTP(ITP)}$$

OAA is reduced to malate in a cytoplasmic malate dehydrogenase reaction (Fig. 6.2, [4]). The OAA reduction is functionally equivalent to the conversion of pyruvate to lactate in conventional glycolysis. This regenerates the NAD^+ for the glyceraldehyde 3-phosphate dehydrogenase step of the Embden–Meyerhof pathway. More importantly, it produces malate, which is transferred into the mitochondrion.

Within the mitochondrion, part of the malate is dehydrated to fumarate

(Fig. 6.2, [5]) which is reduced to succinate (Fig. 6.2, [6]). The NADH-dependent reduction of fumarate is coupled to ATP formation via coupling site 1 of the classical respiratory chain. This anoxic 'oxidative phosphorylation' using fumarate as the terminal oxidant is firmly established in *Ascaris* (Saz, 1971, 1981; Köhler & Bachmann, 1980) as well as other helminths (Barrett, 1976*a*, 1981; Van Vugt, Kalaycioglu & Van den Bergh, 1976; Bryant, 1978; Fioravanti & Saz, 1980; Saz, 1981).

Succinate excretion. The succinate excreted by *Polymorphus minutus* and *Echinorhynchus gadi* is probably formed from malate via fumarate reduction (Crompton & Ward, 1967*a*; Beitinger & Hammen, 1971). Crompton (1965) obtained histochemical evidence for the presence of succinate dehydrogenase (fumarate reductase) in *P. minutus* mitochondria; malate and lactate dehydrogenases appeared to be both mitochondrial and

Fig. 6.2. Pathways leading to metabolic end-product formation in helminths. Compiled from Saz (1971, 1981) and other sources for reactions in *Ascaris suum* and in *Fasciola hepatica.* 1, Pyruvate kinase; 2, Lactate dehydrogenase; 3, Phosphoenolpyruvate carboxykinase; 4, Malate dehydrogenase; 5, Fumarate hydratase; 6, Fumarate reductase; 7, Malic enzyme; 8, Pyruvate dehydrogenase; 9, CoA transferase(s) (Barrett, Coles & Simpkin, 1978; Köhler, Bryant & Behm, 1978; Van Vugt, Van Der Meer & Van den Bergh, 1979); 10, Succinyl-CoA synthetase (Barrett *et al.*, 1978); 11, Methylmalonyl-CoA mutase, methylmalonyl-CoA epimerase, and propionyl-CoA carboxylase (Köhler *et al.*, 1978; Saz & Pietrzak, 1980; Pietrzak & Saz, 1981); 12, Branched volatile fatty acid synthetic pathway in *Ascaris* (Komuniecki, Komuniecki & Saz, 1981*a*, *b*).

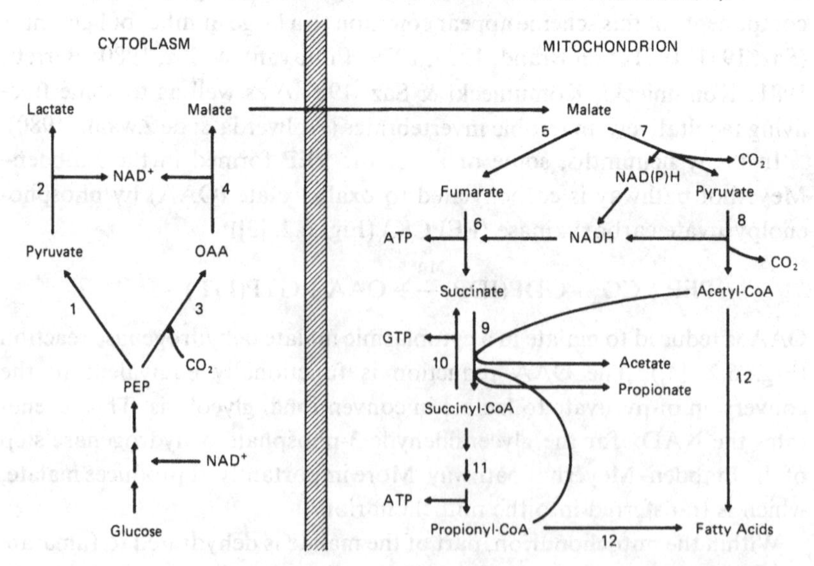

cytosolic. Beitinger & Hammen found that *E. gadi* incorporated [^{14}C]-bicarbonate into non-volatile metabolites, but it is not known whether carbon dioxide fixation occurred via PEPCK or by operation of a malic enzyme (malate dehydrogenase, decarboxylating) reaction in the direction of carboxylation. Either reaction results in the formation of malate for ATP-yielding succinate production in the mitochondrion.

Crompton & Ward (1967*a*) and Beitinger & Hammen (1971) used 100% nitrogen to obtain anaerobic conditions; this would simulate poorly the environment of the vertebrate intestine (§6.6.4). The steady state partial pressure of carbon dioxide (pCO_2) in mammalian and avian intestines averages 30–75 mm Hg (equivalent to a gas phase of 4–10% carbon dioxide), and the pCO_2 can increase significantly as a consequence of postprandial pancreatic secretions (Crompton, 1970; Mettrick & Podesta, 1974; Mettrick, 1980). Although Crompton & Ward and Beitinger & Hammen included bicarbonate in their media, it is likely that significant carbon dioxide was lost to the gas phase during the lengthy incubations used. Since the bicarbonate ion concentration of the incubation medium has a striking effect on the proportions of lactate and succinate excreted by *Hymenolepis diminuta* (Ovington & Bryant, 1981), the end-products detected for both *P. minutus* and *E. gadi* might have been affected by carbon dioxide insufficiency.

The succinate excreted by *Echinorhynchus gadi* and *Polymorphus minutus* requires a source of NADH for fumarate reduction. In helminths which excrete volatile fatty acids + succinate (*Ascaris suum*) or acetate + propionate (*Fasciola hepatica*), reducing equivalents for succinate formation are produced by the oxidative decarboxylations of malate to pyruvate (catalyzed by malic enzyme; Fig. 6.2, [7]) and of pyruvate to acetyl-CoA (catalyzed by pyruvate dehydrogenase; Fig. 6.2, [8]). The ideal stoichiometry for succinate (or propionate) formation in mitochondria in which reducing equivalents derive from dismutation of malate to acetate is 2 succinate (or propionate): 1 acetate. This ratio is realized in *F. hepatica* (Bryant, 1978); in helminths producing volatile fatty acids (Fig. 6.2, [12]), the stoichiometry is less predictable, and excess succinate appears as one of the excretory products. Beitinger & Hammen (1971) did not examine the volatile products formed by *E. gadi* and we do not know whether the reducing equivalents for fumarate reduction may derive from acetate formation in *E. gadi*. It is unlikely that some of the radioactivity which Crompton & Ward (1967*a*) did not recover in their analysis of the organic acids excreted by *P. minutus* was acetate, since the presence of this acid was anticipated in their methodology.

Barrett (1981) suggested that the additional reducing power required for succinate formation in *Polymorphus minutus*, *Echinorhynchus gadi* and other lactate and succinate producers might have come from the oxidation of fatty acids, or from operation of a conventional tricarboxylic acid cycle at low rates, as is the case in anaerobic molluscs (Holwerda & de Zwaan, 1980). However, we do not know at present whether tricarboxylic acid cycle activity (or lipid oxidation) occurs at sufficient rates in *E. gadi* or *P. minutus* to provide the reducing equivalents necessary for fumarate reduction (§6.6.5).

Terminal metabolism in Moniliformis moniliformis. Malate formation via carbon dioxide fixation appears important to metabolism in *Moniliformis moniliformis*. Radioactivity was incorporated into malate, succinate, fumarate, aspartate and alanine when intact *M. moniliformis*, pieces of body wall, or body wall homogenates were incubated aerobically with [^{14}C]-glucose, $NaH^{14}CO_3$, 1,4-[^{14}C]-succinate or [^{14}C]-aspartate (Graff, 1964, 1965; Bryant & Nicholas, 1965; Janssens & Bryant, 1969). Bryant & Nicholas (1966) also demonstrated metabolism of 1,4-[^{14}C]-succinate to

Table 6.14. *Incorporation of* $NaH^{14}CO_3$ *into soluble intermediates by patent female* Moniliformis moniliformis (*after Graff, 1965*)[a]

	CPM/g fresh tissue[b]		
Intermediate	0.5 min	1.5 min	5 min
Malate	17020 (23.6)	46070 (26.2)	88750 (16.3)
Succinate	3190 (4.4)	19680 (11.2)	49320 (9.0)
Fumarate	960 (1.3)	2450 (1.4)	7550 (1.4)
Lactate	420 (0.6)	1290 (0.7)	7370 (1.4)
Unidentified acid or neutral compound	560 (0.8)	970 (0.6)	5020 (0.9)
Pyruvate	—[c]	—[c]	+[c]
α-Ketoglutarate	—[c]	—[c]	0[c]
Oxaloacetate	—[c]	—[c]	+[c]
Aspartate	27320 (37.9)	45450 (25.9)	90310 (16.5)
Alanine	2050 (2.8)	12000 (6.8)	40570 (7.4)
Serine	—[c]	—[c]	1560 (0.3)

[a] Approximately 1 g tissue incubated aerobically for the indicated times in 2 mM $NaH^{14}CO_3$ (1.89 mCi/mmol). Cationic fraction separated from anionic and neutral compounds by ion exchange chromatography. Samples analyzed by paper chromatography

[b] From eluates of spots on paper chromatograms, counted with 9% counting efficiency. Numbers in parentheses are % of total non-volatile radioactivity incorporated by worms

[c] — = not reported; + = radioactivity present in 2,4-dinitrophenylhydrazine derivatives; 0 = dinitrophenylhydrazine derivative not radioactive

malate, fumarate, aspartate and 'CO_2, volatile and insoluble substances' by particulate fractions from *M. moniliformis*. The carbon dioxide fixation observed by Graff (1964) and Bryant & Nicholas (1965) was probably catalyzed by PEPCK, since this enzyme is present in *M. moniliformis* at high levels (Cornish, Wilkes & Mettrick, 1981*a*). Graff (1965) measured [^{14}C]-sodium bicarbonate incorporation by *M. moniliformis* into non-volatile metabolites during brief (0.5–5 min) aerobic incubations (Table 6.14). His results suggest significant equilibration of radioactivity between malate and succinate. More importantly, the substantial radioactivity observed in lactate and alanine indicates that pyruvate was formed from malate.

Despite the evidence that carbon dioxide fixation into malate and succinate occurs readily in *Moniliformis moniliformis*, succinate and other common products of malate dismutation are only minor end-products of metabolism (Ward & Crompton, 1969; Körting & Fairbairn, 1972; Table 6.15). Barrett (1981) has suggested that carbon dioxide fixation in *M. moniliformis* may be related to the fact that *M. moniliformis* alcohol dehydrogenase appears to be NADPH-dependent (Körting & Fairbairn, 1972).

Barrett proposed that carbon dioxide fixation into oxaloacetate and reduction of oxaloacetate to malate provides for regeneration of NAD^+, as is the case in other helminths (Fig. 6.2, [3], [4]). He postulated that the subsequent decarboxylation of malate by an $NADP^+$-dependent malic enzyme (Fig. 6.2, [7]) provided both the NADPH and the pyruvate necessary for ethanol production. Barrett's view appears inconsistent with the fact that Körting & Fairbairn (1972) were unable to demonstrate any malic enzyme activity in preparations from adult *M. moniliformis* (Table 6.11) and that the moderate malic enzyme activity reported by Horvath & Fisher (1971) has been questioned (Cornish *et al.*, 1981*a*). However, Körting & Fairbairn's results may be technical artifacts because the lack of reductants in the buffer appears to have been partly responsible for the failure to demonstrate in *M. moniliformis* fumarate hydratase activity (Wilkes *et al.*, 1982*a*). Furthermore, if malic enzyme is mitochondrial in *M. moniliformis*, a significant amount may have been lost from Körting & Fairbairn's preparations from adults during centrifugation (§6.5.3).

The specific pathway of ethanol formation has not been explored in any ethanol-excreting helminth. The acetaldehyde which serves as a substrate for alcohol dehydrogenase is probably formed from pyruvate via pyruvate decarboxylase, as occurs in yeast. Körting & Fairbairn (1972) found only low levels of pyruvate decarboxylase activity in *Moniliformis moniliformis* homogenates; this may have been caused by poor enzyme preservation, or it may mean that the enzyme is mitochondrial. The possibility that acetaldehyde formation occurs by some other route, for example, reduction of acetyl-CoA by an aldehyde dehydrogenase (Metzler, 1977), should not be overlooked.

Ward & Crompton (1969) found substantial amounts of formate among the metabolites excreted by *Moniliformis moniliformis* (Table 6.15), and Laurie (1959) found comparable or higher amounts of formate produced by *M. moniliformis* incubated aerobically. The formate was a product of helminth and not bacterial metabolism because the non-radioactive formate detected by Ward & Crompton could not have derived from the [^{14}C]-glucose present in the incubation medium.

In Gram-negative bacteria, anaerobic formate production is catalyzed by pyruvate formate lyase (Metzler, 1977):

$$\text{Pyruvate} + \text{CoA} \rightarrow \text{Acetyl-CoA} + \text{Formate}$$

We do not know whether a comparable reaction occurs in *Moniliformis moniliformis* and other formate-excreting helminths, but the possibility deserves examination.

The acetyl-CoA formed in the bacterial pyruvate formate lyase reaction is frequently converted to acetyl phosphate via a reversible phosphotransacetylase reaction, and acetate is formed from acetyl phosphate by acetate kinase, for which ADP is the phosphate acceptor (see Metzler, 1977; Barrett, 1981). Phosphotransacetylase and acetate kinase activities are not present in *Fasciola hepatica* (Barrett, Coles & Simpkin, 1978). Instead, ATP production using the energy available in acetyl-CoA may involve CoA transfer from acetyl-CoA to succinate and cycling of the succinyl-CoA so formed back to succinate via succinate thiokinase (Fig. 6.2, [9], [10]; Barrett, 1981; Ward, 1982). Since acetate is produced by *Moniliformis moniliformis*, it is reasonable to presume that it is derived from acetyl-CoA, regardless of the pathway by which acetyl-CoA is formed. It is also quite possible that conversion of acetyl-CoA to acetate results in ATP formation. However, we have no information about the activities of these or any other enzymes involved in substrate level phosphorylations in acanthocephalans.

6.6.3 *Metabolic compartmentation*

Ward & Crompton (1969) used exogenous [^{14}C]-glucose as a tracer to identify the products excreted by *Moniliformis moniliformis* and determined the chemical levels of excreted products and their specific radioactivities (Table 6.15). Although significant amounts of formate and detectable amounts of propionate and *n*-valerate were excreted during the

2.5-h incubations, no radioactivity was observed in these compounds. Some radioactivity was present in excreted acetate and *n*-butyrate, but the specific activities of these products were very low.

We may assume that the non-radioactive acids excreted by *Moniliformis moniliformis* arose from the metabolism of glycogen or pools of malate,

Table 6.15. *End-products of* Moniliformis moniliformis *metabolism under 100% nitrogen*[a]

Product	μmol/ 100 mg wet mass per 2.5 h	Total (μmol)	Relative specific activity[b]	Percentage of total radioactivity in medium
Experiment N1[c]				
Ethanol	14.35	74	0.30	55.1
Lactate	2.75	14	0.24	12.2
Succinate	0.007	0.034	0.18	0.03
Experiment N4[d]				
Ethanol	6.80	62	0.271	60.0
Formate	7.09	63.5	—[e]	—[e]
Experiment N6[f]				
Ethanol	7.6	69.0	0.25	55.1
Lactate	1.15	10.4	0.24	12.1
Succinate	0.033	0.28	0.05	0.09
Experiment N8[g]				
Ethanol	7.05	69.6	0.30	69.6
Lactate	1.08	11.7	0.22	12.6
Succinate	0.04	0.4	0.38	1.0
Formate	3.08	29.7	0	0
Acetate	0.58	5.6	0.07	1.25
Propionate	0.20	1.9	0	0
n-butyrate	0.12	1.1	0.014	0.1
n-valerate	0.12	1.1	0	0

[a] Calculated from Ward & Crompton (1969). Incubations apparently included both male and female worms. Medium (Tyrode's solution) contained [^{14}C]-glucose, most of which was depleted during the incubations

[b] Specific activity *per carbon atom* in product divided by specific activity *per carbon atom* of [^{14}C]-glucose in incubation medium

[c] 515-mg 53-day-old worms removed 23.9/27.6 μmol glucose from the medium and excreted products equivalent to 44 μmol glucose

[d] 915-mg 72-day-old worms removed 21.6/22.9 μmol glucose from the medium; total products not determined

[e] Not determined

[f] 515-mg 63-day-old worms incubated with 21.4 μmol [^{14}C]-glucose excreted products equivalent to at least 35 μmol glucose. Respiratory gas (CO_2) recovered (DPM):ethanol (DPM) = 1.24:1

[g] 985-mg 108-day-old worms removed 20.6/21.7 μmol glucose from the medium and excreted products equivalent to 46.4 μmol glucose. Respiratory gas (CO_2) recovered (DPM):ethanol (DPM) = 0.83:1

succinate or amino acids (Cornish *et al.*, 1981*b*; Crompton & Ward, 1984). Perhaps the experimental conditions affected the incorporation of radioactive Embden–Meyerhof pathway intermediates into pathways leading to volatile acid production, but the pathways of volatile acid formation might have been restricted to tissue compartments to which the [^{14}C]-glucose absorbed during the incubations did not have immediate access. Such compartmentation has been demonstrated in *Ascaris suum* tissues and is probably a feature common to many species of helminths (Barrett, 1981).

Effects of exogenous sugars and helminth sex on end-products. Laurie (1959) found that fasted *Moniliformis moniliformis* incubated aerobically with exogenous sugars excreted a higher proportion of their acid end-products as lactate than did worms incubated in the absence of exogenous carbohydrate. Similar increases in lactate excretion in the presence of exogenous glucose have been observed for cestodes and trematodes (Barrett, 1981; Ovington & Bryant, 1981). Dunagan (1963, 1964) found that lactate accounted for increasingly less of the products of endogenous glycogen utilization by species of *Neoechinorhynchus* maintained *in vitro* as glycogen stores diminished during the prolonged incubations. Laurie's results indicate that the metabolic products formed by *M. moniliformis* and other acanthocephalans incubated either in the absence of exogenous glucose (Körting & Fairbairn, 1972) or under conditions in which substrate was depleted during the incubations (Crompton & Ward, 1967*a*, *b*; Ward & Crompton, 1969; Beitinger & Hammen, 1971) may not represent the metabolic states that exist during nutrient uptake. Significantly, Ward & Crompton's (1969) data showed that ethanol accounted for a smaller percentage of the radioactive end-products excreted by *M. moniliformis* during shorter incubations, when the length of time that metabolism proceeded under conditions of exogenous glucose depletion seems to have been less. Moreover, Dunagan's observations suggest that the relative proportions of end-products excreted by starving acanthocephalans may shift continuously as a function of the metabolic status of the tissues.

Barrett (1981) suggested that the increased lactate excretion observed when carbon dioxide-fixing helminths are incubated in the presence of exogenous glucose represents 'overflow metabolism' that occurs when carbon dioxide fixation by PEPCK is operating at maximum capacity (Fig. 6.2, [3]). This hypothesis agrees with the fact that the concentration of PEP determined for *Moniliformis moniliformis* tissues (Cornish *et al.*, 1981*b*) is three to six times the apparent K_m(PEP) reported for the PEPCK purified from *M. moniliformis* (Cornish *et al.*, 1981*a*). Alternatively, such overflow metabolism might be a consequence of changes in concentrations of tissue metabolites and the effects of these changes on the activities of regulatory enzymes, rather than simple saturation of a key enzyme in one metabolic pathway.

Laurie (1959) found that male *Moniliformis moniliformis* from fasted hosts excreted significantly more lactate than did female worms when incubated aerobically in exogenous fructose, mannose or maltose, but not glucose. Results obtained by Körting & Fairbairn (1972) suggest that lactate comprised a higher proportion of the metabolic end-products excreted by male *M. moniliformis* incubated anaerobically without exogenous substrate. The significance of these sex differences is not clear, but they may indicate a higher proportion of lactate-producing tissues in male worms. Interestingly, Crompton (1965) found histochemical evidence for very high lactate dehydrogenase activity in the testis and cement gland of *Polymorphus minutus*.

The PEPCK/PK branchpoint. According to Saz (1971), the major factor which determined the extent to which lactate fermentation predominated over the formation of malate dismutation products in most species of helminths was competition between pyruvate kinase (PK) and PEPCK for PEP (Fig. 6.2, [1], [3]). The PEPCK/PK branchpoint is sensitive to a variety of metabolites normally present in tissues because of the regulatory properties each of the enzymes (Bryant, 1975, 1978; Barrett, 1981). Although there have been no studies of the regulatory properties of PK from any acanthocephalan, the mass action ratios observed for the PK substrates and products in *Moniliformis moniliformis* (Table 6.12) indicate that PK was strongly down-regulated in the worms examined by Cornish *et al.* (1981*b*).

The PEPCK isolated from *Moniliformis moniliformis* by Cornish *et al.* (1981*a*) was mildly sensitive to inhibition by succinate (K_i = 7mM), as was the PEPCK isolated by the same investigators from *Ascaris suum* (Wilkes, Cornish & Mettrick, 1982*b*). Although the K_i(succinate) for the acanthocephalan PEPCK is of the same magnitude as the apparent succinate concentration in whole worms (Cornish *et al.*, 1981*b*), Cornish *et al.* (1981*a*) concluded that succinate was probably not a significant inhibitor of PEPCK because of the high affinity of the enzyme for PEP. The acanthocephalan PEPCK has much lower apparent K_m s for GDP and IDP as nucleotide substrates (1.8μM and 16.5μM, respectively) than do the enzymes purified from *Hymenolepis diminuta* by Wilkes, Cornish & Mettrick (1981) and from *A. suum* by Wilkes *et al.* (1982*b*), or the partially purified enzymes (Behm & Bryant, 1975, 1982) from *Moniezia expansa* and

Fasciola hepatica. The enzyme from *Moniliformis moniliformis* is also much more sensitive to inhibition by GTP ($K_i = 3.5\mu M$). None of the purified PEPCK enzymes is directly affected by ATP or other adenine nucleotides, but one might anticipate an indirect effect of adenylate charge as a result of nucleoside diphosphate kinase activity. Substrate concentration and tissue pH are probably the major factors regulating the activities of PEPCK enzymes from helminth tissues (Behm & Bryant, 1982).

Cornish *et al.* (1981*a*) identified α-ketoglutarate ($K_i = 0.15mM$) as a potential negative modulator of PEPCK from *Moniliformis moniliformis*. As they noted, the incorporation of [^{14}C]-sodium bicarbonate into aspartate during a 30-s incubation (Graff, 1965; Table 6.14) undoubtedly occurred by transamination of the oxaloacetic acid produced in the PEPCK reaction, and the transamination reaction probably used glutamate as the amino donor. Cornish *et al.* suggested that the aspartate formation observed by Graff (1965) might be a synthetic reaction, and they implicated the sensitivity of PEPCK to α-ketoglutarate as a feedback mechanism for regulation of amino acid synthesis. However, it is unlikely that amino acid synthesis is a major activity in acanthocephalans. Transamination reactions involving glutamate and aspartate are no doubt involved in intertissue movement of carbon skeletons, but the role of such shuttles and the significance of α-ketoglutarate inhibition of PEPCK in energy metabolism, if any, remains to be demonstrated.

6.6.4 *Oxygen and metabolism*

Crompton & Ward's (1967*b*) preliminary report of anaerobic ethanol production by *Moniliformis moniliformis* indicated that worms incubated aerobically excreted 'a similar quantity of ethanol'. However, the data presented by Ward & Crompton (1969) (Table 6.16) suggest that less ethanol is excreted by *M. moniliformis* under aerobic conditions. It must be emphasized, however, that Ward & Crompton (1969) published the results of only one aerobic incubation (see Table 1 of that paper) and little more than tentative conclusions can be drawn here until more work has been done. Also, in the light of more recent work, interpretation is complicated because Ward & Crompton's incubations included both male and female worms (§6.6.3) and at first their samples were stored frozen before analysis. Later they discovered that some ethanol (and lactate) was lost from frozen samples during storage.

Ward & Crompton (1969) found that the amounts of acetate and formate excreted in the experiment summarized in Table 6.16 were <0.25 and <1.0 μmol, respectively (the lower limits of detection in their analyses). These results differ from those observed in later, more carefully controlled anaerobic incubations reported by Ward & Crompton (Table 6.15) and from Laurie's (1959) observation that *Moniliformis moniliformis* excreted significant amounts of acetate and formate aerobically. It is not clear whether the apparently low amounts of formate and acetate excretion observed by Ward & Crompton in their single aerobic incubation were real effects due to worm age, shorter incubation time, or some other parameter, or whether they were affected by sample storage.

Beitinger & Hammen (1971) reported that *Echinorhynchus gadi* previously incubated in [^{14}C]-glucose excreted nearly equal amounts of radioactive lactate and succinate under anaerobic (100% nitrogen) conditions, but that lactate greatly exceeded succinate among the radioactive products excreted aerobically. Although Beitinger & Hammen's results are not quantitative (§6.6.2), the differences between end-products of anaerobic and aerobic metabolism agree with observations on other helminths. In general, aerobic conditions result in an increase in products formed from pyruvate relative to those formed from succinate and its derivatives (Bryant, 1975, 1978; Behm & Bryant, 1975; Bryant & Behm, 1976; Van Vugt *et al.*, 1976; Körting & Barrett, 1977; Köhler & Bachmann, 1980; Barrett, 1981; Rahman & Mettrick, 1982).

Manometric studies on adult *Moniliformis moniliformis* by Laurie (1959) and Bryant & Nicholas (1966) and on *Echinorhynchus gadi* by Beitinger & Hammen (1971) have detected the consumption of 0.2–0.35 μl oxygen/mg wet mass per h *in vitro*; this rate of oxygen consumption is similar to that observed for adult *Ascaris suum* and other adult intestinal helminths

Table 6.16. *Ethanol production by 42-day-old* Moniliformis moniliformis*: comparison of a single aerobic incubation with incubations under anaerobic (100% nitrogen) conditions*[a]

	Anaerobic incubations	Aerobic incubation
Ethanol production (μmol/100 mg per 1.17 h)	4.83	3.08
Total ethanol produced (μmol)	14.4	9.3
Relative specific activity[b,c]	0.49	0.55

[a] Calculated from Ward & Crompton (1969). Initial glucose concentration = 8.6 mM (0.94 mCi/mmol; 12.6 μmol total); final glucose concentration = 1.3 mM. Samples (sex unspecified) = 300 mg fresh mass. Some ethanol may have been lost during storage before analysis in first incubations, but this was avoided later
[b] Specific activity *per carbon atom* in ethanol divided by specific activity *per carbon atom* of [^{14}C]-glucose added to medium
[c] Radioactivity in ethanol was calculated based on the amount of radioactivity volatilized from medium heated to 80 °C for 10 min

incubated aerobically (von Brand, 1973; Barrett, 1981). Oxygen uptake by *M. moniliformis* acanthors resembles that of adults in its responses to respiratory inhibitors and artificial electron acceptors (Bryant & Nicholas, 1966). Exogenous sugars or succinate did not affect the rate of oxygen utilization by *M. moniliformis* or *E. gadi*, but this observation is characteristic of many other helminths (see von Brand, 1973) and does not preclude a possible role for oxygen in energy metabolism. Dunagan (1962, 1964) could not demonstrate oxygen uptake by species of *Neoechinorhynchus*, but he did not describe the sensitivity of his measurements or the conditions of incubation. Since oxygen tension in the vertebrate intestine may be as high as 50 mm Hg (Mettrick & Podesta, 1974; Mettrick, 1980), oxygen uptake by acanthocephalans may be of physiological consequence *in vivo*. The cautious reports that oxygen may reduce acanthocephalan survival *in vitro* (Van Cleave & Ross, 1944; Nicholas & Grigg, 1965) are of little significance because of the relatively poor survival of the animals under any conditions.

Bryant & Nicholas (1966) found that cyanide ions (10mM) inhibited part of the oxygen uptake by intact *Moniliformis moniliformis* adults and acanthors, and that antimycin A caused a moderate reduction in oxygen uptake. Exogenous cytochrome *c* was reduced by crude *M. moniliformis* homogenates and by particulate fractions supplemented with succinate or NADH, but conventional cytochrome *c* oxidase could not be detected. In the presence of succinate, acanthor homogenates or particulate fractions from adults incubated with redox dyes (phenazine methosulfate or methylene blue) exhibited substantial increases in oxygen uptake which were unaffected by cyanide or antimycin A. Bryant & Nicholas (1966) concluded that *M. moniliformis* mitochondria possess a branched cytochrome chain, with one (minor) branch similar to the mammalian mitochondrial respiratory chain and the other (major) branch leading to a cyanide-insensitive terminal oxidase. Such branched cytochrome chains, with *b*-type cytochrome *o* serving as the terminal oxidase for the cyanide-insensitive branch, are characteristic of several bacterial groups, and branched (or parallel, possibly spatially separated) cytochrome chains using cytochrome $a+a_3$ and cytochrome *o* as separate terminal oxidases have been reported for several helminths, the most thoroughly studied being those of *Ascaris suum*, *Moniezia expansa* and *Fasciola hepatica* (Bryant, 1970, 1978; von Brand, 1973, 1979; Bryant & Behm, 1976; Barrett, 1981).

The functional significance of aerobic respiration in helminths is controversial (von Brand, 1973, 1979; Bryant, 1978; Barrett, 1981; Ward, 1982). The majority of the oxygen utilization involves the 'alternate terminal oxidase' (i.e., putative cytochrome *o*). This component of the succinate oxidase system has a characteristically low affinity for oxygen, so that oxidation rates vary directly with oxygen tension. Furthermore, the amount of oxygen reaching deep tissues, especially in large helminths, is probably severely limited (Bryant, 1970, 1978; von Brand, 1973, 1979; Barrett, 1981).

Malate oxidation by *Ascaris suum* muscle mitochondria has a P:2e$^-$ ratio of about 1.0 under both aerobic and anaerobic conditions. Approximately half of the malate is converted to pyruvate + acetate, both aerobically and anaerobically, but significantly more succinate is produced anaerobically (Köhler & Bachmann, 1980). Köhler & Bachmann found no evidence for significant site 2 or site 3 phosphorylations via the conventional cytochrome chain in ascarid mitochondria. These characteristics suggest anaerobic ATP production via a site 1 phosphorylation, with fumarate reduction coupled to NADH oxidation, supplemented by opportunistic aerobic phosphorylation at coupling site 1, with NADH oxidation coupled to the 'alternate terminal oxidase' rather than to fumarate reduction (Bryant, 1978). Other helminths may behave similarly as opportunistic microaerobes (Barrett, 1976*a*, 1981; Bryant & Behm, 1976; Van Vugt *et al.*, 1976; Körting & Barrett, 1977; Bryant, 1978; Fioravanti & Saz, 1980; Köhler & Bachmann, 1980; Ward, 1982) and the contribution of oxygen-linked phosphorylation to total aerobic energy metabolism may be considerable (Behm & Bryant, 1975; Bryant & Behm, 1976; Körting & Barrett, 1977).

The phosphorylation capacity of acanthocephalan mitochondria has not been examined, but circumstantial evidence suggests that some species may have mechanisms for opportunistic oxidative phosphorylation. Ward (1952) found that *Macracanthorhynchus hirudinaceus* incubated *in vitro* consumed slightly more glycogen under anaerobic conditions than under aerobic conditions, the ratio of anaerobic to aerobic glycogen consumption being about 1.2. This reduced aerobic glycogen depletion would be expected if oxygen can supplant organic oxidants to a limited extent. The amounts of acetate and formate excreted aerobically by *Moniliformis moniliformis* also suggest an oxygen-coupled oxidation of reducing equivalents. Interestingly, Dunagan (1964) reported that 'survival' of species of *Neoechinorhynchus in vitro* was identical under aerobic and anaerobic conditions, but that mobility 'was reduced in the absence of free oxygen'.

6.6.5 *The tricarboxylic acid cycle and related reactions*

The conventional tricarboxylic acid cycle is not generally considered to contribute significantly to energy metabolism in most species of helminth because of the limited activity of one or more enzymes in the tricarboxylic acid limb of the cycle (von Brand, 1973, 1979; Barrett, 1981). However, substrate oxidation to succinate via the conventional tricarboxylic acid cycle may be a valuable source of reducing power for ATP-generating fumarate reduction in small intestinal nematodes (Barrett, 1976*a*) as in anaerobic molluscs (Holwerda & de Zwaan, 1980). Recent evidence suggests that aerobic tricarboxylic acid cycle activity is significant to energy metabolism in several large helminths (Bryant & Behm, 1976; Körting & Barrett, 1977; Ward, 1982).

Beitinger & Hammen (1971) found that *Echinorhynchus gadi* incubated in [^{14}C]-glucose or [^{14}C]-sodium bicarbonate formed significant quantities of [^{14}C]-citrate and small amounts of radioactive α-ketoglutarate. Thus they concluded that *E. gadi* had a complete tricarboxylic acid cycle, but their data do not permit an analysis of the extent of its activity. Körting & Fairbairn (1972) detected low levels of citrate synthase, aconitate hydratase, and isocitrate dehydrogenase in *Moniliformis moniliformis* cystacanths, but failed to find citrate synthase or aconitate hydratase in supernates from sucrose homogenates of adults (Table 6.9). They found small amounts of isocitrate lyase in preparations from adults and malate synthase in cystacanth homogenates, but could not demonstrate malate synthase in adult preparations. These observations suggest that *M. moniliformis* has neither a complete tricarboxylic acid cycle nor a glyoxylate cycle. However, Körting & Fairbairn's failure to identify several mitochondrial enzymes in adults may be an artifact of the use of supernatant fractions or of poor preservation of enzyme activity (Wilkes *et al.*, 1982*a*). Interestingly, Bryant & Nicholas (1965) found that *M. moniliformis* body walls and homogenates incorporated small amounts of radioactivity from 2-[^{14}C]-acetate into malate, and that body wall pieces, but not homogenates, converted some 2-[^{14}C]-acetate into lactate and alanine. It is not known whether the incorporation of acetate into malate and pyruvate derivatives involved reactions of a complete tricarboxylic acid cycle, a glyoxylate cycle or some other pathway.

Dunagan & Scheifinger (1966*a*) concluded that aconitate hydratase and isocitrate dehydrogenase were not present in *Macracanthorhynchus hirudinaceus*, but their data indicate very low activities for these enzymes in particulate fractions. Rasero, Monteoliva & Mayor (1968) implied that *M. hirudinaceus* possessed a γ-aminobutyrate pathway by which succinate could be formed from α-ketoglutarate in the absence of α-ketoglutarate dehydrogenase activity, and α-ketoglutarate transaminase was determined in *M. hirudinaceus*.

Crompton (1965) examined oxidoreductase activities in *Polymorphus minutus* using histochemical techniques and found evidence for lactate, malate and isocitrate dehydrogenase activities in both mitochondrial and cytosolic compartments. Additionally, he indicated that mitochondria contained succinate dehydrogenase, glutamate dehydrogenase, glycerol phosphate dehydrogenase and, interestingly, alcohol dehydrogenase. However, no quantitative information is available on the activities of these or any other mitochondrial enzymes in *P. minutus*.

At present, we have insufficient information to determine whether any acanthocephalan possesses a full complement of tricarboxylic acid cycle enzymes or tricarboxylic acid cycle bypasses which yield succinate and malate by oxidative routes. The possibility should not be dismissed, particularly in view of the apparent lack of redox balance among the acidic end-products of metabolism in *Echinorhynchus gadi* and *Polymorphus minutus* (§6.6.2).

6.6.6 *Metabolism in larvae*

Little information exists about the metabolism of acanthocephalan larvae during development in the invertebrate host. Butterworth (1969*b*, cited in Crompton, 1970) found that carbon dioxide was the only detectable radioactive product released by early, middle and late acanthellae and by non-activated cystacanths of *Polymorphus minutus* incubated *in vitro* in [^{14}C]-glucose. Cystacanths from the gammarid hemocoel utilized oxygen at a rate of 0.1 μl (4.4 nmol)/mg dry mass per h and exhibited an apparent respiratory quotient of 1.16. Based upon these observations, Crompton (1970) suggested that *P. minutus* developing within the gammarid hemocoel metabolize carbohydrate at least partly via aerobic pathways. Crompton (1970) also suggested that *P. minutus* cystacanths might metabolize lipids while within the hemocoel of the intermediate host although Butterworth (1967) had found no evidence for lipid metabolism in cystacanths.

Horvath (1971) showed that cystacanths of *Moniliformis moniliformis* activated by treatment with 0.25% (mass/vol) sodium taurocholate for 30 min (Graff & Kitzman, 1965) consumed 30% of their glycogen reserves during a 2-h incubation in the absence of glucose, but that significant glycogen synthesis occurred in the presence of exogenous glucose (§6.5.1). Freshly activated cystacanths incubated for 30 min in 1mM [^{14}C]-glucose incorporated significantly less radioactivity into glycogen in the presence of 5% carbon dioxide (with the balance of the gas phase being air or nitrogen) than when air or 100% nitrogen formed the gas phase. Net incorporation of radioactivity into glycogen was also greater under aerobic (air) than anaerobic (100% nitrogen) conditions. In the absence of exogenous glucose, glycogen depletion in artificially activated cystacanths was greater when the incubation gas phase was 95% nitrogen–5% carbon dioxide than in the presence of air, 100% nitrogen or air containing 5% carbon dioxide (Horvath, 1971).

Horvath considered that a carbon dioxide fixation step is important to carbohydrate metabolism in activated *Moniliformis moniliformis* cystacanths. He identified this carbon dioxide fixation step as the carboxylation of PEP, based on his demonstration that PEPCK is present in cystacanths (Horvath & Fisher, 1971). Although PEPCK fixation of carbon dioxide may be involved in energy metabolism in activated cystacanths, it is not clear that the results completely segregate requirements for metabolism from those for continued cystacanth activation. The apparent effects of gas phase on glycogenolysis and incorporation of [^{14}C]-glucose into glycogen in Horvath's experiments corresponded generally to the effects of gas phase composition on the rate of *Moniliformis moniliformis* cystacanth activation at suboptimal (0.05%) bile salt concentrations (Graff & Kitzman, 1965). Increasing the carbon dioxide content of the gas phase accelerated the rate of cystacanth activation, while substitution of air for nitrogen as the complementary component of the gas phase tended to slow activation.

Horvath suggested that the presence of oxygen 'interfered with' glycogenolysis in his experiments and with cystacanth activation under suboptimal conditions in Graff & Kitzman's experiments. Although the grounds for this suggestion may be difficult to support, the idea has some merit. In *Moniliformis moniliformis* adults, carbon dioxide fixation apparently leads to mitochondrial succinate formation, but oxygen-linked succinate oxidation might occur aerobically (§6.6.4). If similar reactions occur in activated cystacanths, it is possible that elevated oxygen tensions may slow the rate at which succinate (or some other reduced product of carbon dioxide fixation) accumulates in the tissues with the possibility that the build-up of this metabolite is important to biochemical transitions involved in complete cystacanth activation. Alternatively, the effects of oxygen in reducing glycogenolysis and increasing incorporation of [^{14}C]-glucose into glycogen in activated *M. moniliformis* cystacanths may simply indicate that energy metabolism in activated cystacanths may operate more efficiently aerobically, as would be expected if low rates of oxidative phosphorylation can occur (§6.6.4).

6.6.7 *Miscellaneous enzyme studies*

Dunagan & Yau (1968) examined the oligosaccharidases present in the soluble fraction from *Macracanthorhynchus hirudinaceus* homogenates. Evidence was found for significant hydrolysis of maltose, dextrin, trehalose and turanose, in decreasing order of hydrolytic capacity. Modest hydrolysis of inulin, lactose, raffinose and melizitose occurred at acidic pH, and some lactose and melizitose hydrolysis occurred at neutral pH as well. Dunagan & Yau found no evidence for hydrolysis of sucrose, α-methyl glucoside, α-methyl mannoside, or melibiose. The putative maltase activity from *M. hirudinaceus* had a pH optimum of 5.0, suggesting a lysosomal hydrolase. In the presence of Tris ions, which inhibit maltase and general α-glucosidase activities (Semenza, 1968), the optimal pH appeared to be 4.5. Although these results suggest more than one enzyme acting as a maltase, Dunagan & Yau were able to identify only one band by starch gel electrophoresis using *p*-nitrophenyl-α-glucoside as a substrate (see below). The pH optimum of 5.8 for trehalase activity is in the range characteristic of trehalases from a broad spectrum of sources (Elbein, 1974).

Trehalase activity has been found in soluble, but not particulate, fractions from *Moniliformis moniliformis* body walls (Starling & Martinez, unpublished). Partially purified trehalase has a relative molecular mass of about 77000. The enzyme is highly sensitive to oxidizing conditions, losing activity rapidly in the absence of reducing agents such as dithiothreitol, a property which might explain why Dunagan & Yau (1968) sometimes observed no trehalase in *M. hirudinaceus* preparations if that enzyme were equally sensitive to oxidizing conditions. Trehalase from *M. moniliformis* appears to be highly anionic, binding tightly to DEAE–cellulose at pH 6.5 and migrating anodally in standard polyacrylamide gel electrophoresis systems. The apparent K_m for trehalase is 10.4mM, and maximal activity is seen at pH 6.4 in the presence of 2mM magnesium ions. Phosphate ions stimulate trehalase activity significantly at concentrations within the physiological range, which suggests that cellular phosphate may be involved in regulating trehalose metabolism (§6.5.2).

Moniliformis moniliformis also contains significant α-glucosidase activity in both soluble and particulate preparations, but the properties of the particulate enzyme(s) have not been examined. The soluble fraction contains two distinct enzymes which have widely different electrophoretic mobilities on polyacrylamide gels. Both α-glucosidases can be demonstrated on polyacrylamide gels with *p*-nitrophenyl-α-glucoside as a substrate, but the band migrating more slowly at pH 7.5 is stained more intensely by this reagent. The more slowly migrating band hydrolyzes trehalose, glycogen and starch to a limited extent, and has significant activity against sucrose. None of these substrates is hydrolyzed significantly by the more rapidly migrating band. Both appear to have an acidic pH optimum, but the pH characteristics of the individual enzymes have not been determined. The mixture of partially purified α-glucosidases has 10–15-fold higher activity with maltose than with *p*-nitrophenyl-α-glucoside as a substrate, while the hydrolysis of glycogen is about comparable to that of *p*-nitrophenyl-α-glucoside. The evidence suggests that the band with higher electrophoretic mobility may be a relatively specific maltase, but it has not been technically possible to stain polyacrylamide gels for maltase activity.

Parshad & Guraya (1978*b*) examined the acid and alkaline phosphatases in low-speed supernatant fractions from *Sphaerirostris pinguis* and several other species of helminth, and surveyed the effects of metal ions, metabolic inhibitors and anthelmintic drugs on the phosphatase activities. Because they were able to demonstrate moderate to strong inhibition of phosphatase activities by most of the agents tested, they concluded that anthelmintics may affect transport processes by inhibiting surface phosphatases. However, Parshad & Guraya did not separate soluble and particulate phosphatases and so there is no way of knowing how much of the phosphatase activity was due to hydrolase(s) associated with the absorptive surface and how much to cytoplasmic or organellar phosphatases.

6.7 Conclusions

Moniliformis moniliformis has received more attention in modern biochemical and physiological studies than any other acanthocephalan species because it is easy to maintain in the laboratory and is large enough to be manipulated for the provision of sufficient tissue for biochemical analysis. *Moniliformis moniliformis* and its larger cousin *Macracanthorhynchus hirudinaceus* are the only acanthocephalans from which enzymes have been purified. There have been several good biochemical studies on these archiacanthocephalans in recent years; let us hope that the activity continues.

Studies of physiology of *Moniliformis moniliformis* have the advantage that there is a great deal of information available on the intestinal physiology of its definitive host, in part because of the ubiquity of the laboratory rat in biological research but also because of the extensive research that has been conducted on the cestode *Hymenolepis diminuta* and its interactions with the host intestine. The *M. moniliformis–H. diminuta–*host system offers opportunities for comparative studies and for investigations of interspecific helminth interactions. With rare exceptions, we have been remiss in exploiting those opportunities in experimental design or interpretation.

Acknowledgements

The research cited in this chapter was supported by grant AI-15397 from the National Institutes of Health and by a Weldon Springs Research Grant from the University of Missouri.

7

Reproduction

D.W.T. Crompton

7.1 Introduction

When discussed broadly, reproduction can be defined as the replacement of one pair or population of parents with the next (Cohen, 1977), which means that sexual reproduction will not have been accomplished until parents have become grandparents. This view is rather impractical because it suggests that every aspect of an acanthocephalan parasite's biology can be claimed to fall within the subject of reproduction. It does, however, have some conceptual value by implying that reproduction depends not only on the properties of nucleic acids and behaviour of gametes but also on numerous somatic and environmental factors. In this chapter, reproduction is taken to include all the events and processes occurring from the maturation of the adult worms to the release of the shelled acanthors which may be conveniently called eggs.

As far as is known, all species of Acanthocephala attain sexual maturity in the alimentary tract of vertebrates, usually in a relatively precise site, or microhabitat (Crompton, 1973, 1975). Some qualification of this generalization may soon become necessary because several species are undoubtedly precocious (§7.5.2) and active spermatozoa have been observed by Brattey (1980) in male cystacanths of *Acanthocephalus lucii* in its intermediate host *Asellus aquaticus*. All the evidence shows that sexual reproduction involving separate male and female worms is the only form that occurs; hermaphroditism, parthenogenesis and any other form of asexual propagation are unknown (Van Cleave, 1953; Kennedy, in press). Acanthocephalans appear to be polygamous (Jewell, 1976) and, following gametogenesis, copulation, insemination and internal fertilization occur while the worms are inhabiting the alimentary tract. Much of this reproductive activity is centred on the unique ovarian balls or free-floating ovaries (Atkinson & Byram, 1976; Parshad & Crompton, 1981) from

Table 7.1. *Estimates of the duration of the prepatent periods of acanthocephalans in experimental infections*

Parasite	Definitive host	Approximate duration (days)			Reference
		First insemination[a]	Acanthor formation[b]	Total period	
ARCHIACANTHOCEPHALA					
Macracanthorhynchus hirudinaceus	Domestic pig	n.d.[c]	n.d.	70[d]	Kates (1944)
Mediorhynchus centurorum	*Centurus carolinus*[e]	n.d.	n.d.	35 (28–39)	Nickol (1977)
Moniliformis clarki	*Peromyscus maniculatus*[f] *sonoriensis*	n.d.	n.d.	65	Crook & Grundmann (1964)
Moniliformis moniliformis	Laboratory rat	16	22	38	Abele & Gilchrist (1977); Atkinson & Byram (1976); Burlingame & Chandler (1941); Crompton (1974); Crompton, Arnold & Barnard (1972); Robinson (1965)
PALAEACANTHOCEPHALA					
Acanthocephalus dirus	*Lepomis macrochirus*[g]	1	n.d.	n.d.	Muzzall & Rabalais (1975*a*)
Acanthocephalus lucii	*Perca fluviatilis*[g]	1 at 20 °C	n.d.	n.d.	Brattey (1980)
Echinorhynchus truttae	*Salmo trutta*[g]	2	62	64	Awachie (1966)
Leptorhynchoides thecatus	*Ambloplites rupestris*[g]	21–28	35	56	DeGiusti (1949*a*)
Polymorphus minutus	Domestic duck	5	17	22	Crompton & Whitfield (1968*b*); Nicholas & Hynes (1958); Petrochenko (1958); Romanovski (1964)
Pomphorhynchus bulbocolli	*Catostomus commersoni*[g]	21	49	70	Jensen (1952)
EOACANTHOCEPHALA					
Octospiniferoides chandleri	*Gambusia affinis*[g]	n.d.	n.d.	30	DeMont & Corkum (1982)
Paratenuisentis ambiguus	*Anguilla rostrata*[g]	n.d.	n.d.	28	Samuel & Bullock (1981)

[a] The time of the first insemination is usually based on the first finding of copulation caps on female worms (see Abele & Gilchrist, 1977)
[b] The time taken from the estimate of first insemination until eggs (=shelled acanthors) are first detected in host faeces
[c] n.d., not determined
[d] Kates (1944) wrote 'small numbers of eggs were detected in the feces from seventy to eighty days after infection' and 'the prepatent period...has been calculated to be from two to three months' in the same paper
[e] Red-headed woodpecker
[f] Deer mouse
[g] Species of freshwater fish

which the zygotes are discharged into the body cavity after fertilization. The embryological development of the acanthor from the zygote continues to take place in the body cavity of the female worm during its association with the vertebrate host. Each acanthor becomes enclosed in several envelopes (Chapter 9), and, as has been found in almost all laboratory demonstrations of life cycles (Chapter 8), the acanthors from the body cavity are fully infective to animals susceptible as intermediate hosts (Crompton, 1970). Acanthocephalan worms, therefore, would seem to be ovoviviparous or as equally viviparous as species of filarial nematode (Muller, 1975).

Recent summaries of knowledge of the reproductive biology of the Acanthocephala have been prepared by Parshad & Crompton (1981), Crompton (1983*a*, *b*, and in press) and Kennedy (in press) and additional detailed information about the functional morphology of the acanthocephalan reproductive system has been provided by Miller & Dunagan (Chapter 5).

7.2 Prepatent period

The terms prepatent period and patent period are somewhat artificial because they reflect mainly an observer's ability to detect the presence of a parasitic infection in a host. The first detection of eggs in host faeces is generally assumed to be the beginning of the patent period in an acanthocephalan infection (Crompton, Arnold & Barnard, 1972) and time between this event and the day on which the host was known to have swallowed the infective cystacanths, or juveniles, represents the prepatent period. The processes occurring during prepatency include gametogenesis, copulation, insemination, fertilization and acanthor development. A selection of observations on the duration of the prepatent periods of different species of Acanthocephala is given in Table 7.1. In the case of acanthocephalans, the notion of a prepatent period applies to a population of worms which must consist of at least one male and one female.

Generalizations should not be drawn at present from the information in Table 7.1. It might have been expected that some relation would exist between the length of the prepatent period and the lifespan of an acanthocephalan, the duration of its association with the definitive host or its fecundity (§7.8). Under laboratory conditions, female *Polymorphus minutus* and female *Moniliformis moniliformis* may live for up to about 54 days (Crompton & Whitfield, 1968*b*) and 144 days (Crompton *et al.*, 1972), respectively. Their estimated prepatent periods of 22 and 38 days (Table 7.1) represent about 41% and 26% of their lifespans, which does not suggest that the prepatent period forms a uniform part of the acantho-

Table 7.2. *Chromosomes and sex determination in some species of Acanthocephala (from Parshad & Crompton, 1981)*

Species	Chromosome number (2*n*)	Sex chromosomes	Reference
ARCHIACANTHOCEPHALA			
Macracanthorhynchus hirudinaceus	M = 6, F = 6[a]	XY–XX	Jones & Ward (1950); Robinson (1964)
Mediorhynchus grandis	6	—	Schmidt (1973*b*)
Moniliformis moniliformis	M = 7, F = 8	X0–XX	Robinson (1965)
PALAEACANTHOCEPHALA			
Acanthocephalus ranae	8	XY–XY	John (1957)
Acanthocephalus ranae	16	—	Walton (1959)
Echinorhynchus gadi	16	—	Walton (1959)
Echinorhynchus truttae	M = 7, F = 8[a]	X0–XX	Parenti, Antoniotti & Beccio (1965); Parenti & Antoniotti (1967)
Leptorhynchoides thecatus	M = 5, F = 6	X0–XX	Bone (1974*a*)
Polymorphus minutus	16	—	see Bone (1974*b*)
Polymorphus minutus	6 or 12	—	Nicholas & Hynes (1963*b*)
Pomphorhynchus laevis	8	—	Walton (1959)
EOACANTHOCEPHALA			
Neoechinorhynchus cylindratus	M = 5, F = 6	X0–XX	Bone (1974*b*)

[a] M, male; F, female

cephalan life history. Some encouragement for those attempting to relate the duration of prepatency to the length of the host–parasite association is given by the brief time before copulation that occurred for *Acanthocephalus dirus* in experimental infections of *Lepomis macrochirus* (Table 7.1; Muzzall & Rabalais, 1975*a*). Apparently, *A. dirus* may have a relatively brief association with *L. macrochirus* and Muzzall & Rabalais were of the opinion that gravid female worms passed out of the fish into the water rather than releasing eggs in the typical acanthocephalan manner (§7.9). This tentative suggestion might hold for *A. lucii* (Table 7.1) but not for *Echinorhynchus truttae*, despite the latter's early copulatory phase, because egg release or the beginning of the patent period did not begin until about 9 weeks later (Table 7.1).

A few observations suggest that for some species of Acanthocephala, male and female worms may mature at different rates. Harms (1965*a*), on the basis of field work and limited experimental data, reckoned that male *Octospinifer macilentus* reached maturity in about 8–10 weeks and females in about 16 weeks. Keppner (1974) infected creek chub (*Semotilus atromaculatus*) with cystacanths of *Paulisentis missouriensis* and found 3 weeks later that, although the males were producing spermatozoa and the females contained free ovaries (= ovarian balls), developing eggs were not yet present in the females' body cavities. The estimate of the time taken for the development of the acanthors of *Echinorhynchus truttae* (Table 7.1) cannot, however, be accounted for by differential maturation rates of male and female worms; Awachie (1966) observed copulation caps on female *E. truttae* by the end of the second day in brown trout (*Salmo trutta*) and, as expected, developing eggs were found free in the body cavities of the female worms during the second week of the infection.

7.3 Sex determination and sex ratio

The sex of acanthocephalan worms, like that of other dioecious organisms (Sinnott, Dunn & Dobzhanksky, 1958), is established during fertilization by a process involving sex chromosomes. The males have been identified as the heterogametic sex (X0 or XY) and the females as homogametic (XX). The mechanism involving X0 appears to be more common and is represented in the three classes of the phylum (Table 7.2).

The most detailed description of the karyotype and mechanism of sex determination for an acanthocephalan species is that given for *Acanthocephalus ranae* from toads by John (1957) who examined squash preparations of gametes after the gonads had been ruptured in acetic orcein. John's regrettably neglected paper (for example, it is not mentioned by Robinson (1964, 1965), Parenti, Antoniotti & Beccio (1965) or Parshad & Crompton

(1981)), provided unequivocal evidence that male and female *A. ranae* possess three pairs each of identical chromosomes plus one extra pair of which each was different in males ($2n = 8$). These separate pairs could only be explained in terms of a heteromorphic XY–XX sex-determining system which became operational after fertilization (John, 1957: §7.7). Earlier, Jones & Ward (1950) had suggested that an XY–XX mechanism occurred in *Macracanthorhynchus hirudinaceus* and Robinson (1964) obtained the necessary evidence. In germ cells from mature male and female *M. hirudinaceus* of unknown age, Robinson identified three pairs of homologous chromosomes ($2n = 6$) at both oogonial and spermatogonial mitotic metaphase. In the male, however, while one acrocentric pair and one metacentric pair were similar in appearance to pairs in the female cells, the third pair consisted of a metacentric chromosome with arms of equal length and a subacrocentric chromosome with arms markedly different in length. Robinson concluded that the metacentric and subacrocentric members of the third pair in the male were the X and Y chromosomes, respectively (Table 7.2).

During oogonial mitotic metaphase in 21-day-old *Moniliformis moniliformis*, Robinson (1965) observed four homologous pairs of metacentric chromosomes ($2n = 8$) of which one was a small pair measuring about 4 μm in length while the other three were larger, measuring about 7 μm. On examination of spermatogonial metaphase, Robinson identified seven chromosomes consisting of three pairs and one solitary chromosome ($2n = 7$). Thus, he concluded that sex in *M. moniliformis* was determined by an X0 mechanism with the male as the heterogametic sex. The karyotype of *Echinorhynchus truttae* has been characterized by Parenti *et al.* (1965) and Parenti & Antoniotti (1967) and that of *Leptorhynchoides thecatus* by Bone (1974*a*, *b*), and a similar sex-determining mechanism has been found (Table 7.2). The karyotypes of more species of Acanthocephala should now be determined in order to investigate the taxonomic and phylogenetic significance of the XY–XX and X0–XX mechanisms. The X0 condition could have arisen from the XY (Robinson, 1965); instances of the loss of Y chromosomes from animals are known (Lewis & John, 1968).

Systems of sex determination that depend on one heterogametic and one homogametic parent ought in theory to give rise to a sex ratio of 1:1 at the time of fertilization unless one of the two types of gamete from the heterogametic sex is in some way more favoured or more active during fertilization. Observations on populations of adult acanthocephalans in naturally occurring or laboratory infections generally reveal a sex ratio of 1:1 or else a ratio in which female worms predominate. Examples of this pattern include the following species: *Moniliformis moniliformis* described by Burlingame & Chandler (1941), Graff & Allen (1963) and Crompton & Walters (1972); *Macracanthorhynchus hirudinaceus* studied by Kates (1944); *Acanthocephalus dirus* observed by Camp & Huizinga (1980); *A. ranae* by John (1957); *Echinorhynchus salmonis* by Tedla & Fernando (1970) and Valtonen (1980); *E. truttae* by Parenti *et al.* (1965) and Awachie (1966); *Fessisentis necturorum* by Nickol & Heard (1973); *Polymorphus minutus* by Nicholas & Hynes (1958) and Crompton & Whitfield (1968*b*); *Neoechinorhynchus cristatus* reported by Muzzall (1980); *N. rutili* by Walkey (1967); and *N. saginatus* by Muzzall & Bullock (1978). The observed sex ratios reflect the fact that male and female acanthocephalans develop in equal numbers so that cystacanths occur with a sex ratio of 1:1 in intermediate hosts (Muzzall & Rabalais, 1975*b*; Brattey, 1980) and young worms similarly in definitive hosts. In nearly all cases, however, females live longer than males (Crompton, 1970; Chapter 10) and in 1953, Van Cleave wrote '...but older infections of the definitive host commonly show a preponderance of females...it is not uncommon for an infection of long standing to consist largely or even exclusively of females'. Infections of *Pomphorhynchus laevis* in goldfish appear to be an exception to this rule since, although male *P. laevis* do not become established in the fish as easily as females, they survive longer (Kennedy, 1972).

If worms of all species of Acanthocephala are polygamous, with individuals mating with more than one partner of the other sex (§7.6, §7.9), it is interesting to investigate the reason for the retention of a 1:1 sex ratio. In the context of current behavioural and evolutionary theory, natural selection is considered to act on individual organisms rather than on groups (Krebs & Davies, 1981) so that in an uneven sex ratio the rarer sex would always have an evolutionary advantage. Thus, if male acanthocephalans usually mate with more than one partner, as has been demonstrated with *Moniliformis moniliformis* (Crompton, 1974), a reduction of the sex ratio against males would promote opportunities for the established males to pass on their genes at the expense of the larger number of females. Future work may be expected to show that the sex ratio of acanthocephalans remains at 1:1 until sufficient spermatozoa have been transferred to the females to complete the fertilization of their total oocytes. This proposal, which assumes that an operational sex ratio of 1:1 exists, would ensure that male and female acanthocephalans would have equal evolutionary opportunities. The loss of males from the small intestine before females (Van Cleave, 1953) would facilitate the use of available nutrients for acanthor production, an activity which would also favour individual male and female worms equally. Direct observation of this type of reproductive behaviour would be difficult in naturally established populations where recruitment and loss of individual worms are normal (Chubb, Awachie & Kennedy, 1964), and where competition for resources and various host-mediated factors may occur (§7.10). Perhaps methods should also be considered for determining the operational sex ratio, which is the ratio between the number of sexually active males and sexually receptive females (Krebs & Davies, 1981), rather than the absolute value as is common practice at present.

7.4 Sexual dimorphism

A detailed account is given by Parshad & Crompton (1981) of the differences in various external features which exist between the sexes of adult acanthocephalan worms in their definitive hosts. Sexual dimorphism may be frequently observed in the size and shape of the body, the distribution of body spines, the number, size and shape of proboscis hooks, the occurrence of papillae, and the position of genital openings (Van Cleave, 1920*b*; Ward & Nelson, 1967; see Yamaguti, 1963). Occasionally, striking differences are seen between the proboscides of males and females of the same species, as has been described for *Filicollis anatis* (Van Cleave, 1920*b*) and for *Megapriapus ungriai* by Golvan, Gracia-Rodrigo & Diaz-Ungria (1964). *M. ungriai* is equally unusual in that it was found in the spiral valve of an elasmobranch fish.

The commonest difference between the sexes in the Acanthocephala is that of body size, with the female usually being larger than the male. *Moniliformis moniliformis*, *Mediorhynchus grandis* and *Hexaspiron nigericum* are three species in which mature females have been shown to be as much as five times as long as the males. *Corynosoma hamanni*, studied by Holloway & Nickol (1970), and *Echinorhynchus lageniformis*, by Olson & Pratt (1971), are probably exceptional species in which the males have been found to be bigger than the females. Body size in the Acanthocephala is an extremely variable character which may be influenced by the age of the worms (Crompton, 1972*a*), their reproductive state (Crompton, 1974), their population structure (Graff & Allen, 1963), host species and distribution (Bullock, 1962; Amin, 1975*a*, *b*; Buckner & Nickol, 1979), host sex (Graff & Allen, 1963), host diet (Nesheim *et al.*, 1977, 1978; Parshad, Crompton & Nesheim, 1980) and host environment (Walkey, 1967). Evolutionists have argued that as the female of a species becomes more adapted to receive the male and to process the products of fertilization, and the male becomes more adapted to bring the spermatozoa to the female, differences develop between the sexes and these may be reinforced by sexual selection (Calow, 1978; §7.4.3). At present, however, it is difficult to see how as variable and plastic a character as adult body size could contribute to the reproductive behaviour and successful copulation in the Acanthocephala (Parshad & Crompton, 1981). Van Cleave (1920*b*) was even more cautious about assuming that somatic differences between the sexes might be involved in sexual selection, although at that time relatively few acanthocephalan species were known and even fewer had been studied in detail. Perhaps small morphological differences, which are known to occur at the posterior ends of individuals of the same species, for example, *Fessisentis fessus* described by Nickol (1972), may facilitate sexual recognition and copulation (§7.6.2). This is a highly speculative suggestion and it is not known whether acanthocephalans produce or detect any chemicals which may be involved in sex recognition.

7.5 Gonads and gametogenesis

Sexual reproduction consists basically of meiosis and nuclear fusion, fertilization. The first of these events, gametogenesis, involves the production of anisogamous gametes in the gonads of mature worms and so it seems appropriate to include here a comparative description of current knowledge of the development and structure of acanthocephalan ovaries and testes. There is general agreement that in both male and female acanthocephalans the gonadial and many of the somatic organ primordia arise during development from the central nuclear mass or entoblast located in the middle of the acanthor (Chapter 8). Nothing, however, appears to be known about the processes of primary determination which must be assumed to have occurred to enable particular nuclei or units of nuclei and cytoplasm within the acanthor to be destined for the development of the gonads and the continuation of the germ line. In many invertebrates the nature of the oogonial cytoplasm and its distribution after fertilization between the blastomeres of the early embryo are crucial in the determination of cell fates (Davenport, 1979). The role of cytoplasmic factors in acanthocephalan development would be an intellectually and technically challenging study because the whole embryo appears to become syncytial after a few divisions and the presence of the egg envelopes often impedes observations.

7.5.1 *Terminology*

In preparing this account about the development and structure of the gonads and the processes of gametogenesis, it became apparent that published results, and perhaps those to be written in future years, would be easier to evaluate and compare if some agreement could be reached about terminology. Several terms could be dropped and confusion would be reduced if 'primordium' was used for nuclei or cells that appear to be

Table 7.3. *Observations on the development of acanthocephalan gonads during experimental infections of intermediate hosts*

Parasite	Intermediate host[a]		Temperature (°C)	Days after infection of intermediate host							Reference
				Gonad primordia[b]							
	Taxon	Habitat		M	u.s.	F	Testes	Ovary	Infective to definitive host	Ovarian fragmentation[c]	
ARCHIACANTHOCEPHALA											
Mediorhynchus grandis	or	t	29–33	25	—	—	27	—	27	27	Moore (1962)
Moniliformis clarki	or	t	25	22	—	n.d.[d]	—	n.d.	60	DH	Crook & Grundmann (1964)
Moniliformis moniliformis	di	t	23–26	—	—	—	38	—	49	IH	Moore (1946*a*)[e]
PALAEACANTHOCEPHALA											
Echinorhynchus lageniformis	am	m	23	—	—	—	15	18	30	—	Olson & Pratt (1971)
Echinorhynchus truttae	am	fw	17	—	25	—	37	68	82	79	Awachie (1966)
Leptorhynchoides thecatus	am	fw	20–25	17	—	n.d.	18	n.d.	32	DH	DeGiusti (1949*a*)
Plagiorhynchus cylindraceus	is	t	20–23	22	—	32	25	37	60	IH	Schmidt & Olsen (1964)
Polymorphus minutus	am	fw	17	—	17	—	35	—	60	IH	Hynes & Nicholas (1957)
EOACANTHOCEPHALA											
Neoechinorhynchus emydis	os	fw	—	—	16	—	—	—	21[f]	—	Hopp (1954)
Neoechinorhynchus rutili	os	fw	15	33	—	—	38	48	48	DH	Merritt & Pratt (1964)
Neoechinorhynchus saginatus	os	fw	25	10	—	9	12	12	16	n.d.	Uglem & Larson (1969)
Octospinifer macilentus	cp	fw	21	—	—	—	20	—	33	—	Harms (1965*b*)
Octospiniferoides chandleri	os	fw	23	14	—	n.d.	20	20	20	DH	DeMont & Corkum (1982)
Paratenuisentis ambiguus	am	m	22–25	—	—	—	16	15	27	—	Samuel & Bullock (1981)
Paulisentis fractus	cp	fw	21–23	—	—	—	7	9	13	—	Cable & Dill (1967)
Paulisentis missouriensis	cp	fw	20–23	10	—	11	—	14	14	DH	Keppner (1974)

[a] am, amphipod; cp, copepod; di, dictyopteran; is, isopod; or, orthopteran; os, ostracod; fw, freshwater; m, marine; t, terrestrial

[b] F, female; M, male; u.s., unsexed

[c] DH, during infection of definitive host; IH, during infection of intermediate host

[d] n.d., not discerned; —, no information

[e] See also Yamaguti & Miyata (1942), King & Robinson (1967), Nicholas (1967), Atkinson & Byram (1976) and Asaolu *et al.* (1981)

[f] *N. emydis* was found by Hopp (1954) to be a special case. The ostracods bearing infective juveniles, or cystacanths, are eaten next by snails which are in turn eaten by the definitive host (Chapter 9)

undifferentiated, but are of known developmental fate, and 'rudiment' was reserved for structures that are showing some degree of differentiation (Parshad & Crompton, 1981). A strong scientific case can now be made for dropping the term 'ovarian ball' from the literature because each of these is either a developing or functional ovary (Atkinson & Byram, 1976; Asaolu *et al.*, 1981). The functional ovarian tissue in female Acanthocephala is considered to be organized into ovarian balls which develop from the ovary in what Van Cleave (1953) described as an unexplained form of fragmentation. The term 'ovarian ball' appears to be a direct translation of *ovarial ballen* (Meyer, 1933), although other authors have also used the terms 'ovarial ball', '*Eiballen*', 'ovarian fragments', 'egg balls' and 'egg masses' for the units of functional ovarian tissue. In fact, DeGiusti (1949*a*) clearly referred to ovarian balls in *Leptorhynchoides thecatus* as floating ovaries, which is exactly what they are (§7.5.3), but the value of this terminology was overlooked.

7.5.2 *General pattern of gonad development*

There seems to be general agreement that the gonads develop in close association with the ligament. A summary of observations on the development of the female and male gonads of acanthocephalan worms during their relationships with their intermediate hosts is given in Table 7.3. The timings of events shown in the table are rough estimates based on either text or figure legends of the references cited. Considerable difficulty was experienced in the compilation of the table because it is not always clear whether authors were including the gonads under the term 'genital primordium' and because of the different conventions used to indicate the timing of events. Comparative studies would be simplified if it could be agreed that the start of an experimental infection, for the purposes of a developmental or life cycle study, would be designated 'day 0' and that all other references to the timing of events and processes would be given in the convention of day 1 (=24 h post infection), day 2 and so on.

Nevertheless, two general points can be made from the information in Table 7.3, which is based almost entirely on observations made by light microscopy. First, with the exception of *Neoechinorhynchus saginatus* and *Octospiniferoides chandleri*, it appears that the development of acanthocephalan testes begins before that of the ovaries and is largely completed within the intermediate host. This pattern occurs in representatives of each of the three classes of Acanthocephala. Secondly, the process of ovarian fragmentation (see below) begins while the parasite is located in either the intermediate or definitive host as is known for *Plagiorhynchus cylindraceus* and *Leptorhynchoides thecatus*, respectively (Schmidt & Olsen, 1964; DeGiusti, 1949*a*). In 1964, Schmidt & Olsen wrote, 'No other acanthocephalan is known to have an ovary which develops sufficiently to break into ovarian balls while still in the intermediate host. Thus the cystacanth of *P. formosus* [= *Plagiorhynchus cylindraceus*] appears to be precocious, attaining an advanced stage of development within the intermediary'. However, Yamaguti & Miyata (1942) and Moore (1946*a*) had already observed ovarian fragmentation in *Moniliformis moniliformis* during its development in cockroaches and Merritt & Pratt (1964) have also described a similar process in *Neoechinorhynchus rutili* developing in ostracods. Too few species have been studied to enable the biological significance of such marked differences in the onset of fragmentation to be assessed. It might have been expected that fragmentation would begin during development in the intermediate host if the lifespan of the adult stages, or more particularly the male worms, was relatively short. According to Yamaguti & Miyata (1942), Moore (1946*a*) and Asaolu *et al.* (1981), ovarian fragmentation in *M. moniliformis* begins while the parasite is contained within its cockroach host, but the mean expected lifespan of a female *M. moniliformis* is about 57 days (Crompton, Keymer & Arnold, 1984) under experimental conditions, which does not seem to be short. In marked contrast, ovarian fragmentation in *M. moniliformis* is considered by Atkinson & Byram (1976) to be delayed until the parasite has been established in the rat for at least 9 days. Although this view is difficult to confirm, it is interesting that Crook & Grundmann (1964) in their study of the closely related *M. clarki* (Table 7.3) made virtually no reference to the development of the ovary and the reader must assume for the present that ovarian fragmentation in *M. clarki* occurs after the females have reached their definitive host. Nothing seems to be known about where the gonads continue to develop or whether ovarian fragmentation proceeds during an acanthocephalan worm's association with a paratenic host (Chapter 9).

Fig. 7.1. Diagrammatic representation, based on transmission electron microscopy, of the organization of a free-floating ovary from an inseminated mature female *Moniliformis moniliformis*. *d.o.*, developing oocyte; *e.*, egg; *i.*, inclusion; *mv.*, microvillus; *m.o.*, mature oocyte; *o.*, oogonium; *o.mt.*, oogonial mitochondrion; *o.n.*, oogonial nucleus; *o.s.*, oogonial syncytium; *pf.s.*, postfertilization space; *s.n.*, supporting syncytial nucleus; *s.s.*, supporting syncytium; *s.c.*, cortex (*s.s.*); *s.m.*, medulla (*s.s.*); *s.mt.*, mitochondrion (*s.s.*); *v.*, fertilization membrane; *z.*, zygote. (From Crompton & Whitfield, 1974.)

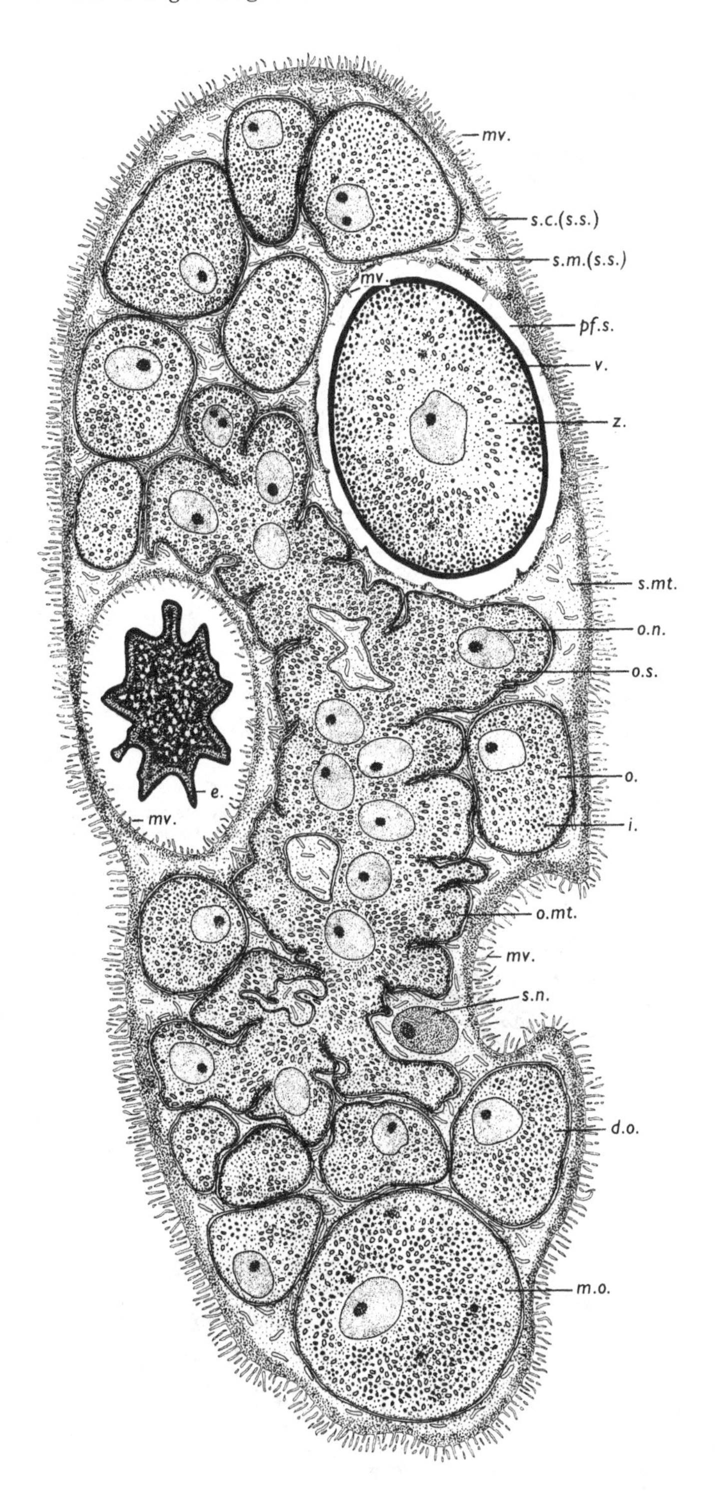
mv.
s.c.(s.s.)
s.m.(s.s.)
mv.
pf.s.
v.
z.
s.mt.
o.n.
o.s.
o.
e.
i.
mv.
o.mt.
mv.
s.n.
d.o.
m.o.

Fig. 7.2. Diagrammatic representation of some of the processes considered to occur in an acanthocephalan ovary. The fates of individual nuclei originating from the oogonial syncytium may be followed through a series of stages. Note that oocyte atresia is shown in the absence of fertilization (nucleus 27(1)). (From Crompton & Whitfield, 1974.)

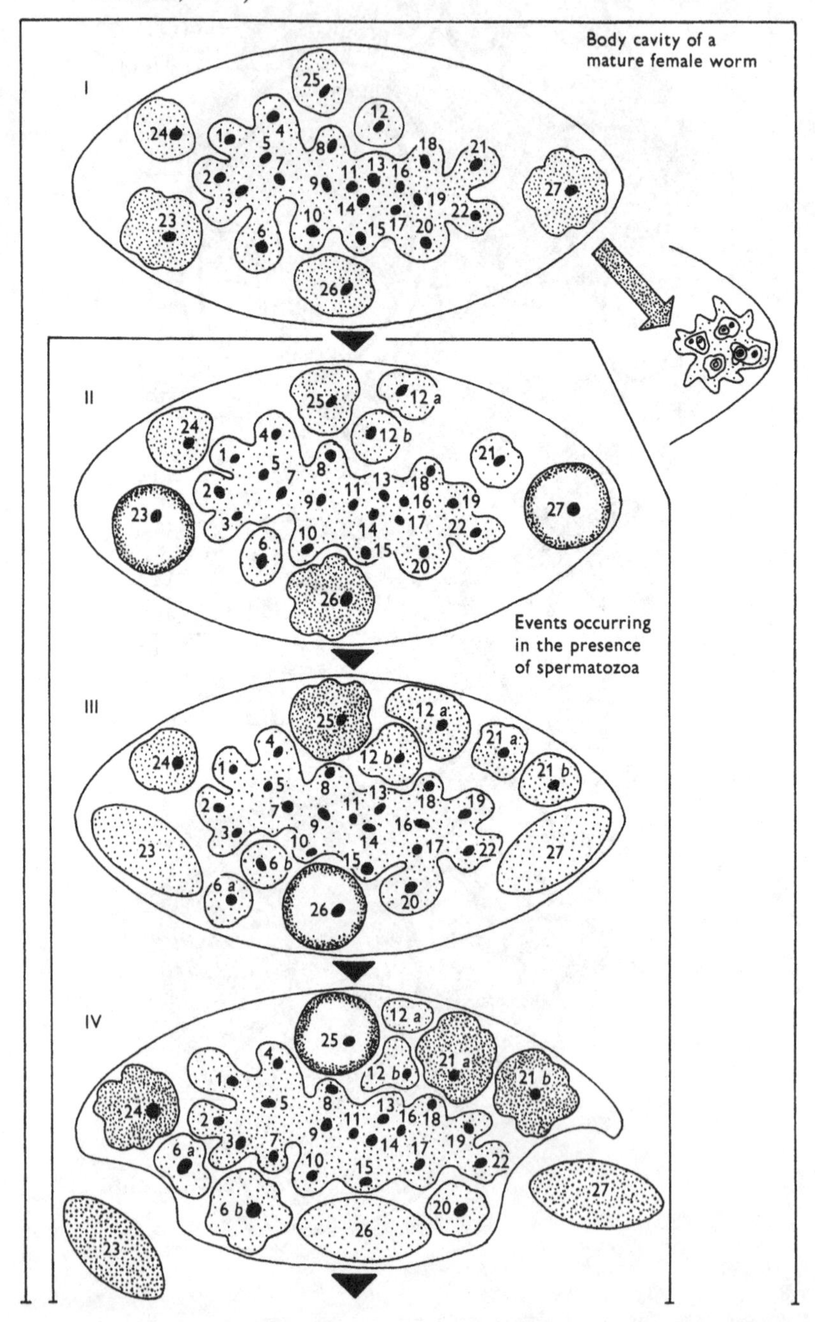

7.5.3 *Oogenesis and oocytes*

Mature acanthocephalan ovaries float freely amongst the developing eggs in the fluid of the body cavity (Meyer, 1933; Van Cleave, 1953; Yamaguti, 1963; Bullock, 1969). In archiacanthocephalans, the eggs and ovaries remain within the persistent ligament sacs, while in palaeacanthocephalans the single ligament sac ruptures and the eggs and ovaries become free in the body cavity (Bullock, 1969). In eoacanthocephalans, there appear to be both dorsal and ventral ligament sacs, but these may not persist throughout adult life as Bullock (1969) found for the genera *Neoechinorhynchus*, *Octospiniferoides* and *Tanaorhamphus*.

A list of publications that are concerned with the detail of ovarian structure, oogenesis and oocytes in species of the three classes of Acanthocephala is given in Table 7.4. These accounts are in various ways founded on the early work of Hamann (1891*a*) and Kaiser (1893) who sought to trace the origins and fate of acanthocephalan germ cells.

Structure of the mature ovaries. Electron microscopy (Table 7.4) has led to the identification of three components in mature free-floating ovaries. These are two separate multinucleate syncytia, named the oogonial syncytium and the supporting syncytium, and a cellular zone. These have been recognized in the archiacanthocephalan *Moniliformis moniliformis* (Fig. 7.1) by Crompton & Whitfield (1974) and Atkinson & Byram (1976), in the palaeacanthocephalans *Breizacanthus* sp. by Marchand & Mattei (1980), *Corynosoma semerme* by Peura, Valtonen & Crompton (1982) and *Polymorphus minutus* by Crompton & Whitfield (1974), and in the eoacanthocephalans *Acanthogyrus tilapiae*, *Neoechinorhynchus agilis* and *Pallisentis golvani* by Marchand & Mattei (1976*d*, 1979). The cells, which consist of oocytes and various stages in their formation, appear to develop from the central germ line or oogonial syncytium (Fig. 7.2).

Two lines of evidence suggest that the cells of the ovary develop from the oogonial syncytium. First, there is no other reasonable explanation for the origin of the cells and, secondly, the oogonial cytoplasm contains electron-dense spherical inclusions which can also be observed in the cells identified as oogonia, primary oocytes and mature oocytes (Crompton & Whitfield, 1974). In *Polymorphus minutus*, *Breizacanthus* sp. and *Corynosoma semerme*, the cytoplasmic inclusions, which appear to increase in number as the oocytes mature, are membrane-bounded, but not homogeneously electron-dense (Fig. 7.3; Crompton & Whitfield, 1974; Marchand & Mattei, 1980; Peura *et al.*, 1982). Their presence in the cytoplasm of the oogonial syncytium serves as a most useful identification marker because they are not found in the cytoplasm of the supporting syncytium.

Earlier workers (Hamann, 1891*a*; Kaiser, 1893; Meyer, 1928, 1933; Nicholas & Hynes, 1963*b*) were satisified about the presence of the cells and the oogonial syncytium but, without the greater resolving power of an electron microscope, were unsure about the nature of the other material. Meyer (1928) thought that the surface of the free ovaries and matrix (='*grundsubstanz*') in which the oogonia and oocytes were embedded in *Macracanthorhynchus hirudinaceus* might be composed from the remains of degenerating oocytes. In *Moniliformis moniliformis* and *Polymorphus minutus*, however, the electron microscope shows that Meyer's '*grundsubstanz*' is syncytial, being made up of cytoplasm and containing nuclei, but apparently no cell membranes (Crompton & Whitfield, 1974; Atkinson & Byram, 1976). The surface of a free ovary from a mature worm is membrane-bounded and gives rise to numerous microvilli (Fig. 7.1), each

Fig. 7.3. Transmission electron micrograph of a section through part of the cytoplasm of a mature oocyte of *Corynosoma semerme*. Note the structure of the membrane-bounded inclusions. (From Peura *et al.*, 1982.)

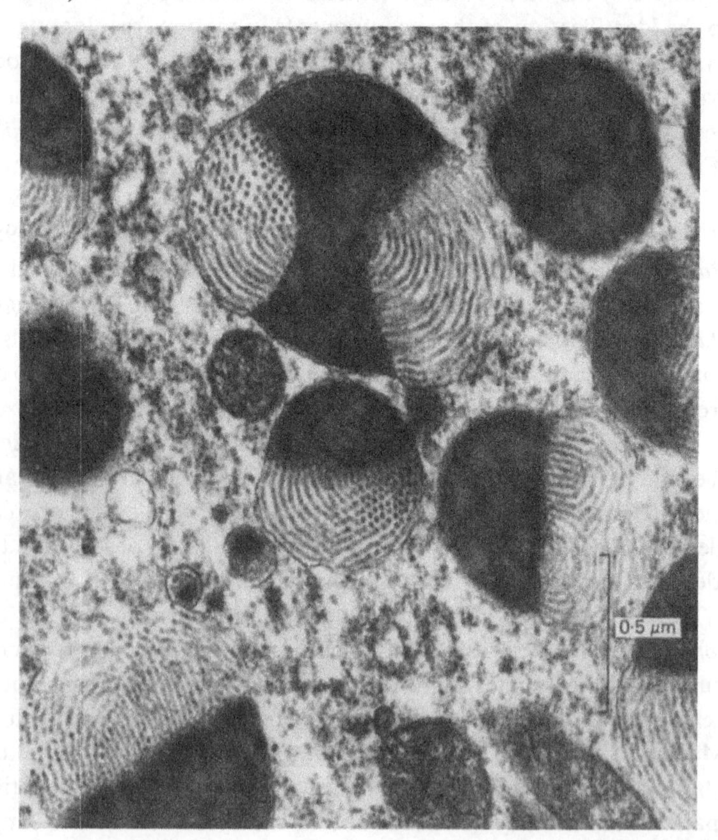

measuring about 50 nm in diameter and from 1–5 μm in length (Crompton & Whitfield, 1974; Parshad, Crompton & Martin, 1980). The nature of the surface formed from the supporting syncytium of a mature ovary suggests that it might have an absorptive function and the intimate interdigitation of its cytoplasm with that of the oogonial syncytium and the cells would be ideal for the transport of nutrients from the body cavity into the ovary (Atkinson & Byram, 1976; Parshad, Crompton & Martin, 1980). Such transport would be facilitated by the ducts and channels which can be seen in electron micrographs of the cortical region of the supporting syncytium. Not all acanthocephalan species possess free ovaries which comply in structure with the general description given above. Each free ovary of *Sphaerirostris pinguis* (Table 7.4) appears to be composed of 24–30 functional units with each unit possessing its own oogonial and supporting syncytia and oogonial cells (Parshad & Guraya, 1977*a*).

Table 7.4. *Studies giving information on ovary structure, oogenesis and oocytes in the Acanthocephala*

ARCHIACANTHOCEPHALA	
Macracanthorhynchus hirudinaceus	Meyer (1928, 1933); Robinson (1964)
Moniliformis moniliformis	Asaolu (1977)[a]; Asaolu *et al.* (1981)[a]; Atkinson & Byram (1976)[a]; Crompton & Whitfield (1974)[a]; Parshad, Crompton & Martin (1980)[b]; Robinson (1965)
Prosthenorchis sp.	Guraya (1969)
PALAEACANTHOCEPHALA	
Breizacanthus sp.	Marchand & Mattei (1980)[a]
Corynosoma semerme	Peura *et al.* (1982)[a]
Echinorhynchus truttae	Parenti & Antoniotti (1967)
Leptorhynchoides thecatus	Bone (1974*a*)
Polymorphus minutus	Crompton & Whitfield (1974)[a]; Nicholas & Hynes (1958)
Sphaerirostris pinguis	Parshad & Guraya (1977*a*, 1978*a*)
Pomphorhynchus laevis	Stranack (1972)[a]
EOACANTHOCEPHALA	
Acanthogyrus oligospinus	Anantaraman & Subramoniam (1975)
Acanthogyrus tilapiae	Marchand & Mattei (1976*d*)[a]
Neoechinorhynchus agilis	Marchand & Mattei (1979)[a]
Neoechinorhynchus cylindratus	Bone (1974*b*)
Pallisentis golvani	Marchand & Mattei (1976*d*)[a]

[a] Transmission electron microscopy
[b] Scanning electron microscopy

Cytological development of the free ovaries. Many complex events occur during the formation of reproductively functional ovaries in the adult female worm from the ovarian primordium in a stage in the intermediate host (Table 7.3). Yamaguti & Miyata (1942), Moore (1946*a*), Asaolu (1977) and Asaolu *et al.* (1981) have identified immature free ovaries in the late developmental stages of *Moniliformis moniliformis* in cockroaches, Sandground (1926) observed them in cystacanths of *M. moniliformis* in paratenic hosts and Asaolu *et al.* (1981) investigated the cytological changes occurring in the still immature ovaries during the first 9 days of the course of the infection in rats. This study was undertaken after Atkinson & Byram (1976) had written, 'Seven to nine days after infection of the definitive host by cystacanths of *M. dubius* [=*M. moniliformis*], the genital primordium of the female acanthocephalan is transformed from a

Fig. 7.4. Transmission electron micrograph of a section through a free ovary from the body cavity of a female *Moniliformis moniliformis* after one day in the rat. Note the prominent surface coat. *mb*, multivesicular body; *n*, nucleus; *nu*, nucleolus; *sc*, surface coat. (From Asaolu *et al.*, 1981.)

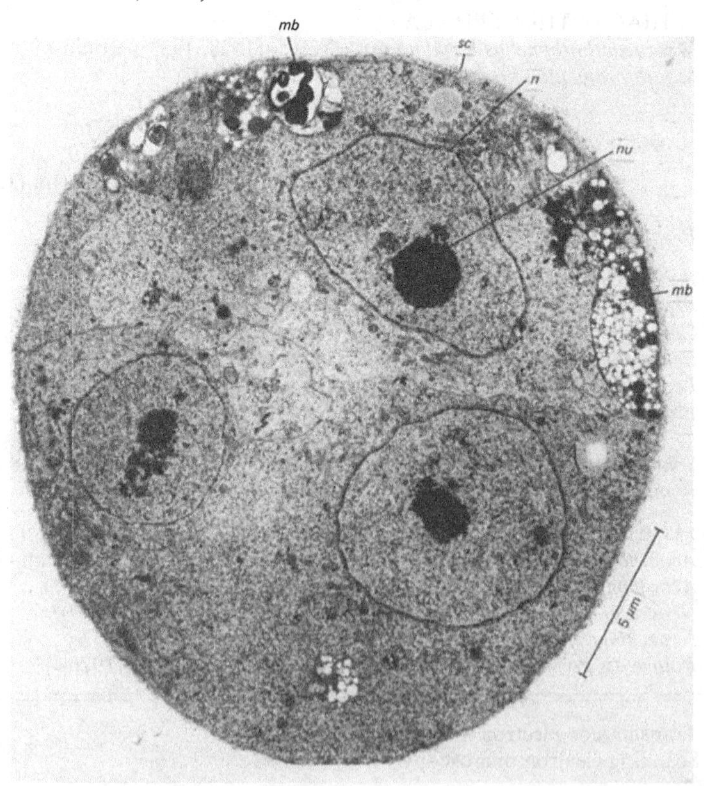

fragmented mass of cells into discrete ovarian balls. This is accomplished by the envelopment of free germinal cells by somatic tissue which originates from the ligament sac primordium.' Their view about the onset of ovarian fragmentation (Van Cleave, 1953) is difficult to accept when free ovaries can easily be seen in female cystacanths (Asaolu *et al.*, 1981). Their view is important, however, because it stimulates ideas about how the supporting and oogonial syncytia might have been formed.

Transmission electron micrographs obtained from sections through structures identified as free ovaries in female *Moniliformis moniliformis* established in rats for 1 and 5 days are shown in Figs 7.4 and 7.5, respectively. Further details of the structures of these ovaries and of others in a time sequence are shown in Fig. 7.6 (Asaolu *et al.*, 1981). Several changes in cytological complexity and development can be observed in this

Fig. 7.5. Transmission electron micrograph of a section through a free ovary from the body cavity of a female *Moniliformis moniliformis* after 5 days in the rat. *mb*, multivesicular body; *n*, nucleus; *nu*, nucleolus; *sc*, surface coat. (From Asaolu *et al.*, 1981.)

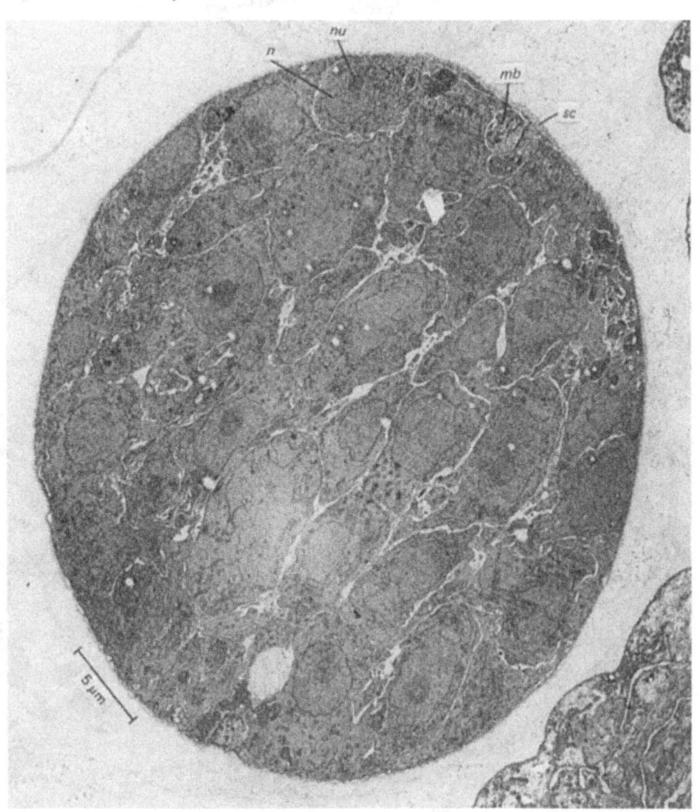

(*a*)

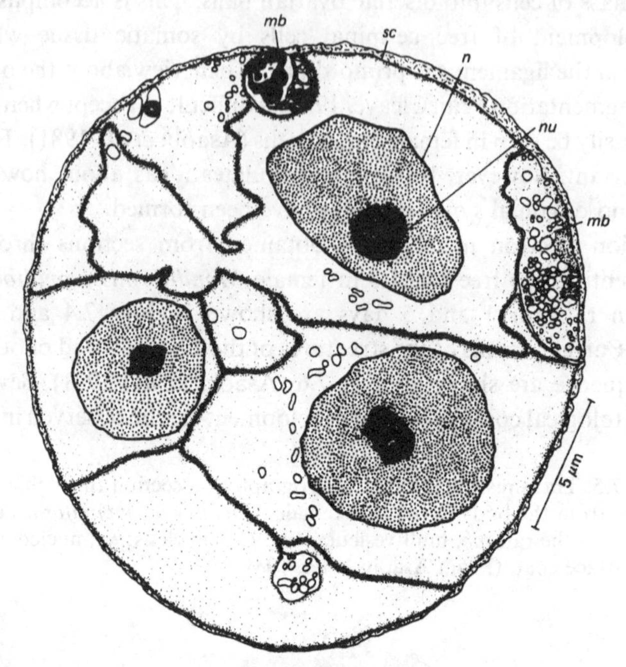

(*b*)

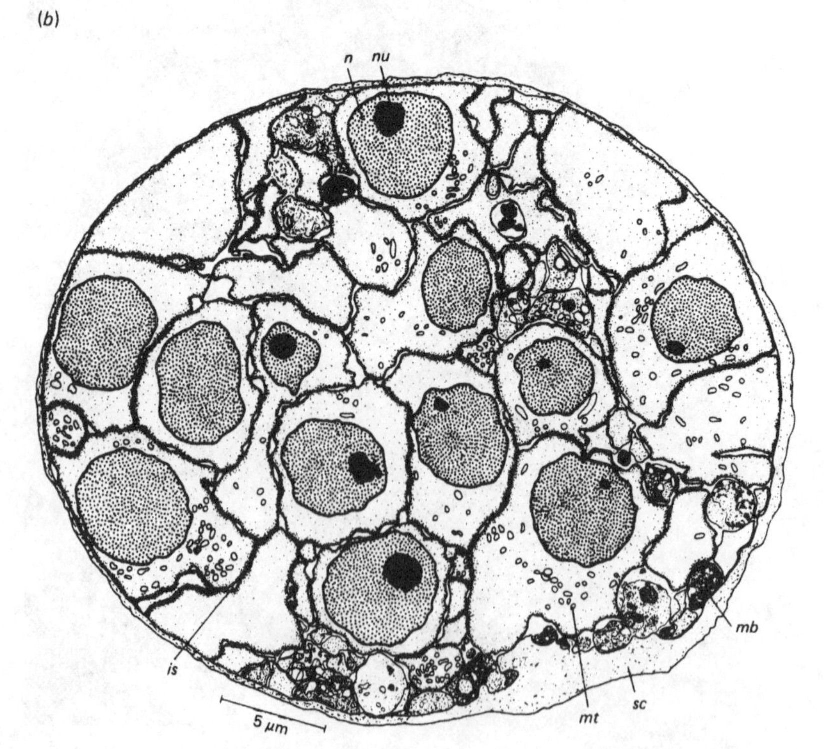

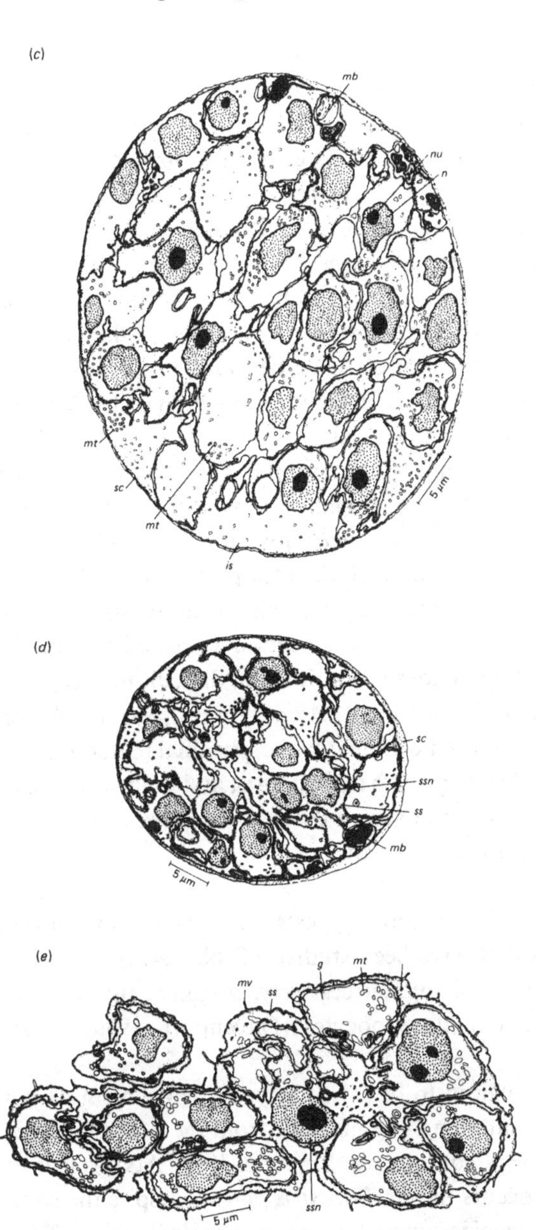

Fig. 7.6. Diagrammatic interpretations based on transmission electron micrographs of the general structure of free ovaries from the body cavities of female *Moniliformis moniliformis* aged (*a*) 1, (*b*) 3, (*c*) 5, (*d*) 6 and (*e*) 9 days. *g*, gonocyte; *is*, intercellular space; *mb*, multivesicular body; *mt*, mitochondrion; *mv*, microvillus; *n*, nucleus; *nu*, nucleolus; *sc*, surface coat; *ss*, supporting syncytium?; *ssn*, supporting syncytium nucleus? (From Asaolu *et al.*, 1981.)

chronological series. First, the early free ovaries contain relatively few cells and perhaps part of the surface of each one of these is in contact with the relatively thick surface coat (Fig. 7.6*a*). Secondly, as the worms age, more cells develop within the ovaries and fewer of these have contact with the surface coat material (Fig. 7.6*a–d*). Thirdly, as the worms age further, the surface coat around each ovary can no longer be detected and microvillous projections which are typical of ovaries in sexually mature *Moniliformis moniliformis* are formed (Figs 7.1 and 7.6*e*). The supporting syncytium has been formed by the time the female *M. moniliformis* are 9 days old (Asaolu *et al.*, 1981), perhaps by fusion of those peripheral cells which retained contact with the surface coat. No evidence was obtained by Asaolu *et al.* to suggest that certain ligament cells enveloped free germinal cells in *M. moniliformis* recently established in rats; it would be rather difficult to explain how any cells of external origin (Atkinson & Byram, 1976) might be incorporated into an immature ovary invested in a thick surface coat. This objection does not conflict with the idea that cells of different origin might have combined to form the ovarian primordium at a stage in development earlier than any that has been investigated to date. If such a combination of cell types does occur, it is not difficult to accept that the oogonial syncytium could also have arisen by a process of cell fusion to give rise to ovaries composed of somatic and germinal material (Atkinson & Byram, 1976). Studies to resolve these points, and to identify the origins of the ovarian primordia and ovarian tissue development in other species of Acanthocephala are urgently required.

Oogenesis. The available evidence suggests that oogenesis in those acanthocephalan species that have been studied (Table 7.4) probably follows the expected pattern involving, successively, oogonia, primary oocytes, secondary oocytes and mature oocytes (Davenport, 1979; Parshad & Crompton, 1981).

The oogonia are believed to be formed by the detachment of portions of cytoplasm, each containing one nucleus, from the oogonial syncytium (Fig. 7.2; Meyer, 1933; Crompton & Whitfield, 1974; Atkinson & Byram, 1976). This process occurs in *Sphaerirostris pinguis* despite the somewhat different structural organization within its free ovaries (Parshad & Guraya, 1977*a*). Nuclei of three sizes have been observed in the oogonial syncytium of *S. pinguis* and it is the largest that are involved in oogonia formation (Parshad & Guraya, 1977*a*). Each free oogonium is observed in electron micrographs to be totally surrounded, and so isolated from all other components of the ovary, by the cytoplasm of the supporting syncytium (Figs 7.2 and 7.7). This condition remains for the remainder of the life of that cell and its division products within the ovary. The fact that the inner as well as the outer surface of the supporting syncytium is bounded by a plasma membrane means that there is also no continuity between this syncytium and any of the germ line cells.

The free oogonia divide mitotically to form primary oocytes and Atkinson & Byram (1976) have tentatively suggested that meiotic prophase may occur at this stage of oogenesis in *Moniliformis moniliformis* while Parshad & Guraya (1977*a*) drew the same conclusion for *Sphaerirostris pinguis*. The structure of the acanthocephalan free ovary makes difficult the precise identification of stages in oogenesis until the mature oocytes are formed. All observers agree, however, that marked cytoplasmic expansion occurs during the maturation of the oocytes. In *Moniliformis*

Fig. 7.7. Transmission electron micrograph of a section through a free ovary from the body cavity of a female *Moniliformis moniliformis*. The supporting syncytium (*s.c.*) is seen to surround the developing oocyte (*d.o.*) with its prominent nucleus (*n*) and inclusions. (From Crompton & Whitfield, 1974.)

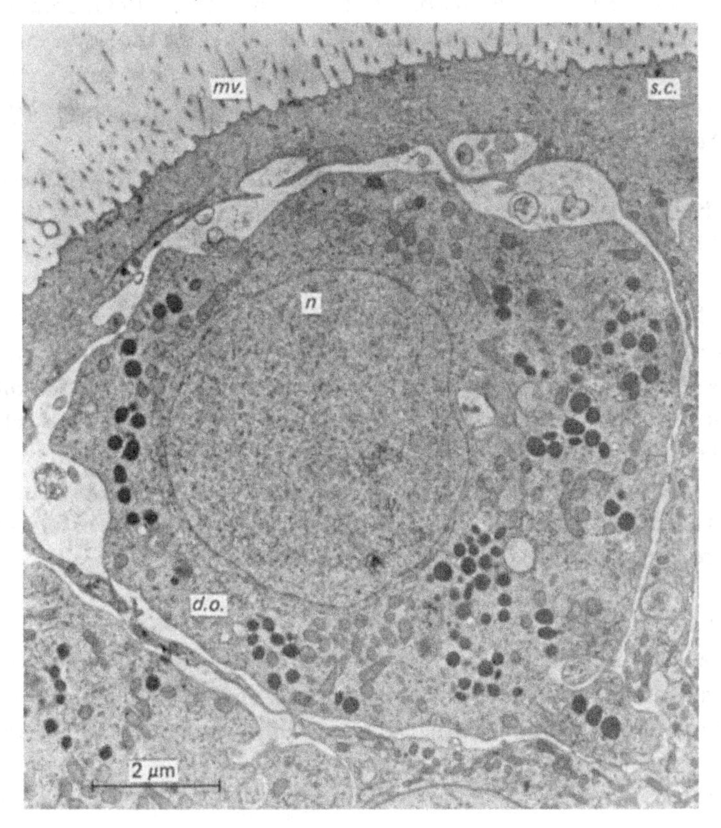

moniliformis, *Polymorphus minutus*, *Acanthogyrus tilapiae* and *Pallisentis golvani*, there is ultrastructural evidence of much metabolic activity in the growing oocytes including the presence of numerous mitochondria, rough endoplasmic reticulum, annulate lamellae, Golgi bodies and lipid droplets (Crompton & Whitfield, 1974; Atkinson & Byram, 1976; Marchand & Mattei, 1976*d*). In *Prosthenorchis* sp., *A. oligospinus* and *Sphaerirostris pinguis*, histochemical tests suggest that lipids, phospholipids, lipoproteins, proteins and RNA accumulate during the growth of oocytes (Guraya, 1969; Anantaraman & Subramoniam, 1975; Parshad & Guraya, 1977*a*).

Mature oocytes usually appear as roughly spherical or elliptical cells which come to lie just below the surface of the free ovary. This is the site of sperm penetration and fertilization (Bullock, 1969). The mature oocytes of *Corynosoma semerme*, which seem typical of others that have been described, measure about 25 μm in diameter after fixation in 3% glutaraldehyde with a prominent spherical nucleus measuring about 9 μm in diameter (Peura *et al.*, 1982). Many mitochondria, electron-dense inclusions and lipid droplets are observed to be dispersed evenly in the cytoplasm. Not all mature oocytes become fertilized and Meyer (1933) drew attention to the presence of 'abortive Keimzellen' in the free ovaries of *Macracanthorhynchus hirudinaceus*. This is probably the first reference to oocyte atresia, the significance of which is still unclear (Cohen, 1977), in the Acanthocephala. Crompton & Whitfield (1974) considered that oocyte atresia occurred in *Moniliformis moniliformis*, Parshad & Guraya (1978*a*) observed the process in ovaries from uninseminated and inseminated *Sphaerirostris pinguis* and Marchand & Mattei (1980) wrote that in *Breizacanthus* sp. the oocytes degenerate without insemination. About 8.5% of oocytes from inseminated and 32% from uninseminated *S. pinguis* were estimated by Parshad & Guraya (1978*a*) to be in some stage of degeneration or atresia. These observations suggest that insemination and/or fertilization, which eventually lead to an extrusion of cells from the ovary, affect the rate of oocyte atresia. In uninseminated worms, cell loss following fertilization is not available to relieve the pressure of cell production from the oogonial syncytium so the rate of atresia, and probable recycling of materials, is increased.

Ovarian fragmentation. There is little reason to doubt that the number of free ovaries increases with age in the body cavities of female acanthocephalan worms, apparently reaching a peak and then declining. Crompton, Arnold & Walters (1976) estimated the numbers of free ovaries in inseminated and uninseminated female *Moniliformis moniliformis* during the course of experimental infections in male rats. Some rats were infected orally with 20 cystacanths each to provide inseminated females while others received surgical transplants of female worms of less than 7 days of age to ensure a supply of uninseminated females. The rats were allowed to feed *ad libitum* on Oxoid 41B diet (Oxoid, 1977) and water. These experimental details are most important because recent work (Keymer, Crompton & Walters, 1983) has shown that host diet and density-dependent factors are associated with striking effects on the fecundity of *M. moniliformis*. A graph of the estimated average numbers of free ovaries in inseminated and uninseminated *M. moniliformis* aged from 1 to 20 weeks is shown in Fig. 7.8. On average, the uninseminated females contained 4238 ovaries per worm which is significantly more ($p < 0.001$) than the average number of 3313 per worm found in inseminated females (Crompton *et al.*, 1976).

The free ovaries of *Moniliformis moniliformis* are prolate spheroids and, after obtaining average measurements of length and width of ovaries from females of different ages, estimates were made by Crompton *et al.* (1976) of the average volume of ovarian tissue per worm. More ovarian tissue was found in the uninseminated worms than in the inseminated, but this may be indicative of less synthetic activity and a slower tissue turnover rate when zygotes are not being produced. Because fertilization in the Acanthocephala involves a reaction between spermatozoa and mature oocytes at the surface of the free ovaries (§7.7), the greater the surface area relative to the mass of ovarian tissue and the greater the rate of mature

Fig. 7.8. The average numbers of free ovaries estimated to be present in uninseminated (filled circles) and inseminated (open circles) female *Moniliformis moniliformis* during the course of the infection in male rats each given 20 cystacanths orally. (From Crompton *et al.*, 1976.)

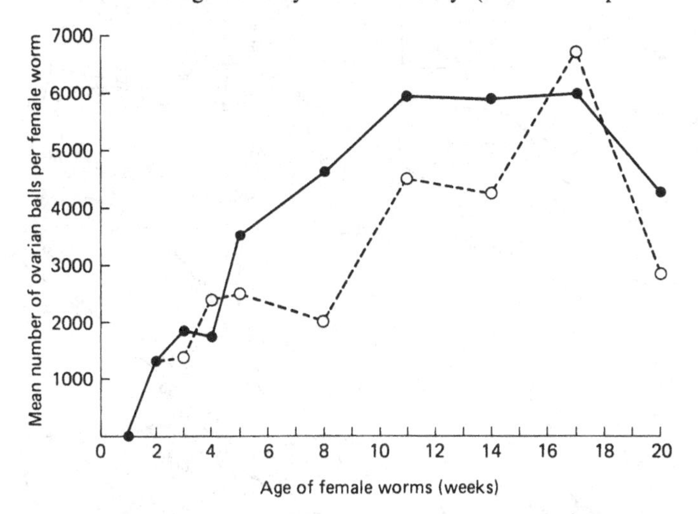

Fig. 7.9. Observations on free ovaries from the body cavities of female *Moniliformis moniliformis* (A–J from uninseminated and K–O from inseminated worms). The ovaries were assumed to be in some stage of division or fission when examined by light microscopy. A, 7 days; B and C, 14 days; D, 21 days; E, 28 days; F and G, 35 days; H, 56 days; I, 77 days; J, 98 days; K, 28 days; L and M, 35 days; N, 56 days; O, 98 days. Dimensions are given in micrometres. *de*, developing egg; *mo*, mature oocyte; *z*, zygote. (From Parshad, Crompton & Martin, 1980.)

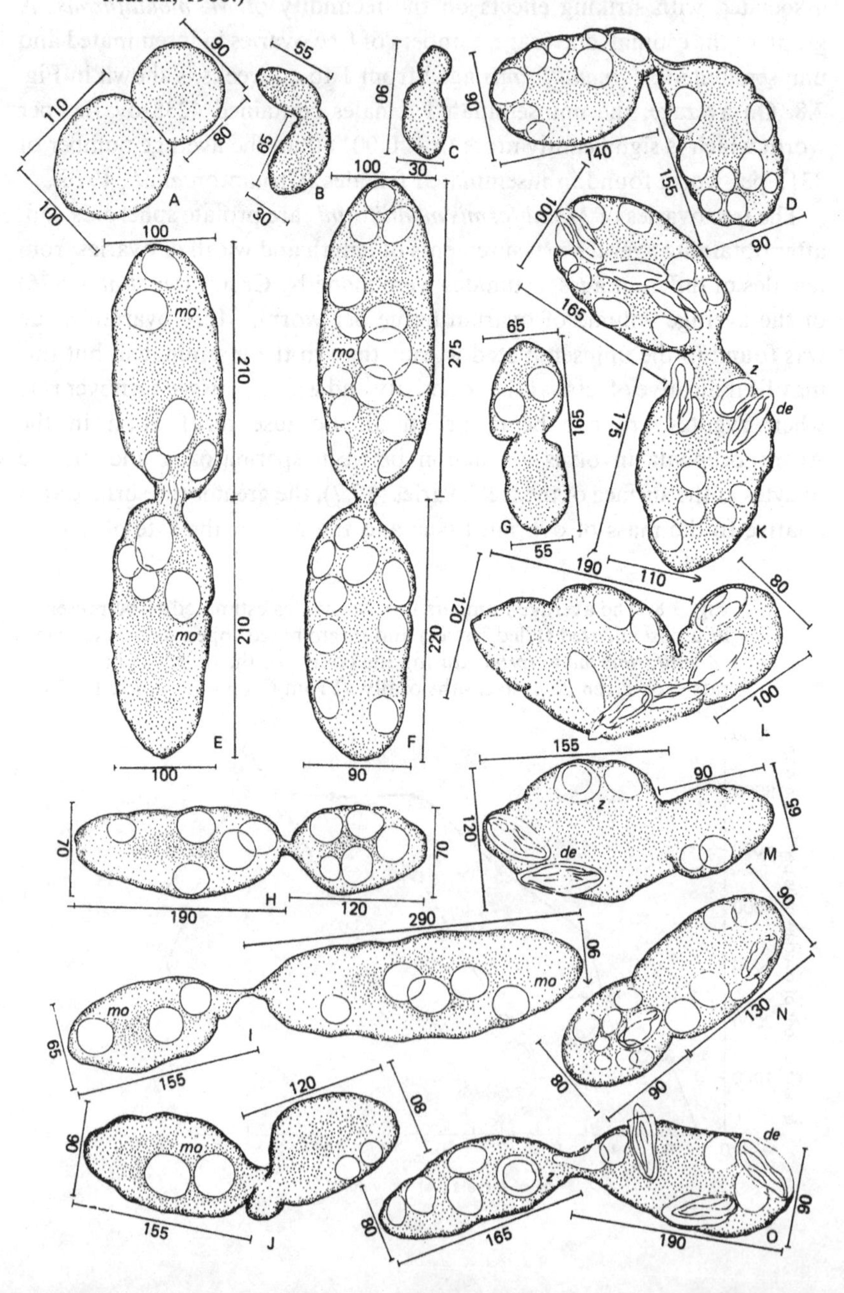

oocyte production, the better will be an individual female worm's chances of passing on its genes to the next generation.

Van Cleave (1953) considered that the production of free ovaries occurred by fragmentation in an unexplained manner. Some clues, however, can now be offered to initiate the much-needed explanation by reference to observations on *Moniliformis moniliformis* made by Crompton & Whitfield (1974), Crompton *et al.* (1976) and Parshad, Crompton & Martin (1980). Fresh and fixed ovaries, when mounted under a coverslip and viewed by light microscopy, are frequently observed to possess furrows, constrictions or creases (Fig. 7.9) as if a gradual process of division were in progress. That division of ovaries occurs can be proposed for free ovaries from uninseminated and inseminated worms, regardless of worm age or parent ovary size (Parshad, Crompton & Martin, 1980). Also the ovaries do not usually appear to divide evenly; fragmentation gives a better concept of the process than division, although fragmentation implies inaccurately that the process occurs quite rapidly. In transmission electron micrographs of the ovarian supporting syncytium of *M. moniliformis*, the cortical cytoplasm contains numerous structures identified tentatively as microfilaments (Crompton & Whitfield, 1974). These potentially contractile elements are located in exactly the right place for initiating ovary division or any other redeployment of the supporting syncytium.

So far in this account no general comparative statements have been made about the dimensions and shapes of the free ovaries. Those illustrated for *Moniliformis moniliformis* (Fig. 7.9) are not untypical while those of other species are much closer to being spherical. A species by species list could easily be compiled from published taxonomic descriptions, but it might serve little purpose since so many factors, including worm age and reproductive state (Crompton *et al.*, 1976), the process of division or fragmentation (Parshad, Crompton & Martin, 1980) and host diet (Keymer *et al.*, 1983), clearly influence the ovaries.

The functional ovary. An outline summary of the cellular and developmental processes that are now believed to occur in a free ovary of a sexually mature female acanthocephalan worm is depicted in Fig. 7.2. The diagram shows the same ovary at four points (I to IV) during its existence in the female body cavity and, by reference to the numbered oogonial nuclei, oogonia formation, oogenesis and oocyte atresia are illustrated, together with fertilization and zygote release. In addition, the ovary is shown to divide with all the cellular processes continuing in the division products. Finally, it should be remembered that the free ovaries grow for at least the first part of the course of infection in the definitive host and that all this activity occurs concurrently in a population of ovaries undergoing similar events. This summary, however, should be considered with some caution since it is based largely on observations made on a single representative species of the phylum.

7.5.4 *Spermatogenesis and spermatozoa*

A list of publications which contains information about the testes, spermatogenesis and spermatozoa, and also draws attention to the major contribution to our knowledge made by B. Marchand and X. Mattei working in Senegal, is given in Table 7.5. Much more work has been done on spermatogenesis and the structure of the spermatozoa than on the testes.

Structure of the testes. The testes of mature male acanthocephalan worms may be located in the anterior (*Sphaerirostris turdi*), middle (*Fessisentis vancleavei*) or posterior (*Moniliformis moniliformis*) part of the body cavity and always in association with the ligament (see Parshad & Crompton, 1981). Usually there are two oval, elliptical or spherical testes arranged in tandem, but some degree of overlap is not uncommon. Nor is it uncommon to find monorchic males as have been described for *Centrorhynchus elongatus*, *Acanthocephalus dirus*, *Corynosoma constrictum*, *Neoechinorhynchus oreini*, *M. moniliformis* and *F. vancleavei* by Kobayashi (1959), Bullock (1962) and Amin (1975*a*,*b*), Schmidt (1965), Fotedar (1968), Crompton (1972*b*) and Buckner & Nickol (1978), respectively. The single testis of a monorchic male is generally larger than either of the testes of a diorchic male of the same age and species and in *F. vancleavei* the monorchic condition appeared to be typical (Buckner & Nickol, 1978). Monorchidism may not be disadvantageous; a single monorchic male *M. moniliformis* was shown by Crompton (1972*b*) to be capable of carrying out its role in copulation, insemination and successful fertilization.

Less is known about the cytological development of the testes than that of the ovaries. Whitfield (1969) recognized a form of cytological organization typical of the early testis of *Polymorphus minutus* towards the end of its development in the intermediate host (Table 7.5) and a form typical of the late testis in young adult worms in domestic ducks. Two cell types were identified by electron microscope examination of the early testis; the gonocyte or primordial germ cell and the supporting cell. Each discrete gonocyte is reasonably regular in outline with a diameter of 5–7 μm and an undifferentiated appearance characterized by a relatively large nucleus and basophilic cytoplasm. The supporting cell, in contrast, appears to be well differentiated, with an irregular outline and cytoplasm containing mitochondria, Golgi bodies and other organelles and inclusions. The outline of each supporting cell was highly irregular and cytoplasmic extensions surrounded and separated the gonocytes. Whitfield suggested tentatively that, since contacts between supporting cells were so common and close, further work might reveal the presence of a syncytium rather than cell population. The most obvious change in cellular organization observed by Whitfield (1969) in the late testis of *P. minutus* was the presence of the division products of the gonocytes and the onset of spermatogenesis. Whitfield concluded that five spermatogonic and spermiogonic divisions occurred in the late testis of *P. minutus* within about 50 h.

Parshad & Guraya (1979) studied the cytology of testes of *Sphaerirostris pinguis* from naturally infected house crows sampled on four occasions during the year. Circumstantial evidence suggested that seasonal changes may occur in the testes of *S. pinguis*, and it is tempting to suggest that the reproductive cycle of the host might have influenced the changes (§7.9; Chapter 9). The testes of *S. pinguis* collected from May to July showed clear signs of spermatogenesis while those from worms collected from November to December did not and tended to be vacuolated. However, this potential example of a relation between host hormones and acanthocephalan reproduction will remain uncertain until the course of infection of *S. pinguis* has been worked out in the laboratory. At present, it is difficult to distinguish between testes which are inactive because of the state of the host's hormonal profile and those which are inactive because the male worms are senescent.

Spermatogenesis. Three phases may be recognized in the orderly sequence of events known as spermatogenesis. These are spermatocytogenesis, during which spermatogonia proliferate through mitosis to form spermatocytes; meiosis, during which the chromosome number is halved and clusters of spermatids are produced; and spermiogenesis, during which the spermatids change into the spermatozoa (Bloom & Fawcett, 1968). This summary is derived mainly from studies on mammalian material, but such evidence as we have indicates that spermatogenesis will be found to be reasonably similar in all species of the Acanthocephala (Table 7.5). Interestingly, spermatogenesis in many species of mammals can only occur normally once the testes have descended into the scrotum where the temperature will be anything from 4 to 7 °C cooler than the deep body temperature (Setchell, 1982) which will be experienced by the testes of their acanthocephalan parasites in the small intestine.

Spermiogenesis was followed in detail in the late testis (Whitfield, 1969) of *Polymorphus minutus* by Whitfield (1971 *a*). Morulae of spermatids were

Table 7.5. *Studies giving information on testis structure, spermatogenesis and spermatozoa in the Acanthocephala*

	Reference	Common microtubule arrangement of spermatozoan flagellum
ARCHIACANTHOCEPHALA		
Macracanthorhynchus hirudinaceus	Kaiser (1893); Meyer (1933); Robinson (1964)	—
Mediorhynchus sp.	Marchand & Mattei (1978*a*)[a]	—
Moniliformis cestodiformis	Marchand & Mattei (1977*b*, 1978*a*)[a]	9+2
Moniliformis moniliformis	Asaolu (1977)[a]; Robinson (1965); Whitfield (1969)[a]	9+2
Prosthenorchis sp.	Guraya (1971)	—
PALAEACANTHOCEPHALA		
Breizacanthus sp.	Marchand & Mattei (1980)[a]	—
Centrorhynchus milvus	Marchand & Mattei (1976*c*, 1978*a*)[a]	9+2
Echinorhynchus gadi	Franzen (1956)	—
Polymorphus minutus	Whitfield (1969, 1971*a*, *b*)[a]	9+2
Porrorchis centropi	Marchand & Mattei (1978*a*)[a]	—
Rhadinorhynchus pristis	Marchand & Mattei (1978*a*)[a]	9+2
Serrasentis sagittifer	Marchand & Mattei (1977*a*, 1978*a*)[a]	9+2
Sphaerirostris pinguis	Parshad & Guraya (1979)	—
Tegorhynchus furcatus	Marchand & Mattei (1976*a*, 1978*a*)[a]	9+0
EOACANTHOCEPHALA		
Acanthogyrus tilapiae	Marchand & Mattei (1976*b*, 1977*b*, 1978*a*)[a]	9+2
Neoechinorhynchus agilis	Marchand & Mattei (1977*b*, 1978*a*, *b*, 1979)[a]	9+3
Pallisentis golvani	Marchand & Mattei (1977*b*, 1978*a*)[a]	9+2
Tenuisentis niloticus	Marchand & Mattei (1977*b*)[a]	9+2

[a] Transmission electron microscopy

identified by means of electron microscopy and each of these was considered to consist of a multinucleate syncytium of as many as 64 spermatids. Each early spermatid contained a relatively large nucleus, concentrations of ribosomes, a few mitochondria and a centriole, but no obvious rough endoplasmic reticulum. Later spermatids, still attached to their partners by a cytophoral stalk, were observed to contain membrane-bounded electron-dense inclusions and a conspicuous Golgi body. The flagellum was observed to develop and to emerge out of the spermatid roughly opposite to the cytophoral stalk with its origins at the centriole. Soon after the flagellum appeared, the nucleus became very elongate and either grew around the extending flagellum, or developed a cleft to accommodate it. Next, the nuclear envelope disintegrated to form the spermatozoan body containing the chromatin and with its characteristic electron-dense inclusions along both sides of the flagellum. The spermatozoon of *P. minutus* was seen as an elongate, filiform cell having neither mitochondria nor any obvious acrosome (Whitfield, 1971*a*). The mitochondria appeared to be extruded into the cytophoral stalks which remained as the spermatozoa broke away to form free cells.

Since Whitfield's study of *Polymorphus minutus*, Marchand and Mattei have investigated aspects of spermatogenesis and spermatozoan structure in 12 other species representing the three classes of the phylum (Table 7.5). Their observations (Marchand & Mattei, 1977*b*, 1978*a*, *b*, 1979), which are perhaps best summarized by their description of spermiogenesis in *Neoechinorhynchus agilis*, have much in common with that of Whitfield (1971*a*) except for the direction of the formation of the flagellum and the mode of elimination of the mitochondria. According to Marchand and Mattei, the centriole migrates to the anterior region of the spermatid at the start of spermiogenesis. The flagellum then extends posteriorly for a while until the nucleus changes shape and extends along the length of the flagellum until its end is reached. At this time, the flagellum extends in the opposite direction while the nuclear envelope disintegrates and the nucleocytoplasmic derivative or spermatozoon body is formed. In this way the anatomy of the spermatozoon is reversed with most of the free flagellum extending anteriorly as illustrated in figure 1 of Marchand & Mattei (1978*a*). This mode of flagellum formation must be considered the typical process in the development of acanthocephalan spermatozoa. When the nucleocytoplasmic derivatives are formed, drops of apparently unrequired cytoplasm containing the mitochondria are extruded from the spermatids (Marchand & Mattei, 1978*b*). In 1965, Austin wrote that the cytoplasmic components of many spermatozoa, and possibly of all, are known to include mitochondria. Not all acanthocephalan species fit this view. Regardless of the high quality of the electron micrographs, all should remember that the authors are using a series of static images to present a description of a continuously changing living process.

Spermatozoa. With the exception of *Tegorhynchus furcatus*, in which the spermatozoa may be unusual in possessing a distinct ovoid structure at one end (Marchand & Mattei, 1976*a*), acanthocephalan spermatozoa appear to be filiform cells measuring from 25 to 65 μm in length and up to about 0.5 μm in width (Fig. 7.10; Parshad & Crompton, 1981). The spermatozoa of *Moniliformis moniliformis* are known to be motile *in vivo* and *in vitro* and, since mitochondria appear to be lacking, glycolysis from carbohydrate to pyruvate may be assumed to be the principle form of energy metabolism.

Spermatozoon motility is effected through the free flagellum which contains the usual axoneme consisting of microtubules. The combination of nine double outer microtubules and two central microtubules is most

Fig. 7.10. Simplified cutaway stereogram of a spermatozoon of *Polymorphus minutus*. *a.*, anterior; *d.*, dense inclusion; *f.*, flagellum; *f.c.*, flagellar centriole; *s.*, spermatozoan body. (From Whitfield, 1971*a*.)

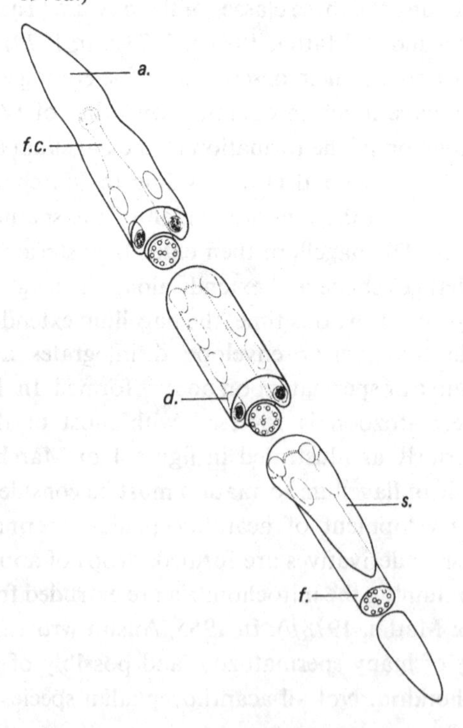

common in the species studied (Table 7.5), but in *Neoechinorhynchus agilis* about 96% of the axonemes showed a 9+3 rather than a 9+2 pattern (Marchand & Mattei, 1977*b*). Combinations of 9+0, 9+1, 9+4 and 9+5 have also been observed, but these rarely form more than a small proportion of the sample examined. In spermatozoa of *Acanthogyrus tilapiae* about 7% showed a 9+0 pattern, 18% showed 9+1, 73% showed 9+2 and 1% showed 9+3 (Marchand & Mattei, 1977*b*). The functional significance of these variations in the axoneme is not known; perhaps motility is affected in those spermatozoa which do possess the 9+2 array of microtubules. Living spermatozoa normally display propagated sine waves along the length of the flagellum. It may be assumed that the motility is associated with ATPase activity in the arms of the A microtubules (Warner, 1974) in the outer circle of nine (Whitfield, 1971*a*).

Guraya (1971) and Whitfield (1971*a*) concluded that there was no acrosome in the spermatozoon of a species of *Prosthenorchis* and *Polymorphus minutus*, respectively. During spermiogenesis, the Golgi body is known to be involved in the formation of the acrosome which subsequently facilitates fertilization (Austin, 1965). Golgi bodies are undoubtedly present during acanthocephalan spermiogenesis (see references cited in Table 7.5) and it is possible, as has been suggested by Marchand & Mattei (1977*a*) for *Serrasentis sagittifer*, that a derivative from the centriole may have taken over the acrosomal function, which is concerned with the attachment of a spermatozoon and its penetration through the oocyte membranes (Austin, 1965).

7.6 Sexual congress, copulation and insemination

Internal fertilization, which depends on copulation and insemination, is essential for the survival of endoparasitic helminths, but difficult to account for in discussions of the origins and evolution of the Acanthocephala from free-living, aquatic ancestors (Conway Morris & Crompton, 1982).

7.6.1 *Sexual congress*

Sexual congress has been defined by Cohen (1977) as the association of males and females for sexually reproductive purposes. Although there is no supporting evidence, it seems reasonable to assume that acanthocephalan worms are likely to be polygamous, which means that each individual may copulate with more than one partner of the opposite sex (Jewell, 1976). When viewed in terms of selfish genes, each adult worm should be considered to have an overriding sexual drive to pass on its genetic material to the next generation. For females this will be achieved

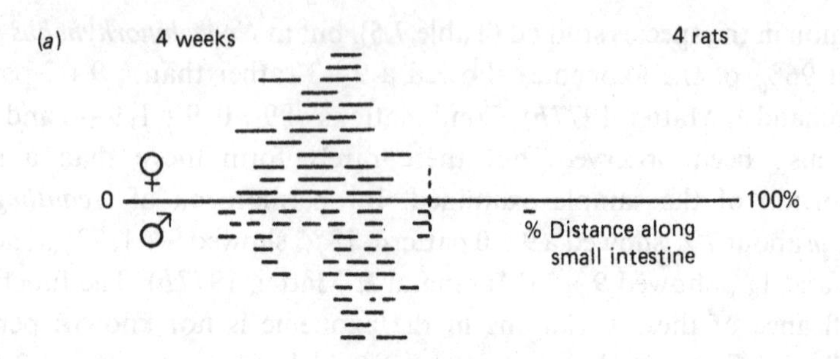

(a)
4 weeks
4 rats
♀
♂
0
100%
% Distance along small intestine

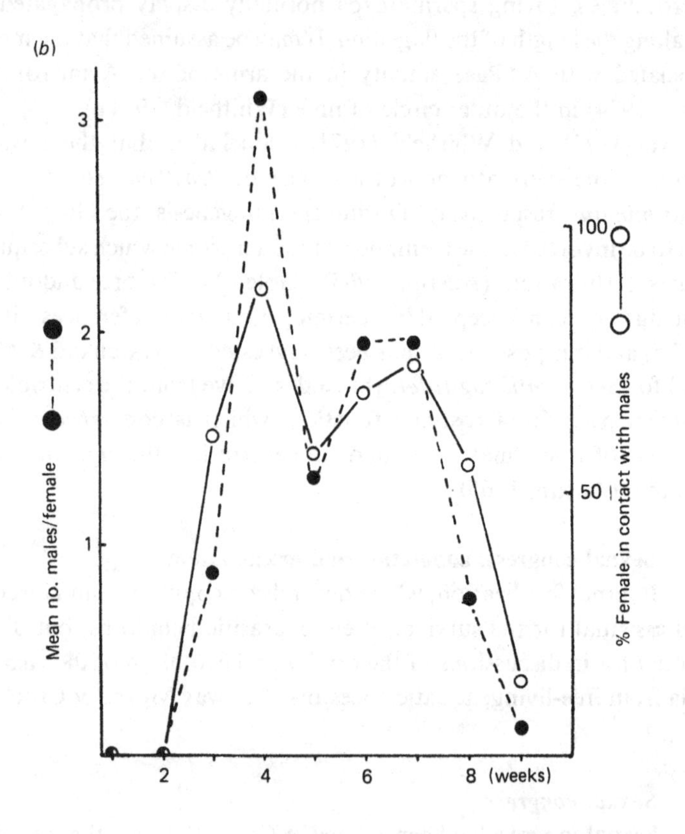

(b)
Mean no. males/female
% Female in contact with males
3
2
1
100
50
2
4
6
8 (weeks)

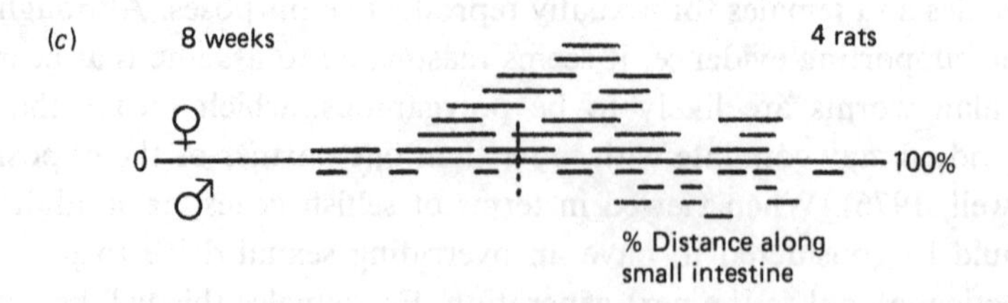

(c)
8 weeks
4 rats
♀
♂
0
100%
% Distance along small intestine

theoretically by (1) an increased longevity without loss of fertility; (2) the production of the number of offspring that contributes most recruits to the next generation; and (3) successful competition with other females. In contrast, a male acanthocephalan will have many opportunities for passing on its genes provided that access can be gained to many fertile females (see Jewell, 1976).

Most information relevant to sexual congress, copulation and insemination in the Acanthocephala can be gleaned from the results of experimental infections of *Moniliformis moniliformis* in laboratory rats (Burlingame & Chandler, 1941; Holmes, 1961; Crompton *et al.*, 1972; Crompton, 1974, 1975; Atkinson & Byram, 1976; Abele & Gilchrist, 1977; Nesheim *et al.*, 1977, 1978; Parshad, Crompton & Martin, 1980).

The adults of *Moniliformis moniliformis*, like those of other species, occupy fairly precise sites in the small intestine of the definitive host (Crompton, 1973, 1975) although these frequently change due to distinct emigrations during the course of the infection. Newly established *M. moniliformis* are found in the posterior half of the rat's small intestine, but 3–4 weeks later, most are observed to be attached to the mucosa in a compact group in the anterior part of the small intestine (Crompton, 1975; Nesheim *et al.*, 1977). Presumably the worms have moved anteriorly by a distance of some 50 cm and have also attained sexual maturity. The anterior emigration and the sexual development of the worms is dependent on the composition of the rat's diet (Nesheim *et al.*, 1977; Parshad, Crompton & Nesheim, 1980; §7.9). Crompton (1975) undertook an analysis of body contacts between male and female *M. moniliformis* during the course of infection in appropriately nourished rats. It was possible to claim (Fig. 7.11) that when the worms were aged 4 weeks and had emigrated anteriorly, each female was, on average, in contact with the maximum number of males. The highest proportion of females experiencing contact with at least one male also occurred at this time. This is the period of the infection when the sex ratio is 1:1 (§7.3), when fertilization can be shown to have begun (Crompton, 1974; Atkinson & Byram, 1976) and when the start of egg release is drawing near (Burlingame & Chandler,

---

Fig. 7.11. Observations on the distribution of male and female *Moniliformis moniliformis* in male rats fed *ad libitum* on a purified diet rich in starch (Nesheim *et al.*, 1977) for 4 (*a*) or 8 (*c*) weeks. Each rat was given 20 cystacanths orally. A graphical representation (*b*) of the results of an analysis of body contacts observed between worms at postmortem examination suggests that most worms will be in contact with each other and most females will be in contact with males 4 weeks after the start of the infection. (From Crompton, 1975.)

1941; Crompton *et al.*, 1972), and therefore the time when copulation would be expected to be occurring frequently. Other opinions, however, show how density-dependent factors may influence reproduction by their effect on sexual congress (§7.9). It is interesting to note (Fig. 7.11) that less contact appears to occur between male and female *M. moniliformis* by the time they are 8 weeks old. By this time, the number of worms has started to decline (Crompton, Keymer & Arnold, 1984) and there has been a gradual posterior movement of the population in the small intestine.

Sexual congress as defined above is not such a useful concept unless attention is drawn to the duration of the association between male and female worms. Under controlled conditions, where the population size of *Moniliformis moniliformis* could not exceed 12 worms per rat, Crompton *et al.* (1972) found that the average patent period of a female worm was about 106 ± 16 days. When female *M. moniliformis* were transplanted into rats after having had varying periods of contact with males from the start of the course of the infection, patent periods similar to those measured by Crompton *et al.* (1972) were not observed unless male and female worms had been together for the first 5 weeks after becoming established in rats (Crompton, 1974). Eggs of *M. moniliformis* were no longer detected in the host's faeces when the female worms reached 53 days old if they had been 17 days old on transplantation, whereas eggs were still present when the worms were 119 days old provided that they were at least 35 days old on transplantation (Crompton, 1974). The young female worms (17 days old on transplantation) were not lost from the rats when egg production stopped under these circumstances. The main conclusion from this work is that the period of 5 weeks is necessary to allow the female worms to acquire sufficient spermatozoa, from one or several males, for the fertilization of the average production of mature oocytes during the lifespan (§7.9). Also, 5 weeks would encompass the probable period of maximum physical contact between male and female worms (Fig. 7.11).

7.6.2 *Copulation*

Although the phylum is comprised of about 1000 species, copulation has rarely been observed in the Acanthocephala. Species with males and females assumed to have been *in copulo* at postmortem examination of the definitive host include *Sphaerechinorhynchus rotundocapitatus* by Johnston & Deland (1929), *Acanthogyrus dattai* by Podder (1938), *A. holospinus* by Sen (1938), *Gorgorhynchus clavatus* by Van Cleave (1940), *Porrorchis rotundatus* by Golvan (1956*c*), *Acanthogyrus partispinus* by Furtado (1963) and *Moniliformis moniliformis* by Atkinson & Byram (1976). Individuals of *Neoechinorhynchus emydis* and *N. pseudemydis* were observed *in copulo* by Dunagan (1962) during an attempt to grow the worms *in vitro*.

Dunagan's experimental system might be extended to provide more knowledge of how copulation occurs. At present a miscellany of observations reveals that the process involves a union of the posterior ends of the partners (Podder, 1938; Sen, 1938) although there has been some disagreement as to whether the copulatory bursa is everted during copulation. Podder (1938) wrote, in describing *Acanthogyrus dattai*, that the everted bursa of the male became attached to the posterior end of the female, while Sen (1938) stated that the vaginal end of the female *A. holospinus* appeared to be inserted into the bursa of the male which did not show any evidence of eversion. Most authors, however, are convinced that the bursa of the male, which is shown in Fig. 7.12 from a recent study of *Pseudoacanthocephalus bufonis* by Kennedy (1982), must have a copulatory function (Van Cleave, 1949; Parshad & Crompton, 1981) and Whitfield (1969) interpreted its morphology in male *Polymorphus minutus* as essential for copulation.

The role in the copulatory process of the so-called genital spines, protruding from the surface of the posterior part of the trunk in worms of some species (Van Cleave, 1920*b*), is less clear than that of the bursa. Those on the female may become embedded in the inner walls of the bursa of the male and so strengthen the copulatory union, albeit at the cost of some tissue damage to the male. Those on the male can have no obvious

Fig. 7.12. The copulatory bursa of a male *Pseudoacanthocephalus bufonis*. The worms were described as being found 'with fully everted bursa and cirrus'. (From Kennedy, 1982.)

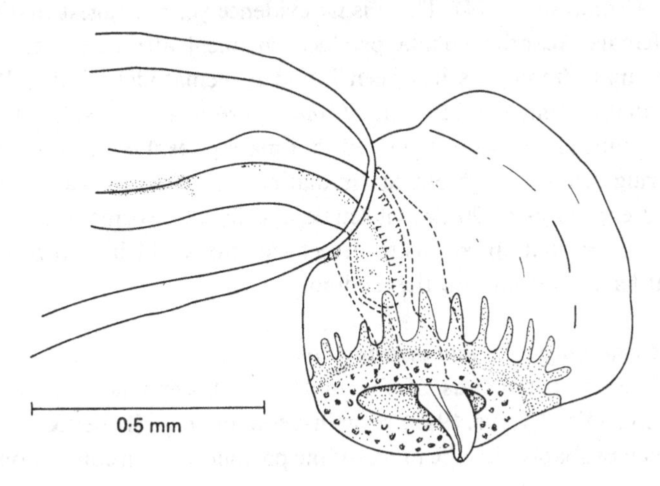

copulatory function while many species copulate successfully without any trace of genital spines. The adjective 'genital' may be misleading; any trunk spines are bound to increase friction against the intestinal mucosa and aid in locomotion or the retention of position.

Some idea of the onset of copulation for several species may be obtained from Table 7.1. It is to be expected that the males will have a more active role in copulation than the females (Parshad & Crompton, 1981). The copulatory anatomy of the male (Fig. 7.12) is more complex than that of the female and the sensory apparatus and nervous system of the male (Chapter 5) are more extensive. Until recently it was known that while both male and female Acanthocephala possessed a cerebral ganglion, the males also possessed a genital ganglion, or pair of ganglia, consisting of two groups of cells joined by at least one commissure (Meyer, 1933). A third ganglion, called the bursal ganglion, has been found by Dunagan & Miller (1977, 1978*a*, 1979) in male *Moniliformis moniliformis* and male *Macracanthorhynchus hirudinaceus*. Dunagan & Miller (1977) suggested that the activities of the bursal ganglion may be coordinated with those of the genital ganglion by the nerves that innervate the muscles responsible for bursa eversion and withdrawal; those movements must require considerable control. The absence of any such neural tissue in females strongly indicates that the males must initiate copulation. That this seems likely is supported indirectly by the common observations that male acanthocephalans are often found free in the host's gut lumen at postmortem examination or attached less firmly to the intestinal mucosa than are the females (Nicholas & Hynes, 1958).

Male and female *Moniliformis moniliformis* invariably find each other and copulate, regardless of age or numbers, when surgically transplanted into rats (Crompton, 1974). There is no evidence yet to suggest that male and/or female Acanthocephala produce chemical attractants to assist successful mate finding as has been found in nematodes (Anya, 1976). Heterosexual mating attraction in *M. moniliformis* was considered to be absent by Bone (1976) who suggested that mating resulted from trial and error pairing through thigmotaxic stimulation. This opinion was based on controlled exposures of 20-day-old male and female worms *in vitro* in a specially constructed apparatus to detect secretions which each or either sex might have released into the medium.

7.6.3 *Copulatory caps and insemination*

The cement glands of the male acanthocephalan discharge a mucilaginous (Van Cleave, 1949), proteinaceous material (Haley & Bullock, 1952) which probably aids the union of the partners during copulation and insemination, but soon after the act is over becomes hardened to form a copulatory cap around the genital extremity of the female (Van Cleave, 1949). The presence of copulatory caps on females is a useful indicator to the observer that copulation and probably insemination must have occurred. Copulatory caps may also show the observer that the acanthocephalans under study were located in a suitable and supportive host species.

The significance of copulatory caps is not straightforward. If the caps become hardened around the genital spines, as has been described by Van Cleave (1949) for young mature females of the genus *Corynosoma*, further insemination may be prevented for a time and egg release through the efferent duct system (Whitfield, 1968, 1970) may be delayed. In a general discussion of sexual selection, Krebs & Davies (1981) referred to the copulatory caps of *Moniliformis moniliformis* as chastity belts, a suggestion reminiscent of that made by Devine (1975) in a discussion of copulatory plugs in snakes. If female *M. moniliformis* copulate only once or rarely as the chastity belt might suggest, sufficient spermatozoa would have to be transferred at that time, and stored subsequently to fertilize enough mature oocytes for an average egg production per female of about 600000 eggs during a period of about 106 days (Crompton *et al.*, 1972). That possibility seems unlikely since male *M. moniliformis* aged 16 and 17 days can copulate and inseminate females of the same age, but nothing like full egg production is achieved if the worms are not allowed to associate for at least 19 more days in the rat (Crompton, 1974). A copulatory cap that remained attached to a female worm for a few days, however, could have many advantages for the genes of the individual male that secreted the cement. Brief closure of the female gonopore would not only minimize losses of that male's spermatozoa, but might also ensure that at least some of them would encounter mature oocytes without competition from spermatozoa from another male. In the case of *M. moniliformis* established in laboratory rats, a picture is emerging to suggest that most copulatory activity and insemination probably occurs when the worms are between the ages of 16 and 35 days (Crompton, 1974, 1975), perhaps with maximum activity taking place at the lower end of this range. Thereafter, copulatory caps, which might be expected to survive for a few days at most, would be a serious impediment to egg release which can first be detected when the worms are on average 38 days old (Burlingame & Chandler, 1941; Crompton *et al.*, 1972).

Some observers have described the presence of copulatory caps over the posterior ends of male worms. Nicholas & Hynes (1958), Awachie (1966), Amin (1975*c*), Abele & Gilchrist (1977) and Moore & Bell (1983) spotted copulation caps on males of *Polymorphus minutus*, *Echinorhynchus truttae*, *Acanthocephalus dirus*, *Moniliformis moniliformis* and *Plagiorhynchus cylindraceus*, respectively. Abele & Gilchrist interpreted the capping of the males in terms of sexual selection in which some males or possibly one male would effectively and temporarily eliminate other males from the competition for available females. This suggestion is fully in accord with the current view that the combination of gamete dimorphism and a sex ratio of 1:1 results in competition between males for females (Krebs & Davies, 1981). Abele & Gilchrist conceded that the caps on the males might have resulted from poor sex recognition, but they obviously felt that the male *M. moniliformis* could distinguish between the sexes because they stated that they could find no evidence that spermatozoa had been transferred into the males that had been capped. In marked contrast, Moore & Bell (1983) concluded that the copulatory caps they found on male *Plagiorhynchus cylindraceus* might have been a response of the worms to the deteriorating environment in dead hosts. Further information about copulatory caps is given by Fisher (1960), Dunagan & Miller (1973) and Amin (1975*c*).

In most animals, spermatozoa are transferred to the female in semen, a fluid secreted by various glands (Cohen, 1977). Presumably acanthocephalan spermatozoa must also be transferred in semen, but its origin is unknown. Atkinson & Byram (1976) suggested tentatively that spherical objects, measuring about 55 μm in diameter and situated in the body cavities of 18-, 19- and 154-day-old female *Moniliformis moniliformis*, might have been spermatophores. These objects might also have been a rare form of abnormal or degenerating free ovary; no obvious organ for spermatophore production in a male *M. moniliformis* has been described.

7.7 Fertilization

The process of fertilization appears to involve, for most species of animal which have been studied in detail, a sequential series of at least five identifiable stages (Cohen, 1977). These are (1) penetration of the oocyte envelopes by the spermatozoon; (2) some form of membrane fusion between the spermatozoon surface and the oocyte to bring male and female cytoplasm into contact; (3) the entry of at least the spermatozoon nucleus, and often other components, into the oocyte; (4) the activation of the oocyte which often releases a prominent cortical reaction resulting in a block to polyspermy, metabolic and synthetic changes and the formation of polar bodies; and (5) syngamy or the fusion of nuclear materials (see Parshad & Crompton, 1981). This scheme cannot yet be claimed with certainty to apply to fertilization in the Acanthocephala, but in the account that follows some of these events can be recognized as occurring in the same order as set out above for the few species which have been observed.

Fertilization almost certainly begins while the mature oocytes are located beneath the surface of the free ovaries. Meyer (1928, 1933) demonstrated that spermatozoa became attached to the surfaces of the free ovaries of female *Macracanthorhynchus hirudinaceus* and more recent evidence obtained from *Acanthocephalus ranae* by John (1957), *Polymorphus minutus* by Crompton & Whitfield (1974), *Moniliformis moniliformis* by Crompton & Whitfield (1974), Atkinson & Byram (1976) and Parshad, Crompton & Martin (1980), *Acanthogyrus tilapiae* and *Pallisentis golvani* by Marchand & Mattei (1976*d*), *Neoechinorhynchus agilis* and a species of *Breizacanthus* by Marchand & Mattei (1979, 1980) and from *Corynosoma semerme* by Peura *et al.* (1982) has confirmed Meyer's observation and those of the workers listed above have shown that fertilization in these species, which represent the three classes of the phylum, begins while the mature oocytes are contained within the free ovaries. In this context, 'mature oocyte' means an egg cell that is able to participate in fertilization; no indication of its phase of cell division during oogenesis (§7.5.3) is intended.

Some doubt about the site of fertilization for all species of Acanthocephala must remain since Van Cleave (1953) wrote 'fertilization takes place as the ova leave the egg balls'. This view was shared by Guraya (1969) who stated that in a species of *Prosthenorchis*, the ripe oocytes escaped from the ovarian balls and were then fertilized, and by Nicholas & Hynes (1963*b*) who concluded that in *Polymorphus minutus* the eggs were budded off from the ovarian balls and were fertilized in the body cavity. In the same paper, however, Nicholas & Hynes reported the presence of twisted, rather spindle-shaped, brilliantly stained (Feulgen's method) bodies in the ovarian balls and these were assumed to be spermatozoan heads. Anantaraman & Subramoniam (1975) observed cells identified by them as free oocytes in the body cavity of *Acanthogyrus oligospinus* and Marchand & Mattei (1976*d*) also noticed free oocytes in the body cavity of *Pallisentis golvani*; in neither case was fertilization considered to be occurring.

The onset of fertilization in the Acanthocephala depends on spermatozoa, or at least on their hereditary material, first passing through the supporting syncytium before penetration of the oocyte can occur. Spermatozoa undoubtedly make an intimate and strong attachment to the microvillous surface of the free ovaries and some attachments survive the rigors of fixation and processing for light and electron microscopy. According to Marchand & Mattei (1976*d*, 1977*a*, 1979, 1980), who investigated three eoacanthocephalan and two palaeacanthocephalan species, the free flagellum, which they consider to be the anterior extremity of the spermatozoon, becomes attached to the ovarian surface. Marchand & Mattei (1979) described, from transmission electron micrographs, a swelling at the flagellar apex at the time of contact of the spermatozoa of *Neoechinorhynchus agilis* with the ovarian surface. A spermatozoon with a swollen portion in close contact with an ovarian surface of *Moniliformis moniliformis* was observed with the scanning electron microscope by Parshad, Crompton & Martin (1980). These and other aspects of the behaviour of acanthocephalan spermatozoa in the body cavity of the female worm are extremely difficult to follow and interpret because the timing of events is almost impossible to determine *in vivo*. Acanthocephalan fertilization and early embryology would soon be better understood if a short-term *in vitro* system could be developed for the maintenance of spermatozoa and ovaries.

Changes also appear to occur at the ovarian surface of some species in response to the close presence of spermatozoa. In *Neoechinorhynchus agilis*, the surface microvilli seem to become much reduced in number and are replaced by relatively large vesicles which may facilitate the entry of the spermatozoa into the ovarian tissues (Marchand & Mattei, 1979) and bring them nearer to the mature oocytes. Earlier, Atkinson & Byram (1976) published a transmission electron micrograph showing longitudinal and transverse sections through either a spermatozoon or spermatozoa enclosed in a sizeable 'vacuole' of supporting syncytium near the surface of an ovary from a 106-day-old *Moniliformis moniliformis*. They considered that some fusion might have occurred between the microvilli and the spermatozoa plasmalemma. Scattered observations like these show how much more work needs to be done on the interesting relationship between the spermatozoa and the ovaries.

There is, however, abundant evidence to show that the spermatozoa become incorporated into the tissues of the free ovaries. For example, more than 100 sections of spermatozoan flagella were counted by Marchand & Mattei (1976*d*) in a single free ovary of *Acanthogyrus tilapiae*, but it appeared that the spermatozoa were enclosed in supporting syncytium which separated them from the germ cells. In ovaries of a species of *Breizacanthus*, Marchand & Mattei (1980) interpreted transmission electron micrographs as showing that the spermatozoa could move freely in any part of the ovary, including the germ cells, but that a cytoplasmic jacket of supporting syncytium was always present. However, when a spermatozoon was observed to have penetrated a mature oocyte, the cytoplasmic sheath disappeared so that the entire and much-folded spermatozoon (60 μm in length as compared with an oocyte of 10 μm diameter) must have been in contact with the cytoplasm of the oocyte. They also concluded that earlier observations (Marchand & Mattei, 1976*d*, 1979), where only spermatozoan flagella and their accompanying sheaths of supporting syncytium were found in germ cells, indicated that the nucleocytoplasm and its chromatin did not penetrate unless the oocyte was mature. Following the disintegration of the cytoplasmic sheath around the newly penetrated spermatozoon in the mature oocyte of a species of *Breizacanthus*, Marchand & Mattei (1980) observed the separation of the flagellum from the nucleocytoplasmic derivative and an amalgamation of the derivative with elements of the oocytic cytoplasm which resulted in a 'reticulum saccule'. It seems probable that the saccule will be found to be the male pronucleus, but confirmation of this view will depend on the identification of the female pronucleus and the formation of the zygote nucleus by fusion. Earlier, Crompton & Whitfield (1974) had published a transmission electron micrograph of a section through a zygote from a 28-day-old *Polymorphus minutus* in which several sections could be seen of a spermatozoan flagellum (Fig. 7.13). There was absolutely no trace of any sheath of supporting syncytium around the flagellum but, since the cell under observation was unquestionably an established zygote (see below), the intermingling of gamete components described by Marchand & Mattei (1980) had probably occurred leaving the discarded flagellum.

The entry of a spermatozoon into a mature oocyte, which is probably in the premeiotic condition (Robinson, 1965; Crompton & Whitfield, 1974; Marchand & Mattei, 1980; Peura *et al.*, 1982), initiates a complex series of processes. One of the most obvious and best studied of these is a cortical reaction resulting in a block to polyspermy and the production of a fertilization membrane which becomes the outer envelope of the egg or shelled acanthor. Until spermatozoon entry has occurred, the cytoplasm of the mature oocytes contains numerous randomly dispersed membrane-bounded electron-dense inclusions (Stranack, 1972; Crompton & Whitfield, 1974; Atkinson & Byram, 1976; Marchand & Mattei, 1980; Peura *et al.*, 1982; see §7.5.3). Some of these inclusions, like those seen in *Polymorphus minutus* by Crompton & Whitfield (1974), *Corynosoma semerme* (Fig. 7.3) by Peura *et al.* (1982) or *Breizacanthus* sp. by Marchand & Mattei (1980), have a complex structure. By means of enzyme histochemistry, Marchand & Mattei (1980) showed that the inclusions contained a protein and a polysaccharide fraction. On spermatozoon entry, the cytoplasmic distribution of oocyte inclusions alters so that they either disappear or become arranged at the periphery of the oocyte; these changes occur at about the same time as an amorphous, electron-dense envelope is formed around the oocyte or zygote as it may be by this time (Figs 7.13 and 7.14). This amorphous deposit, which is assumed to be produced from the contents of the inclusions, is a fertilization membrane (Marchand & Mattei, 1980; Parshad & Crompton, 1981). Histochemical observations made with the light microscope on a species of *Prosthenorchis* by Guraya (1969) and on *Sphaerirostris pinguis* by Parshad & Guraya (1977*a*) suggest that a similar reaction must follow penetration of the mature oocytes by the spermatozoa in these species.

Two other obvious early changes which are associated with spermatozoon penetration of the mature oocyte are the alteration in shape of the penetrated cell, usually from spherical to ovoid, and the development of an obvious postfertilization space between the fertilization membrane and the inner surface of the supporting syncytium which still surrounds the cell (see Parshad & Crompton, 1981; Figs 7.13 and 7.14). It is less easy to generalize about the other changes which accompany spermatozoon penetration and indicate that fertilization is in progress or has recently

Fig. 7.13. Transmission electron micrograph of a section through a zygote from a free ovary of an inseminated 28-day-old female *Polymorphus minutus*. The cytoplasm appears to be devoid of electron-dense inclusions (compare Fig. 7.3), a fertilization membrane (*f.m.*) has been formed and the postfertilization space (*pf.s.*) is prominent. Note the profiles of the axoneme of a spermatozoan flagellum (arrowed). (From Crompton & Whitfield, 1974.)

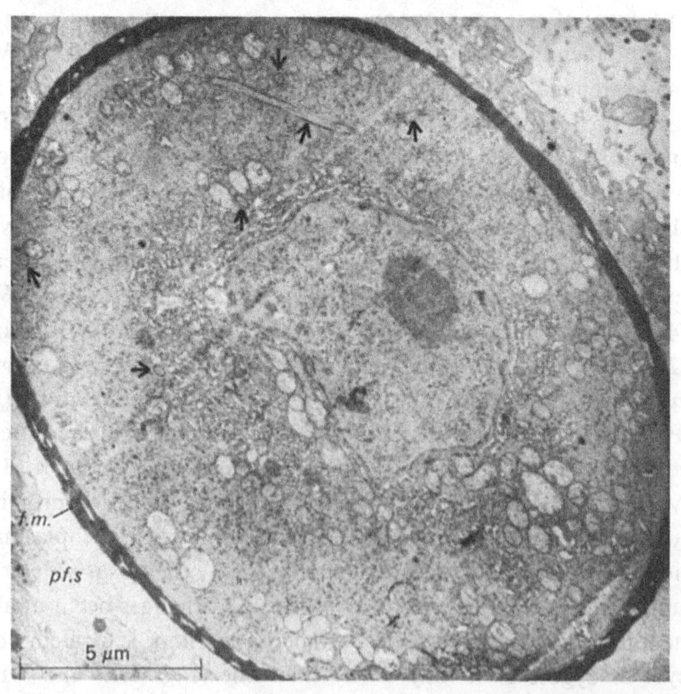

occurred. Striking reorganizations, compared to the mature oocyte, in the distributions of mitochondria and rough endoplasmic reticulum were detected in the zygote (or penetrated oocyte) of *Corynosoma semerme* (Fig. 7.14) by Peura *et al.* (1982). In inseminated female *Moniliformis moniliformis*, in which fertilization was occurring, the somatic tissues of the worms increased, presumably to accommodate the developing eggs. A comparison of the mean amount of nitrogen in female *M. moniliformis* aged from 5 to 15 weeks, in controlled infections, showed that there was always more nitrogen in the bodies alone of inseminated females than in the bodies plus gonads of uninseminated females of the same age (Crompton, 1972*a*, 1974).

According to Hopper & Hart (1980), each primary or, in the context of this chapter, mature, oocyte does not normally undergo meiotic divisions until after spermatozoon penetration when, in theory, a zygote and three polar bodies would be expected to be produced. Apparently,

Fig. 7.14. Transmission electron micrograph through a section of a zygote from a free ovary of an inseminated female *Corynosoma semerme*. Note the postfertilization space (*pf.s.*), the fertilization membrane (*f.m.*), the apparent absence of electron-dense inclusions (compare Fig. 7.3) and the aggregations of mitochondria (*mt.*) and rough endoplasmic reticulum (*er.*). (From Peura *et al.*, 1982.)

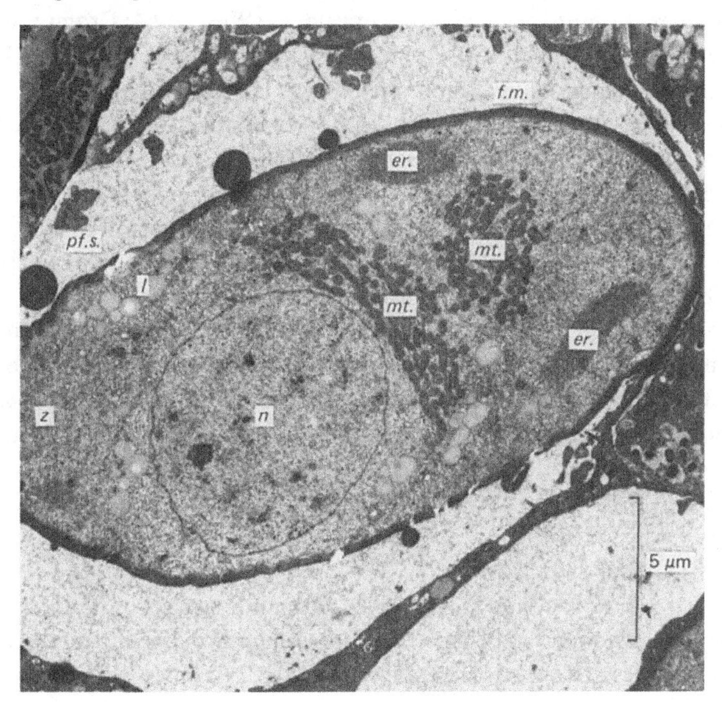

there are some exceptions in that oocytic meiosis may occur before spermatozoon penetration in some species of invertebrate. In the Acanthocephala that have been studied, most of the evidence suggests that spermatozoon penetration initiates meiosis and the formation of polar bodies. Two polar bodies per developing embryo were depicted by Nicholas & Hynes (1963*b*) for *Polymorphus minutus*, three for *Moniliformis moniliformis* by Robinson (1965) and two for *Mediorhynchus grandis* by Schmidt (1973*b*). It would be interesting to discover which was the usual number for the phylum and whether any chromosomes are discarded as the polar bodies are formed and extruded (Austin, 1965, 1982). Such a study might also help to elucidate the interpretation made by Atkinson & Byram (1976) that meiotic prophase might have been occurring in the oocytes of free ovaries from 12-day-old *M. moniliformis*, even though insemination is not known to occur in this species until they are about 16 days old (Crompton, 1974; Atkinson & Byram, 1976).

Fertilization may be assumed to have been completed when the zygote starts to cleave and the development of the embryo begins (Chapter 8).

7.8 Egg release and egg production

The term 'egg' is used by most workers on the Acanthocephala to describe the fully-formed acanthor larva (Chapter 8) enclosed in a series of three, or more probably four, envelopes (West, 1964; Crompton, 1970; Wright, 1971; Whitfield, 1973; Parshad & Crompton, 1981). Further information about acanthocephalan eggs is given in Table 7.6.

The fact that the outer egg envelope (number 1, Crompton, 1970) is produced by a form of cortical reaction when the inclusions of the oocytes release their contents during fertilization (§7.7) means that the developing embryo must synthesize and lay down the other complex envelopes from precursor molecules obtained from the female worm's body cavity fluid. The early embryo contained within envelope 1 is released from the ovary to join the assorted population of eggs in different stages of development and free ovaries and spermatozoa in the body cavity. No one has described any organ or structure that might assist in the formation of egg envelopes, but the possibility that some additions might be made while the eggs are located in the uterine bell or other parts of the efferent duct system (Chapter 5) should not be overlooked.

7.8.1 *Egg release*

Fully developed females of all known species of Acanthocephala possess a uterine bell (Chapter 5) whose function has been a matter of some controversy (Parshad & Crompton, 1981). In theory, some form of egg sorting must take place, otherwise incompletely developed eggs would be lost from the body cavity and unnecessary wastage would occur. Usually only shelled acanthors are found in the host's faeces and since eggs have been observed to pass in and out of the bell (Kaiser, 1893; Whitfield, 1968; Dunagan & Miller, 1973), the evidence in favour of egg sorting by the uterine bell seems stronger than the arguments against (see Parshad & Crompton, 1981). Generally, fully developed eggs are larger than other stages, and Whitfield (1968) concluded that since fully developed eggs of *Polymorphus minutus* were always 100 μm in length (Table 7.6), size could serve as the distinguishing feature which would prompt the uterine bell to pass them on into the uterus and out into the host's intestine or to return them to the body cavity for further development. If so, taxonomists should be careful to state whether dimensions apply to eggs taken from the host's faeces, the worm's body cavity or the worm's efferent duct system. Whitfield also suggested that the finding of developing eggs and even free ovaries in the efferent duct system (Meyer, 1933; Yamaguti, 1963; Asaolu, 1977) could be attributed to abnormal bell activity on the death of the host. The studies that have been made of the onset of patency by acanthocephalan worms (Table 7.1; §7.2) and of the patent period (see below) have not been concerned to see whether any rhythmical release of eggs occurs.

7.8.2 *Egg production*

Information about the duration of patency and the egg production of the few species of Acanthocephala which have been investigated to date is summarized in Table 7.7. The main reason that probably accounts for this paucity of knowledge is the failure to overcome the technical difficulty of obtaining acanthocephalan eggs from host faeces for quantitative counts (Crompton *et al.*, 1972). Routine flotation and concentration methods or modifications of them for archiacanthocephalan and palaeacanthocephalan species do not seem to lead to reproducible results (Kates, 1944; Crompton *et al.*, 1972; Holland, 1983*a*).

The observations made on the egg production of *Macracanthorhynchus hirudinaceus*, *Moniliformis moniliformis* and *Polymorphus minutus* (Table 7.7) can be criticized for several reasons, from the use of too few experimental animals to the little consideration given to the efficiency of the faecal processing systems used. In the case of *M. moniliformis*, the estimates of egg production are at least 10% lower than the actual values (Crompton *et al.*, 1972). Peak values are shown for the egg production of *M. hirudinaceus* and *P. minutus* (Table 7.7), but this pattern was not so obvious for *M. moniliformis* where egg production rose steadily and then seemed to hold a plateau phase for much of the patent period.

Table 7.6. *Observations on the eggs (shelled acanthors) of Acanthocephala (based on Parshad & Crompton, 1981)*

Parasite	Dimensions[a,b] (μm)	Number of envelopes	Reference
ARCHIACANTHOCEPHALA			
Macracanthorhynchus hirudinaceus	l. 80–100 w. 46–65	4	Meyer (1933); Kates (1943)
Mediorhynchus centurorum	l. 55 w. 36	4(3)[c]	Nickol (1969, 1977)
Mediorhynchus grandis	l. 58–64 w. 35–38	4	Moore (1962); Schmidt (1973*b*)
Moniliformis clarki	l. 60–75 w. 35–40	4	Crook & Grundmann (1964)
Moniliformis moniliformis	l. 112–120 w. 56–60	4	Moore (1946*d*); Edmonds (1966)[a]; Wright (1971)[d]
PALAEACANTHOCEPHALA			
Acanthocephalus dirus	l. 65–103 w. 14–19	4(3)[c]	Bullock (1962); West (1963, 1964)[d]; Oetinger & Nickol (1974)
Echinorhynchus lageniformis	l. 90 w. 20	4	Olson & Pratt (1971)
Echinorhynchus truttae	l. 110–140 w. 23–27	3(4)[e]	Awachie (1966)
Fessisentis fessus	l. 111–124 w. 14–19	4	Nickol (1972); Dunagan & Miller (1973); Buckner & Nickol (1979)
Fessisentis vancleavei	l. 68–83 w. 13–15	4	Buckner & Nickol (1978)

Leptorhynchoides thecatus	l. 85 w. 21	4(3)[c]	DeGiusti (1949*a*); Uznanski & Nickol (1976)
Plagiorhynchus cylindraceus	l. 75 w. 30	3	Schmidt & Olsen (1964)
Polymorphus minutus	l. 96–109 w. 18–20	3(4)[e]	Monné & Hönig (1954)[d]; Whitfield (1973)[d]
EOACANTHOCEPHALA			
Acanthogyrus oligospinus	n.g.[f]	3	Anantaraman & Ravindranath (1976)[d]
Neoechinorhynchus cristatus	l. 56 w. 27	4	Uglem (1972*b*)
Neoechinorhynchus emydis	l. 25 w. 18–22	3	Hopp (1954)
Neoechinorhynchus rutili	l. 27 w. 17	3	Merritt & Pratt (1964)
Neoechinorhynchus saginatus	l. 44–46 w. 16–20	4	Uglem & Larson (1969)
Octospinifer macilentus	l. 51–60 w. 23–28	3	Harms (1965*a*)
Pallisentis nagpurensis	l. 92 w. 48	3	George & Nadakal (1973)

[a] Measurements were made on eggs obtained from various locations including host faeces, the body cavities of both living and fixed worms and the efferent duct system. l, length; w. width
[b] Egg sizes vary according to location (see Nickol, 1972; Buckner & Nickol, 1978, 1979)
[c] Outer envelope may be lost once the egg has been passed out of the host
[d] References give details of the structure and composition of the egg envelopes, which contain protein, keratin, chitin and other materials
[e] Middle envelope may be further resolved into two components
[f] n.g. = measurements not given

Table 7.7. *Estimates of the duration of the patent periods and egg (shelled acanthor) production of acanthocephalan species in controlled infections*

Parasite	Cystacanths per host	Patent period	Mean daily egg production per female	Estimate of fecundity[a] ($\times 10^{-3}$)	Reference
ARCHIACANTHOCEPHALA					
Macracanthorhynchus hirudinaceus	24, 48, 59 and many (4 pigs)	Possibly 10 months	260000 (peak)[b]	24000	Kates (1944)
Moniliformis moniliformis	12 (23 rats)	106 ± 16 days	5500[c]	600	Crompton *et al.* (1972)
PALAEACANTHOCEPHALA					
Polymorphus minutus	41, 42 and 46 (3 ducks)	25	1700 (peak)	13	Crompton & Whitfield (1968*b*)

[a] Fecundity defined here as the total number of shelled acanthors estimated to be produced per female
[b] Wolfhügel (1924) estimated that about 82000 eggs were released on average per day by a female *M. hirudinaceus*
[c] Sita (1949), without giving details, wrote 'as many as 6500 eggs were expelled with the faeces of a rat in a day'

Any measurement of acanthocephalan egg production or fecundity must be interpreted with care, particularly if extrapolations are to be made from a controlled laboratory infection to a naturally occurring host–parasite relationship (see Chapters 9 and 11). Caution is needed because many host-mediated factors and other species of parasite may operate in the intestinal environment in conjunction with density-dependent constraints (Anderson, 1978) to influence the fecundity of an acanthocephalan parasite (§7.9). If fecundity is a measure of the number of offspring produced (Cohen, 1977), an estimate of the mean number of shelled acanthors produced per female acanthocephalan during a mean patent period would seem to be the appropriate parameter to consider (Table 7.7). This type of estimate, which has to be made while following the course of a controlled infection, gives mean values from the egg production of a population (Crompton *et al.*, 1972). A more informative indicator of fecundity can be obtained by examining the body contents of females by sampling during the course of an infection and determining the proportions of free ovaries, developing and fully developed eggs (Keymer, Crompton & Walters, 1983). This procedure enables the contribution of individual females to the reproduction potential of the whole population to be evaluated. The same data provide a measure of the variability of an individual female's fecundity; this is considerable as was found by Crompton *et al.* (1972) for *M. moniliformis*. Both approaches, however, are limited because they are based on destructive sampling of the host population if the numbers of female worms involved are to be determined reliably.

7.9 An ecological perspective of acanthocephalan reproduction

In this section, acanthocephalan reproduction, which has arbitrarily been defined here as the processes occurring from the maturation of the adult worms to the release of shelled acanthors, is considered with more emphasis on the role of ecological factors. A simplified scheme showing how acanthocephalan fecundity may be affected by both physicochemical and biotic factors is given in Fig. 7.15. Several of the relations are no more than suggestions based on observations made on other organisms while others are founded on the results of laboratory experiments. Generally, the ecological factors likely to affect acanthocephalan reproduction may be conveniently discussed under five headings: (1) parasite population structure; (2) host nutrition; (3) immune responses; (4) interactions with other parasites; and (5) seasonal and climatic variability. There is, however, considerable overlap and feedback between factors, regardless of this scheme of classification.

Fig. 7.15. A schematic representation of some of the actual and potential factors that influence the fecundity of acanthocephalan worms.

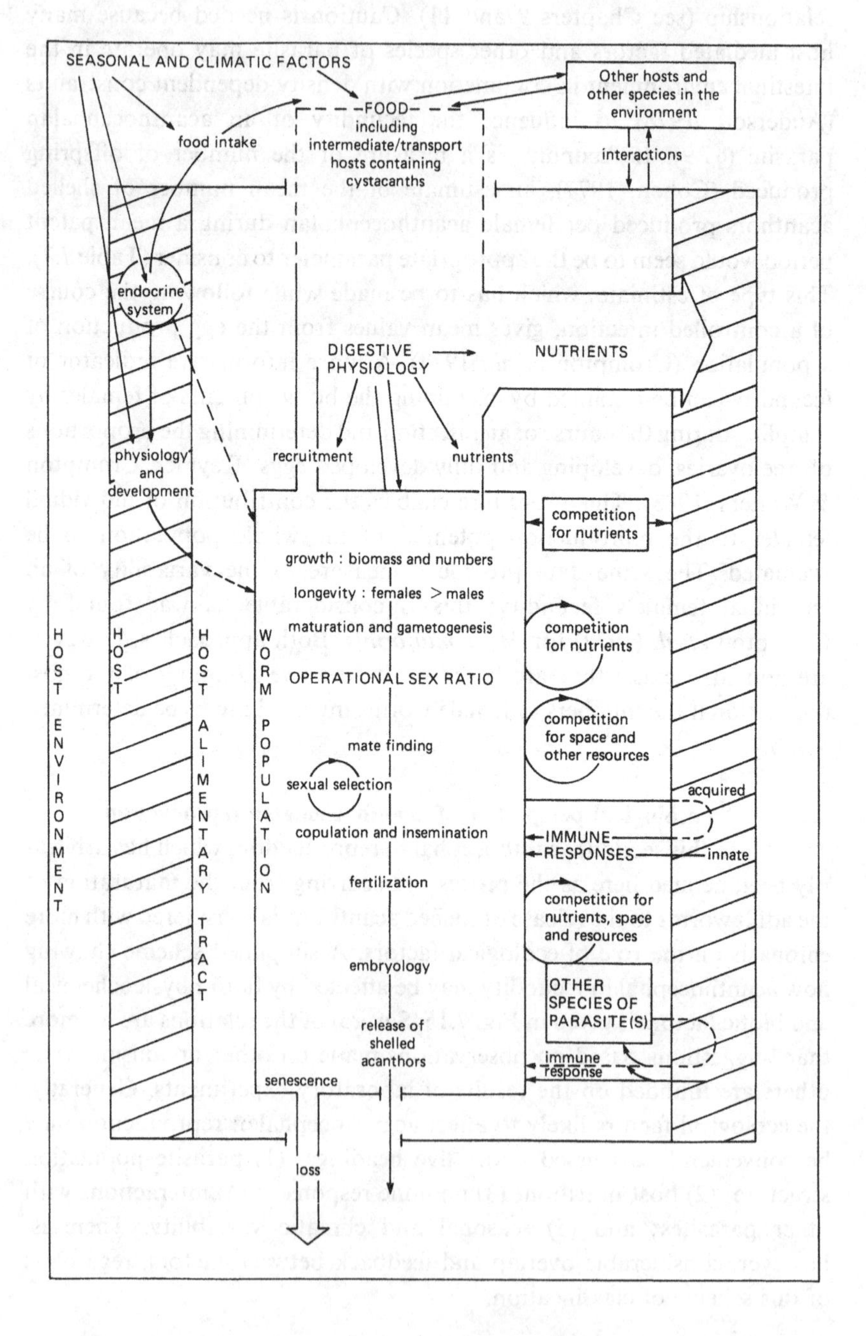

7.9.1 *Parasite population structure*

Density-dependent constraints on fecundity may affect and contribute to the regulation of helminth populations in their hosts (Anderson, 1978). For a dioecious acanthocephalan species, the notion of constraint should not be assumed to apply only as the numbers of worms increase in a host. If the structure of the population of worms under study is such that males are absent or in short supply, the chances of copulation and full insemination occurring (§7.6) for all the mature females present may be restricted. This type of constraint on fecundity could be operative at a lower population density than would cause competition between worms for nutrients or space (Fig. 7.15) to have any measurable effect (see below).

Anderson (1982*a*) argued that for a parasitic species mating probability is dependent on whether the parasite is monogamous or polygamous and on the form of the frequency distribution of parasite numbers per host. The distribution of adult acanthocephalans in their definitive hosts appears to be greatly overdispersed (Pennycuick, 1971*c*; Kennedy & Rumpus, 1977; Lee, 1981; Crompton, Keymer & Arnold, 1984) and this means, provided that acanthocephalans are indeed polygamous (see Jewell, 1976), that more females in a given population will be likely to be inseminated than would have occurred had the parasites been distributed at random. Population in this context means all the subpopulations of the same species in a community of hosts (Anderson, 1982*a*). Elsewhere, in this discussion and in Fig. 7.15, Anderson's subpopulation is referred to as a population. Incidentally, Anderson (1982*a*) did not discuss the impact of polygyny or polyandry on mating probability and it would now be useful to discover which form of sexual partnership is typical of the Acanthocephala.

The common observation that female acanthocephalan worms usually live longer than males (Parshad & Crompton, 1981; Chapter 10) enables density-dependent effects to influence fecundity in two ways, at least during the course of an experimental infection when the entire worm population is acquired by the host at the same time. The higher population density during the early part of the infection will increase mating probability (Anderson, 1982*a*) and will favour sufficient insemination to support maximum egg production (Crompton *et al.*, 1972; Crompton, 1974). Thereafter the greater rate of loss of males before females will serve to reduce competition and will tend to optimize the use of available resources for embryological development (Fig. 7.15). In a naturally occurring infection, such processes may not be so effective as immature worms become established in the intestine and other regulatory effects of density dependence are expressed (Fig. 7.15). It is worth noting that one significant consequence of the overdispersal of developmental stages of acanthocephalans in their intermediate hosts (Crofton, 1971*a*; Lackie, 1972*a*; Holland, 1983*a*), will be the occasional simultaneous arrival of several or even many parasites in a host. A population of worms of the same age, or a reinforcement in this way of an existing population, should enable the advantage of the differential longevities of the male and female worms to operate.

This hypothetical advantage to acanthocephalan reproduction of over-dispersal of the parasites in their intermediate hosts is not without some foundation. One hundred and sixty five cystacanths of *Moniliformis moniliformis* from a single *Periplaneta americana* were shown to be capable of infecting rats (Crompton *et al.*, 1984). More than 300 *Macracanthorhynchus hirudinaceus* were found in one *Popilla japonica* by Miller (1943), 63 *Echinorhynchus truttae* in one *Gammarus pulex* by Awachie (1966), 18 *Polymorphus minutus* in another *G. pulex* by Hynes & Nicholas (1963), and 31 *Acanthocephalus dirus* in an *Asellus intermedius* by Seidenberg (1973). These cases, and others of relative magnitude, could enhance the reproduction of the species concerned. It would probably be important for the heavily parasitized individual intermediate host to be eaten by a definitive host having the potential to harbour many adult worms (see Crompton *et al.*, 1984).

Some evidence from the study of naturally occurring infections of acanthocephalan worms in wild hosts also points to density-dependent influences on fecundity. Chubb (1964) related the greater fecundity of *Acanthocephalus clavula* in eels, as compared with that in grayling, roach and pike, to the greater intensities and the increasing probability of mating. Holmes, Hobbs & Leong (1977) collected data for infections of *Echinorhynchus salmonis* in 10 species of fish in Cold Lake, Alberta. They found that *E. salmonis* attained sexual maturity only in the four species of salmonid fish and, after a computer-simulated test of a mathematical model based, as they stated, on myriad assumptions, they were able to show that the entire population of *E. salmonis* in the lake could have been regulated simply by density-dependent constraints on its fecundity in whitefish, *Coregonus clupeaformus*.

7.9.2 *Host nutrition*

Adult acanthocephalan parasites, which are largely confined to the small intestine of their definitive hosts and lack an alimentary tract of their own, are likely to be highly sensitive to changes in the host's diet (Chapter 6). Several experiments have shown not only how the growth, longevity and reproductive development of *Moniliformis moniliformis* respond to variations in the composition of the rat's diet (Nesheim *et al.*,

1977; Parshad, Crompton & Nesheim, 1980; Crompton, Singhvi & Keymer, 1982), but also how the growth of the rat may be affected by the presence of the parasite (Crompton, Singhvi, Nesheim & Walters, 1981). In the latter case, the growth of young rats had been made dependent on the low fructose content of their diets; apparently competition for fructose between the absorptive surface of the *M. moniliformis* and the intestinal mucosa of the rats could account for the depressed rate of gain of rat body mass.

This background means that density-dependent effects due to competition rather than immunity are likely to be very evident and easy to analyse as *Moniliformis moniliformis* challenge each other and their hosts's mucosa for limited amounts of nutrients, particularly carbohydrate (Fig. 7.15). Recently, Keymer *et al.* (1983) exposed groups of rats, feeding *ad libitum* on purified diets containing 1%, 2%, 3% or 12% (mass/mass) fructose, to oral infections of either 10, 20, 40, 60 or 80 cystacanths of *M. moniliformis*. This investigation showed how density-dependent mechanisms might affect the parasite's reproductive potential; difficulties in quantifying acanthocephalan egg production (§7.8.2) were overcome by examining the state of the free ovaries in the female worms' body cavities (Keymer *et al.*, 1983). The experimental treatments were stopped when the *M. moniliformis* were 5 weeks old and it was found that: (1) the number of free ovaries per female worm increased linearly with increasing dietary fructose concentration, but was independent of the numbers of worms present; (2) the size of the free ovaries increased in response to increasing fructose concentration, but also responded to the worm burden; and (3) at the low dietary fructose concentrations, the magnitude of the density-dependent effects on ovarian tissue was greatly reduced. Furthermore, eggs containing fully developed acanthors were not observed in the body cavities of female worms retrieved from rats fed on diets containing either 1% or 2% fructose. No other information appears to be available at present to show how host nutrition may affect the reproduction of other acanthocephalan species by influencing density-dependent processes.

7.9.3 *Immune responses*

Density-dependent regulation of parasite reproduction could be mediated through the host's immune responses. In theory, and there is some supportive evidence from studies of nematode infections (see Keymer, 1982*a*), an increase in worm burden may represent an increased antigenic stimulation leading to an increased immune response from the host (Fig. 7.15). It would be unusual if such a density-dependent process did not impair egg production or curtail the patent period.

At present, the immunology of acanthocephalan infections in definitive hosts is not well understood and urgently needs more investigation. Harris (1972) detected antibody production by chub *Leuciscus cephalus*, to the presence of *Pomphorhynchus laevis* but could not show that protection or resistance resulted from the response. Andreassen (1975*a,b*) used infections of up to 400 cystacanths of *Moniliformis moniliformis* in rats to show that the loss of worms from the hosts occurred earlier at the higher than at the lower doses. Since expulsion in this way could be prevented by the use of an immunosuppressive agent (cortisone) and since rats given secondary infections of *M. moniliformis* appeared more refractory to their establishment, Andreassen concluded that an immune response was involved. More recently, Miremad-Gassmann (1981*a, b*) also proposed that an immune response affected the course of an infection of *M. moniliformis* in the rat. These findings indicate that the host's immune response cannot be ruled out as a density-dependent mechanism for curbing acanthocephalan fecundity.

7.9.4 *Interactions with other parasites*

Results from experimental studies have shown that interactions occur between *Moniliformis moniliformis*, *Hymenolepis diminuta* (Cestoda) and *Nippostrongylus brasiliensis* (Nematoda) in the small intestine of laboratory rats (Holmes, 1961, 1962*a*; Stock, 1978; Holland, 1983*a, b*). Responses between species of *Neoechinorhynchus* and populations of cestodes in fish have been described by Cross (1934) and Chappell (1969*a*). None of these reports claims that acanthocephalan fecundity is affected (Fig. 7.15), but the possibility should not be ignored.

7.9.5 *Seasonal and climatic variability*

As early as 1916, Van Cleave was impressed by the varying degrees of acanthocephalan infestation of certain vertebrate hosts at different times of year. More recently, Walkey (1967), Tedla & Fernando (1970), Muzzall & Bullock (1978) and Jilek (1978) have described seasonal fluctuations in the prevalences, intensities and reproduction of acanthocephalan worms from fish hosts. Although these aspects are considered elsewhere (Chapters 9 and 11), their bearing on reproduction (Fig. 7.15) justifies further discussion here.

The cases cited above involve poikilothermic hosts from temperate regions and so it is plausible to assume that environmental temperature might be involved. Gonadal function in poikilothermic vertebrates and in invertebrates is regularly found to be controlled by temperature change (Austin, 1965). In fish, body temperature cannot fall below that of the water because there is no opportunity for heat loss through evaporation, the highly vasculated gills ensure the rapid equilibration of the blood temperature with that of the surrounding water, while significant rises in body temperature through intense metabolic activity are probably rare (Schmidt-Nielson, 1975). Any effects of temperature change on the course of an acanthocephalan infection or on acanthocephalan reproduction could easily be indirect and be mediated through effects on the food supply, the numbers and condition of intermediate hosts, and so on (Fig. 7.15).

No seasonal periodicity was observed in the prevalence and intensity of infections of *Neoechinorhynchus rutili* in sticklebacks, *Gasterosteus aculeatus*, although a marked rise and fall in the numbers of egg-producing female worms was demonstrated (Fig. 7.16; Walkey, 1967). Changes in temperature or nutrition seem unlikely to have been responsible because other aspects of the course of infection continued undisturbed throughout the years of observation. In contrast, the seasonal effects on the reproduction of *Echinorhynchus salmonis* in yellow perch, *Perca fluviatilis* (Fig. 7.17), discovered by Tedla & Fernando (1970), showed that gametogenesis and egg production, which must have stopped at some point during the late summer, were at a peak when the water temperature would have been cold. The physical properties of water, however, are such that temperature changes occur gradually. Seasonal effects on populations of *E. salmonis* in yellow perch might have been related in some way to spawning (Fig. 7.17). Amin (1978*a*) considered that peak maturation and, therefore, fecundity of *E. salmonis* infecting chinook salmon, *Oncorhynchus tshawyt-*

Fig. 7.16. The maturation cycle of *Neoechinorhynchus rutili* in a population of *Gasterosteus aculeatus* in Great Britain from January 1960 to January 1962. The percentage composition of immature (stippled), mature (unshaded) and gravid (shaded) worms is shown for each monthly sample. (From Walkey, 1967.)

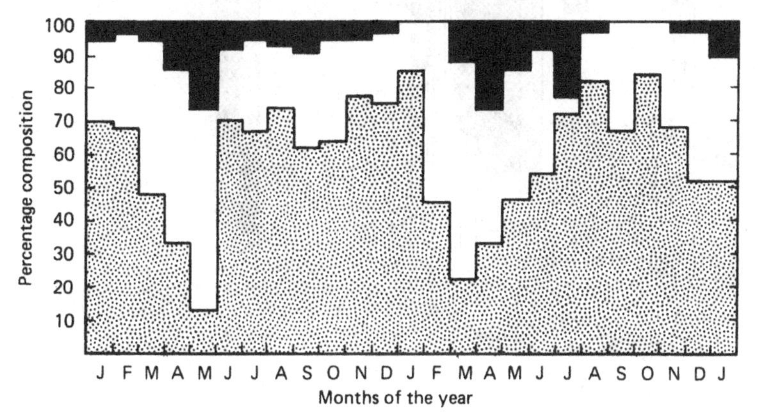

Fig. 7.17. Relations between the seasons and the reproductive biology of *Echinorhynchus salmonis* in *Perca fluviatilis* from the Bay of Quinte, Lake Ontario, Canada. (Constructed from data of Tedla & Fernando (1970).)

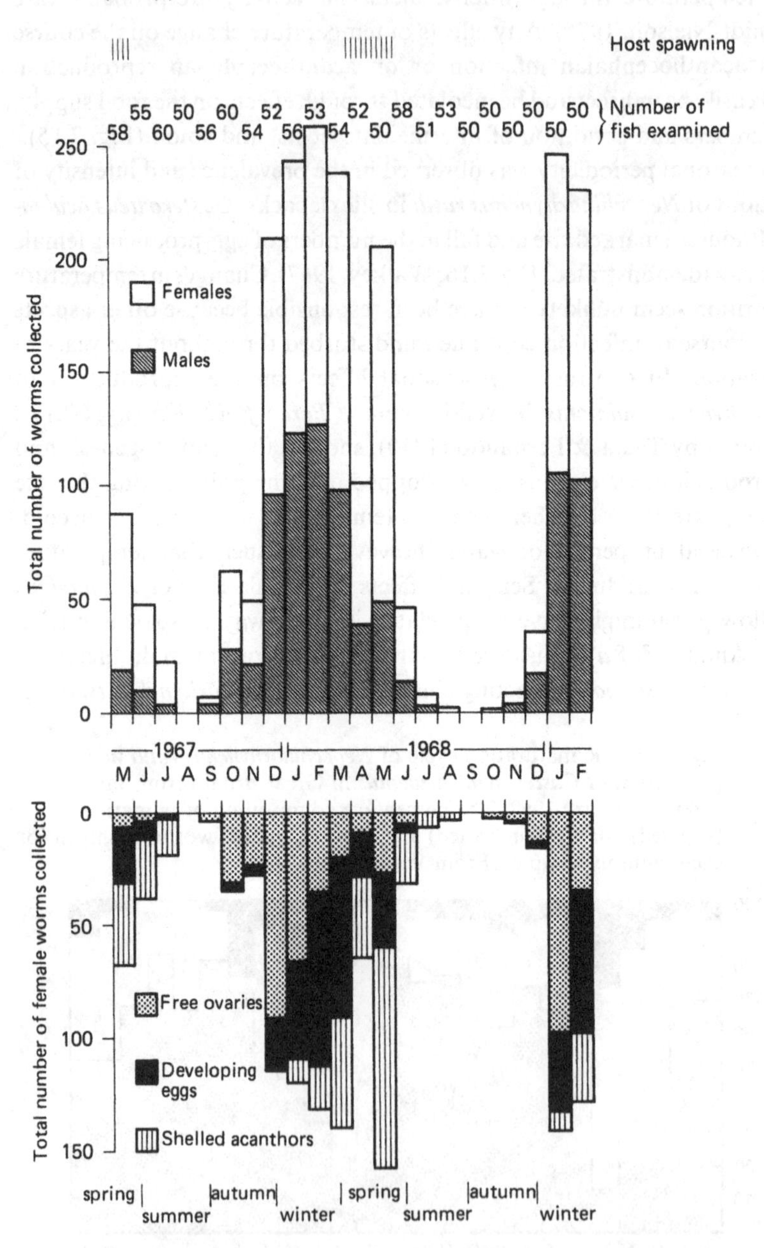

scha, occurred during spawning. This implies that an acanthocephalan's reproductive cycle could be regulated by its host's endocrine system (Fig. 7.15). Since the replication of a form of herpes virus, the formation of cysts by the protozoon *Opalina ranarum* and the production of eggs by the monogenean fluke *Polystoma integerrimum* can each be closely correlated with the advent of sexual activity in frogs (Newton, 1982; El Mofty & Smyth, 1964; Tinsley, 1983), similar influences on acanthocephalan reproduction cannot yet be discounted. Other evidence for the role of host hormones on acanthocephalan reproduction is either equivocal (Parshad & Guraya, 1979) or against their involvement. For example, gravid female *Neoechinorhynchus saginatus* were obtained from immature *Semotilus corporalis* during most months of the year and a strong positive relation between prevalence and worm burden, as fish length increased, was interpreted by Muzzall & Bullock (1978) as showing that host hormonal levels did not affect the maturation of *N. saginatus*.

8

Development and life cycles

Gerald D. Schmidt

8.1 Introduction

Embryological development and biology of the Acanthocephala occupied the attention of several early investigators. Most notable among these were Leuckart (1862), Schneider (1871), Hamann (1891 *a*) and Kaiser (1893). These works and others, including his own observations, were summarized by Meyer (1933) in the monograph celebrated by the present volume. For this reason findings of these early researchers are not discussed further, except to say that it would be difficult to find more elegant, detailed and correct studies of acanthocephalan ontogeny than those published by these pioneers.

Since Meyer's monograph, many studies have been published on most aspects of the biology of the Acanthocephala. When comparing these, recurring patterns in development are observed. A discussion of these patterns is the main content of this chapter. The postzygotic development of the acanthor, acanthella and cystacanth stages are described first, followed by examples from the three classes, Archiacanthocephala, Palaeacanthocephala and Eoacanthocephala. Finally, a table of all post-1932 literature references to intermediate and paratenic hosts that I have been able to find is presented.

8.2 Postzygotic development

8.2.1 *Acanthor*

Following fertilization and reorganization of the zygote, described in the previous chapter, cell division begins the formation of the acanthor, the stage in the life cycle that is infective to the intermediate host. The first published study of this early embryology, that of Hamann (1891 *a*), remains the definitive work on general acanthocephalan embryology to this day (Schmidt, 1973 *b*). In this landmark paper Hamann clearly demonstrated the early embryology of *Acanthocephalus ranae*, and *Echinorhynchus gadi*, both in the class Palaeacanthocephala. Nicholas & Hynes (1963*b*) studied *Polymorphus minutus*, also a palaeacanthocephalan.

In the Eoacanthocephala, Meyer (1931*a*) studied *Neoechinorhynchus rutili*, the only investigation so far reported for this class.

In the Archiacanthocephala, Meyer (1928, 1937, 1938*a*, *b*) studied *Macracanthorhynchus hirudinaceus*, and Nicholas (1967, 1973) worked with *Moniliformis moniliformis*. Schmidt (1973*b*) described the developing acanthor of *Mediorhynchus grandis*.

Thus, examples from all three classes of Acanthocephala have been examined and can be compared.

Cell division begins and continues while the embryo is still attached to the ovarian ball. The diploid number varies among species from 5 to 16 (Parshad & Crompton, 1981). Polar bodies can be seen as two darkly staining spheres at one end of the embryo (Fig. 8.1). The polar bodies mark the future anterior end of the animal, which is opposite to the condition known for other phyla. This is one of the two unusual aspects of polarity seen in the development of Acanthocephala.

The first division forms two blastomeres of about equal size, the nuclei of which lie one behind the other along the long axis of the embryo. Second divisions occur rapidly with approximately equal blastomere nuclei arranging in a spiral around the axis. Meyer (1936) considered acanthocephalan cleavage to be a distortion of typical spiral cleavage, and even considered that of *M. hirudinaceus* to be determinate, exhibiting the 4D origin of the mesoderm. It is not clear whether this conclusion is justified.

The first four cells are equal in size in all species studied except for *N. rutili* where the posterior cell, called D by Meyer (1931*a*), is larger and denser than the others.

Fig. 8.1. Early embryology of *Mediorhynchus grandis*. 1, Two cell stage. Prophase of division of posterior blastomere. 2, Four-nuclei stage, showing spiral cleavage and absence of apparent cell membranes. 3, Six-nuclei stage. 4, Nine-nuclei stage, with posterior quadrant of micromeres. 5, Twelve-nuclei stage. 6–7, Macromeres and micromeres, showing anterior quadrant of micromeres (some nuclei omitted). 8, Formation of condensed nuclei. 9, Inward migration of condensed nuclei, forming primordium of central nuclear mass. 10, Mature acanthor, with enclosing membranes (diagrammatic). bs, body spines; c, chromosomes; cn, condensed nucleus; cnm, central nuclear mass; fm, fertilization membrane; im, inner membrane; ma, macromere; mi, micromere; om, outer membrane; p, primordium of central nuclear mass; Pb, polar bodies; s, second membrane; vn, vesicular nucleus. (From Schmidt, 1973*b*.)

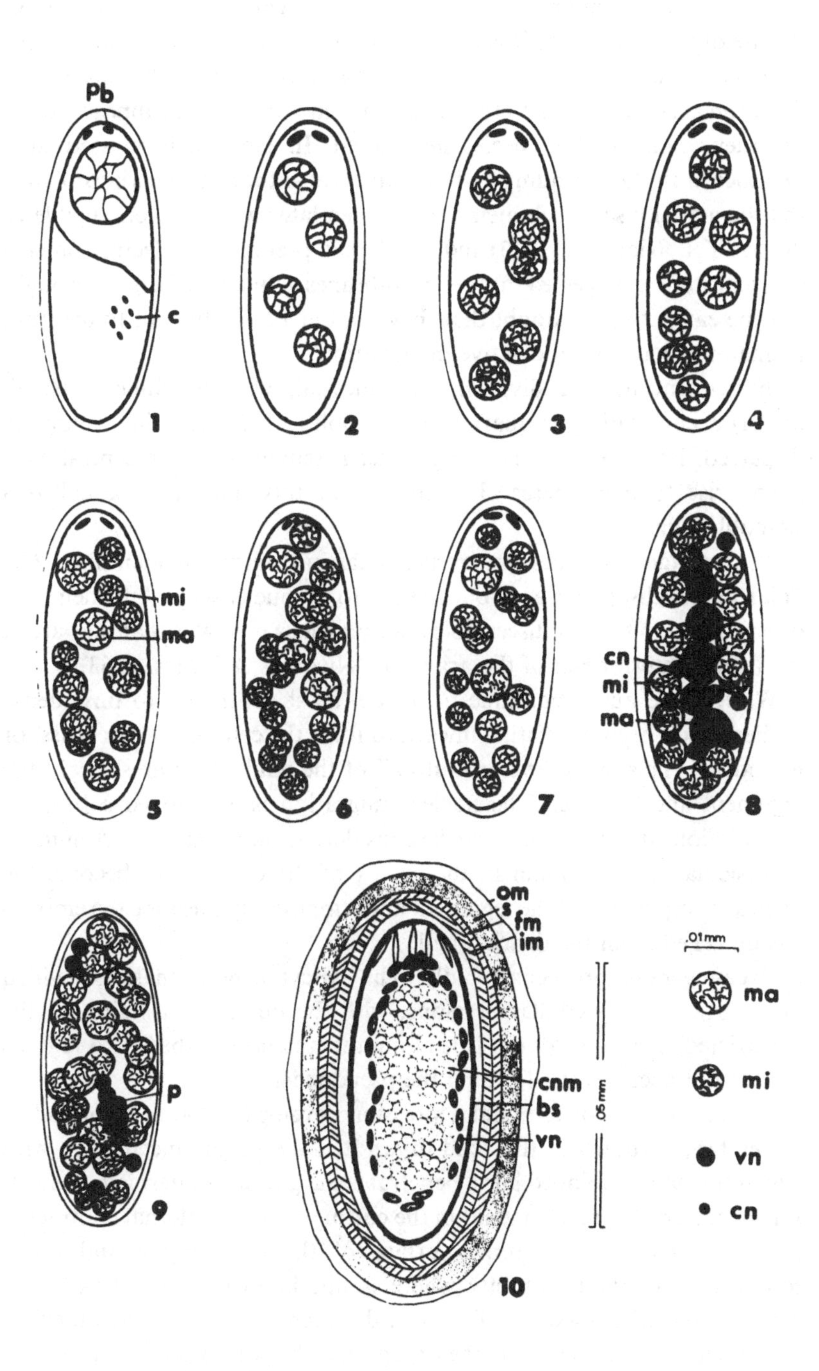
Pb
c
1
2
3
4
mi
ma
5
6
7
cn
mi
ma
8
p
9
om
s
fm
im
cnm
bs
vn
.05 mm
10
.01mm
ma
mi
vn
cn

The body of the embryo becomes a syncytium which remains throughout the life of the worm. The loss of cell membranes may occur as early as the four-cell stage, in the case of *M. grandis*, or 36-cell stage in *P. minutus* and *M. hirudinaceus*. Hamann (1891*a*) did not mention the disappearance of cell membranes in the species he studied. In fact, he illustrated their presence in all stages leading to the complete acanthor. It seems improbable that he actually saw such membranes when later workers were unable to do so. Yet, Nicholas (1973) indicated that preliminary electron microscopical studies suggested that cell membranes actually do persist throughout the early stages. I doubt that they do, but I am reluctant to discount Hamann as having a too active imagination.

Subsequent nuclear divisions are unequal, with micromeres (small nuclei) accumulating as quartets at the poles, then becoming randomly dispersed. Four macromeres (large nuclei) remain near the central axis, giving rise to micromeres. The embryo at this time may be called a stereoblastula.

As cleavage continues, macromeres can no longer be identified. The polar bodies disappear at about the 18- to 36-nuclei stage, but their lack of proximity to the last discernible macronucleus indicates that they are at the future anterior end of the acanthor (Nicholas & Hynes, 1963*a*).

While some nuclei continue to divide, most contract into tiny, dense bodies which begin migrating inward to form the central nuclear mass, or mesoderm. This mass develops into all of the adult structures except the tegument and lemnisci. The nuclear migration is considered a form of gastrulation, and the embryo can be considered a stereogastrula. A number of vesicular nuclei remain in the cortex of the embryo, to become the nuclear components of the tegument and lemnisci. By this time the embryo has detached from the ovarian ball.

Organogenesis proceeds rapidly. The syncytial body mass is divided into cortical and medullary regions, with the nuclear mass girdling the constricted, central medulla. It is along the junction of these two regions that a split occurs later, forming the pseudocoel.

A small, ventral or terminal depression develops at the anterior end of the embryo. Known as the aclid organ it is the boring mechanism with which the fully developed larva penetrates the gut of its intermediate host. It is bordered by spines that aid in the cutting process. Internal contractile fibers originate on the inner surface of the aclid organ and insert somewhere internally; exactly where is not known. When these fibers contract, the aclid organ is pulled inward, which points the spines anteriad. Upon relaxation of the fibers the organ snaps back into place, causing the spines to have a sweeping, cutting action. Aclid organ spines vary considerably in number, size and arrangement among species. For a review see Grabda-Kazubska (1964). They are said to be absent in the eoacanthocephalan genus *Neoechinorhynchus*, but Hopp (1954) observed minute spines on the acanthor of *N. emydis*. They may, in fact, be present in all. Most species have additional circles of spines over most of the body surface. Acanthor spines persist on the acanthella and often have been seen late in the development of the worm.

By the time the embryo has developed as described above, it has also surrounded itself with a series of envelopes, or membranes. Differences in the number of membranes can be found in the literature, with most authors describing three or four. The subject was admirably discussed by Whitfield (1973) who demonstrated electron microscopically that there are three in *Polymorphus minutus*. The outermost envelope probably is the fertilization membrane. The next inner membrane is comprised of two layers, which may account for the reports of a total of four layers. Closely investing the acanthor is the innermost envelope. Only 20–30 nm thick it cannot be seen with the light microscope, except that when the acanthor is observed while hatching a displacement of surrounding fluids clearly outlines its presence. Whitfield's paper should be consulted for further structural and compositional features of the egg membranes.

Further development cannot occur until the acanthor, enclosed within its membranes, exits its mother's genital pore to be swept out of the definitive host along with other intestinal contents. Then it must be eaten by the proper intermediate host, where it hatches in the intestine. Mechanisms of hatching have been studied by Edmonds (1966).

Without a known exception the first intermediate hosts of acanthocephalans are arthropods, either a crustacean in aquatic species or an insect (or rarely a myriopod) in terrestrial species (see Table 8.1). Reports of cystacanths in other invertebrates, such as snails or annelids probably represent accidental or paratenic hosts.

Differing terminologies have been used in the past for subsequent developmental stages. The term 'acanthella' was proposed by Van Cleave (1937) for all stages between the acanthor and the infective juvenile, or cystacanth. This name has been nearly universally accepted and is used here.

8.2.2 *Acanthella*

Upon hatching, the spindle-shaped acanthor immediately increases in size and begins cutting actions with its aclid organ. After penetrating into the wall of the mesenteron of the intermediate host it may cease activities for hours or even days, or it may proceed through the wall of

Fig. 8.2. Development of *Echinorhynchus truttae*. 1, Mature egg. 2, Acanthor removed from the hemocoel of *Gammarus pulex* 11.5 h after infection. 3, Twelve-day embryo. 4, Eighteen-day acanthella. 5, Twenty four-day acanthella. 6, Thirty two-day acanthella. 7, Thirty

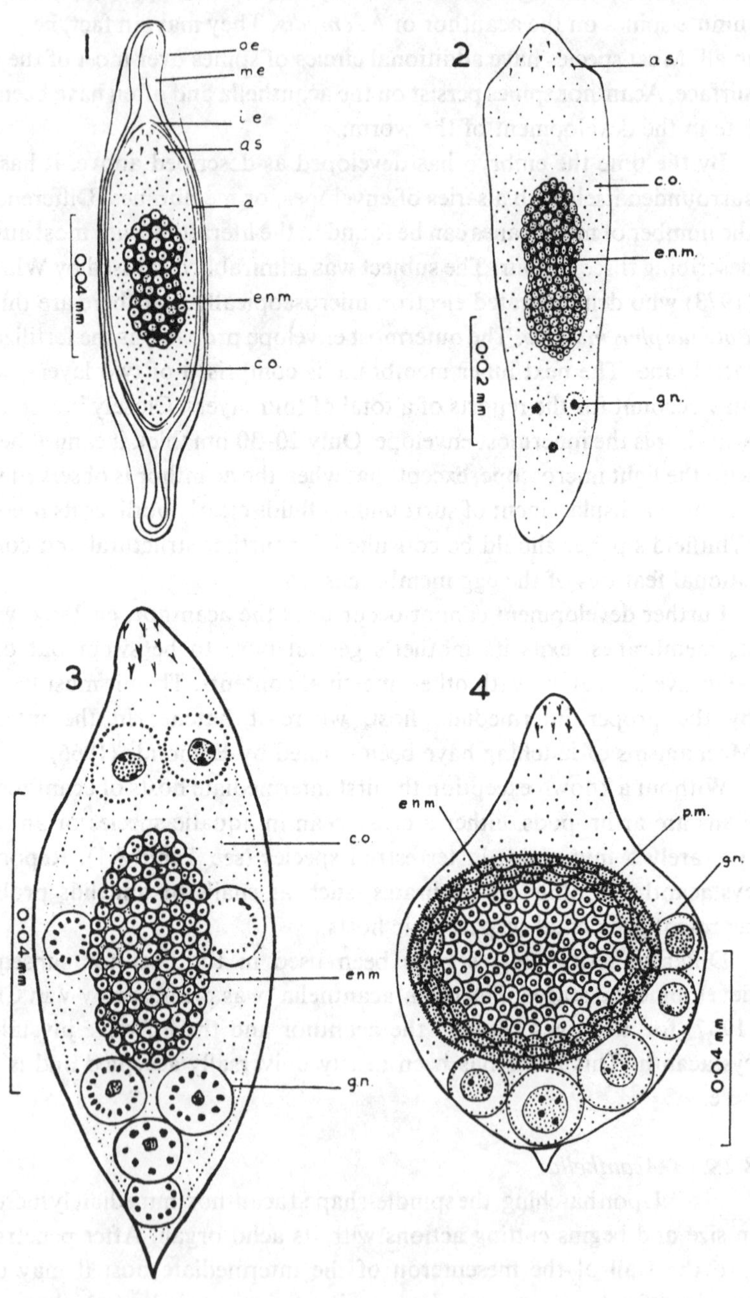

seven-day acanthella. 8, Posterior half of male acanthella, 39 days after infection. a, acanthor; a.s., spine of acanthor; a.t., anterior testis; b.m., muscle layer of body wall; c, cuticula; c.g., cement gland; c.o., cortex; e.n.m., embryonic nuclear mass; g., ganglion; g.n., cortical giant nucleus; i.e., inner envelope of egg; m.e., middle shell of egg; n.r., nucleus of proboscis retractors; o.e., outer envelope of egg; p.b., primordium of copulatory bursa; p.c., primordium of cement gland; p.g., primordium of genitalia; p.g.g., primordium of gonad and genitalia; p.m., primordium of muscles of body wall; p.p., primordium of definitive proboscis; p.p.a., primordium of proboscis apparatus; p.p.r., primordium of proboscis retractors; p.s., proboscis sheath; r.a., remains of acanthor; s., Saefftigen's pouch; s.c., subcuticula; s.d., primordium of sperm duct; s.f., space filled with orange fluid; s.r., retractor muscle of proboscis sheath; t., testis. (From Awachie, 1966.)

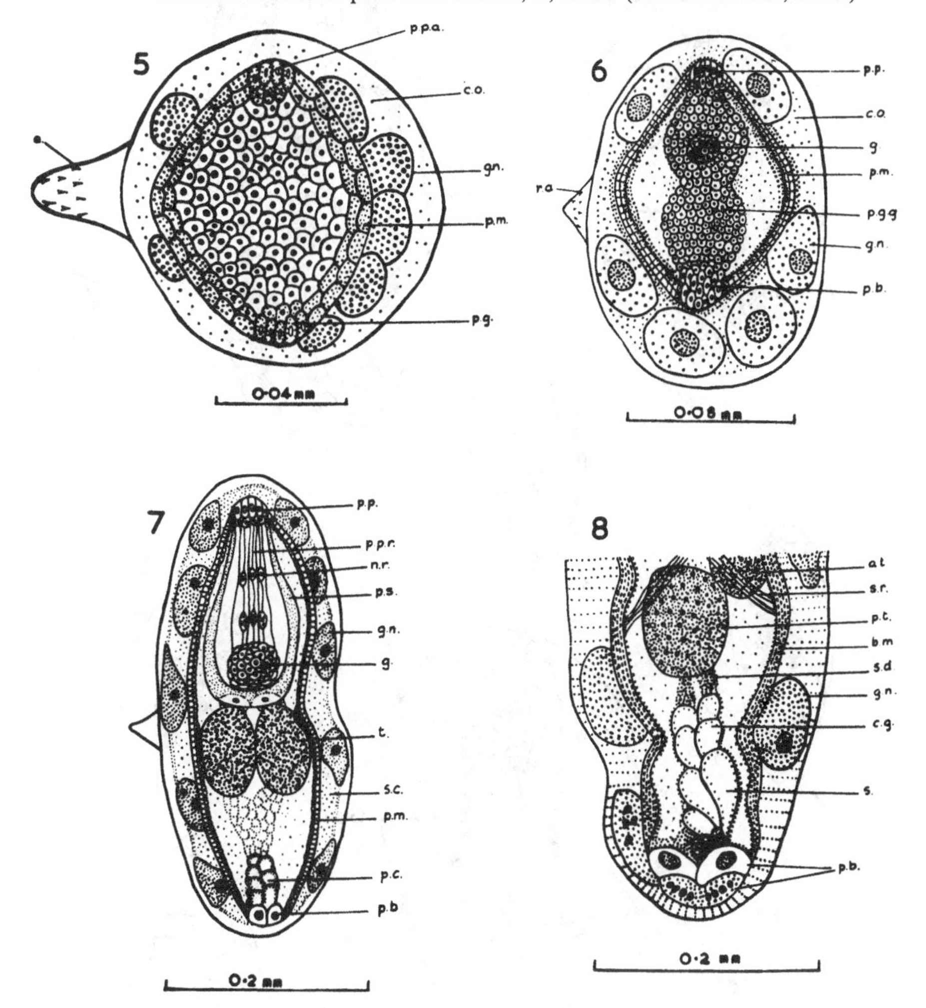

Fig. 8.3. Development of *Plagiorhynchus cylindraceus*. 1, Egg, containing mature acanthor. 2, Acanthor after escape from egg. 3, Acanthella, 22 days. 4, Acanthella, 25 days. 5, Acanthella, 27 days. 6, Anlagen of female reproductive system, 32 days. 7, Developing brain with principal nerve trunks, 28 days. 8, Cross-section at level of developing proboscis, 25 days. 9, Acanthella, 30 days. 10, Cystacanth, 37 days. 11, Cystacanth, 60 days, proboscis artificially forced to evert. AC, aclid organ; AN, apical nuclei; B, bursa; BA, brain anlage; BM, anlagen of body wall musculature; C, cortex; CG, cement glands; EH,

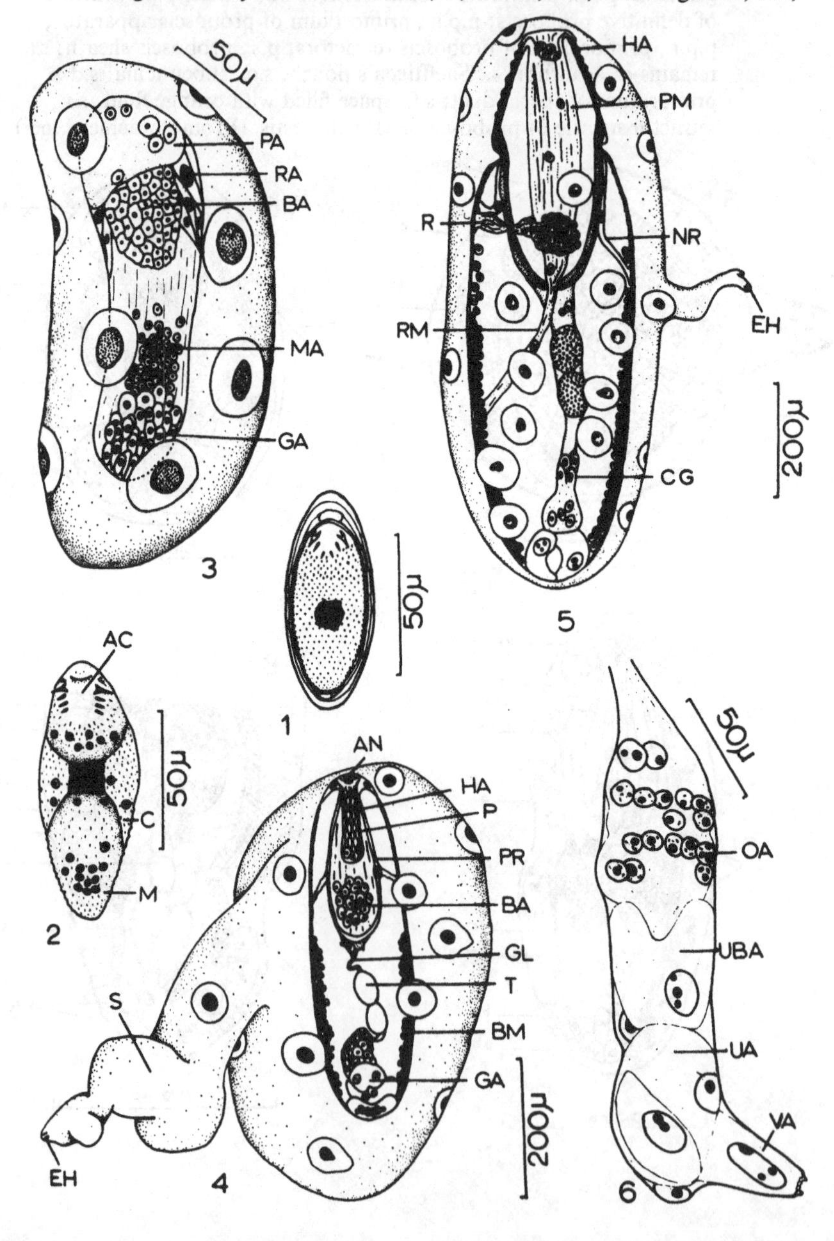

embryonal hooks; GA, genital anlage; GL, genital ligament; HA, hook anlagen; L, lemniscus; LR, nucleus of lemniscal ring; M, medulla; MA, anlagen of body wall musculature, gonads, and genital ligament; NA, lateroanterior nerve trunk; NM, anteromedial nerve trunk; NP, lateroposterior nerve trunk; NR, neck retractor; O, ovarian balls; OA, ovary anlage; P, proboscis; PA, proboscis anlage; PM, proboscis retractor; PR, proboscis receptacle; R, retinaculum; RA, anlagen of proboscis receptacle and retractor muscles; RM, retractor muscle; S, stalk; T, testis; U, uterus; UA, uterus anlage; UB, uterine bell; UBA, uterine bell anlage; V, vagina; VA, vagina anlage. (From Schmidt & Olson, 1964.)

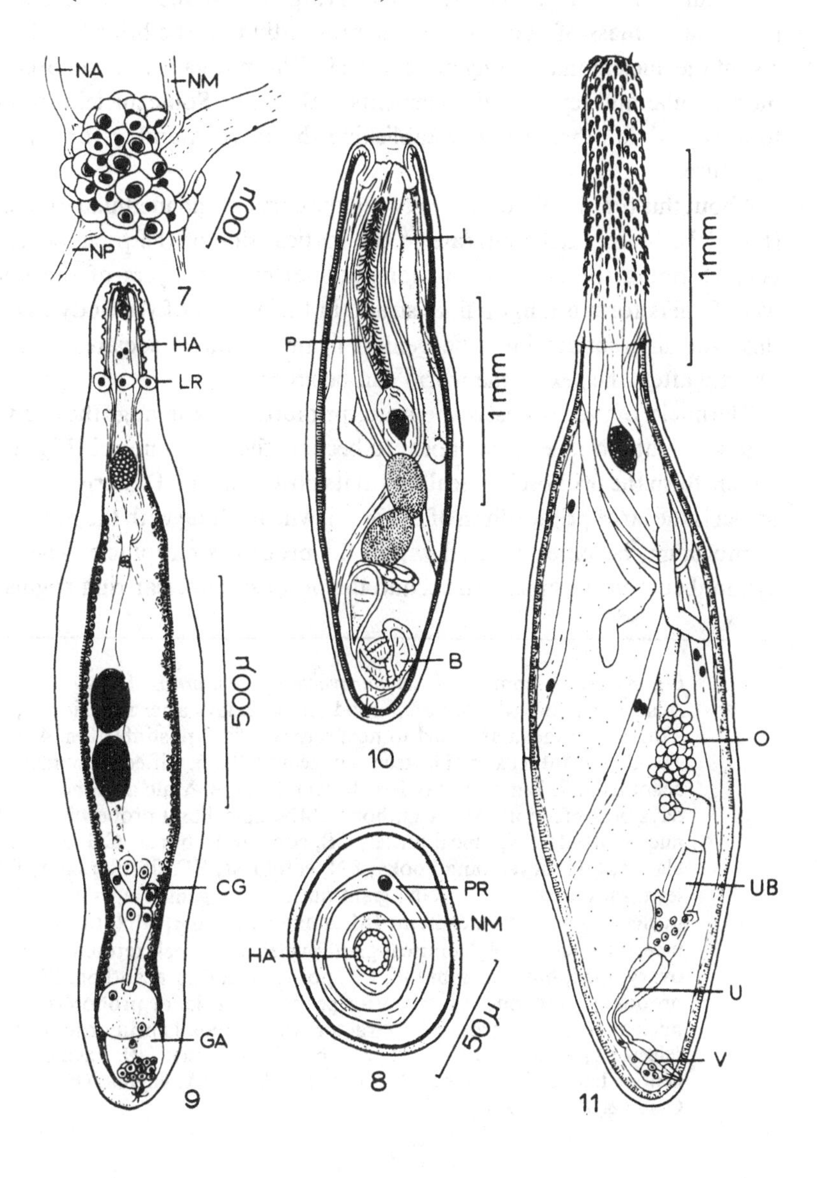

the intestine. In most cases, if not all, the entire acanthor stops under the serosa and begins to grow. Growth first appears as a swelling outward, toward the hemocoel of the host, in a direction 90 degrees from the longitudinal axis of the acanthor. The cortical nuclei enlarge and become vesicular (Fig. 8.2). Their number may increase or not, depending on the species. As growth proceeds, the central nuclear mass begins to organize itself into distinct regions which are the primordia of the internal structures of the adult worm (Fig. 8.2). The proboscis apparatus primordium is a syncytium containing a number of nuclei (Fig. 8.3). Immediately posterior to it, a large mass of nuclei forms the primordium of the brain, while the rest of the nuclear mass elongates behind it. This nuclear mass will become the muscular and reproductive elements of the body. Some nuclei continue to divide while others cease, establishing the eutelic number for a given structure.

About this time a split occurs between the cortical and medullary regions (Fig. 8.3). Many nuclei migrate to the cortical side of the pseudocoel to begin forming the circular and longitudinal muscle layers of the body wall. Others form a longitudinal strand in the middle of the body cavity; this will differentiate into the genital ligament and associated organs. Shortly after, the sex of the worm can be ascertained.

The nuclei of the proboscis apparatus primordium rearrange themselves. A few migrate to its extreme apex, to become the apical nuclei (Fig. 8.4). Others form the proboscis apical ring at its posterior end. Posterior to these nuclei is the large primordium of the brain, which, along with the proboscis primordium, becomes encased within the proboscis receptacle. The syncytium between the apical nuclei and the proboscis nuclear ring begins to

---

Fig. 8.4. Development of *Echinorhynchus lageniformis*. 1, Egg. 2, Acanthor removed from amphipod serosa 2 days after infection. 3, Young acanthella attached to host serosa 5 days postinfection. 4, Ten-day acanthella. 5, Thirteen-day acanthella. 6, Fifteen-day male acanthella. 7, Eighteen-day female acanthella. 8, Male acanthella 20 days postinfection. AC, acanthor; AMN, apical and proboscis inverter nuclei mixed; AN, apical nuclei; CB, copulatory bursa; CG, cement glands; DH, developing hooks; EN, entoblast; FC, fibrillar coat; FM, fertilization membrane; GN, giant nuclei; GO, genital opening; GP, nuclei of genital primordium; IM, inner membrane; IPN, proboscis invertor nuclei; LN, lemniscal giant nuclei; NR, neck retractor; O, ovary; OM, outer membrane; PA, posterior end of acanthor; PG, proboscis ganglion; PN, posterior proboscis nuclei or proboscis nuclear ring; PR, proboscis retractor; PP, proboscis primordium; SA, selector apparatus; SP, Saefftigen's pouch; T, testis; UB, uncinogenous bands; UT, uterus; UTB, uterine bell; VA, vagina. (From Olson & Pratt, 1971.)

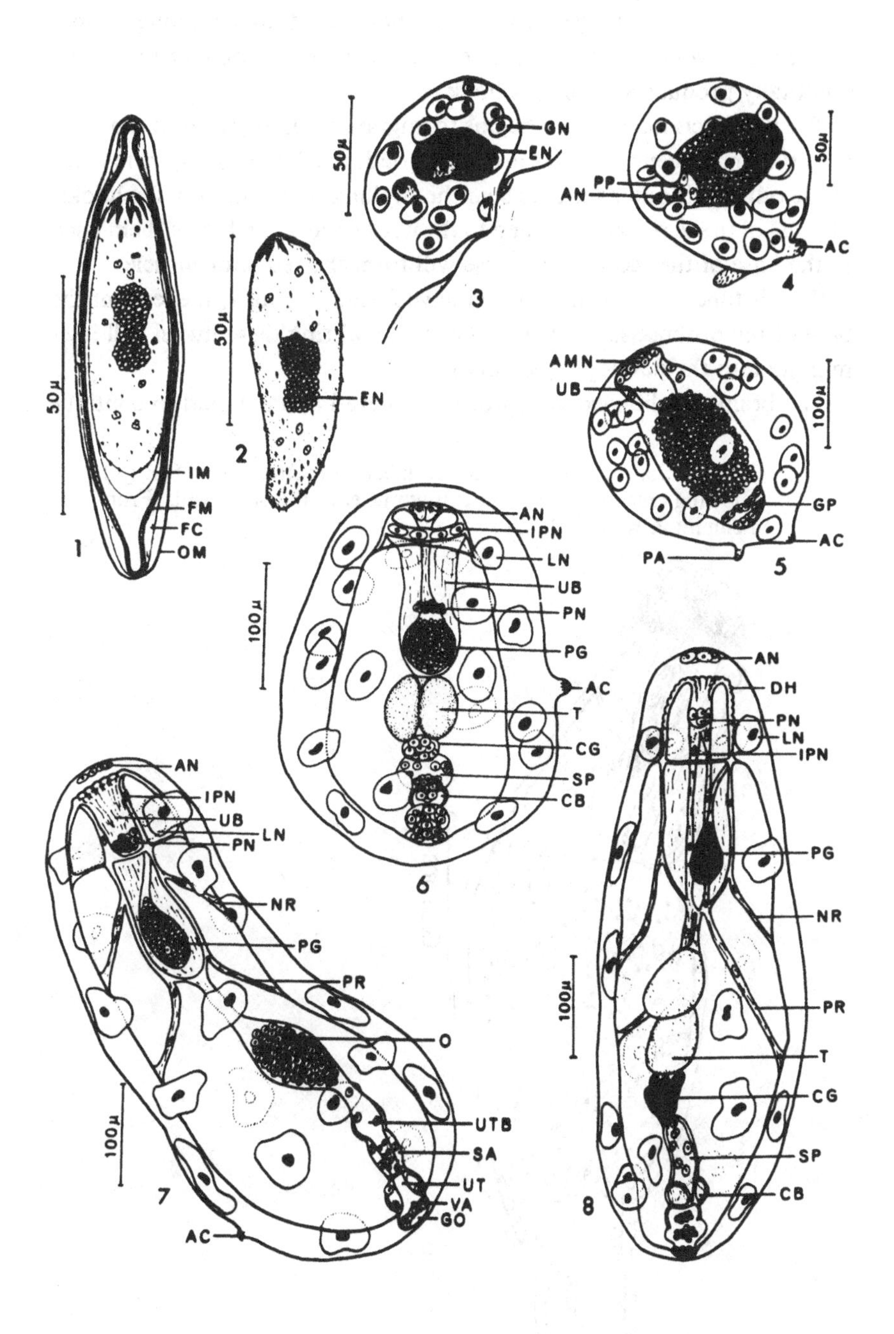
50μ
IM
FM
FC
OM
1
50μ
EN
2
50μ
GN
EN
3
50μ
PP
AN
AC
4
AMN
UB
100μ
GP
AC
PA
5
AN
IPN
LN
UB
PN
PG
100μ
AC
T
CG
SP
CB
6
AN
IPN
UB
LN
PN
NR
PG
PR
O
100μ
UTB
SA
UT
VA
GO
7
AC
AN
DH
PN
LN
IPN
PG
NR
100μ
PR
T
CG
SP
CB
8

differentiate into uncinogenous bands. These are strandlike, longitudinal thickenings, which alternately swell and constrict. Each swelling will eventually produce a proboscis hook.

The proboscis nuclear ring begins to migrate anteriad, while at the same time the uncinogenous bands arch posteriad, in effect turning themselves inside out. Their outer surfaces bear the swellings that will form the hooks. Strands of tissue remain attached to the nuclei and extend from them back to the base of the receptacle. These will form the retractor muscles.

By this time a few cortical nuclei have formed a ring at the level of the base of the proboscis. Known as the lemniscal ring these nuclei will later migrate into the developing lemnisci.

The brain ganglion extends processes laterad and anteriad to continue

Fig. 8.5. Late acanthella (natural infection in *Macrobrachium* sp.), showing ameboid and fragmenting nuclei of the body wall. (From Schmidt & Kuntz, 1967*a*.)

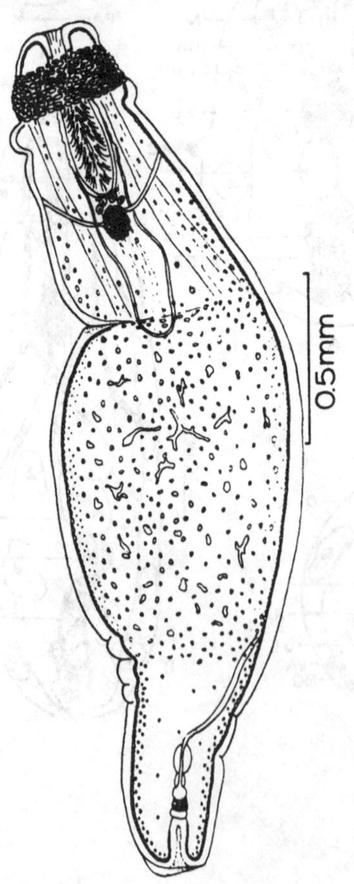

the development of the nervous system. Gonads increase in size within the genital ligament, and associated genital organs are rapidly differentiating near the posterior end.

As internal organs continue growing, the cortical layer gets thinner. In several species a stalk, bearing the hooks of the aclid organ, has been described, extending from the side of the embryo (Figs 8.2, 8.3, 8.4). They eventually disappear, but demonstrate two important phenomena: the axis of the adult worm is rotated 90 degrees from that of the acanthor, and the outer tegument of the acanthella is the stretched tegument of the acanthor.

If the adult of the species has giant nuclei in the body wall, these nuclei continue to enlarge and migrate to their definitive locations at this time. If the adult has nuclear fragments in its body wall, the cortical nuclei become ameboid (Fig. 8.5), with long processes that break up and move to their proper locations.

After the proboscis anlagen has everted, hook development proceeds rapidly, soon piercing the tegument and growing to the final size found in the adult worm. Morphogenesis and an analysis of hook structure of *Moniliformis* was presented by Hutton & Oetinger (1980).

By this time the lemnisci have begun their inward growth, with the nuclei of the lemniscal ring following them in. This growth process must be very rapid, for few researchers have reported seeing it.

Neck retractor muscles, proboscis retractor and protruser muscles, nervous system, and all other organs are now essentially complete and in place. The proboscis then invaginates, at least in most species. In some species the neck retractor muscles pull the entire praesoma into the body, while the posterior end is also pulled in. This is particularly true of parasites of birds, where they must be able to withstand the grinding action of the host's ventriculus.

8.2.3 *Cystacanth*

The worm now has all of the structures of the adult and is capable of infecting the definitive host. Essentially a juvenile, it is most commonly called a cystacanth. Cable & Dill (1967) made a point for calling this stage a juvenile rather than a cystacanth, because it is unencysted and differs from the adult only in size and degree of sexual development. The name cystacanth would be more appropriate to juveniles encysted in the tissues of a paratenic host, according to those authors. Even so, I prefer to follow common usage in this case and use this name for the stage that has completed development in the intermediate host and is infective to the definitive host. In most, if not all, species the cystacanth in the intermediate host is surrounded by a thin membranous envelope. Whether this membrane is of host or parasite origin has received considerable attention. The literature was reviewed by Wanson & Nickol (1973), who concluded that the envelope is in fact the stretched body covering of the acanthor. It is quite possible that in some cases this layer is thickened by a contribution from the host's defensive cells. Factors that influence the activation of cystacanths were reviewed by Graff & Kitzman (1965).

Paratenic hosts often are important ecological bridges between the intermediate and definitive hosts, such as the ostracod–bluegill sunfish–black bass cycle of *Neoechinorhynchus cylindratus*, as outlined by Ward (1940*a*) (see below). But because no further development occurs in the paratenic host, this stage is not considered further here, except to note that a possible exception is seen in the case of *Sphaerechinorhynchus serpenticola*. Worms of this species have been found in the intestinal mesenteries of several species of snakes and without exception the individuals were nearly adults, with active sperm formation in the males. Inseminated females have not been found.

8.3 Life cycles

Too many life cycles are known for all of them to be reviewed here. Table 8.1 lists records of known intermediate and paratenic hosts, with references, that have accumulated since Meyer's (1933) list. An illustrative sampling of life cycles within the three classes of Acanthocephala are listed below.

8.3.1 *Class Archiacanthocephala*

Moore (1946*b*) produced laboratory infections of *Macracanthorhynchus ingens* in the scarabaeid beetles *Phyllophaga crinita*, *P. hirtiventris* and *Ligyrus* sp. Crites (1964) found a natural infection of this species in

---

Fig. 8.6. Development of *Macracanthorhynchus hirudinaceus*. 1, Egg. 2, Acanthor. 3, Very early acanthella. 4, Early acanthella. 5, Acanthella. 6, Late acanthella. 7, Early cystacanth. 8, Cystacanth before invagination. 9, Invaginated cystacanth. A, infective acanthor; BSN, branched skin nuclei; C, cuticle; EN, embryonic nuclear mass (condensed nuclei); GM, gonad and muscle primordia; Gg, ganglion; H, hooks; L, lemnisci; L-1, L-2, L-3, L-4, four shell layers; Lg, ligament; LH, larval hooks; LN, lemnis nuclei; LR, lemniscus nuclear ring; LS, larval spines; MB, muscle cell band; PA, proboscis primordium; PM, proboscis musculature; PN, protonephridia; PR, proboscis nuclear ring; R, raphe; Re, receptaculum; RO, rostellum; RP, retracted proboscis; RS, radial skin fibers; S, skin; SS, scalloped shell; SN, skin or subcuticular nuclei; SP, sensory papilla; T, testes; UA, urogenital primordium (except gonads); VE, vasa efferentia. (After Kates, 1943.)

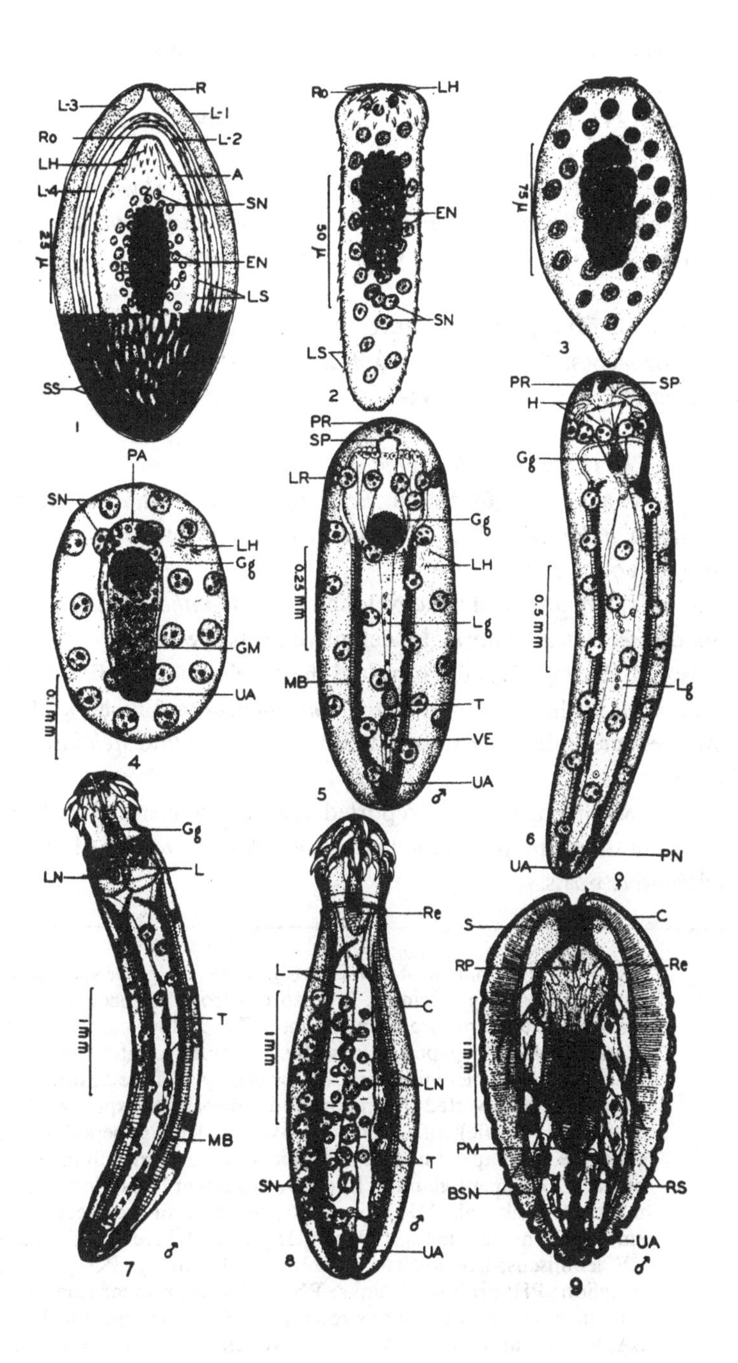
R
L-3
L-1
Ro
L-2
LH
A
L-4
SN
25 μ
EN
LS
SS
1
Ro
LH
EN
50 μ
SN
LS
2
75 μ
3
PR
SP
H
Gg
PA
SN
LH
Gg
GM
0.1 mm
UA
4
PR
SP
LR
Gg
LH
0.25 mm
Lg
MB
T
VE
UA
5
0.5 mm
Lg
6
UA
PN
Gg
LN
L
1 mm
T
MB
7
Re
L
C
1 mm
LN
T
SN
UA
8
S
C
RP
Re
1 mm
PM
BSN
RS
UA
9

a millipede, and Elkins & Nickol (1983) found *M. ingens* cystacanths in woodroaches (*Parcoblatta pensylvanica*). Definitive hosts are racoons and black bears.

Kates (1943) repeated earlier studies on the development of *M. hirudinaceus* in dung beetles (Fig. 8.6).

Moore (1962) traced the development of *Mediorhynchus grandis* in grasshoppers (*Chortophaga viridifasciata*, *Orphulella pelidna*, *Arphia luteola* and *Schistocerca americana*). Cystacanths were infective to birds in 27–30 days.

Nickol (1977) reported the development of *Mediorhynchus centurorum* in woodroaches, *Parcoblatta pensylvanica*. Cystacanths were infective to woodpeckers in about 47 days.

8.3.2 *Class Palaeacanthocephala*

Awachie (1966) found *Echinorhynchus truttae* to develop in *Gammarus pulex* (Fig. 8.2). Cystacanths were infective to trout, *Salmo trutta*, in about 82 days.

Amin (1982) described the development of *Acanthocephalus dirus* in the isopod *Caecidotea militaris*. Its definitive hosts are fishes.

Olson & Pratt (1971) demonstrated the development of *Echinorhynchus lageniformis* in the amphipod *Corophium spinicorne* (Fig. 8.4). The cystacanth becomes infective to starry flounders, *Platichthys stellates*, in about 30 days.

Hynes & Nicholas (1957) reported the development of *Polymorphus minutus* in the amphipod *Gammarus pulex*. A wide variety of ducks serve as definitive hosts.

Fig. 8.7. Development of *Paulisentis fractus*. 1, Egg or shelled embryo drawn from living specimen. 2, Acanthor from hemocoel of copepod one hour after exposure to infection. 3, Three-day acanthor. 4, Young acanthella from copepod 7 days after exposure to infection. One nucleus of future lemnisci omitted for clarity. 5, Late acanthella (male) with proboscis everted; from copepod 9 days after exposure. 6, Late acanthella (female) with proboscis inverted; from copepod 9 days after exposure. AN, apical nuclei; AP, apical organ primordium; BU, bursa; CG, cement gland; CL, central ligament; DN, dorsal giant nuclei of trunk wall; DR, dorsal retractor of proboscis receptacle; EN, embryonal nuclear mass; GP, genital pore; IP, inverter of proboscis; LE, lemniscus; LN, nucleus of lemnisci; OV, ovary; PG, proboscis ganglion; PH, proboscis hooks; PN, proboscis nuclear ring; PS, muscular sheath of proboscis receptacle; RS, uncinogenous bands; SA, selector apparatus; SV, seminal vesicle; TE, testes; UB, uterine bell; UT, uterus; VN, ventral giant nucleus of trunk wall; VR, ventral retractor of proboscis receptacle. (From Cable & Dill, 1967.)

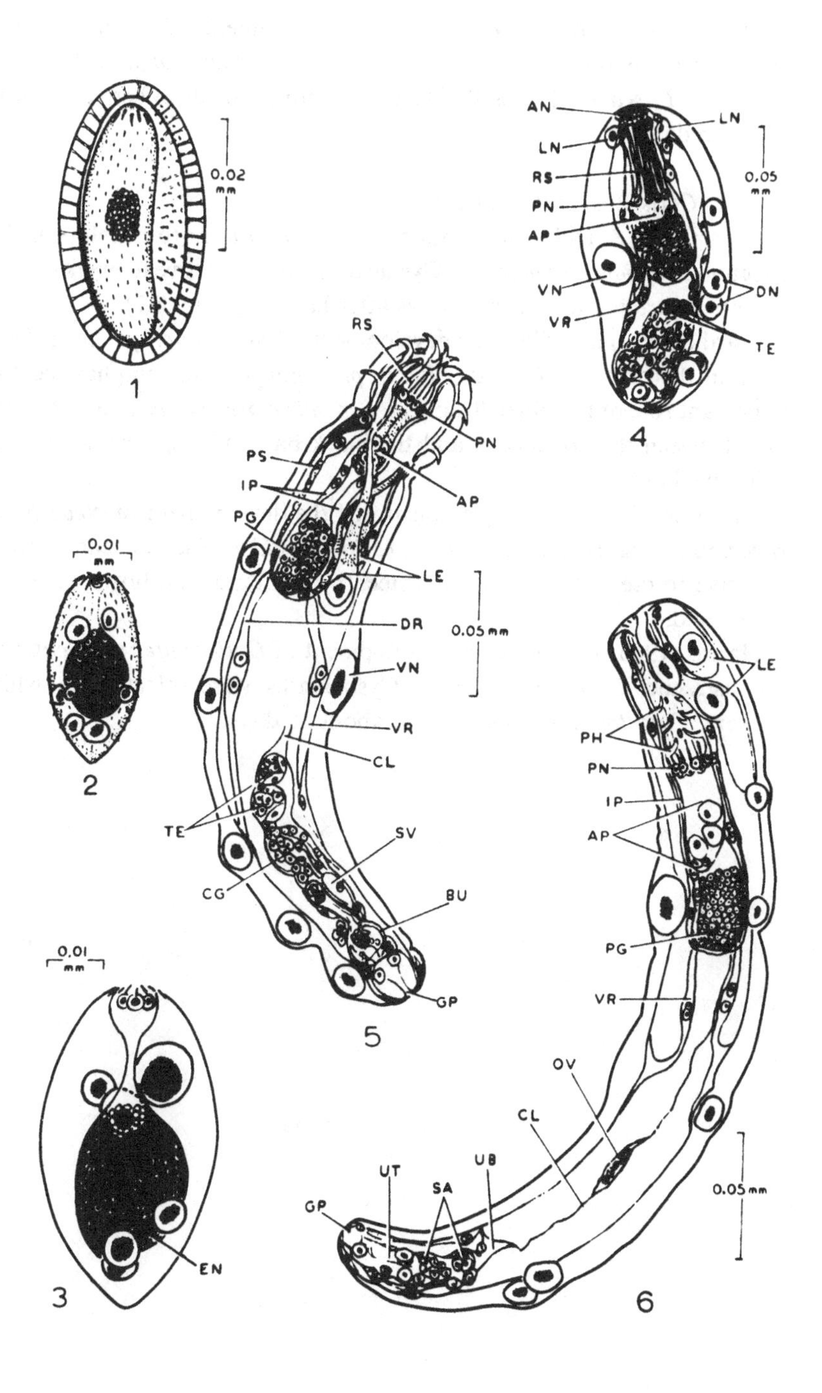
0.02 mm
1
0.01 mm
2
0.01 mm
EN
3
AN
LN
LN
RS
PN
AP
VN
VR
DN
TE
0.05 mm
4
RS
PN
PS
IP
AP
PG
LE
DR
0.05 mm
VN
VR
CL
TE
SV
CG
BU
GP
5
LE
PH
PN
IP
AP
PG
VR
OV
CL
UB
UT
SA
GP
0.05 mm
6

Schmidt & Olson (1964) traced the development of *Plagiorhynchus cylindraceus* in the terrestrial isopods *Armadillidium vulgare*, *Porcellio laevis* and *P. scaber* (Fig. 8.3). The cystacanth becomes infective to birds in 60–65 days.

8.3.3 *Class Eoacanthocephala*

Cable & Dill (1967) found *Paulisentis fractus* to develop in the copepod *Tropocyclops prasinus*. Cystacanths were infective to creek chubs, *Semotilus atromaculatus*, in 13 days after hatching (Fig. 8.7).

Ward (1940*a*) described the development of *Neoechinorhynchus cylindratus* in the ostracod *Cypria* (*Physacypria*) *globula*. She emphasized the important role of the bluegill sunfish, *Lepomis macrochirus*, as a paratenic host between the crustacean and the black bass, *Micropterus salmoides*, definitive host.

Uglem & Larson (1969) demonstrated the development of *Neoechinorhynchus saginatus* in the ostracod *Cypridopsis vidua*. The cystacanths were infective to creek chubs on the sixteenth day. Paratenic hosts were not discovered.

Harms (1965*a*) reported the development of *Octospinifer macilentus* in the ostracod *Cyclocypris serena*. Cystacanths were infective to white suckers, *Catostomus commersoni*, in about 30 days.

Table 8.1. *Intermediate and paratenic hosts for Acanthocephala, with references that have accumulated since Meyer's list of 1933*

Species	Intermediate host	Paratenic host	Reference
CLASS ARCHIACANTHOCEPHALA			
ORDER GIGANTORHYNCHIDA			
Family Gigantorhynchidae			
Mediorhynchus centurorum	*Parcoblatta pensylvanica*		Nickol (1977)
Mediorhynchus grandis	*Arphia luetola*		Moore (1962)
	Chortophaga viridifasciata australior		Moore (1962)
	Gryllus sp.		Moore (1962)
	Orphulella pelidna		Moore (1962)
	Schistocerca americana		Moore (1962)
		Blarina brevicauda	Collins (1971)
Mediorhynchus micracanthus	*Adesmia gebleri*		Rizhikov & Dizer (1954)
Mediorhynchus papillosus	Coleoptera		Kabilov (1969)
	Pimelia subglobosa		Ivashkin & Shmitova (1969)
	Stalagmoptera staundingera		Gafurov (1975)
	Tenthyria taurica		Ivashkin & Shmitova (1969)
	Zophosis punctata		Gafurov (1975)
Mediorhynchus sp.	*Gammarus lacustris*		Borgarenko (1975)
Gigantorhynchidae	*Gammarus lacustris*		Borgarenko, Bronshpits & Solodenko (1975)
Gigantorhynchus sp.	*Orchestoidea trinitatis*		Tsimbalyuk, Kulikov, Ardasheva & Tsimbalyuk (1978)
ORDER MONILIFORMIDA			
Family Moniliformidae			
Moniliformis clarki	*Ceuthophilus fusiformis*		Buckner & Nickol (1975)
	Ceuthophilus utahensis		Crook (1964); Crook & Grundmann (1964)
	Eleodes tuberculata patrulis – incomplete development		Crook (1964)
Moniliformis merionis	*Blaps* sp.		Golvan & Theodorides (1960)
Moniliformis moniliformis	*Blaps deplanata reichardti*		Gafurov (1970)
	Blaps holofila		Ivashkin (1956)
	Blatella germanica		Gonzalez & Mishra (1976)
	Geotrupes impressus		Sultanov, Kabilov & Davlatov (1974)
	Periplaneta americana		Yamaguti & Miyata (1942); Sita (1949); Coronel Guevara, (1953); Bonfonte, Faust & Giraldo

Table 8.1. (*cont.*)

Species	Intermediate host	Paratenic host	Reference
Moniliformis moniliformis	*Periplaneta americana*		(1961); Moore (1962); King & Robinson (1967); Mercer & Nicholas (1967); Robinson & Strickland (1969); Lackie (1972*a*, *b*); Rotheram & Crompton (1972); Acholonu & Finn (1974); Brennan & Cheng (1975); Anvar & Paran (1976); Hutton & Oetinger (1980)
	Prosodes biformis		Gafurov (1970)
	Prosodes vincens		Gafurov (1970)
	Scarabaeus sacer		Nazarova (1959)
Moniliformis sp.	*Periplaneta americana*		Bhamburkar, Garde & Shastri (1970)
ORDER OLIGACANTHORHYNCHIDA			
Family Oligacanthorhynchidae			
Macracanthorhynchus catulinus	*Adesmia gebleri*		Rizhikov & Dizer (1954); Gafurov (1970)
	Dissonomus sp.		Gafurov (1970)
	Pachyscelis banghaasi		Gafurov (1970)
	Stalagmoptera inocostata		Gafurov (1970)
	Tentyria tessulata		Farzaliev & Petrochenko (1980)
	Trigonoscelis gemmulata		Gafurov (1970)
		Agama caucasica	Gafurov (1970)
		Coluber jugularis	Gafurov (1970)
		Citellus dauricus	Dubinin (1948)
		Eremias pleskei	Farzaliev & Petrochenko (1980)
		Erinaceus dauricus	Dubinin (1948)
		Eumeces schneideri	Farzaliev & Petrochenko (1980)
		Lacerta strigata	Farzaliev & Petrochenko (1980)
		Marmota siberica	Dubinin (1948)
		Meles meles raddei	Dubinin (1948)
		Mustela nivalis	Dubinin (1948)
		Naja oxiana	Markov, Bogdanov, Makeev & Khutoryanski (1968)
		Ophisaurus apodus	Farzaliev & Petrochenko (1980)
		Putorius eversmanni	Dubinin (1948)
		Rana ridibunda	Farzaliev & Petrochenko (1980)

Table 8.1. (*cont.*)

Species	Intermediate host	Paratenic host	Reference
Macracanthorhynchus catulinus		*Uromastix hardwicki*	Barus & Tenora (1976)
		Varanus benghalensis	Barus & Tenora (1976)
		Vipera lebetina	Markov, Zinyakova & Lutta (1967)
		Vulpes korsak	Dubinin (1948)
Macracanthorhynchus hirudinaceus	*Anomala mongolica*		Leng, Huang & Liang (1981)
	Aphodius subterranus		Chebotarev (1954)
	Bricoptis variolosa		Daynes (1966)
	Catonia aurata		Shcherbovich (1950)
	Copris lunaris		Trifonov (1963); Sadaterashvili (1970)
	Cotinus nitida		Kates (1943)
	Cryotes nasicornis		Chebotarev (1954)
	Dorcadion pedestrae		Trifonov (1961)
	Dorysthenes hydropicus		Leng, Huang & Liang (1981)
	Dorysthenes paradoxus		Leng, Huang & Liang (1981); Wang, Li & Cai (1981)
	Geotrupes stercorarius		Sadaterashvili (1970); Kashnikov (1972)
	Geotrupes sp.		Chebotarev (1954); Morozov (1959)
	Gnaptor spinimanus		Trifonov (1961)
	Gymnopleurus mopsus		Ono (1933)
	Harpatus tridens		Ono (1933)
	Heteoligus sp.		Simmonds (1960)
	Holotrichia titanus		Leng, Huang & Liang (1981)
	Liocola brevitarsis		Oparin (1962)
	Melolontha hippocastani		Shcherbovich (1950)
	Melolontha melolontha		Shcherbovich (1950)
	Mimela splendens		Leng, Huang & Liang (1981)
	Oryctes nasicornis		Sadaterashvili (1970); Kashnikov (1972)
	Periplaneta americana		Robinson & Strickland (1969)
	Phyllodromia germanica		Ono (1933)
	Phyllophaga anxia		Swales & Gwatkin (1948)
	Phyllophaga futilis		Swales & Gwatkin (1948)
	Phyllophaga rugosa		Swales & Gwatkin (1948)
	Poecilos sp.		Trifonov (1961)
	Polyphylla laticolis		Leng, Huang & Liang (1981)
	Polyphylla olivieri		Sadaterashvili (1970)
	Popillia japonica		Miller (1943)
	Popillia sp.		Leng, Huang & Liang (1981)
	Rhizotrogus aestivus		Sadaterashvili (1970)

Table 8.1. (*cont.*)

Species	Intermediate host	Paratenic host	Reference
Macracanthorhynchus ingens	*Floridobolus penneri*		Bowen (1967)
	Ligyrus sp.		Moore (1946*b*)
	Narceus americanus		Crites (1964)
	Narceus gordanus		Bowen (1967)
	Parcoblatta pennsylvanica		Elkins & Nickol (1983)
	Phyllophaga crinita		Moore (1946*b*)
	Phyllophaga hirtiventris		Moore (1946*b*)
		Agkistrodon piscivorus	Elkins & Nickol (1983)
		Coluber constrictor	Elkins & Nickol (1983)
		Dasypus novemcinctus	Elkins & Nickol (1983)
		Elaphe obsoleta	Elkins & Nickol (1983)
		Lampropeltis getulus	Elkins & Nickol (1983)
		Nerodia cyclopion	Elkins & Nickol (1983)
		Nerodia fasciata	Elkins & Nickol (1983)
		Rana pipiens	Moore (1946*b*)
		Thamnophis sirtalis	Elkins & Nickol (1983)
Macracanthorhynchus larvae	*Amphimallon solstitialis*		Sadaterashvili (1978*a*)
	Amphimallon solstitialis setosus		Sadaterashvili (1978*b*)
	Anoxia pilosa		Sadaterashvili (1978*b*)
	Rhyzotrogus aestivus		Sadaterashvili (1978*b*)
Oncicola canis		*Dasypus novencinctus texanus*	Chandler (1946*a*)
Oncicola oncicola		*Gallus gallus domesticus*	Zeledón & Arroyo (1960)
		Leptoptila verreauxi ochroptera	Pereira (1936)
Oncicola pomatostomi		*Aphelocephala leucopsus*	Schmidt (1983)
		Climacteris leucophaea	Schmidt (1983)
		Climacteris picumnus	Schmidt (1983)
		Climacteris wellsi	Schmidt (1983)
		Cynclosoma castaneum	Schmidt (1983)
		Cynclosoma cinnamoneum	Schmidt (1983)
		Cynclosoma rufiventris	Schmidt (1983)
		Excalfactoria lineata	Schmidt (1983)
		Hylacola pyrrhopgyia	Schmidt (1983)
		Hypotaenidia philippensis	Schmidt (1983)
		Oreocicha lunulata	Schmidt (1983)
		Pachycephala inornata	Schmidt (1983)
		Pedionomus torquatus	Schmidt (1983)
		Pomatostomus rubeculus	Schmidt (1983)
		Pomatostomus ruficeps	Schmidt (1983)
		Pomatostomus supercilious	Schmidt (1983)
		Pomatostomus temporalis	Schmidt (1983)
		Pyrrholaemus brunnus	Schmidt (1983)
		Seriocornis maculatus	Schmidt (1983)

Table 8.1. (*cont.*)

Species	Intermediate host	Paratenic host	Reference
Oncicola schacheri		*Meles meles*	Schmidt (1972*c*)
Oncicola spirula	*Blabera fusca*		Brumpt & Desportes (1938)
	Blatella germanica		Brumpt & Urbain (1938); Dollfus (1938); Van Thiel & Wiegand-Bruss (1945); Eckert (1961)
	Periplaneta orientalis		Brumpt & Urbain (1938)
	Rhyparobia maderae		Brumpt & Desportes (1938)
Oncicola sp.		*Coturnix coturnix*	Padmavathi (1967)
		Francolinus pondicerianus	Padmavathi (1967)
Prosthenorchis elegans	*Blabera fusca*		Brumpt & Desportes (1938)
	Blatella germanica		Brumpt & Urbain (1938); Dollfus (1938); Eckert (1961); Wanson & Nickol (1975)
	Lasioderma serricorne		Stunkard (1965)
	Rhyparobia maderae		Brumpt & Desportes (1938)
	Stegobium paniceum		Stunkard (1965)
CLASS PALAEACANTHOCEPHALA			
ORDER ECHINORHYNCHIDA			
Family Cavisomidae			
Neorhadinorhynchus atlanticus	*Stenoteuthis pteropus*		Naidemova & Zuev (1978); Gaevskaya & Nigmatullin (1981)
Family Echinorhynchidae			
Acanthocephalus anguillae	*Asellus aquaticus*		Shtein (1959); Andryuk (1974); Andryuk (1979*a*, *c*)
Acanthocephalus anthuris	*Proasellus coxalis*		Batchvarov (1974)
Acanthocephalus clavula	*Asellus meridianus*		Chubb (1964); Rojanapaibul (1976)
	Chaetogammarus ischnus		Kurandina (1975)
	Dikerogammarus haemobaphes		Komarova (1969); Kurandina (1975)
	Gammarus balcanicus		Ivashkin (1972)
	Gammarus (Rivulogammarus) balcanicus		Yalynskaya (1980)
	Gammarus (Rivulogammarus) kischeneffensis		Yalynskaya (1980)
	Pallasea quadrispinosa		Shtein (1959)
	Pontogammarus obesus		Kurandina (1975)

Table 8.1. (*cont.*)

Species	Intermediate host	Paratenic host	Reference
Acanthocephalus dirus	*Asellus intermedius*		Seidenberg (1973); Oetinger & Nickol (1981)
	Asellus sp.		Bullock (1962); Camp & Huizinga (1980)
	Caecidotea communis		Muzzall (1981)
	Caecidotea militaris		Amin (1980, 1982); Amin, Burns & Redlin (1980)
	Lirceus garmani		Oetinger & Nickol (1981)
	Lirceus lineatus		Muzzall & Rabalais (1975*a*, *b*, *c*); Oetinger & Nickol (1981)
	Pontoporeia affinis		Amin (1978*b*)
Acanthocephalus galaxii	*Paracalliope fluviatilis*		Hine (1977)
Acanthocephalus lucii	*Asellus aquaticus*		Shtein (1959); Andryuk (1979*b*, *c*); Brattey (1980)
	Asellus sp.		Copland (1956)
Acanthocephalus minor	*Asellus hilgendorfi*		Nagasawa, Egusa & Ishino (1982)
Acanthocephalus ranae	*Asellus aquaticus*		Kirbanov (1978*a*)
Acanthocephalus larvae	*Asellus aquaticus*		Andryuk (1976)
Echinorhynchus gadi	*Caprella septentrionalis*		Val'ter (1976); Val'ter, Kondrashkova & Popova (1980)
	Gammarus duebeni		Kulachkova & Timofeeva (1977)
Echinorhynchus lageniformis	*Corophium spinicorne*		Olson (1970); Olson & Pratt (1971)
Echinorhynchus leidyi	*Mysis relicta*		Prychitko & Nero (1983)
Echinorhynchus salmonis	*Hyalella azteca*		DeGiusti (1949*b*)
	Pontoporeia affinis		Shtein (1959); Brownell (1970)
Echinorhynchus truttae	*Gammarus balcanicus*		Ivashkin (1972)
	Gammarus (*Rivulogammarus*) *balcanicus*		Yalynskaya (1980)
	Gammarus fossarum		Van Maren (1979*a*)
	Gammarus (*Rivulogammarus*) *kischeneffensis*		Yalynskaya (1980)
	Gammarus pulex		Awachie (1963, 1966, 1967)
	Gammarus pulex fossarum		Schütze & Ankel (1976)
Echinorhynchus sp.	*Pontogammarus*		Komarova (1969)

Table 8.1. (*cont.*)

Species	Intermediate host	Paratenic host	Reference
Family Illiosentidae			
Dentitruncus truttae	*Echinogammarus roco*		Orecchia, Paggi, Manilla & Rossi (1978)
	Echinogammarus tibaldii		Orecchia *et al.* (1978)
	Gammarus italicus		Orecchia *et al.* (1978)
Dollfusentis chandleri	*Corophium lacustre*		Buckner, Overstreet & Heard (1978)
	Grandidierella bonnieroides		Buckner, Overstreet & Heard (1978)
	Lepidactylus sp.		Buckner, Overstreet & Heard (1978)
Tegorhynchus furcatus	*Haustorius* sp.		Buckner, Overstreet & Heard (1978)
	Lepidactylus sp.		Buckner, Overstreet & Heard (1978)
Family Fessisentidae			
Fessisentis fessus	*Asellus forbesi*		Buckner (1977); Buckner & Nickol (1979)
	Lirceus lineatus		Buckner (1977); Buckner & Nickol (1979)
Fessisentis necturorum	*Asellus scrupulosus*		Nickol & Heard (1973)
Fessisentis tichiganensis		*Umbra limi*	Amin (1980)
Family Pomphorhynchidae			
Pomphorhynchus bulbocolli	*Gammarus* sp.		Muzzall (1981)
	Hyalella azteca		Schmidt (1964*a*); Muzzall (1981)
		Ameiurus nebulosus	Bangham (1955)
		Ictalurus melas	Sutherland & Holloway (1979)
		Notropis hudsonius	Bangham (1955)
		Osmerus mordax	Bangham (1955)
		Perca flavescens	Bangham (1955)
		Percopsis omiscomaycus	Bangham (1955)
		Umbra limi	Bangham (1955)
Pomphorhynchus dubious		*Rana cyanophylctis*	Kaw (1941)
Pomphorhynchus laevis	*Corophium volutator*		Engelbrecht (1957)
	Gammarus bergi		Chibichenko & Mamytova (1978)
	Gammarus fossarum		Van Maren (1979*a*, *b*)
	Gammarus lacustris		Chibichenko & Mamytova (1978)
	Gammarus pulex		Marshall (1976); Rumpus & Kennedy (1974)
	Gammarus sp.		Engelbrecht (1957)
	Pontogammarus robustoides		Komarova (1969)

Table 8.1. (*cont.*)

Species	Intermediate host	Paratenic host	Reference
Pomphorhynchus perforator	*Gammarus bergi*		Chibichenko & Mamytova (1978)
	Gammarus lacustris		Chibichenko & Mamytova (1978)
Pomphorhynchus rocci	*Gammarus tigrinus*		Johnson & Harkema (1971)
Family Rhadinorhynchidae			
Australorhynchus tetramorphacanthus		*Paratrigla papilo*	Lebedev (1967)
Golvanacanthus problematicus	*Gammarus olivii*		Mordvinova & Parukhin (1978)
Leptorhynchoides plagicephalus	*Gammarus pulex*		Rasin (1949)
Leptorhynchoides thecatus	*Hyalella azteca*		DeGiusti (1949*a*); Uznanski & Nickol (1976, 1980*a*, *b*)
	Hyalella knickerbockeri		DeGiusti (1939)
		Ambloplites rupestris	DeGiusti (1949*a*)
		Lepomis cyanellus	Samuel, Nickol & Mayes (1976)
		Lepomis gibbosus	Samuel, Nickol & Mayes (1976)
		Micropterus salmoides	Samuel, Nickol & Mayes (1976)
Serrasentis sagittifer		*Pagellus erythrinus*	Orecchia, Paggi & Hannuna (1970)
ORDER POLYMORPHIDA			
Family Centrorhynchidae			
Centrorhynchus amphibius		*Ptyas mucosus*	Das (1950)
		Rana tigrina	Das (1950)
Centrorhynchus batrachus		*Rana tigrina*	Das (1952)
Centrorhynchus crocidurus		*Crocidura caerulea*	Das (1950)
Centrorhynchus falconis		*Ptyas mucosus*	Das (1957*a*)
Centrorhynchus longicephalus		*Lycodon* sp.	Das (1950)
Centrorhynchus magnus		*Rana tigrina*	Schmidt & Kuntz (1969)
Centrorhynchus microcervicanthus		*Naia tripudians*	Das (1950)
Centrorhynchus mysentri		*Rana tigrina*	Gupta & Fatma (1983)
Centrorhynchus ptyasus		*Ptyas mucosus*	Gupta (1950)
Centrorhynchus spilornae		*Agkistrodon acutus*	Schmidt & Kuntz (1969)
		Dinodon rufozonatum	Schmidt & Kuntz (1969)
		Psammodynastes pulverulentis	Schmidt & Kuntz (1969)
		Trimeresurus mucrosquamatus	Schmidt & Kuntz (1969)
		Trimeresurus stejnegeri	Schmidt & Kuntz (1969)

Table 8.1. (*cont.*)

Species	Intermediate host	Paratenic host	Reference
Centrorhynchus spinosus		*Dryophis mycterizans*	Pujatti (1952)
		Lycodon falavomaculatus	Pujatti (1952)
		Simotes arnensis	Pujatti (1952)
		Thamnophis sirtalis	Read (1950*b*)
		Zamenis gracilis	Pujatti (1952)
Centrorhynchus teres		*Coluber jugularis*	Sharpilo & Sharpilo (1969)
		Coronella austriaca	Sharpilo & Sharpilo (1969)
		Emys orbicularis	Sharpilo & Sharpilo (1969)
		Eremias arguta	Sharpilo & Sharpilo (1969)
		Lacerta agilis	Sharpilo & Sharpilo (1969)
		Lacerta saxicola	Sharpilo & Sharpilo (1969)
		Lacerta taurica	Sharpilo & Sharpilo (1969)
		Natrix natrix	Sharpilo & Sharpilo (1969)
		Natrix tessellata	Sharpilo & Sharpilo (1969)
		Rana ridibunda	Kirbanov (1978*b*)
		Vipera ursini	Sharpilo & Sharpilo (1969)
Centrorhynchus sp.	*Cantatops quadratus*		Golvan & Ormières (1957)
		Agkistrodon acutus	Schmidt & Kuntz (1969)
		Anolis cristatellus	Acholonu (1976)
		Lycodon subcinctus	Schmidt & Kuntz (1969)
		Natrix sipedon	Ward (1940*b*)
		Natrix stolata	Schmidt & Kuntz (1969)
		Psammodynastes pulverulentis	Schmidt & Kuntz (1969)
		Rhacophorus robustus	Schmidt & Kuntz (1969)
		Trimeresurus mucrosquamatus	Schmidt & Kuntz (1969)
		Trimeresurus stejnegeri	Schmidt & Kuntz (1969)
Sphaerirostris lanceoides		*Hemilepistus pectinatus*	Sultanov, Kabilov & Siddikov (1980)
Sphaerirostris pinguis		*Melogale moschata subaurantiaca*	Schmidt & Kuntz (1969)
		Natrix annularis	Schmidt & Kuntz (1969)
		Paguma larvata taivanus	Schmidt & Kuntz (1969)
		Viverricula indica pallida	Schmidt & Kuntz (1969)
Sphaerirostris sp.		*Rana tigrina rugolosa*	Schmidt & Kuntz (1969)

Table 8.1. (*cont.*)

Species	Intermediate host	Paratenic host	Reference
Family Plagiorhynchidae			
Lueheia inscripta	*Periplaneta americana*	*Anolis cristatellus*	Acholonu (1976)
Plagiorhynchus cylindraceus	*Armadillidium vulgare*		Dollfus & Dalens (1960); Schmidt (1964*b*); Schmidt & Olsen (1964); Wanson & Nickol (1975); Dappen & Nickol (1981); Nickol & Dappen (1982)
	Porcellio laevis		Schmidt (1964*b*); Schmidt & Olsen (1964)
	Porcellio scaber		Schmidt (1964*b*); Schmidt & Olsen (1964)
		Blarina brevicauda	Nickol & Oetinger (1968)
		Eliomys quercinus	Dollfus (1957)
		Erinaceus europaeus	James (1954)
Porrorchis hylae		*Boiga trigonata*	Gupta & Jain (1975)
		Gekko monarchus	Schmidt & Kuntz (1967*b*)
		Hemidactylus frenatus	Schmidt & Kuntz (1967*b*)
		Hyla aurea	Johnston & Edmonds (1948)
		Hyla caerulea	Johnston & Edmonds (1948)
		Japalura swinhonis	Schmidt & Kuntz (1967*b*)
		Limnodynastes dorsalis	Johnston & Edmonds (1948)
		Psammodynastes pulverulentus	Schmidt & Kuntz (1967*b*)
		Rana limnocharis	Schmidt & Kuntz (1967*b*)
		Rana tigrina rugulosa	Schmidt & Kuntz (1967*b*)
		Trimeresurus stejnegeri	Schmidt & Kuntz (1967*b*)
		Zaocys dhumnades	Schmidt & Kuntz (1967*b*)
Porrorchis indicus		*Lycodon* sp.	Das (1957*b*)
Porrorchis leibyi		*Dinodon rufozonatum*	Schmidt & Kuntz (1967*b*)
		Gekko monarchus	Schmidt & Kuntz (1967*b*)
		Japalura swinhonis	Schmidt & Kuntz (1967*b*)
		Natrix stolata	Schmidt & Kuntz (1967*b*)
		Rana latouchi	Schmidt & Kuntz (1967*b*)

Table 8.1. (*cont.*)

Species	Intermediate host	Paratenic host	Reference
Porrorchis leibyi		*Rana tigrina rugulosa*	Schmidt & Kuntz (1967*b*)
		Rhacophorus robustus	Schmidt & Kuntz (1967*b*)
		Sphenomorphus indicus	Schmidt & Kuntz (1967*b*)
		Trimeresurus stejnegeri	Schmidt & Kuntz (1967*b*)
Porrorchis oti		*Rana temporaria ornativentris*	Yamaguti (1939)
Pseudolueheia pittae		*Lycodon subcinctus*	Schmidt & Kuntz (1967*b*)
'Acanthocephaline larvae'	*Porcellio*		Thompson (1934)
Family Polymorphidae			
Arhythmorhynchus comptus	Freshwater isopods		Atrashkevich (1975*a*)
Arhythmorhynchus petrochenkoi	*Asellus* sp.		Atrashkevich (1979*a*)
Arhythmorhynchus uncinatus		*Archosargus probatocephalus*	Bullock (1960)
Corynosoma australe		*Genypterus chilensis*	Vergara & George-Nascimento (1982)
Corynosoma bullosum	*Nototheria coriiceps*		Edmonds (1955)
Corynosoma clavatum	*Platycephalus fuscus*		Johnston & Edmonds (1952)
Corynosoma constrictum	*Hyalella azteca*		Podesta & Holmes (1970)
Corynosoma hadweni		*Oncorhynchus nerka*	Margolis (1958)
		Osmerus mordax	Van Cleave (1953)
Corynosoma hamanni		*Notothenia rossi*	Markowski (1971)
		Rhigophila dearborni	Holloway & Bier (1967)
Cornyosoma obtuscens		*Mycteoperca pardalis*	Van Cleave (1953)
		Umbrina roncador	Ward & Winter (1952)
Corynosoma semerme		*Acerina cernua*	Van Cleave (1953); Dubnitski (1957)
		Anguilla anguilla	Van Cleave (1953)
		Blicca bjoerkna	Van Cleave (1953)
		Clupea harengus	Helle & Valtonen (1981)
		Clupea harangus membras	Van Cleave (1953)
		Coregonus albula	Van Cleave (1953)
		Coregonus fera	Van Cleave (1953)
		Cottus quadricornis	Van Cleave (1953)
		Cottus scorpius	Van Cleave (1953)
		Cyclopterus lumpus	Van Cleave (1953)
		Gadus callorias	Van Cleave (1953)
		Genypterus blacodes	Grabda & Slosarczyk (1981)
		Lota lota	Helle & Valtonen (1981)
		Lota vulgaris	Van Cleave (1953)
		Macruronus novaezelandiae	Grabda & Slosarczyk (1981)

Table 8.1. (*cont.*)

Species	Intermediate host	Paratenic host	Reference
Corynosoma semerme		*Myoxocephalus quadricornis*	Helle & Valtonen (1981)
		Myoxocephalus scorpius	Helle & Valtonen (1981)
		Oncorhynchus nerka	Margolis (1958)
		Onos cumbrinus	Van Cleave (1953)
		Osmerus dentex	Neiland (1962)
		Osmerus eperlanus	Van Cleave (1953); Dubnitski (1957); Jarling (1983)
		Perca fluviatilis	Van Cleave (1953)
		Pleuronectes flesus	Van Cleave (1953)
		Pleuronectes limanda	Van Cleave (1953)
		Pleuronectes platessa	Van Cleave (1953)
		Rhombus maximus	Van Cleave (1953)
		Zoarces viviparus	Van Cleave (1953)
Corynosoma similis		*Osmerus dentex*	Neiland (1962)
Corynosoma strumosum		*Clupea harsangus*	Van Cleave (1953); Helle & Valtonen (1981)
		Conger conger	Van Cleave (1953)
		Coregonus fera	Van Cleave (1953)
		Coregonus laveratus	Van Cleave (1953)
		Cottus quadricornis	Van Cleave (1953)
		Cottus scorpius	Van Cleave (1953)
		Cyclopterus lumpus	Van Cleave (1953)
		Gadus callarias	Van Cleave (1953)
		Gadus macrocephalus	Van Cleave (1953)
		Gasterosteus aculeatus	Van Cleave (1953)
		Lepidopsetta bileneata	Van Cleave (1953)
		Leptocottus armatus	Van Cleave (1953)
		Lophius piscatorius	Van Cleave (1953)
		Lota lota	Helle & Valtonen (1981)
		Lota vulgaris	Van Cleave (1953)
		Myoxocephalus quadricornis	Van Cleave (1953); Helle & Valtonen (1981)
		Myoxocephalus scorpius	Helle & Valtonen (1981)
		Oncorhynchus gorbuscha	Margolis (1958)
		Oncorhynchus nerka	Margolis (1958)
		Osmerus dentex	Neiland (1962)
		Osmerus eperlanus	Van Cleave (1953)
		Osmerus lanceolatus	Van Cleave (1953)
		Perca fluviatilis	Van Cleave (1953)
		Platichthyes stellatus	Van Cleave (1953)
		Pleuronectes flesus	Van Cleave (1953)
		Pleuronectes limanda	Van Cleave (1953)
		Rhombus maximus	Van Cleave (1953)
		Sciaena schlegeli	Van Cleave (1953)
		Trachinus draco	Van Cleave (1953)
		Zoarces viviparus	Van Cleave (1953)

Table 8.1. (*cont.*)

Species	Intermediate host	Paratenic host	Reference
Corynosoma sp.	*Homarus americanus*		Uzmann (1970)
Filicollis anatis	*Asellus aquaticus*		Kotelnikov (1954); Styczynska (1958)
	Asellus sp.		Atrashkevich (1979*b*)
	Astacus astacus		Golvan (1961)
Filicollis trophimenkoi	*Asellus tschaunensis*	*Hynobius keiserlingii*	Astrashkevich (1982)
Hexaglandula mutabilis		*Cichlasoma tetracantha*	Moravec & Barus (1971)
Polymorphus actuganensis	*Gammarus bergi*		Chibichenko & Mamytova (1978)
	Gammarus lacustris		Chibichenko & Mamytova (1978)
Polymorphus biziurae	*Chevax distructor*		Johnston & Edmonds (1948)
Polymorphus botulus	*Carcinus moenas*		Rayski & Garden (1961); Garden, Rayski & Thom (1964)
	Hyas araneus		Rayski & Garden (1961)
	Pagurus pubescens		Uspenskaja (1960)
Polymorphus contortus	*Gammarus lacustris*		Denny (1969); Podesta & Holmes (1970)
	Hyalella azteca		Podesta & Holmes (1970)
Polymorphus formosus	*Macrobrachium* sp.		Schmidt & Kuntz (1967*a*)
Polymorphus kenti	*Emerita analoga*		Reish (1950)
Polymorphus magnus	*Gammarus bergi*		Chibichenko & Mamytova (1978)
	Gammarus lacustris		Petrochenko (1949); Logachev, Bruskin & Kesten (1961); Klesov & Kovalenko (1967); Chibichenko & Mamytova (1978)
	Gammarus maeoticus		Kovalenko (1960); Klesov & Kovalenko (1967)
	Gammarus pulex		Fotedar, Raina & Dhar (1977); Atrashkevich (1979*b*)
	Gammarus wilkitzkii		Atrashkevich (1979*b*)
Polymorphus major	*Cancer irroratus*		Schmidt & MacLean (1978)
Polymorphus marilis	*Gammarus lacustris*		Denny (1969); Tokeson & Holmes (1982)
Polymorphus minutus	*Cambarus affinis*		Golvan (1961)
	Carinogammarus roeselii		Scheer (1934)
	Gammarus duebeni		Hynes (1955)
	Gammarus fossarum		Van Maren (1979*a*)
	Gammarus lacustris		Hynes (1955); Romanovski (1964); Spencer (1974)

Table 8.1. (*cont.*)

Species	Intermediate host	Paratenic host	Reference
Polymorphus minutus	*Gammarus limnaeus*		Schmidt (1964*a*)
	Gammarus pulex		Scheer (1934); Florescu (1936); Crompton (1964); Awachie (1967); Butterworth (1969*a*); Atrashkevich (1979*b*)
	Gammarus pulex fossarum		Noll (1950)
	Gammarus roeseli		Lukacsovics (1959)
	Gammarus wilkitzkii		Atrashkevich (1979*b*)
	Gammarus sp.		Hynes & Nicholas (1958); Crompton (1967)
Polymorphus paradoxus	*Gammarus lacustris*		Bethel & Holmes (1974); Denny (1969)
Polymorphus strumosoides	*Gammarus pulex*		Atrashkevich (1975*b*, 1979*b*)
	Gammarus wilkitzkii		Atrashkevich (1979*b*)
Polymorphus trochus	*Hyalella azteca*		Podesta & Holmes (1970)
Polymorphus sp.	*Chasmagnathus granulata*		Holcman-Spector, Mañe-Garzón & Dei-Cas (1977); Keithly & Ulmer (1965)
	Pagurus longicarpus		Reinhard (1944)
Southwellina dimorpha	*Astacus astacus*		Lantz (1974)
	Procambarus clarkii		Schmidt (1973*a*)
Southwellina hispida	*Macrobrachium* sp.		Schmidt & Kuntz (1967*a*)
		Cyprinus carassius	Yamaguti (1935)
		Elaphe quadrivirgata	Yamaguti (1935)
		Fundulus grandis	Bullock (1957*a*, *b*)
		Mogurnda obscura	Yamaguti (1939)
		Paralichthys lethostigmus	Chandler (1935)
		Rana nigromaculata	Yamaguti (1939)
		Sciaenops ocellatus	Overstreet (1983)
Southwellina macracanthus		*Umbrina roncador*	Ward & Winter (1952)
Polymorphidae sp.	*Orchestoidea trinitatis*		Tsimbalyuk, Kulikov, Ardasheva & Tsimbalyuk (1978)
CLASS EOACANTHOCEPHALA			
ORDER GYRACANTHOCEPHALA			
Family Quadrigyridae			
Acanthogyrus dattai	*Mesocyclops leuckharti*		Sharma & Wattal (1976)
Pallisentis basiri		*Trichogaster chuna*	Hasan & Qasim (1960)
Pallisentis nagpurensis	*Cyclops stenuus*	*Aplocheilus melastiga*	George & Nadakal (1973)
		Barbus sp.	George & Nadakal (1973)

Table 8.1. (*cont.*)

Species	Intermediate host	Paratenic host	Reference
Pallisentis nagpurensis		*Heteropneustes fossilis*	George & Nadakal (1984)
		Macropodus cupanus	George & Nadakal (1973)
		Ophiocephalus gachua	George & Nadakal (1973)
		Rana tigrina	George & Nadakal (1984)
		Wallago attu	George & Nadakal (1984)
ORDER NEOECHINORHYNCHIDA			
Family Neoechinorhynchidae			
Atactorhynchus verecundus	*Cletocamptus deiresi*		Dill (1975)
Neoechinorhynchus cristatus	*Cypridopsis helvetica*		Uglem (1972*b*)
Neoechinorhynchus cylindratus	*Cypria globula*		Ward (1940*a*)
		Ictalurus melas	Samuel, Nickol & Mayes (1976)
		Lepomis gibbosus	Samuel, Nickol & Mayes (1976)
		Lepomis pallidus	Ward (1940*a*)
Neoechinorhynchus doryphorus		*Fundulus majalis*	Van Cleave & Bangham (1949)
		Lucania parva	Bullock (1960)
		Notropis sp.	Bullock (1960)
Neoechinorhynchus emydis	*Cypria maculata*		Hopp (1954)
	Pleurocerea acuta	*Campeloma decisum*	Whitlock (1939)
		Campeloma rufum	Hopp (1946, 1954); Lincicome & Whitt (1947)
		Ceriphasia semicarinata	Hopp (1954); Lincicome & Whitt (1947)
Neoechinorhynchus rutili	*Cypria turneri*		Merritt & Pratt (1964)
	ostracods probably the intermediate host; *Sialis* and *Erpobdella* are the facultative second intermediate hosts		Walkey (1962)
Neoechinorhynchus saginatus	*Cypridopsis vidua*		Uglem & Larson (1969)
Octospinifer macilentus	*Cyclocypris serena*		Harms (1963, 1965*a*)
Octospiniferoides chandleri	*Cypridopsis vidua*		DeMont & Corkum (1982)
	Physocypria pustulosa		DeMont & Corkum (1982)
Paulisentis fractus	*Tropocyclops prasinus*		Cable & Dill (1967)
Paulisentis missouriensis	*Cyclops vernalis*		Keppner (1974)
Family Tenuisentidae			
Paratenuisentis ambiguus	*Gammarus tigrinus*		Samuel & Bullock (1981)
Tenuisentis niloticus		*Hydrocyon brevis*	Dollfus & Golvan (1956)

9

Epizootiology

Brent B. Nickol

9.1 Introduction

In practice, epizootiology deals with how parasites spread through host populations, how rapidly the spread occurs and whether or not epizootics result. Prevalence, incidence, factors that permit establishment of infection, host response to infection, parasite fecundity and methods of transfer are, therefore, aspects of epizootiology. Indeed, most aspects of a parasite could be related in some way to epizootiology, but many of these topics are best considered in other contexts. General patterns of transmission, adaptations that facilitate transmission, establishment of infection and occurrence of epizootics are discussed in this chapter.

When life cycles are unknown, little progress can be made in understanding the epizootiological aspects of any group of parasites. At the time Meyer's monograph was completed (1933), intermediate hosts were known for only 17 species of Acanthocephala, and existing descriptions are not sufficient to permit identification of two of those. Laboratory infections of intermediate hosts had apparently been produced for only two species. Study at that time was primarily devoted to species descriptions, host and geographical distribution, structure and ontogeny. Little or nothing was known about adaptations that promote transmission and the concept of paratenic hosts was unclear.

In spite of the paucity of information, Meyer (1932) summarized pathways of transmission among principal groups of hosts, visualized the relationships among life cycle patterns for the major groups of Acanthocephala, and devised models for the hypothetical origin of terrestrial life cycles from aquatic ones. Nevertheless, most of our knowledge regarding epizootiology has been recently acquired.

9.2 Transmission

9.2.1 *General patterns*

Hosts and developmental stages. All organisms must disperse from sites of propagation to other microhabitats if overcrowding is to be avoided and ranges extended. Parasites must disperse not only in space but also from one host to another. Among the acanthocephalans, only eggs exist outside of a host, free in the environment, and transmission from one definitive host to another requires that eggs be ingested by invertebrate animals appropriate to serve as intermediate hosts and that the final ontogenetic stages achieved there find their way to vertebrate animals where maturation can occur.

Acanthocephalan species for which life cycles have been confirmed by laboratory infections require vertebrates for definitive hosts and arthropods as intermediate hosts. Occurrences of adults in carnivorous invertebrates, such as *Neorhadinorhynchus atlanticus* in squid (Gaevskaya, 1977; Naidenova & Zuev, 1978; Gaevskaya & Nigmatullin, 1981), probably represent transitory infections acquired by transfer of adults from prey. There is no laboratory evidence that adulthood is reached in any invertebrate.

Acanthocephalans parasitic in terrestrial definitive hosts usually have insects, frequently species of Coleoptera or Orthoptera, for intermediate hosts. Microcrustaceans, usually species of Amphipoda, Copepoda, Isopoda or Ostracoda, are generally intermediate hosts for those parasitic in aquatic definitive hosts. In the life cycle of some species, another host occurs between the arthropod intermediate and vertebrate definitive hosts. In such hosts, forms parasitic in intermediate hosts penetrate the intestinal wall and localize in mesenteries or visceral organs where maturity does not occur. Although such intercalated hosts may be required to complete transfer of acanthocephalans from intermediate hosts to the trophic level at which potential definitive hosts feed, there is no evidence that they are essential for achievement of infectivity to the final host. The term 'paratenic host', proposed by Baer (1951) and elaborated by Beaver (1969), has attained wide usage for such animals in which ontogeny does not proceed.

No species of acanthocephalan has been demonstrated to require more than the arthropod intermediate host in order to develop infectivity to vertebrates in which maturity can occur. *Neoechinorhynchus emydis*, which matures in turtles, might be an exception. Whitlock (1939) discovered an unidentified species of *Neoechinorhynchus* in snails from Michigan and Lincicome & Whitt (1947) found *N. emydis* in snails from Kentucky. Hopp (1954) demonstrated that *N. emydis* from *Campeloma rufum* was infective to turtles, but he was unable to infect these snails with eggs of *N. emydis*.

Ostracods, however, were readily infected. In preliminary trials, two turtles fed *N. emydis* from laboratory-infected ostracods did not harbor acanthocephalans at necropsy (Hopp, 1954). Ward (1940*b*) concluded that bluegill sunfish, *Lepomis pallidus* (= *L. macrochirus*), were true second intermediate hosts for *N. cylindratus*, but forms from ostracods are now known to reach maturity directly in several species of the Centrarchidae.

Acanthocephalan ovaries fragment early in life and ova are produced in the resulting masses of cells (§7.5.3). Fertilization is internal and within the female a series of membranes develops around the zygote. Upon discharge from the female, cleavage has produced a larva surrounded by a series of membranes and a shell. Strictly, this stage should, perhaps, be called a shelled embryo. However, for convenience, the term 'egg' is usually applied. The term acanthor designates the larval stage that emerges from the egg upon its ingestion by an arthropod (Van Cleave, 1937). After penetration of the wall of the alimentary canal, the acanthor undergoes a series of ontogenetic stages in the body cavity of its intermediate host. Van Cleave (1937) proposed the term acanthella for each in this series of stages. Misuse of the term prompted him later to reaffirm and clarify this term (Van Cleave, 1947). The final ontogenetic stage in intermediate hosts, which is infective to potential definitive hosts, was termed the juvenile (Van Cleave, 1937, 1947) until Chandler (1949) introduced the term cystacanth. As originally proposed, cystacanth designated the final, infective stage found in invertebrates and juvenile was retained to designate re-encysted forms in paratenic hosts. Cystacanth has achieved general usage as a name for the stage infective to a final host regardless of whether it is found in the arthropod intermediate host (Van Cleave, 1953) or in a vertebrate paratenic host (Van Cleave, 1953). Some objection can be made to use of the term for juveniles in paratenic hosts because not all such stages actually 'encyst'. However, argument can be made for the consistent use of cystacanth to designate a stage of ontogeny, i.e. infective to a potential definitive host, regardless of whether the form is actually encysted.

Transmission from intermediate host directly to definitive host. Many, if not most, acanthocephalans utilize only an intermediate host and a definitive host in their life cycles. Eggs of some species, *Acanthocephalus dirus*, *Polymorphus marilis* and *P. minutus*, might not always be shed individually from females and may leave the definitive host only in the body of passed females, to be released upon deterioration of the adult (Muzzall & Rabalais, 1975*a*; Denny, 1968; Nicholas & Hynes, 1958, respectively). However, eggs of most species are passed free in the feces of the definitive host and are ingested by arthropods in which larval development occurs.

Cystacanths developed in these intermediate hosts attain adulthood in definitive hosts upon ingestion of the intermediate host. In these cases, intermediate hosts are fed upon directly by vertebrates that will become definitive hosts and the path of transmission is clear. Many laboratory studies have confirmed such life cycles for a variety of species.

Paratenic hosts. Cystacanths have long been known to occur in mesenteries and visceral organs of vertebrates, usually poikilotherms, but an understanding of the relationships of these hosts to intermediate and definitive hosts has been slow in developing. Meyer (1932) listed vertebrates harboring cystacanths together with invertebrates as '*Zwischenwirte*' or intermediate hosts.

The suggestion by Van Cleave (1920*a*) that larvae of some acanthocephalans locate extraintestinally in vertebrates when ingested before being fully developed is generally accepted to explain the occurrence of adults of some species in the intestine and cystacanths of the same species in the viscera of conspecific hosts. Late acanthellae of *Leptorhynchoides thecatus* cannot establish in fishes. Those that have developed in amphipods for less than 26 days are unable to infect rock bass, *Ambloplites rupestris*. If development has been from 26 to 29 days in amphipods, the cystacanths localize extraintestinally when fed to rock bass. When at least 30 days of development has occurred before ingestion by rock bass, the worms attach to the intestine and mature (DeGiusti, 1949*a*). This suggests that there is a period in development during which some acanthocephalans are able to survive in vertebrates but are unable to maintain themselves in the intestine and mature. Such adaptations would clearly increase survivorship for those species that mature in predators by giving larvae that were ingested before being completely developed a second chance to reach a definitive host.

Other species of Acanthocephala occur viscerally in certain species of vertebrates in which they are not known to occur intestinally, regardless of the age at which ingestion occurs. Perhaps Mingazzini (1896) was the first to illustrate the epizootiological role of this kind of paratenic host when he produced laboratory infections in falcons, *Falco tinnunculus*, by feeding them cystacanths of *Centrorhynchus aluconis* and *C. buteonis* taken extraintestinally from *Zamenis gemonensis* (Reptilia: Colubridae). Use of paratenic hosts to bridge trophic levels between predatory vertebrates and arthropods is now considered to be an important adaptation that enables acanthocephalans to utilize as definitive hosts groups of animals to which transmission would otherwise be unlikely.

Predatory fishes frequently acquire acanthocephalans from smaller paratenic hosts, usually other fishes, that constitute prey. Hamann (1891*b*) found cystacanths of *Pomphorhynchus laevis* in the viscera of six small fishes and noted that these were frequently prey for larger ones. Riquier (1909) demonstrated that cystacanths of *P. laevis* are indeed, infective to carnivorous fish (*Esox lucius*), but paratenic hosts are not required for development. Cystacanths from intermediate hosts, *Gammarus pulex*, attain adulthood when fed directly to flounder (Engelbrecht, 1957) or goldfish (Kennedy, 1972). Other species of *Pomphorhynchus* also use paratenic hosts. *Pomphorhynchus rocci* occurs in the intestine of several piscine species, including *Morone saxatilus*, striped bass, and encysted in the viscera of striped bass (Paperna & Zwerner, 1976). *Pomphorhynchus bulbocolli* occurs in the mesenteries of several species of fish (Ward, 1940*a*). *Leptorhynchoides thecatus* and *Neoechinorhynchus cylindratus* are other species that occur as adults in carnivorous fishes and commonly use smaller fishes as paratenic hosts. At least two species of *Pallisentis*, *P. basiri* and *P. nagpurensis*, occur viscerally in coarse fishes that are often eaten by larger fishes (Hasan & Qasim, 1960; George & Nadakal, 1973, respectively). Species of the genus *Serrasentis* occur as adults in marine fishes and use various species of other marine fishes as paratenic hosts (Van Cleave, 1924).

Piscivorous birds frequently acquire acanthocephalans of the genera *Arhythmorhynchus*, *Corynosoma*, *Hexaglandula* and *Southwellina* from poikilothermic vertebrate hosts. *Arhythmorhynchus frassoni* occurs in the viscera of Brazilian fishes (Travassos, 1926; Golvan, 1956*a*) and *A. uncinatus* in mesenteries of sheepshead, *Archosargus probatocephalus*, from the coast of Florida (Bullock, 1960). *Corynosoma clavatum* occurs as adults in several species of cormorants, *Phalacrocorax*, in the Southern Hemisphere (Edmonds, 1957*a*) and as cystacanths in mesenteries of *Platycephalus fuscus*, flathead fish (Johnston & Edmonds, 1952). Cormorants, herons and kingfishers in Brazil are definitive hosts for *Hexaglandula mutabilis*, cystacanths of which occur in marine fishes of several genera (Travassos, 1926). *Southwellina hispida* has a broad geographical distribution in *Nycticorax nycticorax*, black-crowned night heron (Schmidt, 1973*a*), and occurs in mesenteries of fishes, frogs and snakes in Japan (Van Cleave, 1925; Yamaguti, 1935, 1939), and in the mesenteries of fishes in Texas (Chandler, 1935; Bullock, 1957*a*). Species of the genus *Andracantha* parasitize a variety of piscivorous birds (Schmidt, 1975), including the American bald eagle (Nickol & Kocan, 1982) and, although no life cycle is known for any species of this genus, it is likely that some, if not all, use fishes as paratenic hosts.

Amphibians and reptiles also serve as paratenic hosts for some acanthocephalan species that mature in flesh-eating birds. Species of *Centrorhynchus* and the related *Sphaerirostris* from around the world, are well known as cystacanths in frogs, lizards and snakes. Adults occur in raptors and other kinds of carnivorous birds. Golvan (1956*b*) and Schmidt & Kuntz (1969) listed many of the definitive and paratenic hosts for species of these genera. Many species of *Oligacanthorhynchus* occur as adults in birds of prey, and the literature abounds with worldwide reports of cystacanths in the viscera and mesenteries of reptiles, usually snakes.

Cystacanths of some species of *Porrorchis* that occur as adults in coucals (*Centropus*), owls (*Bulbo* and *Tyto*), and kites (*Milvus*) in Australia, India, the Philippines and Taiwan are found in the mesenteries of amphibians and reptiles. *Porrorchis hylae* has long been known from the viscera of Australian frogs (Johnston, 1914). Southwell & MacFie (1925) described the adult from *Centropus phasianus*, collected in northern Australia, as *Echinorhynchus bulbocaudatus*, and specimens from *C. viridis*, red-winged coucal, collected in the Philippines were later named *E. centropusi* by Tubangui (1933). *Echinorhynchus bulbocaudatus* and *E. centropusi* are regarded (Edmonds, 1957*b*; Schmidt & Kuntz, 1967*b*) as synonyms of *P. hylae*. Coucals are known to feed upon the species of frogs in which cystacanths of *P. hylae* occur (M'ackness, 1977). *Porrorchis hylae* occurs also in India, but snakes, rather than frogs, are paratenic hosts (Gupta & Jain, 1975). Schmidt & Kuntz (1967*b*) found many species of flesh-eating birds to be definitive hosts in Taiwan where snakes, lizards and frogs are paratenic hosts. *Porrorchis indicus* from India, *P. leibyi* from Taiwan and Palawan, and *P. oti* from Japan are other species of the genus that occur as adults in flesh-eating birds and are known to occur in the viscera of snakes, lizards and frogs (Das, 1957*b*; Schmidt & Kuntz, 1967*b*; Yamaguti, 1939, respectively). The presence of cystacanths of another member of the Porrorchinae, *Lueheia lueheia*, in mesenteries of amphibians, reptiles and even birds of Brazil (Travassos, 1926) is difficult to assess. Adults are only known to occur in insectivorous birds of the families Formicariidae and Furnariidae. Similarly, cystacanths of *L. inscripta* occur in the body cavity of lizards, *Anolis cristatellus*, in Puerto Rico (Acholonu, 1976), but adults occur in passerine birds. Perhaps a yet to be discovered predator is the usual definitive host for these species or perhaps the cystacanths occur in these paratenic hosts but have no epizootiological significance.

Aquatic mammals, especially seals and whales, are frequently definitive hosts for acanthocephalans acquired from fishes. Species of *Bolbosoma* and *Corynosoma*, cosmopolitan parasites of cetaceans and pinnipeds, have been reported many times from a seemingly endless number of piscine paratenic hosts. In spite of the fact that host specificity for fishes is apparently low, the prevalence and intensity of the acanthocephalans seem to vary according to the species of fish. Some acanthocephalan species are more abundant in certain fishes and others more abundant in different species of fish. This is demonstrated by the fact that seasonal variation in the structure of *Corynosoma semerme* and *C. strumosum* populations in ringed seals, *Pusa hispida*, from the Bothnian Bay of the Baltic Sea reflects the migratory habits of herrings. Among fishes that are paratenic hosts in the Bothnian Bay, *C. strumosum* is more predominant than *C. semerme* only in *Clupea harengus*, Baltic herring, and *Lota lota*, burbot (Helle & Valtonen, 1980), but Helle & Valtonen (1981) cited references to indicate that the seals do not feed on burbot. In all seasons *C. semerme* is more predominant in seals than is *C. strumosum*, but the difference, which is great in the spring, is much less in the autumn when herrings are available to the seals (Helle & Valtonen, 1981). Similar relations between paratenic hosts and acanthocephalan population structures in aquatic mammals are likely to be evident when parasite distribution within piscine paratenic hosts is analyzed with the same care at other localities.

Acanthocephalans that occur as adults in carnivorous mammals and are known to use paratenic hosts belong primarily to the genera *Macracanthorhynchus* and *Oncicola*. Adults of *M. catulinus* occur in many species of carnivores throughout the eastern European and Asian portion of the USSR, and cystacanths are found in smaller mammals. Petrochenko (1958) listed definitive and paratenic hosts to which Barus, Kullmann & Tenora (1970) added rodents as paratenic hosts in Afghanistan. *Macracanthorhynchus catulinus* also uses poikilothermic vertebrates as paratenic hosts: *Varanus benghalensis* and *Uromastix hardwicki* (Sauria) in Afghanistan (Barus & Tenora, 1976); frogs, snakes and lizards in Azerbaidzhan (Farzaliev & Petrochenko, 1980); snakes, *Vipera lebetina*, in central Asia (Markov, Zinyakova & Lutta, 1967); and *Naja oxiana* in Turkmenistan and Tadzhikistan (Markov, Bogdanov, Makeev & Khutoryanski, 1968). Tenebrionid beetles serve as intermediate hosts (Rizhikov & Dizer, 1954) for *M. catulinus*.

In North America, *Macracanthorhynchus ingens* parasitizes carnivores, primarily *Procyon lotor*, raccoon (Chandler & Melvin, 1951), and cystacanths occur in a variety of frogs and snakes (Moore, 1946*b*; Elkins & Nickol, 1983). Moore (1946*b*) demonstrated that eggs fed to beetles of the genera *Phyllophaga* and *Ligyrus* developed into cystacanths; however, only *Narceus americana*, a species of millipede (Crites, 1964), and *Parcoblatta pensylvanica*, a species of woodroach (Elkins & Nickol, 1983), are known to serve as intermediate hosts in nature.

Several pathways are apparently feasible for transmission of *Macra-*

canthorhynchus ingens to raccoons (Fig. 9.1). Cystacanths from woodroaches develop into adulthood when fed to raccoons or penetrate the intestine and re-encyst in the viscera when fed to snakes. Cystacanths from snakes develop into adults when fed to raccoons (Elkins & Nickol, 1983). Additionally, it is probable that frogs may be intercalated as paratenic hosts between intermediate hosts and reptilian paratenic hosts or between intermediate hosts and raccoon definitive hosts.

Adults of species of *Oncicola* also parasitize carnivorous mammals, primarily the Canidae and Felidae. The role of paratenic hosts in the epizootiology of these species is a little less clear because the infectivity of encysted forms has not been verified by laboratory infections. Except for *O. spirula*, a parasite of primates, no intermediate host is known for species of this genus, but cystacanths occur in several avian and mammalian species.

In southwestern North America, *Oncicola canis* occurs in the viscera of *Dasypus novemcinctus*, armadillo (Van Cleave, 1921; Chandler, 1946*a*), beneath the epithelial lining of the esophagus of turkey poults (Price, in Christie, 1929), and in the connective tissue and on the outer surface of the esophagus and crop of *Colinus virginianus*, bob-white quail (Cram, 1931). Cystacanths of *O. canis* from the armadillo are frequently calcified (Chandler, 1946*a*), suggesting that armadillos may not be significant paratenic hosts. *Oncicola oncicola* occurs in the connective tissue and musculature of armadillos (*Tatus* sp.) in Brazil (Travassos, 1917) and subcutaneously and in the musculature of domestic chickens in Costa Rica (Zeledón & Arroyo, 1960).

Australian and Asian species of *Oncicola* also use birds for paratenic hosts. *Oncicola pomatostomi*, which occurs intestinally in canine and feline definitive hosts in Australia, Borneo, Malaysia and the Philippines, has been reported from under the skin of 19 species of birds (Schmidt, 1983).

Fig. 9.1. Schematic diagram of pathways of transmission for *Macracanthorhynchus ingens* demonstrated feasible by laboratory infections and field studies. (From data in Elkins & Nickol, 1983.)

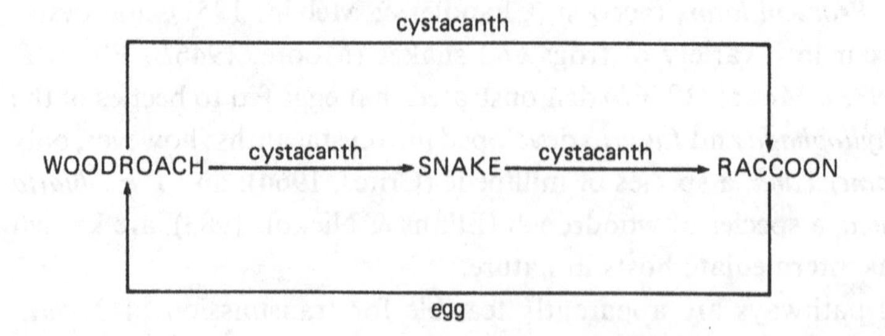

An unidentified species of *Oncicola* has also been reported (Padmavathi, 1967) from the musculature of gallinaceous birds in Madras.

It is clear that paratenic hosts play an important role in transmission of many acanthocephalans, but frequently they lead to termination of the life cycle before reproduction can occur. More or less frequently, encysted forms are found in viscera of animals from which transmission is unlikely or impossible. This seems to be the case for *Oncicola schacheri.* Schmidt (1972*c*) reported that adults of this species occur in Lebanese fox and cystacanths in the mesenteries of badgers, *Meles meles*, but completion of the life cycle by use of badgers as paratenic hosts seems improbable.

At two locations in the USA, Erie County, New York (Nickol & Oetinger, 1968) and near Lincoln, Nebraska, cystacanths of *Plagiorhynchus cylindraceus* occur commonly in mesenteries of shrews. Adults parasitize passerine birds, usually robins or starlings, and although robins are known to kill shrews (Penny & Knapton, 1977) and feed them to their nestlings (Weeden & Weeden, 1973), the epizootiological significance is probably negligible. *Mediorhynchus grandis* has also been reported from the mesenteries of a shrew (Collins, 1971), but because adults occur in non-predatory birds, usually icterids, transmission from the shrew and subsequent attainment of maturity seem unlikely. Cystacanths in hosts from which transmission leading to maturation probably cannot occur, have been reported for many other acanthocephalan species. Such instances could provide a means by which distribution to new kinds of hosts may occur through continued exposure and adaptation, or they may simply represent retention of a general adaptation that is useful only to certain species. In either event, they reflect the frequency with which acanthocephalans, as a group, successfully incorporate paratenic hosts into their pathways of transmission and thereby achieve distribution to groups of animals that would be inaccessible otherwise.

Postcyclic parasitism. When ingested as adults within their definitive hosts, some acanthocephalans survive and parasitize the predator. This phenomenon is termed postcyclic parasitism and hosts that are parasitized as a result are either eupostcyclic or parapostcyclic hosts, depending on whether the predator is conspecific with the prey (Bozkov, 1976). Little is known regarding postcyclic parasitism by Acanthocephala, but it is possible for individuals of at least five species, *Acanthocephalus ranae*, *Echinorhynchus salmonis*, *Moniliformis moniliformis*, *Neoechinorhynchus cristatus* and *Octospiniferoides chandleri*, to be transferred from definitive host to definitive host in this manner (Bozkov, 1980; Hnath, 1969; Moore, 1946*a*; Uglem & Beck, 1972; DeMont & Corkum, 1982, respectively).

DeMont & Corkum (1982) theorized that postcyclic transmission could explain some of the 'mysteries of acanthocephalan transmission', for example, the occurrence of acanthocephalans without known paratenic hosts in definitive hosts that do not feed on the appropriate intermediate hosts.

If postcyclic transmission of some species occurs more frequently in nature than is generally assumed, hosts in which maturation and reproduction of parasites are not at a rate sufficient to maintain the population or in which enteric survival occurs without maturation may not always exert inhibitory effects on acanthocephalan populations.

9.2.2 *Adaptations that increase probability of transmission*

Most species of parasite have high levels of fecundity and are towards the *r* end of the *r–K* continuum (Esch, 1977), and it is often argued that *r*-selection results from the uncertainties of transmission. In spite of this, Croll (1966) suggested that because of the great demands of reproduction the number of eggs produced should be the bare minimum required to overcome the natural toll of transmission and successfully propagate the next generation.

Croll's view was challenged by Jennings & Calow (1975) who contended that high fecundity is a natural consequence of the stable, nutrient-rich environment of adult parasites. They believed that in such an environment energy supplies are not limiting so that accumulation of energy reserves to buffer against competition and possible reductions in food supply or allocation of energy to other *K*-strategies would not be at the expense of egg production. Selection pressure to reduce fecundity to the minimum suggested by Croll would not be present. In such circumstances, parasites could readily produce eggs in excess of the minimum number required to insure successful transmission. Whether or not parasites are under selective pressure to hold egg production to a minimum, it is clear that characteristics of potential hosts are commonly exploited in a manner to facilitate transmission, and elaborate adaptations that help insure larval success have evolved.

Seasonal distribution. Van Cleave (1916) was among the earliest to report that acanthocephalans are not distributed similarly among hosts throughout the year (§7.9.5). He found that *Gracilisentis gracilisentis* occurs in gizzard shad, *Dorosoma cepedianum*, of the Illinois River only from October through April and that *Tanaorhamphus longirostris* is present in the same piscine species only from June through December. Seasonal distribution of acanthocephalans was still largely unstudied by 1932, and Meyer's monograph (1932, 1933) contains little information regarding seasonal occurrence. The rapidity with which studies on the subject have accumulated is evidenced from a recent review (Chubb, 1982) of seasonal occurrence in freshwater fishes. The seasonal distribution is summarized for 34 species and is discussed in relation to the world's climatic zones.

Obviously periods of seasonal activity, including feeding by potential intermediate and definitive hosts, must coincide long enough for transmission to definitive host, maturation and reinfection of intermediate hosts to occur. Eggs could lie dormant at the end of the cycle with reinfection of the intermediate host population being postponed, but there is little indication that they do so. Petrov (1973), however, reported that eggs of a species of *Polymorphus* retained their infectivity to invertebrates for 90 days in the salt lake Kuspek, Kazakhstan, USSR. Kennedy (1972) found that feeding response was an important factor in controlling the level of *Pomphorhynchus laevis* parasitism in dace, *Leuciscus leuciscus*, and that water temperature influenced this response. Unless negated by another factor, such as temperature-dependent rejection (Kennedy, 1972), prevalence and intensity should clearly be greatest during times when intermediate hosts with infective larvae constitute an appropriate portion of the diet of potential definitive hosts. When this occurs throughout the year, there may be no seasonal periodicity in the occurrence of acanthocephalans.

Echinorhynchus salmonis shows no seasonal periodicity in prevalence, intensity or development and maturation in rainbow smelt, *Osmerus mordax*, from Lake Michigan where the amphipod intermediate host, *Pontoporeia affinis*, is available to fish throughout the year (Amin, 1978*a*, 1981). Likewise, availability of intermediate hosts in all seasons has been postulated to explain the uniform seasonal prevalence of *Acanthocephalus clavula* in the fish of Llyn Tegid (Chubb, 1964), *Echinorhynchus truttae* in brown trout, *Salmo trutta*, from North Wales (Awachie, 1965) and the lack of seasonal periodicity in both prevalence and intensity of *Neoechinorhynchus saginatus* in fallfish, *Semotilus corporalis*, of New Hampshire (Muzzall & Bullock, 1978).

It may not follow, however, that absence of seasonal periodicity implies that rates of recruitment and mortality are constant throughout the year. Hine & Kennedy (1974*a*, *b*) found infected *Gammarus pulex* in the River Avon during each month of the year. In that river *Pomphorhynchus laevis* shows no seasonal cycle in incidence or intensity in fishes (Kennedy, 1974; Rumpus, 1975; Kennedy & Rumpus, 1977). Because high water temperatures in the laboratory reduce the success with which *P. laevis* establishes in *Carassius auratus*, goldfish, Kennedy (1972) believed that increased feeding by fish in summer balanced a lower rate of parasite establishment. Thus, seasonal differences in rates of recruitment and mortality (from failure to establish in fish) could occur even if parasite occurrence shows no periodicity.

There are instances in which species of invertebrates appropriate as intermediate hosts are consumed by potential definitive hosts throughout the year, but definite seasonal cycles in acanthocephalan distribution occur. Awachie (1965) demonstrated that even though infective larvae of *Echinorhynchus truttae* occurred in *Gammarus pulex* throughout the year and that the amphipods formed an important part of the food of brown trout all year, seasonal differences in densities of infective larvae led to cycles of intensity in fish. Cases such as this suggest that fish of a different species, one for which seasonal cycles are significant, might be more important hosts.

Seasonal cycles in occurrence of acanthocephalans are often linked directly with seasonal changes in the external environment, especially changes in water temperature. Variations in habits of hosts during the year, for instance changes in diet, also cause seasonal differences in prevalence and intensity of infections (Halvorsen, 1972). While obviously true (intermediate hosts must be available and eaten), this view more or less implies that parasite activity remains constant and alterations in host habits result in distribution of parasites. Parasites are frequently not so passive. Acanthocephalans often show adaptations that are directly related to seasonal events in the life histories of definitive and intermediate hosts in a way that helps insure transmission.

Reproductive effort by parasites may be limited to coincide with breeding cycles in host populations. Muzzall (1978) described seasonal cycles of *Fessisentis friedi* in isopods, *Caecidotea communis*, from New Hampshire that suggest peak acanthocephalan egg production occurs as pickerel move into shallow water to spawn. Maximum intensity and egg production of *Pomphorhynchus bulbocolli* in white suckers, *Catostomus commersoni*, have also been related (Muzzall, 1980) to migration and spawning. In Lake Michigan, *Echinorhynchus salmonis* reaches peak maturity in chinook salmon, *Oncorhynchus tshwytscha*, during spawning (Amin, 1978*a*). These seasonal patterns result in peak egg production by the acanthocephalans when fishes are in shallow waters where appropriate invertebrates are most abundant.

A similar potential endocrine relation may exist between the reproductive cycles of acanthocephalans and the maturation of amphibian hosts. In Georgia, aquatic isopods, *Asellus scrupulosus*, are intermediate hosts for *Fessisentis necturorum*. Prevalence and intensity in salamanders,

Ambystoma opacum, increase from January through March while the salamanders are larvae. Parasitism declines in April and May during metamorphosis and the worms are absent from adults (Nickol & Heard, 1973). Adult *A. opacum* are found in water much less frequently than are adults of other species of salamander and do not even enter standing water to breed. Instead, eggs are laid in moist litter that is later inundated. In Louisiana and Illinois, *F. necturorum* and *F. fessus* occur in *Necturus beyeri* and *Siren intermedia*, respectively (Nickol, 1967, 1972). These amphibians, which retain gills and remain aquatic throughout their lives, are parasitized into summer, long after acanthocephalans are gone from *A. opacum*. Avery (1971) found no seasonal difference in levels of *Acanthocephalus anthuris* infection in newts, *Triturus helveticus*, which remain in the water for the entire year in England. Gravid females were present in adults and larvae throughout the year. No seasonal difference was detected in larval or adult *T. vulgaris*, but adults were surveyed only during months in which they were aquatic. If acanthocephalans were lost from these species at metamorphosis, the aquatic existence of adults could have resulted in their reinfection.

Kennedy (1970) concluded that when maturation of helminths from freshwater fishes showed seasonal cycles, peak egg production would almost always occur late in spring or early in summer. Many species of Acanthocephala might conform to this generalization. Prevalence, density and maturity of *Acanthocephalus dirus* in Wisconsin fishes are greatest in the spring. Eighty per cent of the female acanthocephalans contain fully formed eggs in May (Amin, 1975*c*). Walkey (1967) showed this cycle for *Neoechinorhynchus rutili* in England. There was no seasonal difference in prevalence or intensity in three-spined sticklebacks, *Gasterosteus aculeatus*, but the proportion of immature worms plunged by May so that gravid worms were most common during the late spring and early summer and rarest in winter. This finding is consistent with a later report by Tesarcik (1972) who noted that *N. rutili* was recruited into carp populations from late February to early March in Czechoslovakia. Egg production then occurred from March through July (Tesarcik, 1970). Although he did not note periods of maturity, Bibby (1972) reported a seasonal distribution in Wales of *N. rutili* in minnows, *Phoxinus phoxinus*, that fits with this cycle.

There are many other reports, including those by von Möller (1974, 1975) for *Echinorhynchus gadi*, by Muzzall & Rabalais (1975*a*) for *Acanthocephalus dirus*, by Paperna & Zwerner (1976) for *Pomphorhynchus rocci*, and by Jilek (1978) for *Tanaorhamphus longirostris*, of acanthocephalan infections of vertebrates increasing in prevalence and intensity in the spring. Egg production then occurs late in spring and perhaps throughout the summer. Numbers decline in the fall and reach a minimum during winter. Komarova (1950) found that the prevalence and intensity of *A. lucii* in perch, *Perca fluviatilis*, of the Dneiper River rose sharply in March and April, but that the worms were immature until summer. Towards the end of summer, acanthocephalans decreased in perch until they were absent in the fall. Anderson (1978) described a similar seasonal cycle for *A. lucii* in Norway, although peaks were a little later. In England, the prevalence of *A. lucii* in perch also rises sharply in March. At the same time the percentage of *Asellus aquaticus* and *Gammarus pulex*, intermediate hosts, is at its yearly highest in perch stomachs (Mishra, 1978). *Dentitruncus truttae* has nearly the same seasonal dynamics in Italy. Prevalence of detectable cystacanths in gammarid intermediate hosts is highest from February through March. It falls during the remainder of the year. Prevalence in trout, *Salmo trutta*, begins to rise after April and continues to rise through October, after which it falls (Orecchia, Paggi, Manilla & Rossi, 1978).

Although most studies of seasonal distribution have been concerned with parasites of poikilothermic hosts, there is indication that some of those from homeotherms conform to the general pattern. Hair (1969) found *Plagiorhynchus cylindraceus* more numerous in starlings, *Sturnus vulgaris*, of South Carolina during spring and summer. They were absent from birds during five winter months. *Moniliformis clarki* is most prevalent and produces more eggs in voles, *Microtus ochrogaster* and *M. pennsylvanicus*, in Indiana during the summer than in spring and winter (Fish, 1972). Sadaterashvili (1977) reported egg production and development of *Macracanthorhynchus hirudinaceus* in swine to be slower in winter than in summer.

These studies of seasonal distribution show that annual cycles in vertebrates frequently begin in spring and that most transmission between vertebrates and invertebrates occurs during summer and early fall. Development in intermediate hosts infected late in the season is slow or non-existent until temperatures rise again in the spring. Petrochenko (1958) reported that at 9–12 °C *Polymorphus magnus* developed in *Gammarus lacustris* at only one third the rate exhibited at 18–25 °C. Others (including DeGiusti, 1949*a*; Awachie, 1966; Lackie, 1972; Nickol, 1977; Tokeson & Holmes, 1982) have also observed halted or retarded larval development in invertebrates at low temperatures. During late winter at Cooking Lake, Alberta, less than 3% of the *G. lacustris* contained cystacanths of *P. marilis*. After these amphipods were held in aquaria at temperatures greater than 15 °C, 20–30% of them developed cystacanths (Tokeson & Holmes, 1982).

Ability to resume development is not always related to temperature alone. Tokeson & Holmes (1982) found that the developmental rate of *Polymorphus marilis* in amphipods placed in temperatures favoring development, after being held at temperatures below the developmental threshold, depended upon the time spent at low temperatures. Because development was faster after longer periods below the threshold, they postulated a diapause condition that might prevent development to the cystacanth stage during sporadic intervals of warm weather in early fall.

A common type of cycle, then, consists of spring recruitment by vertebrates after the environment has warmed and potential vertebrate and invertebrate hosts have become active; summer transmission between vertebrates and invertebrates; slow development in invertebrates during the winter; and completion of larval development with rising temperatures in the spring.

Development of parasites geared to the same stimuli that promote increased activity in hosts insures that larvae reach infectivity during times when transmission is most likely. Fully developed, infective larvae probably increase the mortality of hosts even in the absence of predation. Larger forms might make greater metabolic demands of their hosts and increase stress in an already stressful season. Further, alterations of host behavior that may accompany parasite infectivity (Bethel & Holmes, 1974) could result in the intermediate host responding in an unadvantageous manner that could increase stress at a time when transmission is unlikely.

Deviations from the common 'spring recruitment cycle' are often related to deviations in vertebrate–invertebrate relationships and environmental disturbances. Prevalence of *Echinorhynchus salmonis* in yellow perch, *Perca flavescens*, from Ontario rises in the fall after which it declines until the parasite is absent from fish during the summer (Tedla & Fernando, 1970). Watson & Dick (1979) also found that *E. salmonis* was most common in whitefish, *Coregonus clupeaformis*, in late fall in Manitoba when the amphipod intermediate host, *Pontoporeia affinis*, was a prominent whitefish food item.

In instances where invertebrates are rare or absent during winter months, the cycle may be altered also, resulting in recruitment to the definitive host population in the fall. Maturation is then slow with egg production occurring in the spring. Eure (1976) found this to be true for *Neoechinorhynchus cylindratus* in a heated reservoir in South Carolina. Recruitment into largemouth bass, *Micropterus salmoides*, occurred in the fall while ostracod numbers decreased. By November ostracods were scarce and most acanthocephalans were in bass. Egg production was delayed until spring when ostracods were again abundant. Ice cover might affect seasonal distribution of acanthocephalans of waterfowl similarly. Spencer (1974) found no *Gammarus lacustris* infected with *Polymorphus minutus* in a Colorado lake from December through March. Prevalence in amphipods peaked in July and started down in August. Similarly, Hynes (1955) reported that *G. lacustris* infected with *P. minutus* began die-offs during fall. Apparently, during the winter *P. minutus* is found in mallards or occurs in amphipods as small, undetected acanthors. A similar species, *P. magnus*, is rare or absent in waterfowl during the spring and summer, but becomes common in the fall (Okorokov, 1953; Moskalev, 1976).

Environmental disturbances may not affect host relationships of all parasites in the same manner. Contrary to Eure's (1976) findings, Boxrucker (1979) detected no seasonal difference in prevalence or intensity of *Pomphorhynchus bulbocolli* in a thermally heated sample area in Wisconsin, but found that in a non-heated reference area this species displayed the familiar pattern of increasing prevalence and intensity during the spring followed by declining numbers in the fall. Elevated winter temperatures may have resulted in more rapid parasite development in amphipods and in fish feeding more extensively on them than would occur during winter in non-heated environments.

Egg fibrils. There has been considerable attention devoted to the number (see Table 7.6), composition and structure (von Brand, 1940; Monné, 1964; Wright, 1971; Stranack, 1972; Whitfield, 1973), and homologies (West, 1964) of egg envelopes of acanthocephalans. Even early observers frequently illustrated eggs with fibrils contained in the envelopes surrounding acanthors. The illustrations were reproduced without comment by Meyer (1932). Yamaguti (1935) may have been the first to mention fibrils specifically, but he associated them with the outer envelope as did Petrochenko (1953) who described enveloping threads on the surface of *Pseudoacanthocephalus caucasicus* eggs. Monné & Hönig (1954) recognized the fibrils as a separate, distinct envelope, under the outermost covering and first applied the term fibrillar coat. West (1964) incorporated this term in his scheme for naming acanthor envelopes and associated the filamentous nature with aquatic forms.

Monné & Hönig (1954) noted that crushing or treatment with sulphuric acid caused unraveling of the fibrillar coat due to destruction of the outer envelope, but Whitfield (1973) was the first to postulate a function for egg fibrils. He recognized three envelopes in eggs of *Polymorphus minutus*. His envelope II comprises a refractile inner zone, IIb (fertilization membrane in West's scheme), and an outer layer, IIa (fibrillar coat of West), of fibrillar threads. Whitfield then suggested that keratin-destabilizing conditions in the alimentary canal of intermediate hosts could cause lengthening and softening of envelope IIb. Such destabilization could burst envelope I causing release of the fibrillar envelope IIb. The looping threads could serve to slow down passage through the gut giving acanthors better opportunity to complete hatching and to initiate penetration of host tissue. A similar view was adopted by Oetinger & Nickol (1974) in explaining the fibrils of *Acanthocephalus dirus* eggs (Fig. 9.2). They observed, however, that fibrils could be released by the action of bacteria and protozoans on the outer envelope before eggs were ingested.

Uznanski & Nickol (1976) observed that eggs of *Leptorhynchoides thecatus* in fish feces lack outer membranes and possess free fibrils. Scanning electron microscopy revealed the fibrils to be edges of a membranous band wrapped like a bandage around the egg (Fig. 9.3). The band is often torn to produce ribbons of various lengths and numbers. These fibrils entangle in algae and suspend eggs among algal filaments

Fig. 9.2. Eggs of *Acanthocephalus dirus* with freed filaments viewed at 500X by scanning electron microscopy. (Micrograph by David F. Oetinger.)

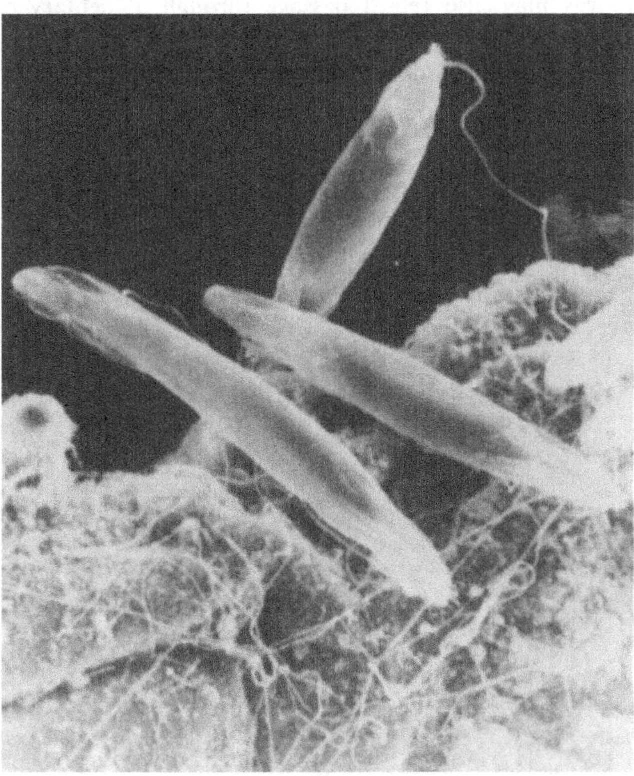

where *Hyalella azteca*, potential intermediate hosts, feed. Feeding experiments resulted in significantly greater prevalences and intensities of infection in amphipods held in containers to which eggs were added after algae (permitting entanglement) than in those held in containers to which eggs were added before the algae (prohibiting entanglement).

Adaptations resulting in egg production at the proper place and time to promote likelihood of successful transmission have been considered. Because eggs are the only free-existing stage in acanthocephalan life cycles, they are of considerable importance in dispersal and transmission. In spite of the fact that eggs are frequently viewed as passive stages disseminated randomly from definitive hosts, it appears that adaptations occur that help to facilitate transmission. Fibrils on egg envelopes represent one such adaptation and there may be others. For example, the outer membrane of *Pallisentis nagpurensis* eggs swells upon contact with water allowing the eggs to float (George & Nadakal, 1973). Planktonic copepods, *Cyclops strennus*, are the intermediate hosts and floating eggs might be more accessible to them. It is clear that fibrils, and perhaps devices for floating, can promote retention of eggs in microhabitats of potential intermediate hosts. Fibrils may also retard passage through alimentary canals of intermediate hosts.

Fig. 9.3. Eggs of *Leptorhynchoides thecatus* with freed fibrillar bands viewed at 610X by scanning electron microscopy. (Reproduced from Uznanski & Nickol, 1976, with permission of the editor of the *Journal of Parasitology*.)

Pigmentation. There are many reports showing that animals that contrast with their backgrounds are more frequently taken by predators. Animals that differ from conspecifics of their population may also be more vulnerable (Holmes & Bethel, 1972). Infective stages of acanthocephalans are known to modify intermediate hosts in ways that may result in increased probability of consumption by potential definitive hosts. These modifications are frequently morphological, involving parasite and host pigmentation, or behavioral.

Although cystacanths of most species are colorless, some of those occurring in aquatic crustaceans are brightly colored. Among the best known and most widely distributed of these colored cystacanths is *Corynosoma constrictum* in central North America. Cystacanths of *C. constrictum* appear through the body wall of *Hyalella azteca* as conspicuous, bright orange spheres. Cystacanths of *Polymorphus contortus*, *P. marilis*, and *P. paradoxus* in *Gammarus lacustris* are also orange (Denny, 1969). Among European species, *Echinorhynchus truttae* and *P. minutus* cystacanths are bright orange (Barrett & Butterworth, 1968) as are those of *Pomphorhynchus laevis* in amphipod intermediate hosts (Kennedy, Broughton & Hine, 1978; Van Maren, 1979*a*, *b*).

Analysis of the carotenoids of *Polymorphus minutus* cystacanths by Barrett & Butterworth (1968) revealed that in spite of the fact that a variety of carotenoids was present in *Gammarus pulex*, only one, esterified astaxanthin, occurred in the cystacanths. They concluded that developing *P. minutus* either selectively absorb this pigment or convert all the others into astaxanthin.

Pigmentation of cystacanths may be independent of that in adults. While all known pigmented cystacanths are shades of orange, adults of several species are red, orange, yellow or brown, and pigmented individuals may occur alongside unpigmented worms of the same species. Ravindranathan & Nadakal (1971) found adults of *Pallisentis nagpurensis* to be orange, red, brown, yellow or colorless. There is apparently no correlation between the type of intermediate host and the carotenoid found in adult parasites (Barrett & Butterworth, 1973).

Although no metabolic role has been demonstrated for pigmentation in acanthocephalans, pigmented cystacanths certainly render infected crustaceans more conspicuous. Such animals readily stand out from uninfected conspecifics and thus may be more likely selected by sight-feeding vertebrates. Likelihood of transmission could be enhanced by making infected invertebrates more obvious or by allowing a predator to single out an individual from the group. Holmes & Bethel (1972) developed a scenario comparing unidirected efforts at prey capture to 'flockshooting' by hunters. Theoretical approaches to predator–prey relationships suggest advantages in transmission for pigmented cystacanths, but little experimental verification has been attempted.

Alterations of intermediate host pigmentation may also promote acanthocephalan transmission. By interfering with normal pigmentation of intermediate hosts, cystacanths of some species cause an abnormal contrast with backgrounds. Munro (1953) was the first to note a relation beween isopod pigmentation and infection with acanthocephalans. In Scotland, he found more than 90% of the *Asellus aquaticus* to be darker than normal when infected with larval polymorphids. Individuals of *A. aquaticus* infected with *Acanthocephalus anguillae* are also more darkly colored (Balesdent, 1965). In Italy, *A. coxalis* infected with *A. clavula* have been described as more darkly pigmented than uninfected individuals (Fresi, 1967*a*, *b*).

Other species of developing acanthocephalans have the opposite effect on crustacean pigmentation. Hindsbo (1972) found that 94.9% of the adult *Gammarus lacustris* from two ponds in Denmark were of the normal gray-to-brown color, but that the remainder were blue. Dissection showed that all blue amphipods, but less than 4% of the brown ones, were infected with a species of *Polymorphus*, probably *P. minutus*. Hindsbo attributed the abnormal color of infected amphipods to the blue hemolymph showing through unpigmented cuticle. Seidenberg (1973) reported 'dichromatism' of the isopod *Asellus intermedius* in Illinois in which all light-colored isopods were infected with *Acanthocephalus dirus* and most dark individuals were not. He related infection and pigmentation in a manner to suggest that developing acanthocephalans promoted depigmentation. *Lirceus lineatus*, when infected with *A. dirus*, are also non-pigmented (Muzzall & Rabalais, 1975*b*, *c*).

Integumental pigmentation of *Asellus intermedius*, *Lirceus garmani* and *L. lineatus* differs with species, sex, reproductive state and size of the isopod. Comparison of pigmentation between uninfected and infected isopods requires consideration of these factors. Normal pigmentation is altered in each of these species when isopods harbor *Acanthocephalus dirus* cystacanths (Oetinger & Nickol, 1981). Acanthocephalan-infected isopods fail to develop pigmentation comparable to uninfected forms, rather than becoming depigmented (Oetinger & Nickol, 1981). Spectrophotometric study of integumental pigment extracts from *A. dirus*-infected and uninfected *A. intermedius* and methanolic hydrochloric acid extracts from *A. dirus* suggests that competition between developing acanthocephalans and developing isopods for amino acids may cause the pigmentation dystrophy (Oetinger & Nickol, 1982*a*). Only small isopods can be infected routinely in the laboratory (Oetinger & Nickol, 1982*b*). Occasional acanthocephalan infections in normally pigmented isopods apparently are the result of infrequent infection of older individuals in which pigmentation has already developed (Oetinger & Nickol, 1981, 1982*a*, *b*). Some epigean isopods possess reduced pigmentation when associated with hypogean habitats (Lisowski, 1979), and thus not all isopods with reduced pigmentation are infected. However, it is clear that developing acanthocephalans of some species cause pigmentation dystrophy that renders infected isopods more conspicuous against the dark background of their habitat (Fig. 9.4).

Experimental demonstration that acanthocephalan-induced conspicuousness actually leads to increased vulnerability to predation has been attempted. Feeding experiments in which unpigmented, blue amphipods (*Gammarus lacustris*) infected with *Polymorphus minutus* were offered to ducks along with the normal brown, uninfected forms revealed that the chance of unpigmented individuals being eaten was 2.5 times greater than that for the pigmented forms (Hindsbo, 1972). Similarly, significantly more light-colored isopods (*Asellus intermedius*) infected with *Acanthocephalus dirus* were eaten than uninfected, dark forms when offered to creek chubs, *Semotilus atromaculatus*, in aquaria (Camp & Huizinga, 1979). However,

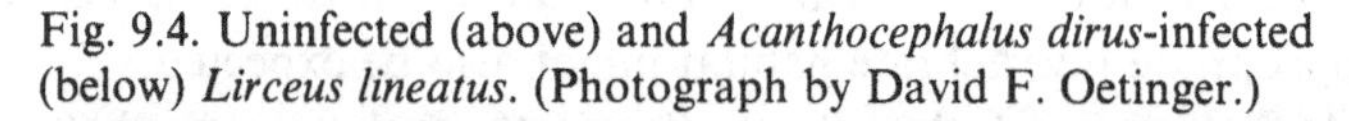

Fig. 9.4. Uninfected (above) and *Acanthocephalus dirus*-infected (below) *Lirceus lineatus*. (Photograph by David F. Oetinger.)

P. minutus and *A. dirus* evoke behavioral changes (Hindsbo, 1972; Camp & Huizinga, 1979) in their intermediate hosts. The importance of altered pigmentation as opposed to modified behavior in promoting differences in predation is unclear. Bethel & Holmes (1977) attempted to distinguish between the effects in promoting increased vulnerability of amphipods to predation by ducks. They painted oval marks, about the size and color of cystacanths, on carapaces of uninfected amphipods. Upon exposure to mallards, the proportions of marked and unmarked amphipods that were eaten were not significantly different. However, mallards are surface feeders and although cystacanths of *P. paradoxus*, for which mallards are a principle definitive host, are brightly colored, other species with orange cystacanths (*Corynosoma constrictum* and *P. marilis*) occur much more frequently in diving ducks, such as scaup, for which brightly colored prey might be more significant.

Alteration of host behavior. Holmes & Bethel (1972) pointed out that the degree of overlap between the habitats or feeding niches of definitive and intermediate hosts could influence the evolution of transmission mechanisms. They suggested that when the habitats only overlap partially, alterations might occur in responses of intermediate hosts so as to move them into the area of overlap with potential definitive hosts. Modifications in definitive host behavior that would result in release of parasite eggs in the appropriate invertebrate habitat might also be expected.

Little is known of the effects of acanthocephalans on behavior of definitive hosts. In man, however, *Moniliformis moniliformis* may cause nearly unbearable tinnitus (Grassi & Calandruccio, 1888) and violent turning of the head (Al-Rawas, Mirza, Shafig & Al-Kindy, 1977). Rabies-like symptoms associated with *Oncicola canis* in dogs and coyotes (Parker, 1909) led Van Cleave (1921) to speculate that acanthocephalan-induced pain could drive a host 'mad'. Intestinal parasitism is the principle cause of pinniped strandings on Australian beaches where cestodes and acanthocephalans are the main parasites seen (Bergin, 1976). Although acanthocephalan-induced behavioral responses have not been studied critically in definitive hosts, it is conceivable that transmission could be influenced by alterations of host habitat selection and distribution.

On the other hand, modifications of behavior that promote transmission have been demonstrated clearly in intermediate hosts of Acanthocephala. Differences in response to light between infected and uninfected amphipods were studied thoroughly by Holmes & Bethel (1972) and by Bethel & Holmes (1973). Their results showed that uninfected *Gammarus lacustris* and *Hyalella azteca* are strongly photophobic and negatively phototactic. *G. lacustris* becomes photophilic, without showing a differential response to different light intensities, but negatively phototactic when infected with cystacanths of *Polymorphus marilis* and photophilic and positively phototactic when infected with *P. paradoxus*. *H. azteca* are strongly photophilic and select regions of highest illumination when infected with *Corynosoma constrictum*. Holmes & Bethel demonstrated that differences in responses to light by these amphipods placed cystacanths of *C. constrictum*, *P. marilis* and *P. paradoxus* in different microhabitats from each other and from uninfected amphipods. This microhabitat difference was magnified by differences in evasive response displayed by amphipods. For example, uninfected amphipods found at the surface among leaves and emergent vegetation dove and vanished immediately when surrounding water or vegetation was disturbed. However, upon disturbance, amphipods infected with *P. paradoxus* cystacanths never dove but instead clung persistently to surrounding floating material, even when it was shaken or removed from the water. If material on which to cling was not present, infected amphipods swam to the top of the water and began 'skimming' along the surface rapidly digging or grasping at the air–water interface with their gnathopods. On the other hand, *G. lacustris* harboring cystacanths of *P. marilis* were photophilic, but displayed normal evasive responses (Bethel & Holmes, 1973).

Other species of amphipod infected with different species of Acanthocephala display similar behavioral modifications. *Gammarus lacustris* and *G. fossarum* infected with cystacanths of a species of *Polymorphus* (probably *P. minutus*) and *Pomphorhynchus laevis*, respectively, display a much higher degree of positive phototropism than do uninfected individuals (Hindsbo, 1972; Van Maren, 1979*a*), and *P. laevis*-infected *G. pulex* spend more time in open water and less on the substrate, move more often towards the surface and rest more in surface vegetation than do uninfected *G. pulex* (Kennedy, Broughton & Hine, 1978).

Certain species of isopods and ostracods also display modifications of behavior when infected by acanthocephalans. Non-parasitized *Lirceus lineatus* spend significantly more time under leaves and less time 'wandering' than do those harboring *Acanthocephalus dirus* cystacanths (Muzzall & Rabalais, 1975*c*), and similar behavioral differences occur between *Asellus intermedius* infected with cystacanths of *Acanthocephalus dirus* and uninfected conspecifics (Camp & Huizinga, 1979). *Cypridopsis vidua* and *Physocypria pustulosa* form aggregations milling about at the water surface when infected with cystacanths of *Octospiniferoides chandleri*, while uninfected ostracods concentrate at the bottom (DeMont & Corkum, 1982).

Evidence that behavioral differences are acanthocephalan-induced comes not only from comparisons of behavior of infected and uninfected invertebrates but also from study of the relation between modified behavior and acanthocephalan development. Not all acanthocephalan species are infective to vertebrates immediately upon reaching the cystacanth stage. Some, such as *Plagiorhynchus cylindraceus* in pillbugs (Schmidt & Olsen, 1964), *Polymorphus trochus* in amphipods (Podesta & Holmes, 1970) and *Mediorhynchus centurorum* in woodroaches (Nickol, 1977), require a period of development as a cystacanth before attaining infectivity. *Polymorphus paradoxus* is not infective to vertebrates during the first week after becoming a cystacanth. The behavioral modifications of infected amphipods are not manifested until 15–20 days after the cystacanth stage is reached. By this time, cystacanths are infective (Bethel & Holmes, 1974).

Experiments to test whether acanthocephalan-induced modifications of intermediate host behavior actually increase vulnerability to predation have been conducted. Bethel & Holmes (1977) showed that amphipods infected with cystacanths of *Polymorphus paradoxus* were significantly more vulnerable to predation by mallards and to accidental ingestion by muskrats, and demonstrated how altered amphipod behavior relates to feeding techniques of these two types of definitive host. Likewise, *Gammarus pulex* harboring *Pomphorhynchus laevis* are eaten by predatory fishes in significantly greater proportions than are uninfected individuals, even when infected amphipods constitute a much smaller proportion of the amphipod population (Kennedy, Broughton & Hine, 1978).

9.3 Establishing infection

9.3.1 *Host recognition and activation*

Acanthocephalans, like most parasites, reach an ontogenetic stage at which no further development occurs until an external stimulus is received. Such stimuli are provided by potential hosts and result in activation and resumption of development. Without these stimuli there is no spread of the parasite.

Hatching of eggs. The egg and cystacanth represent acanthocephalan stages at which development is suspended until external stimuli are provided after ingestion by suitable invertebrates and vertebrates, respectively. However, cystacanths of some species are activated after ingestion by certain vertebrates (which become paratenic hosts), but become parenteric and do not resume development until ingested by a different vertebrate (which becomes a definitive host). Eggs, free in the environment after being passed from the definitive host, and cystacanths, in the body cavity of intermediate or paratenic hosts, represent the stages of epizootiological significance. These stages must achieve transmission while viable, respond to host stimuli and establish infections.

There are numerous records concerning the duration of infectivity of some eggs for various species, but little information on the effect of aging on infectivity of a population of eggs. Retention of viability of at least some eggs ranges from less than 3 weeks (Hynes & Nicholas, 1957) for *Polymorphus minutus* to more than 3 years (Kates, 1942) for *Macracanthorhynchus hirudinaceus*. Environmental factors are a major influence on retention of viability (Kates, 1942), but freezing does not seem to be harmful and may, in fact, extend viability. *M. hirudinaceus* eggs are viable 140 days after freezing (Kates, 1942), *P. minutus* eggs are infective after 6 weeks at −22 °C (Hynes & Nicholas, 1963) and *Plagiorhynchus cylindraceus* eggs are viable for at least an additional 9 months after 30 h at −80 °C (Nickol & Dappen, 1982). As time passes, however, there appears to be a decrease in the percentage of eggs that is viable. The greatest percentage of *Moniliformis moniliformis* eggs hatches after storage for between 3 and 5 days. After 5 days of storage, the percentage that hatches gradually decreases (Edmonds, 1966).

Eggs are not known to hatch under natural circumstances until ingested by an appropriate invertebrate. Alternate drying and wetting results in 'artificial' hatching among eggs removed from body cavities of some species; however, hatching does not occur when eggs from host feces are subjected to such treatment (Manter, 1928; Moore, 1942). The mechanism of such artificial hatching is apparently different from that of 'natural' hatching and fundamental differences occur between acanthors freed artificially and naturally (Uglem, 1972*a*,*b*; Nickol, 1977). After artificial hatching, no species has been demonstrated to be infective to invertebrates and acanthors of *Neoechinorhynchus cristatus* are known to be non-infective (Uglem, 1972*b*). Further, the process of hatching induced artificially is passive (Manter, 1928) and Uglem (1972*b*) interpreted it as osmotic release of dead material. Acanthors stimulated to hatch by more natural means play an active role in their release. Mechanics of hatching have been detailed by Crook & Grundmann (1964) for *Moniliformis clarki* and by Uglem (1972*b*) for *N. cristatus*. In each of these cases, acanthors used hooks of their aclid organs (Schmidt & Olsen, 1964) to assist in penetration of embryonic membranes. Others (DeGiusti, 1949*a*; Merritt & Pratt, 1964) have also noted active participation by the acanthor during hatching. Hatching induced by alternate drying and rewetting, which could be prompted by periods of sunshine followed by rain (Manter, 1928), is of no apparent epizootiological significance.

Upon ingestion by appropriate invertebrates, hatching begins within 15 min (Schmidt & Olsen, 1964) for *Plagiorhynchus cylindraceus*, 45 min (DeGiusti, 1949*a*) for *Leptorhynchoides thecatus*, 60 min (Awachie, 1966; Uglem, 1972*b*; Hynes & Nicholas, 1957, respectively) for *Echinorhynchus truttae*, *Neoechinorhynchus cristatus*, and *Polymorphus minutus*, 90 min (Nickol, 1977) for *Mediorhynchus centurorum*, 2 h (Edmonds, 1966) for *Moniliformis moniliformis*, 4 h (Harms, 1965*a*) for *Octospinifer macilentus*, 6 h (Merritt & Pratt, 1964) for *N. rutili*, and 24 h (Moore, 1942) for *Mediorhynchus grandis*. Hatching of *M. grandis* eggs continues for at least 48 h (Moore, 1962).

Attempts to discover stimuli for 'natural hatching' were begun when Hynes & Nicholas (1957) were unable to induce hatching of *Polymorphus minutus* eggs by placing them in macerated portions of suitable *Gammarus* gut. Schmidt & Olsen (1964) achieved hatching of *Plagiorhynchus cylindraceus* eggs by mixing them with crushed digestive glands of an appropriate invertebrate. *In vitro* hatching of *Moniliformis moniliformis* eggs is prompted by solutions of a variety of electrolytes provided that the molarity is between 0.2 and 0.4 and that the pH is greater than 7.5. In 0.25M sodium bicarbonate, hatching occurs over the temperature range of 10 °C to 37 °C. In the presence of carbon dioxide, hatching occurs at pH as low as 6 and the percentage of those hatching is greater over the entire range of suitable pHs (Edmonds, 1966). Lackie (1972*b*) found that when eggs removed from body cavities and stored more than 48 h in 60% sucrose were placed in a solution of 0.3M sodium bicarbonate at room temperature, more than 10% hatched within 24 h. Because these aliquots included a proportion of eggs not fully developed, hatching success under natural conditions should be considerably greater. During hatching, acanthors release chitinase (Edmonds, 1966). Because chitin reportedly occurs in the embryonic membranes of some acanthocephalans (von Brand, 1940; Monné & Hönig, 1954) and because external chitinase has no effect on eggs (Edmonds, 1966), it appears that hatching requires contributions from both the host and parasite. Crompton (1970) compiled information to demonstrate that the physicochemical conditions under which Edmonds (1966) achieved hatching occur in invertebrate alimentary canals, suggesting that the stimulus is physiological. There may be a tendency to view feeding habits and presence or absence of a hatching stimulus in digestive tracts of various invertebrates as the basis for intermediate host specificity. This is not, however, the entire explanation. Eggs of *Neoechinorhynchus cristatus* are ingested and hatch in four species of Ostracoda, but acanthors penetrate the intestine in only two of them and develop to cystacanths in only one species (Uglem, 1972*a*, *b*).

Initiation of infection in intermediate hosts. Factors other than ingestion and hatching must be considered in the epizootiology of acanthors. Success in initiating infection, the effect of previous infection and the ultimate ability to develop cystacanths are important considerations in the spread of parasites through host populations. It is difficult, if not impossible, to know what percentage of eggs is viable in nature. In cockroaches, *Periplaneta americana*, approximately 25% of the fully developed eggs of *Moniliformis moniliformis* given orally in small numbers succeed in reaching the cystacanth stage. Although the percentage of success decreases as the number of eggs administered to each cockroach increases, there is no evidence of a saturation level and previous infection does not seem to affect the success of superimposed infections (Hynes & Nicholas, 1957). Various life cycle studies (Ward, 1940*b*; Hynes & Nicholas, 1957; Harms, 1965*a*; Nickol & Dappen, 1982) reveal that under laboratory conditions, where exposure and numbers of eggs are great, it is usually possible to infect between 70% and 80% of the appropriate invertebrates. Of course, wild-caught invertebrates are infected at much lower levels, probably due to lower exposure rates.

Cystacanth activation. After reaching the cystacanth stage, acanthocephalan development is again suspended until an external stimulus is supplied by an appropriate vertebrate. Little information is available regarding the length of time cystacanths remain infective in intermediate or paratenic hosts, but those of *Moniliformis moniliformis* remain infective to rats after at least one year in *Periplaneta americana* (Buckner & Nickol, 1975). It is generally assumed that cystacanths remain infective throughout the life of the intermediate host.

Although cystacanths of *Moniliformis moniliformis*, *Echinorhynchus truttae* and *Polymorphus minutus* are known to be freed from their intermediate hosts in the vertebrate stomach, activation does not appear to occur until they reach the small intestine (Graff & Kitzman, 1965; Awachie, 1966; Lingard & Crompton, 1972, respectively). *Polymorphus minutus* cystacanths introduced directly into the small intestine of ducks established infections (Lingard & Crompton, 1972), demonstrating that all necessary conditions for activation are present there.

Bile salts are important in cystacanth activation. Graff & Kitzman (1965) achieved *in vitro* activation of at least 80% of the *Moniliformis moniliformis* cystacanths treated with sodium taurocholate, sodium glycocholate and sodium cholate in a pH range of 7.4 to 8.5. Sodium taurocholate was the most effective bile salt, giving activation more rapidly and at lower concentrations. Activation was stimulated by carbon dioxide and inhibited by oxygen. Cystacanths activated by this treatment were infective to rats. Pretreatment of cystacanths with proteolytic enzymes was unnecessary for activation and any effect questionable. Rats in which the bile duct openings were surgically transferred to the cecum were refractory to infection.

Cystacanths of *Polymorphus minutus* can be activated *in vitro* in a basic salt solution, but addition of duck bile or any of several bile salts markedly enhances activation. Temperatures between 42 °C and 44 °C and a pH of 7.0 are optimal for activation (Lackie, 1974). Kennedy *et al.* (1978) studied effects of a variety of stimuli on activation of *Pomphorhynchus laevis* cystacanths. They found that natural bile (either piscine or mammalian) was the most effective of the media tested.

Conditions under which cystacanth activation has been achieved *in vitro* are consistent with those expected to exist *in vivo*. The temperature, pH, osmotic pressure and bile salt concentrations in the environment of *Polymorphus minutus* in ducks (Crompton, 1969; Crompton & Edmonds, 1969) are similar to those found by Lackie (1974) to be favorable for activation of *P. minutus* cystacanths. The speed with which activation occurs *in vitro* is also consistent with *in vivo* observations (Lackie, 1974).

The role in host specificity played by stimuli for cystacanth activation apparently varies. Although Kennedy *et al.* (1978) concluded that stimuli for some species are more specific than for others, the requirements for activation are comparatively general. As Crompton (1970) pointed out, the fact that cystacanths of many species are activated by conditions in vertebrates destined to become paratenic hosts suggests that activation stimuli may not be a principal factor in determining the host specificity of many species.

Initiation of infection in definitive hosts. After activation, cystacanths encounter obstacles in addition to defensive responses by hosts. Several laboratory studies (Kennedy, 1974; Buckner & Nickol, 1975) report great intraspecific differences in cystacanth success. The success of *Pomphorhynchus laevis* cystacanths depends upon water temperatures and the nutritional state of the host but not on the number of cystacanths administered (Kennedy, 1972). On the other hand, survival of *Leptorhynchoides thecatus* cystacanths is density-dependent at high levels of infection. Green sunfish, *Lepomis cyanellus*, fed 10, 20, or 40 *L. thecatus* cystacanths retain only 10 to 15 of them. When fed 40 cystacanths, parasite survivorship declines until, after 2 weeks, the worm level is similar to that in fish fed 20 cystacanths (see Fig. 11.6). Resources in green sunfish may limit *L. thecatus* to about 15, or approximately two per pyloric cecum (Uznanski & Nickol, 1982).

In addition to environmental temperature, nutritional state of the host and the number of cystacanths ingested, the sex of the host may also affect the success rate of cystacanths. For example, male rats retain a greater number of *Moniliformis moniliformis* cystacanths than do females (Crompton & Walters, 1972). The monograph by Crompton (1970) should be consulted for a more complete review of these and other aspects of the infection process.

9.3.2 *Host response*

Site location. After activation, acanthocephalans must locate in the proper site for continued development. Frequently this site is not that at which activation occurs. In intermediate hosts, acanthors are freed from embryonic membranes in the alimentary canal and must penetrate surrounding tissue and continue development in a parenteric site. Even in definitive hosts, the initial site of attachment is not always that required for continued growth (Holmes, 1962*a*; Uglem & Beck, 1972). Site location does not bear directly on epizootiology, but defensive response by the host during migration and development does, when spread of the parasite is prevented by stopping egg production or by killing infective stages.

Morphological considerations. Structural features do not seem to be important defensive devices against acanthocephalans. There is no evidence that anatomical factors such as villus length or the size and shape of the crypts of Lieberkühn, suspected (Smyth & Smyth, 1968) of rendering some carnivores refractory to *Echinococcus granulosus*, serve as deterrents to acanthocephalan attachment.

Various grinding organs found in alimentary canals of many potential hosts pose possible risks to invading acanthocephalans, but direct evidence of damage to eggs, acanthors or cystacanths is lacking. Indeed, mechanical damage by amphipod mouth parts is suspected to assist hatching of *Polymorphus minutus* eggs (Hynes & Nicholas, 1957) and grinding by the gastric mill promotes liberation of *Echinorhynchus truttae* acanthors (Awachie, 1966). Chitinous peritrophic membranes of many insects and some crustaceans represent possible barriers to infection. The toll on invading acanthors is unmeasured, but the presence of viable cystacanths in these organisms attests to the fact that at least some are successful in penetration.

Absence of evidence that morphological features inhibit successful transmission may be due to lack of study. Certainly many acanthors succumb to cellular responses in the epithelium (Moore, 1962; Robinson & Strickland, 1969; Nickol & Dappen, 1982) or in the serosa (DeGiusti, 1949*a*; Hynes & Nicholas, 1958). The extent to which prior damage contributes to these fatalities is unknown. In *Gammarus pulex* dead or damaged acanthellae of *Polymorphus minutus* are readily encapsulated and melanized (Crompton, 1967). Future studies could reveal that structural features are important, just as are feeding preferences and physiological traits, in determining host specificity and intensity of parasitism.

Cellular response. There are many reports describing pathology of acanthocephalans in definitive hosts, but there is little evidence that host responses retard or prevent egg production. It is well known, though, that individuals of some species parasitize the alimentary canal of vertebrates in which they cannot attain maturity (Holmes, 1979). The exact mechanisms, or missing requirements, that prevent egg production are unknown. Nutritional studies such as those of Nesheim, Crompton, Arnold & Barnard (1977), Parshad, Crompton & Nesheim (1980) and Crompton, Singhvi & Keymer (1982) illustrate that this failure to attain maturity, at least in some cases, is not a result of a defensive response by the host.

Invertebrates, on the other hand, frequently display cellular responses that result in destruction of invading acanthors. In some cases these responses serve to limit intensity of infection in normal intermediate hosts. A portion of the invading *Moniliformis moniliformis* acanthors is encapsulated, melanized and killed in *Periplaneta americana* (Robinson & Strickland, 1969; Schaefer, 1970) and the intensity of the hemocyte response may be greater when large numbers of acanthors are present (Robinson & Strickland, 1969). Some *Leptorhynchoides thecatus* are destroyed by cellular responses in *Hyalella azteca*, the usual intermediate host, especially at low temperatures when development is slowed (DeGiusti, 1949*a*).

In other instances, defensive responses may account for the host distribution of acanthocephalan species. Larvae of *Macracanthorhynchus hirudinaceus* develop successfully in many species of scarabaeid beetles (Kates, 1943), but not all scarabaeids are equally satisfactory hosts. Defensive reactions by some result in encapsulation and death of many invading acanthors (Miller, 1943). In Great Britain all three native species of *Gammarus* can be found infected by *Polymorphus minutus*. Hynes & Nicholas (1958), however, described strains of the parasite adapted to each of the amphipod species. In the 'wrong' amphipod, development is slowed and many parasites are encapsulated and melanized (Hynes & Nicholas,

1958). Similarly, *Mediorhynchus grandis* is capable of development in dock beetles, but development is delayed and many parasites are partially or wholly chitinized; development is much more satisfactory in grasshoppers (Moore, 1962). At sites where *Plagiorhynchus cylindraceus* acanthors penetrate the gut walls of isopods (*Armadillidium vulgare*), there is an immediate accumulation of hemocytes that may completely engulf the parasites leading to their degeneration (Schmidt & Olsen, 1964). This response (Fig. 9.5) is made by adult isopods and contributes to their great resistance to infection; juvenile isopods display no such response and are highly susceptible (Nickol & Dappen, 1982).

Acanthocephalans of many species are invested in a thin, transparent envelope as they develop in the hemocoel of arthropod intermediate hosts (Fig. 9.6). Perhaps influenced by the knowledge that arthropods make defensive hemocyte responses to acanthors and encouraged by Salt's (1963) review of defensive reactions by insects to metazoan parasites that frequently result in parasite encapsulation and death, the envelope frequently has been viewed as originating from host hemocytes as a defensive reaction (Bowen, 1967; Mercer & Nicholas, 1967; Poinar, 1969; Ravindranath & Anantaraman, 1977). After penetrating epithelial tissue of the host, acanthors undergo development beneath the serosa before entering the hemocoel (DeGiusti, 1949*a*; Crompton, 1964) leading Crompton

Fig. 9.5. Photomicrograph of a cross-section through the mid-hindgut of an adult *Armadillidium vulgare* 47 hours after feeding of *Plagiorhynchus cylindraceus* eggs, showing an acanthor in the epithelium and the hemocyte response. (Reproduced from Nickol & Dappen, 1982, with permission of the editor of the *Journal of Parasitology*.)

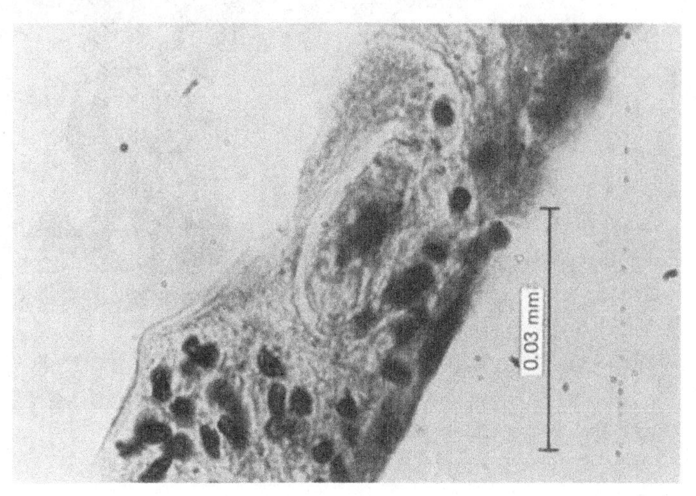

(1964) to the initial conclusion that the envelope originates as a result of a wound-healing response by the host to its stretched or damaged serosa. Subsequent descriptions (Bowen, 1967; Denny, 1969) were consistent with that view. Robinson & Strickland (1969), however, injected *Moniliformis moniliformis* acanthors directly into the hemocoel of cockroaches and demonstrated that envelope formation does not require parasite development in the serosa. Earlier workers (Meyer, 1938*a*; Pflugfelder, 1956; Moore, 1962) attributed the envelope's origin to the parasite. Rotheram & Crompton (1972) and Lackie & Rotheram (1972) concluded that the

Fig. 9.6. Early cystacanth of *Moniliformis moniliformis* removed from the hemocoel of *Periplaneta americana*, showing surrounding envelope.

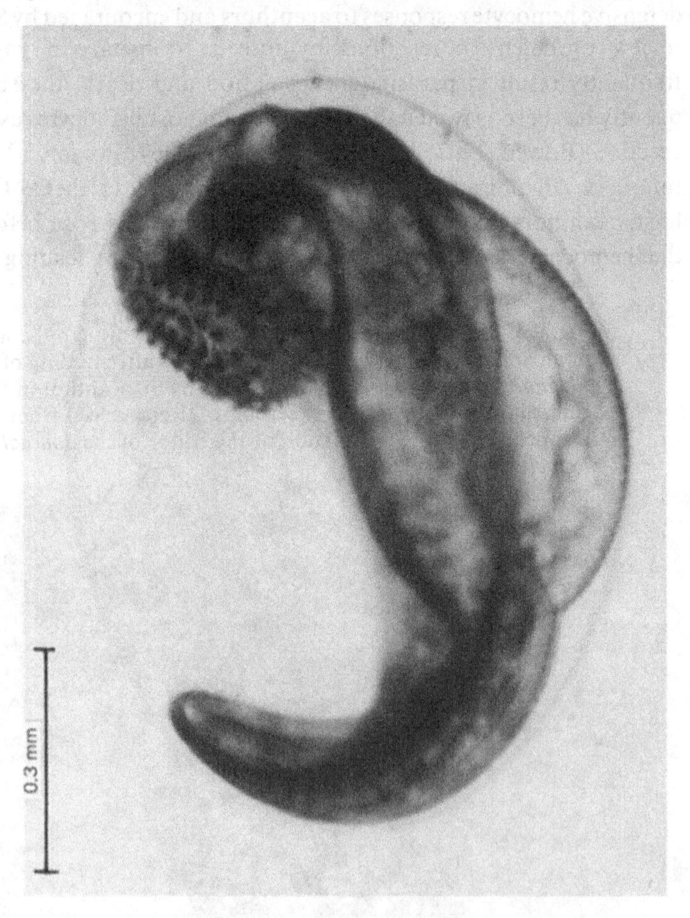

envelope comprises a membranous coat formed from microvillate projections of the early acanthella and later separated from the parasite's surface. Based on observations of *M. moniliformis*, *Plagiorhynchus cylindraceus* and *Prosthenorchis elegans* every second or third day until cystacanths were fully formed, Wanson & Nickol (1973) noted that the membranous coat around the early acanthella bears armature of the acanthor and concluded that as the acanthella develops from the central nuclear mass of the acanthor, remnants of the acanthor's tegument remain as the envelope. Lackie & Lackie (1979) provided additional evidence of parasite origin for the envelope by demonstrating that most cystacanths of *M. moniliformis* grown in locusts and transplanted within their envelopes to cockroaches did not evoke hemocyte response, while locust tissue was readily encapsulated by cockroaches.

Hemocytes collect on the surface of the envelope during early stages of acanthella development but are soon dispersed or destroyed (Robinson & Strickland, 1969; Lackie & Rotheram, 1972; Wanson & Nickol, 1973) and the envelope apparently provides protection from further hemocyte response. Lackie (1975) examined the envelope's role in determining intermediate host specificity for *Moniliformis moniliformis* and observed that the degree of protection afforded by the envelope varies depending upon the species of cockroach. He suggested that avoidance of hemocytic encapsulation resulted from quantitative variation of some unknown parameter of surface properties rather than from the roach's all-or-none discrimination of a qualitative difference. Brennan & Cheng (1975) provided evidence that mucins in the envelope ultimately prevent melanization of *M. moniliformis* by blocking required enzymatic activity in the cockroach hemolymph. It is possible that the degree to which mucins of an acanthocephalan species retard enzymatic activity of various invertebrates represents the quantitative variation hypothesized by Lackie (1975).

It is apparent that the envelope surrounding developing acanthocephalans of some species is a product of the parasite rather than a result of a defensive response by the host hemolymph. Although some acanthors may be under hemocyte assault in the gut wall while others in the same host are forming envelopes under the serosa (Rotheram & Crompton, 1972), hemocytic encapsulation is independent of envelope formation. Both of these phenomena, in turn, are distinct from cystacanth encapsulation that sometimes occurs in visceral organs of vertebrates. Upon ingestion by vertebrates in which they cannot grow to adulthood, cystacanths of some species localize in mesenteries or visceral organs where they frequently are encapsulated. *Neoechinorhynchus cylindratus* is the only species for which structure of the cyst has been studied carefully. The walls of these cysts are of host origin and comprise an inner layer of collagenous connective tissue and a thinner outer layer of fibroblastic cells (Bogitsh, 1961). While the cystacanths remain viable, the vertebrates serve as paratenic hosts but, frequently, encapsulated cystacanths are necrotic and appear to have been destroyed. The duration of viability is unknown but Crompton (1970) suggested that encapsulation follows the pattern of chronic inflammatory reactions; thus, it is likely that degeneration is an extended process.

Humoral response. There is evidence that humoral responses by invertebrate and vertebrate hosts play a role in limiting acanthocephalan infections. Juvenile isopods, *Armadillidium vulgare*, are highly susceptible to infection by *Plagiorhynchus cylindraceus*, but adults are nearly refractory. In addition to an effective cellular response to invading acanthors, electrophoresis reveals that adult isopods produce an increase in one of the protein components of their hemolymph when subjected to penetrating acanthors. Hemolymph of juvenile isopods shows no such change. This finding is interpreted as indirect evidence for humoral resistance (Nickol & Dappen, 1982).

Evidence that poikilothermic vertebrates are capable of responding immunologically to acanthocephalans is clearer, but whether the response results in resistance to parasitism is unresolved. Numbers of a cestode, *Proteocephalus exiqus*, and of an unidentified species of *Neoechinorhynchus* show an inverse relation in ciscoes, although the two species occupy different intestinal sites. A nonspecific immunity may limit either tapeworms or acanthocephalans when one of them is present in large numbers (Cross, 1934). *Pomphorhynchus laevis* matures in chub, *Leuciscus cephalus*, and evokes production of a precipitating antibody (Harris, 1970). Antibodies are not produced in four other piscine species in which *P. laevis* occurs but does not mature (Harris, 1972). These findings led Harris (1972) to speculate that the antibodies are produced only in response to excretory or secretory products of mature worms. The role of antibody response in limiting reinfection was not considered. In goldfish, *Carassius auratus*, the rate of *P. laevis* establishment is not affected by the presence of an existing infection (Kennedy, 1974), but *P. laevis* is not known to mature in goldfish and may not evoke an antibody response in them. Thus, the role of antibody response in protection against infection is still undetermined.

The possibility that acquired immunity limits acanthocephalan infections in homeotherms is better documented. Little information regarding immunity in birds is available although protein changes occur in the serum of *Polymorphus*-infected ducks (Petrov & Nikitin, 1975). Burlingame &

Chandler (1941) showed that rats infected with *Moniliformis moniliformis* possess decided resistance to superinfection. They interpreted their results as evidence of crowding or environmental effects rather than immunological reactions. This conclusion was based on the facts that survival of worms in the initial infection was not dose-dependent and that survival of superimposed worms that succeeded in establishing within the 'zone of viability' was equal to survival in primary infections. However, localization of the secondary worms was different from those of similar age in primary infections because 15–40% of them were found posterior to the zone of viability.

Others have attributed the resistance described by Burlingame & Chandler (1941) to immunological causes. Density-dependent expulsion in primary infections occurs at doses of 100 cystacanths, far more than the 20–25 used by Burlingame and Chandler. Recovery in primary infections of this magnitude has been shown to drop from 66% to 26% after 2 weeks (Miremad-Gassmann, 1981*a*) and from about 85% to about 15% between the fourth and eighth weeks (Andreassen, 1975*a*). At lower initial doses, expulsion is delayed (Andreassen, 1975*a*; Miremad-Gassmann, 1981*a*). Worms from secondary and tertiary infections are recovered in smaller percentages, are smaller, and are located more posteriorly in the intestine than are those from primary infections (Andreassen, 1975*a*; Miremad-Gassmann, 1981*a*).

Serum from rats fed 100 *Moniliformis moniliformis* cystacanths contains antibody capable of sensitizing homologous skin for at least 16 weeks after infection (Andreassen, 1975*b*) and Miremad-Gassmann (1981*b*) found specific antibodies in *M. moniliformis*-infected rats. The lemnisci and tegument are suspected either to form or to store metabolic antigens, because indirect immunofluorescence revealed that specific circulating antibodies bound predominantly to them (Miremad-Gassmann, 1981*b*).

There seems little question that acanthocephalans of at least some mammals, and perhaps of birds, poikilothermic vertebrates and invertebrates, induce resistance potentially capable of limiting the number of parasites and reducing the likelihood of epizootics. The extent to which immunity actually influences populations in nature is largely unknown.

9.4 Epizootics

9.4.1 *Epizootics in wild animals*

Definitive hosts. Helminth prevalence is seldom monitored regularly in wild populations, making detection of sudden increases unlikely. However, the particular number of eggs produced by most acanthocephalans (even if the number is minimal for survival of the species) and the variety of adaptations that promote transmission present an ever-present risk of hyperinfestation for some animals.

Isolated instances of morbidity and death are frequently attributed to acanthocephalans. A dog with rabies-like symptoms that died in San Antonio, Texas, harbored about 300 specimens of *Oncicola canis*, some of which had perforated the intestine (Parker, 1909). *Plagiorhynchus cylindraceus* has occasionally been linked with paralysis of robins and starlings (Jones, 1928; Webster, 1943; Holloway, 1966) and death of bluebirds (Thompson-Cowley, Helfer, Schmidt & Eltzroth, 1979). The pathological significance of acanthocephalans in dying or dead animals is difficult to interpret because the link is frequently circumstantial. Further, densities are often no greater in dead animals supposedly killed by the acanthocephalans than in conspecific individuals that showed no adverse effects. Whether or not acanthocephalans cause or contribute to isolated morbidity and death, these cases probably do not indicate epizootics. More likely they represent instances of acanthocephalans supplementing effects of other stresses or of hyperinfestation in single animals rather than being a reflection of sudden increases of prevalence in the population.

On the other hand, sudden local increases in morbidity and mortality, including epizootics, do occasionally occur. From 1966 through 1970, mottled sculpin, *Cottus bairdi*, found dead in a creek in La Salle County, Illinois, suffered unusual pathology and dense infections from *Acanthocephalus dirus*. The intestine of one of the fish was detached from the stomach and in another instance a portion of sculpin intestine with 22 worms attached was found in the bottom of the stream. Many other sculpin harbored large numbers of *A. dirus*. After this apparent epizootic, no sculpin could be found in that region of the creek possibly due to extermination by acanthocephalans (Schmidt, Walley & Wijek, 1974).

Polymorphus magnus and *P. mathevossianae* were among the helminths deemed responsible for the 1968 death of 40% of the cygnets inhabiting a lake in the Kurgal'dzhin (USSR) nature reserve (Maksimova, 1972). Sanford (1978) described gross and histopathological findings in cygnets from two Canadian flocks of mute swans, *Cyngus olor*, whose deaths were attributed to heavy infections of *P. minutus*. In one case cygnets from one of seven ponds were noticeably smaller than those on the other ponds. One of the small cygnets died and was found to harbor more than 300 acanthocephalans. Seven of eight of the 1976 and at least six of seven 1977 cygnets from the other flock also died, apparently from large numbers of *P. minutus*. The birds had gradually lost weight and body condition over the summer and died in close succession from August through mid-October.

Eiders, *Somateria mollissima*, seem especially vulnerable to sudden outbreaks of acanthocephaliasis. *Polymorphus minutus* was linked to a 1947 epizootic with high mortality in eiders in Denmark (Christiansen, 1948), and *P. botulus* was regarded as the major contributing factor to 1956 and 1957 epizootics during which numerous eider ducks were found dead and dying in Maine and Massachusetts (Clark, O'Meara & Van Weelden, 1958). During the summers of 1956 and 1957, eiders from breeding colonies in the Netherlands remained in large numbers near a coastal island. Many were apparently ill and some were found dead. This epizootic was also attributed (Swennen & Van Den Broek, 1960) to *P. botulus*. Infections with as many as 3500 individuals of *P. minutus* may have been partially responsible for unusually high mortality of juvenile eiders in an archipelago of southwestern Finland during the summers of 1976 through 1979 (Itämies, Valtonen & Fagerholm, 1980).

Several factors, including the physical characteristics of the landscape, appear to initiate acanthocephalan-induced epizootics. Van Maren (1979*b*) described a locality on the river Merloux, in France, at which 48% of a dense population of *Gammarus fossarum* harbored *Polymorphus minutus*. At this site, in a meadow with ducks and chickens, the stream is narrow and shallow. It is an excellent habitat for amphipods and provides an unusually good opportunity for them to become infected. Away from this site, at a point not far downstream, only 2% of the amphipods were infected and farther downstream infected amphipods were not found.

There is evidence that constant exposure to acanthocephalans leads to a relatively small, stabilized density of worms among a population of vertebrates, but that sudden exposure to large numbers of intermediate hosts results in severe infections (Hynes & Nicholas, 1963). When a new or escaped flock of ducks encounters an environment such as the one described by Van Maren (1979*b*), an epizootic may ensue.

Environmental disturbances due to human activities or climatic conditions may also induce epizootics. *Polymorphus minutus* was first present in wildfowl on a reserve in Kent, England, after activities of a sand and ballast company altered the course of a traversing river. Although infected birds appeared healthy, in the succeeding 3 years the prevalence of *P. minutus* in mallards rose steadily to 39%, then 45%, and finally 71% (Crompton & Harrison, 1965). Often, environmental changes, which are frequently only temporary, or human activity produce a rapid increase of acanthocephalans routinely present in small or moderate numbers. Increasingly frequent outbreaks of disease have accompanied growing numbers of birds along the Swedish coast since spring hunting of seafowl was outlawed in the mid-1950s (Peresson, Borg & Fält, 1974). These epizootics, in which approximately 90% of the ducklings died during the summer of 1970, are attributed to serious invasion by endoparasites, including *P. botulus* and *P. minutus*, brought about by overpopulation (Peresson *et al.*, 1974). Ice-covered waters during late winter often limit the places where eiders search for food. Congregation of large numbers in small areas leads to heavy infections. Grenquist (1970) believed these conditions to have caused a high level of mortality from *P. minutus* or *P. botulus* in eiders in Finland. Similarly, concentration of swans in summer on a reduced level of water is blamed for the Kurgal'dzhin reserve epizootic (Maksimova, 1972).

Intermediate hosts. Attempts to produce laboratory infections frequently result in the death of invertebrates, probably as a result of simultaneous penetration of the intestinal wall by large numbers of acanthors. Some of the species known to cause death under these conditions are *Macracanthorhynchus hirudinaceus* in June beetle larvae (Miller, 1943), *Neoechinorhynchus cylindratus* in ostracods (Ward, 1940*b*) and *N. rutili* in ostracods (Merritt & Pratt, 1964). Other species interfere with reproduction by intermediate hosts by retarding development of ovaries (Schmidt & Olsen, 1964), preventing appearance of secondary sex characteristics (Munro, 1953) or by suppressing both oogenesis and secondary sex characteristics (LeRoux, 1931). Hynes (1955) suggested that such interference with reproduction could account for the sudden, unaccountable disappearances of amphipods observed in certain lakes.

Studies of the seasonal occurrence of acanthocephalans in invertebrates frequently reveal different prevalences throughout the year, usually in relation to reproduction by adults in definitive hosts. While it is clear that pathological effects could result in unusually high mortality, acanthocephalan-induced 'die-offs' of invertebrates are unknown outside the laboratory.

9.4.2 *Epizootics in captivity*

Animals in zoological gardens, fish hatcheries and other confinements repeatedly suffer pathogenic effects of acanthocephaliasis and frequently die as a result. Most of the reported cases of acanthocephalan-induced epizootics among captive animals are of primates harboring *Prosthenorchis elegans* and, occasionally, *Oncicola spirula*. It is not unusual for recently captured, imported animals to die from acanthocephalan infections that were probably acquired before confinement. Four of 10 newly acquired squirrel monkeys, *Saimiri oerstedii*, died of *P. elegans* at the Gorgas Hospital in the Panama Canal Zone; the other six were uninfected (Takos & Thomas, 1958). Nine of 30 marmosets acquired by the Medical College of Virginia and Dartmouth Medical School died of massive infections of *P. elegans* or from a combination of the parasite and other infections from 4 to 122 days after acquisition (Richart & Benirschke, 1963). Deaths of a marmoset, *Hapale jacchus*, and a mongoose, *Atilax paludinosus*, within 8 months of arrival at Regent's Park, London, were attributed to acanthocephalans (T-W-Fiennes, 1966). Heavy infections with a species of *Prosthenorchis* were deemed responsible for the death of 12 of 13 squirrel monkeys that died within 18 days of arrival at a primate colony in the USA (Morin, Renquist, Johnson & Strumpf, 1980).

Acanthocephalans do not always cause death directly, but they may cause lesions that enable other pathogens to become established (Schmidt, 1972*a*). A baboon, *Papio papio*, that harbored many *Oncicola spirula* died of generalized tuberculosis in a Paris zoo (Brumpt & Urbain, 1938), suggesting that acanthocephalans may act synergistically with other pathogens.

Circumstances of confinement usually result in the concentration of definitive hosts in an environment abounding with invertebrates suitable as intermediate hosts. This is especially true of primate colonies, where roaches suitable for development of *Prosthenorchis elegans* and *Oncicola spirula* are plentiful (Brumpt & Desportes, 1938). Under such conditions parasites of introduced animals may spread rapidly among others of the colony. *Prosthenorchis elegans* and *O. spirula* occur naturally only in the New World. Importation, however, has permitted spread to both Old and New World primates in zoos and primate colonies around the world, where it commonly results in loss of entire stocks (Schmidt, 1972*a*). *Prosthenorchis elegans* or *O. spirula* epizootics occurred among Old World primates in Paris (Brumpt & Urbain, 1938) and Rotterdam (Van Thiel & Wiegand-Bruss, 1945). Moore (1970) described an epizootic among the great apes at the Hogle Zoo in Salt Lake City, Utah, that involved five gibbons, three orang-utans and one gorilla. Three of the gibbons died. It was concluded that the epizootic began with the introduction of a gibbon that had been acquired from a source in Florida known to handle large numbers of South American monkeys. The gibbon had diarrhea upon arrival and grew progressively weaker during the following months. The other animals subsequently became ill. Importation of *O. spirula*-infected gibbons purchased from a New York dealer had previously (Chandler, 1953) been blamed for an epizootic during which many animals became ill and three gibbons, two squirrel monkeys and a pig-tailed macaque died at the Houston Zoological Gardens.

Among confined animals, epizootics in mammals are most conspicuous and receive the most attention. However, an epizootic caused by *Acanthocephalus dirus* among trout at the New Hampshire State Fish Hatchery at New Hampton (Bullock, 1963) attests to the fact that under the right circumstances poikilotherms are also subject to severe outbreaks of acanthocephaliasis.

9.5 Regulation of numbers

Fecundity of acanthocephalans, like most groups of parasites, is great relative to that of their hosts. Even if egg production is the minimum required for survival of the species, as Croll (1966) suggested, the number of eggs produced by each female is considerable (Kates, 1944; Crompton & Whitfield, 1968*a*; Crompton *et al.*, 1972). The large number of infectious agents and various adaptations to enhance the probability of successful transmission provide potential for massive infections in individual hosts. The relative infrequency of epizootics is due, in part, to mechanisms that regulate numbers (see Chapters 10 and 11).

10

Life history models

A. P. Dobson and A. E. Keymer

10.1 Introduction

Our aim is to develop a further understanding of the population dynamics of acanthocephalans by means of the construction of simple mathematical models. A basic theoretical framework has been proposed by Anderson & May (1978) to describe the population dynamics of a hypothetical macroparasite with a simple life cycle involving the production of directly infective transmission stages by hermaphroditic adult parasites in the definitive host. This framework is extended in the present chapter to encompass the salient features of the acanthocephalan life history. In this way, we attempt to examine the factors which control the population densities of the parasite in its intermediate and definitive hosts, to elucidate the features of the life history which are important in maintaining the parasite at these densities and to investigate how these factors interact to determine lifetime reproductive success. Initially, we will consider an acanthocephalan life history which involves one species of intermediate and one species of definitive host. Once the dynamic properties of this simple model have been examined, we shall extend the basic framework and briefly consider the evolutionary and dynamic consequences of the inclusion in the life history of a second intermediate host, and the utilization by the parasite of more than one definitive host species.

10.2 Model framework

The life history of acanthocephalan parasites may be represented in a generalized manner by a flow chart (Fig. 10.1). All known species of Acanthocephala are dioecious and reproduce sexually in the vertebrate alimentary tract. All are assumed to have indirect life histories with transmission from intermediate to definitive host being effected by predator–prey interactions across trophic levels. Eggs produced by the adult female worms are expelled from the definitive host, and intermediate hosts are infected as a result of the ingestion of eggs present in the environment. The model consists of coupled differential equations representing the dynamics of the intermediate host and parasite populations. It is assumed that the definitive host population possesses a stable age distribution and is of constant size, L, on the time-scale appropriate to changes in parasite population numbers. The assumptions incorporated in the model are detailed in the following sections.

10.2.1 *The distribution of parasite numbers per host*

For most macroparasitic species, the statistical distribution of parasite numbers per host has been found to be aggregated or 'overdispersed', so that most hosts are uninfected or harbour relatively few parasites and a few hosts harbour disproportionately heavy worm burdens.

Fig. 10.1. Flow chart representing a simple acanthocephalan life history. The boxes represent populations and the arrows represent the rate parameters which determine the dynamics of the populations.

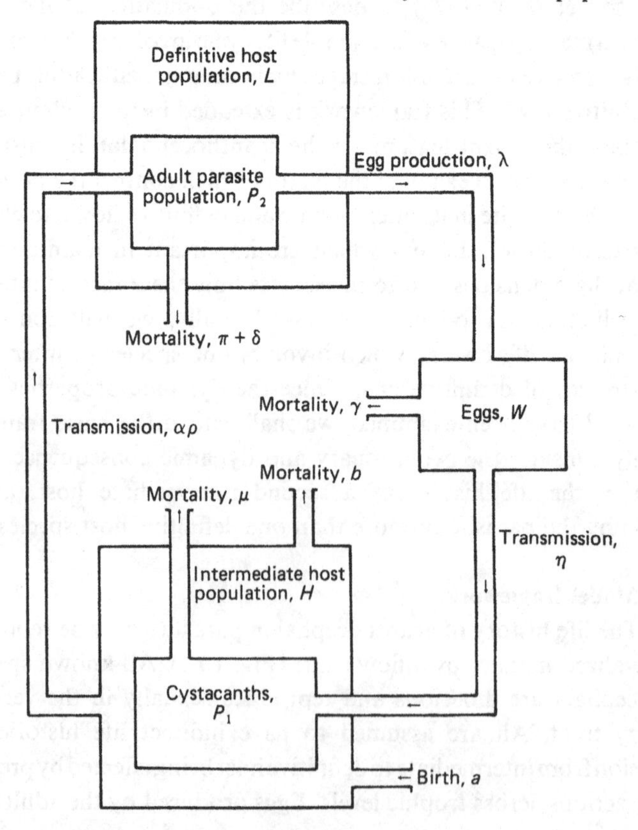

The form of the observed distribution may in most cases be described by the negative binomial probability model (Crofton, 1971*a*; Anderson & May, 1978). This distribution is defined by the mean and a single parameter, usually denoted k, which varies inversely with the degree of aggregation. As illustrated by Fig. 10.2, it provides a good description of the distribution of many acanthocephalan species in both their intermediate and definitive host populations. Here we will adopt the convention of using k to indicate the degree of aggregation of parasites in the definitive hosts and j that in the intermediate hosts.

10.2.2 *Transmission to the intermediate host*

Infection of the intermediate host occurs through the ingestion of eggs and so is governed by the dynamics of a predator–prey interaction. A functional response (i.e. a non-linear relation between the rate of egg

Fig. 10.2. Frequency distributions of the numbers of acanthocephalan parasites per host. (*a*) *Acanthocephalus clavula* in *Gasterosteus aculeatus*, $k = 0.29$ (Pennycuick, 1971*c*). (*b*) *Polymorphus minutus* in *Gammarus pulex*, $j = 1.14$ (Crofton, 1971*a*). (*c*) *Moniliformis moniliformis* in experimentally infected *Periplaneta americana*, $j = 1.46$ (Holland, 1983*a*). (*d*) *Pomphorhynchus laevis* in *Leuciscus leuciscus*, $k = 0.62$ (Kennedy & Rumpus, 1977). The histogram columns show observed frequencies and the fitted curves indicate the predictions of the negative binomial probability model.

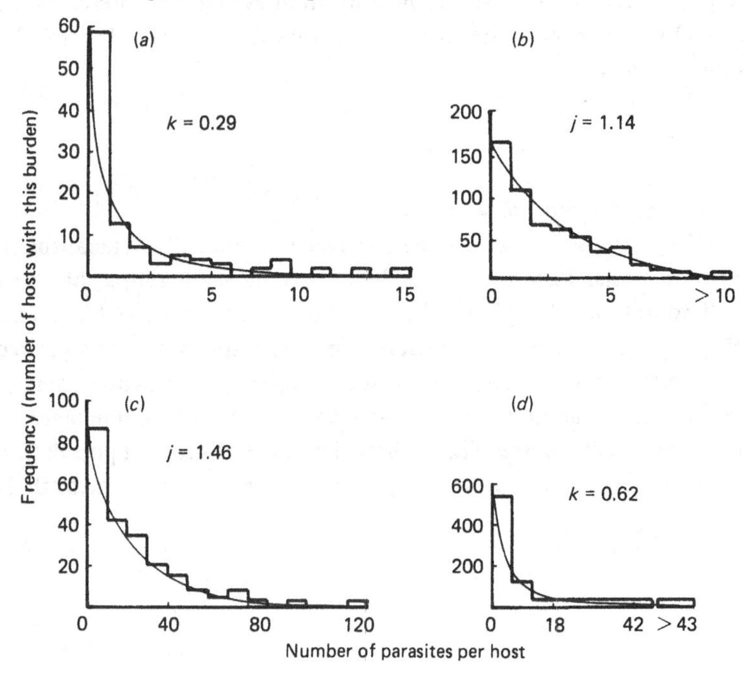

ingestion and egg density) has been demonstrated for the interaction between *Periplaneta americana* and *Moniliformis moniliformis* eggs under laboratory conditions (Lackie, 1972*a*, *b*; Keymer, 1982*a*; Chapter 11). Under most natural conditions, however, the density of acanthocephalan eggs relative to the density of the intermediate host population is unlikely to reach high enough levels for satiation or handling-time effects (see Murdoch & Oaten, 1975) to come into play. Functional responses are therefore not considered to be of importance in the dynamics of transmission of acanthocephalan parasites to the intermediate host. In the present model it is assumed that intermediate hosts acquire parasites at a rate proportional to the density of hosts, H, and the density of infective parasite eggs, W. The net rate of parasite acquisition is thus ηWH, where η is a coefficient representing the rate of transmission. Of those parasites acquired by an individual host, only a proportion, D_2, survive to reach infectivity. If the average prepatent period in the intermediate host (the time period from ingestion of the egg by the intermediate host to the development of the infective cystacanth) is Υ_2 time units, then the proportion D_2 may be expressed as

$$D_2 = \exp[-\Upsilon_2(b+\mu+\rho L)] \quad (10.1)$$

where μ, b and ρL represent losses of larval parasites during the prepatent period due to natural parasite and host mortalities and host predation as defined below. The net rate of gain of infective cystacanths to the population is thus

$$\eta D_2 HW(t-\Upsilon_2)$$

10.2.3 *Cystacanth mortality*

Very little is known of the survival potential of cystacanths in intermediate hosts. Parasite encapsulation occurs in incompatible host species (Crompton, 1975; Uznanski & Nickol, 1980*a*), but to what extent mortality results from either senescence or host responses in compatible hosts remains unknown. In the present model, larval parasites are assumed to have a constant *per capita* instantaneous mortality rate, μ, independent of either their density or age. The net loss of cystacanths from a population of size P_1 is thus μP_1 and the expected lifespan of a cystacanth is $1/\mu$.

10.2.4 *Loss of cystacanths as a result of intermediate host mortality*

Intermediate hosts are assumed to die from causes other than ingestion by a potential definitive host at a *per capita* rate b, such that the net loss of parasites from a host population of size H is

$$bH \sum_{i=0}^{\infty} i \cdot p(i)$$

where $p(i)$ represents the probability that an individual host contains i parasites.

10.2.5 *Loss of cystacanths as a result of transmission to the definitive host*

Infection of the definitive host takes place by means of a second predator–prey association between potential definitive and infected intermediate hosts. It is assumed, however, that variations in intermediate host density do not cause a reduction in ingestion rate due to satiation or handling-time effects. This assumption is likely to be valid since there generally exists much disparity between the sizes of the two host species, and the interaction is often one involving harvesting or grazing rather than the capture and consumption of individual items of prey. Ingestion is thus assumed to be directly proportional to the population densities of definitive and intermediate hosts, L and H, respectively.

Much qualitative evidence is available to suggest that the cystacanths of many acanthocephalan species affect either the pigmentation or the

Table 10.1. *Comparative vulnerability of uninfected and infected intermediate hosts of acanthocephalan parasites to predation by potential definitive hosts*

Acantho-cephalan species	Inter-mediate host	Definitive host	Ratio of hosts predated (infected: un-infected)		Reference
Polymorphus paradoxus	*Gammarus lacustris*	Duck	3.52 4.19		Bethel & Holmes (1977) Holmes & Bethel (1972)
Acantho-cephalus dirus	*Asellus inter-medius*	Creek chub	1.68 7.50	light dark	Camp & Huizinga (1979) (2 different substrates)
Pompho-rhynchus laevis	*Gammarus pulex*	Trout	4.89		Kennedy, Broughton & Hine (1978)
Plagiorhynchus cylindraceus	*Armadil-lidium vulgare*	Starling	1.62		Moore (1983)

behaviour of their intermediate hosts in such a way as to render the infected individuals more vulnerable to predation. Some examples in which quantitative measurements have been made are given in Table 10.1. The net rate of loss to the cystacanth population as a result of definitive host infection may thus be represented by

$$\rho\alpha LH \sum_{i=0}^{\infty} i^2 \cdot p(i)$$

where ρ represents the rate of predation by the definitive host on uninfected intermediate hosts, $p(i)$ is the probability of an intermediate host harboring i cystacanths and α is a coefficient representing the proportional increase in the rate of predation on infected intermediate hosts (per parasite) caused by parasite-induced increase in susceptibility to capture.

10.2.6 *Transmission to the definitive host*

As stated above, the rate of ingestion of intermediate hosts by prospective final hosts (ρ) is assumed to be directly proportional to their densities, L and H, and also dependent on the degree of parasite-induced increase in susceptibility to predation, α, such that the net rate of transmission is

$$\rho\alpha LH \sum_{i=0}^{\infty} i^2 \cdot p(i)$$

Of those parasites acquired by an individual definitive host, only a proportion, D_1, survive to reach sexual maturity, since some are lost due to parasite and host mortalities during the prepatent period (the time period from ingestion of cystacanth to first egg production by mature worms). If the average length of the prepatent period in the definitive host is Y_1 time units, then the proportion D_1 may be expressed as

$$D_1 = \exp[-Y_1 \pi] \tag{10.2}$$

where π is the instantaneous *per capita* rate of adult parasite mortality. The net rate of gain of sexually mature adult parasites is thus

$$\rho\alpha LHD_1 \sum_{i=0}^{\infty} i^2 \cdot p(i)\,(t - Y_1)$$

10.2.7 *Adult worm mortality*

From the experimental evidence available, survival of adult acanthocephalans in the definitive host appears to be age dependent (Fig. 10.3), with the expected lifespan of male worms being less than that of females (Parshad & Crompton, 1981). Although both the fecundity and survival of intestinal helminths have in many cases been found to be dependent on the density of mature parasites within the host (Anderson,

1978), no specific experimental investigations of density dependence in acanthocephalans in their natural definitive hosts have been carried out. This may perhaps partly explain the reported absence of density dependence in some acanthocephalan life histories (Chapter 11); the results of several recent studies indicate the possible general significance of constraints on the population growth of adult worms. Uznanski & Nickol (1982), for example, recorded a reduction in survival through the prepatent period in high-density infections of *Lepomis cyanellus* with *Leptorhynchoides thecatus*, whilst the percentage recovery of *Moniliformis moniliformis* from laboratory rats appears to be reduced when high numbers of cystacanths are administered (Andreassen, 1975*a*; Miremad-Gassmann, 1981*a*; Crompton, Keymer & Arnold, 1984). In the absence of experimental evidence for the existence of density-dependent reductions in fecundity, we limit the operation of density-dependent constraints in our basic model

Fig. 10.3. The numbers of *Moniliformis moniliformis* recovered from 1230 male (open circles) and female (filled circles) rats on postmortem examination at varying intervals postinfection. (*a*) Each rat infected orally with 12 cystacanths; (*b*) Each rat infected orally with 20 cystacanths. (From Crompton *et al.*, 1984.)

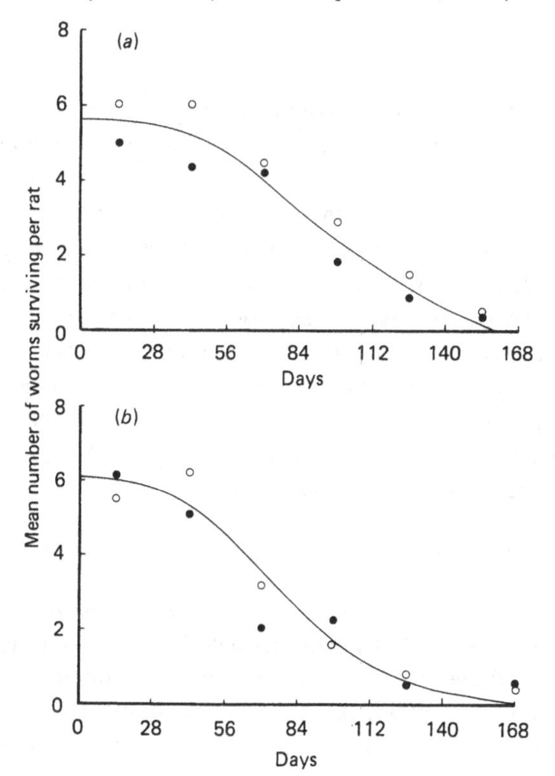

to variations in the level of parasite mortality during the prepatent period. We note that this is a key area of acanthocephalan ecology which would benefit considerably from further experimental research. For the purposes of the present chapter, the *per capita* rate of parasite mortality is assumed to be linearly related to the parasite burden, l, such that the net rate of loss of parasites from the definitive host population is

$$\pi L \sum_{i=0}^{\infty} l \cdot p(l) + \delta L \sum_{i=0}^{\infty} l^2 \cdot p(l)$$

where π is the rate of mortality per worm per unit time, $p(l)$ is the probability of a definitive host harboring l worms and δ is a coefficient measuring the severity of density-dependent constraints on worm survival and establishment. At moderate and low worm burdens, acanthocephalans are generally considered to be without serious pathogenic effects on the definitive host (§10.6), although they may cause considerable morbidity in immature animals (Bendell, 1955). Parasite-induced reductions in definitive host survival or fecundity are therefore not included in our basic model, although their dynamical consequence would be to compound the effects of other density-dependent constraints on the population growth of the adult worms.

10.2.8 *Egg production*

Acanthocephalans are dioecious and so the net rate of egg production is dependent on the burden and statistical distribution of male and female worms within the definitive host population. The mating function, Φ, is defined as the probability that any individual female worm copulates successfully with a mature male worm and produces fertile eggs. At low mean worm burdens, the frequency of concomitant infections with both male and female worms may fall to such low levels as to influence adversely the net transmission success of the parasite population. Such an effect may create a breakpoint, or critical density of parasites per host, below which mating success is too infrequent to ensure the persistence of the parasite within the host population (Macdonald, 1965). A suitable form for the function Φ has been suggested by May (1977):

$$\Phi(N, k) = \left[1 - \left(1 + \frac{N}{2k}\right)^{-(1+k)}\right] \tag{10.3}$$

Here, N and k are the parameters of the negative binomial distribution describing the mean numbers of adult worms per definitive host and their degree of contagion (see above). The relation between the probability of a female worm being mated (the mating function, Φ) and the mean parasite burden per host for three different values of k are shown in Fig. 10.4. High levels of overdispersion in the statistical distribution of worm numbers per host (low k) increase the proportion of female worms mated at low average worm burdens. When the mean worm burden (N) is high, almost all worms have the opportunity to mate, irrespective of whether their distribution within the host population is overdispersed or random in form. A nomogram (Fig. 10.5) shows contours for various proportions of the total female worm population which are likely to mate under various conditions, for a realistic range of k and N values. Data for a number of acanthocephalan species have been superimposed onto this diagram, illustrating the importance of the mating function under conditions where the mean parasite burden is either continually low (e.g. *Acanthocephalus dirus* or *Neoechinorhynchus rutili*), or occasionally low as a result of seasonal variation in transmission rates.

Available evidence suggests that there is normally equality in the sex ratio of cystacanths in the intermediate host (Amin, Burns & Redlin, 1980; Brattey, 1980; Muzzall & Rabalais, 1975*b*; Parenti, Antoniotti & Beccio, 1965). The preponderance of females which is sometimes observed in acanthocephalan populations in the definitive host may result simply from disparity in the longevity of male and female worms (§7.9.1), such that the equilibrium sex ratio is ($\pi♂/\pi♀$) where $\pi♂$ and $\pi♀$ are the instantaneous mortality rates of male and female worms, respectively. There thus seems no reason to suggest any difference in the statistical distribution of male

Fig. 10.4. Predicted influence of the degree of overdispersion of adult parasites in the definitive host (k) on the relation between the mean worm burden per host and the proportion of female worms mated. It is assumed that polygamous mating takes place randomly between male and female worms in the intestine. When $k = 10$, the pattern of worm numbers per host is effectively random and follows the Poisson distribution. (Adapted from Anderson, 1982*a*.)

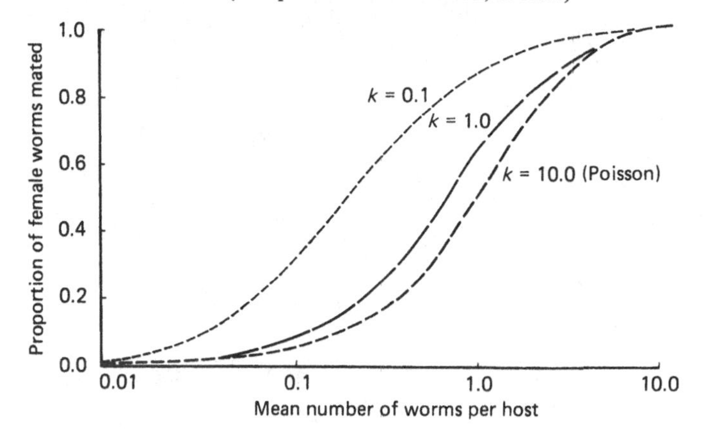

and female worms within the host population (Crompton *et al.*, 1984) which might affect the form of the mating probability function (May, 1977).

Some observations on the mating behaviour of *Moniliformis moniliformis* in laboratory rats have been made (Abele & Gilchrist, 1977; Crompton, 1974) and, in the absence of any evidence to suggest that the worms may be monogamous, it is assumed that polygamous mating takes place randomly between male and female worms in the definitive host (§7.6). Assuming that the *per capita* rate of egg production, λ, is constant and independent of worm age or density, the net rate of egg production by a

Fig. 10.5. Nomogram showing predicted contours for various proportions of the female worm population likely to be mated for various combinations of k (degree of overdispersion in distribution of adult worm numbers per host) and N (average worm burden per host) values. Data points for the following acanthocephalan species have been superimposed: (1) *Acanthocephalus dirus* in *Semotilus atromaculatus* (Camp & Huizinga, 1980); (2) *Acanthocephalus lucii* in *Perca fluviatilis* (Lee, 1981); (3) *Echinorhynchus truttae* in *Salmo trutta* (Awachie, 1965); (4) *Leptorhynchoides thecatus* in *Lepomis macrochirus* (Esch *et al.*, 1976); (5) *Echinorhynchus salmonis* in *Osmerus mordax* (Amin, 1981); (6) *Pomphorhynchus laevis* in *Leuciscus leuciscus* (Kennedy & Rumpus, 1977); (7) *Polymorphus minutus* in *Aythya fuligula* (Crompton & Harrison, 1965); (8) *Polymorphus minutus* in *Anas platyrhynchos* (Crompton & Harrison, 1965); (9) *Polymorphus botulus* in *Somateria mollissima* (Thompson, personal communication); (10) *Neoechinorhynchus rutili* in *Gasterosteus aculeatus* (Walkey, 1967); (11) *Neoechinorhynchus emydis* in *Graptemys geographica* (Hopp, 1954).

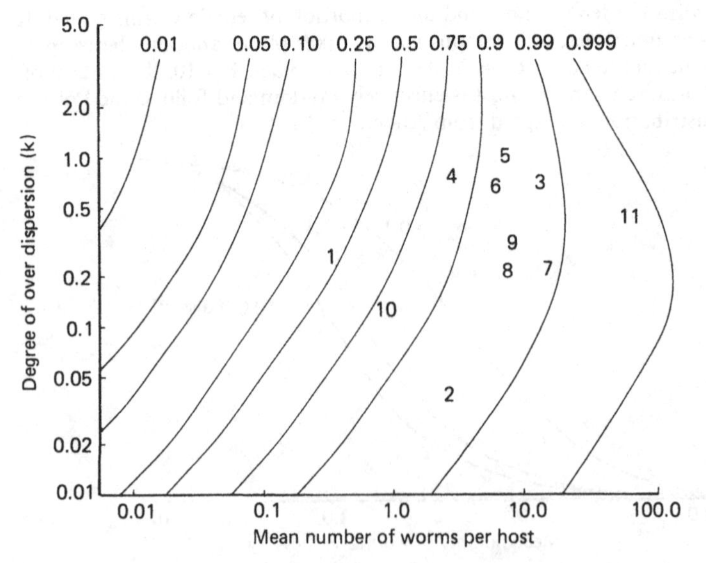

population of P_2 worms may be denoted by $\lambda P_2 \Phi$. Since acanthocephalans are, in effect, ovoviviparous (§7.1), there exist no time delays in egg maturation equivalent to those which occur in the life histories of some nematode and cestode parasites.

10.2.9 *Egg mortality*

The population of eggs, or shelled acanthors, is subject to considerable mortality from a variety of causes outside the host. Assuming that the instantaneous rate of mortality is γ per egg per unit time, the net rate of loss from a population of W eggs is γW. Each egg thus has an expected lifespan of $1/\gamma$.

10.2.10 *Loss of eggs as a result of host infection*

Infective eggs are removed from the population, W, as a consequence of ingestion by potential intermediate hosts. The rate of loss is equal to ηWH, as described in §10.2.2.

10.2.11 *Dynamics of the intermediate host population*

Intermediate hosts are assumed to have constant *per capita* rates of birth and death, a and b, respectively. Intermediate hosts are also lost from the population, H, as a result of predation by definitive hosts (§10.2.5). Although some species of Acanthocephala may reduce the reproductive capacity of their intermediate host (Brattey, 1980; Reinhard, 1956; Munro, 1953), the extent of the reduction varies and many species appear to have no effect at all upon host fertility and fecundity (Chapter 11). For simplicity in the construction of our basic model, it is assumed that parasite-induced reductions in intermediate host fecundity do not occur in the life history of the hypothetical acanthocephalan species under consideration.

10.2.12 *Structure of the basic model*

The assumptions detailed above give rise to the following differential equations describing the changes through time of the adult parasite (P_2), cystacanth (P_1), egg (W) and intermediate host (H) populations:

$$\frac{dP_2}{dt} = \rho\alpha L D_1 H \sum_{i=0}^{\infty} i^2 \cdot p(i)(t - Y_1) - \pi P_2 - L\delta \sum_{i=0}^{\infty} l^2 \cdot p(l) \quad (10.4)$$

$$\frac{dP_1}{dt} = \eta WHD_2(t - Y_2) - (b + \mu) P_1 - \rho\alpha LH \sum_{i=0}^{\infty} i^2 \cdot p(i) \quad (10.5)$$

$$\frac{dW}{dt} = \lambda P_2 \Phi - \gamma W - \eta WH \quad (10.6)$$

$$\frac{dH}{dt} = (a - b) H - \rho LH - \rho\alpha LH \sum_{i=0}^{\infty} i \cdot p(i) \quad (10.7)$$

It is clear from the structure of the above equations that the dynamical behaviour of the model is dependent on the nature of the statistical distribution of parasites within the host populations. It is assumed that the distribution of both adult and larval acanthocephalans may be described by the negative binomial probability model (see §10.2.1). The statistical moments of a negative binomial (parameter k) describing the distribution of P parasites in H hosts may be written as follows:

$$\sum_{i=0}^{\infty} i \cdot p(i) = P/H \tag{10.8}$$

and

$$\sum_{i=0}^{\infty} i^2 \cdot p(i) = \frac{P^2(k+1)}{H^2 k} + P/H \tag{10.9}$$

By incorporating these moments into eqns 10.4 to 10.7, the basic model may be simplified to give

$$\frac{dP_2}{dt} = \alpha\rho L\left[\frac{P_1}{H}+\frac{P_1{}^2}{H^2}\left(\frac{j+1}{j}\right)\right]D_1 H(t-Y_1) - \pi P_2 - \delta L\left[\frac{P_2}{L}+\frac{P_2{}^2}{L^2}\left(\frac{k+1}{k}\right)\right] \tag{10.10}$$

$$\frac{dP_1}{dt} = \eta WHD_2(t-Y_2)-(b+\mu)\,P_1 - \alpha\rho LH\left[\frac{P_1}{H}+\frac{P_1{}^2}{H^2}\left(\frac{j+1}{j}\right)\right] \tag{10.11}$$

$$\frac{dW}{dt} = \lambda P_2 \Phi - \gamma W - \eta WH \tag{10.12}$$

$$\frac{dH}{dt} = (a-b)\,H - \rho LH - \alpha\rho LP_1 \tag{10.13}$$

where k and j are the parameters of the negative binomial distributions describing parasites in the definitive and intermediate hosts, respectively.

10.3 Empirical estimates of model parameters

In order to investigate how various life history features of acanthocephalan parasites interact to produce the observed patterns of population behaviour in natural host–parasite associations, empirical estimates of all the parameters included in eqns 10.10 to 10.13 are necessary. Parameter values for the 14 most intensively studied species of Acanthocephala have been estimated from published results of field and experimental investigations (Table 10.2). Most of these studies were not specifically designed for parameter measurement and the estimates given are thus in many cases only order-of-magnitude values obtained from semi-quantitative information. It is apparent from Table 10.2 that the available data on which to base interspecific studies of acanthocephalan population dynamics is comparatively scarce. A full complement of parameter values is not attainable for any single species. Conspicuously, it appears that no quantitative studies have been carried out on rates of transmission, either to intermediate (η) or definitive ($\alpha\rho$) hosts. Other parameters for which data are lacking include egg production (λ), the relation between worm survival and worm density (δ), egg survival (γ), cystacanth-induced increase in host susceptibility to predation (α), and the population dynamics of intermediate hosts (a and b). Much further information is also required on the ways in which these life history parameters change in response to other biological or physical variables, including the relations between cystacanth burden per host and cystacanth infectivity and survival, and those between cystacanth development and mortality and environmental temperature. Until further quantitative information is available on which to base theoretical studies, the formulation of model frameworks to describe acanthocephalan life histories remains to some extent a conjectural exercise. We feel that this does not detract, however, from its usefulness in identifying the areas in which further research would be of interest.

10.4 Basic reproductive rate

The manner in which the various population processes interact to influence the probable persistence of acanthocephalan infections within host populations may be determined from the structure of eqns 10.10 to 10.13. For parasite persistence to occur, the rate of reproduction of the parasite per generation must balance or exceed the cumulative losses occurring at each stage of its life history. Anderson (1980) has defined an index, R_0, for comparing the basic reproductive rates of different parasite species. This index is directly analogous to the R_0 first derived by Fisher (1930) to describe the net reproductive rate of free-living populations. Today, it is still widely used by ecologists, population geneticists and demographers, as well as more recently by epidemiologists (see Warren, 1982) who regard R_0 as a useful concept from which to develop and compare different control strategies for diseases of medical and veterinary importance.

For acanthocephalan parasites, R_0 may be defined as the average number of female offspring, produced throughout the lifetime of a mature female worm, which would achieve reproductive maturity in the next generation in the absence of density-dependent constraints on establishment, survival or reproduction. R_0 is therefore a dimensionless parameter which encapsulates the biological details of the parasite's life history. For

Table 10.2 *Estimated values of acanthocephalan life history parameters*[a]

	Species	Parameter[b]														
		π	T_1	δ	k	λ	γ	η	μ	T_2	j	α	ρ	a	b	r
A[c]	*Macracanthorhynchus*															
	hirudinaceus	0.003	75	—	—	2.6×10^5	0.001	—	—	75	—	See Table 10.1	—	—	—	—
	(*Cotinus nitida* + pig)[d]	(38)[e]	(38)			(38)	(39)			(26; 34; 36)						
	Moniliformis moniliformis	0.009	38	—	0.60	5.5×10^3	—	—	—	40–58	1.46		—	—	0.010	—
	(*Periplaneta americana* + rat)	(14)	(11; 34)		(13)	(11)				(30; 29)	(21)				(18)	
P	*Acanthocephalus dirus*	0.011	60–90	—	0.23	—	—	—	0.011	150	0.33		—	—	—	—
	(*Asellus intermedius* + fish)	(7)	(7)		(7)				(7)	(7)	(7)					
	Acanthocephalus lucii	—	—	—	0.04	—	—	—	—	40	—		—	0.033	0.010	0.023
	(*Asellus aquaticus* + fish)				(31)					(6)				(40)	(40)	
	Acanthocephalus dirus	0.003	—	—	0.38	—	—	—	0.003	—	0.18		—	—	—	—
	(*Caecidotea militaris* + fish)	(3)			(3)				(3)		(1; 3)					
	Echinorhynchus truttae	0.014	70	—	0.72	—	—	—	0.006	82	> 10		—	0.045	0.020	0.025
	(*Gammarus pulex* + fish)	(5)	(5)		(4)				(5)	(5)	(4)			(23)	(23)	
	Echinorhynchus salmonis	0.017	120	—	1.01	—	—	—	—	—	—		—	—	0.006	—
	(*Pontoporeia affinis* + fish)	(41)	(41)		(2)										(32)	
	Leptorhynchoides thecatus	0.033	56–70	0.002	0.42–0.77	—	—	—	—	25–31	1.31		—	0.080	0.040	0.040
	(*Hyallela azteca* + fish)	(43)	(16; 43)	(43)	(17)					(16; 43)	(43)			(9)	(9; 43)	
	Plagiorhynchus cylindraceus	—	—	0.180	0.16	—	0.004	—	—	60–65	—		4.8	0.174	0.002	0.172
	(*Armadillidium vulgare* + birds)			(45)	(46)		(48)			(47; 50)			(46)	(49)	(49)	
	Pomphorhynchus laevis	0.031	60	—	0.62	—	—	—	—	150	0.44–0.66		—	0.045	0.020	0.025
	(*Gammarus pulex* + fish)	(28)	(19)		(27)					(19)	(19; 27)			(23)	(23)	
	Polymorphus minutus	0.023–0.073	22	—	0.19	1.7×10^3	0.006	—	—	63	0.20–3.00		—	0.045	0.020	0.025
	(*Gammarus pulex* + duck)	(12; 15; 37)	(15)		(12)	(15)	(15; 24)			(24; 25)	(10; 20)			(23)	(23)	
	Polymorphus botulus	0.025	41	—	0.20–0.30	—	—	—	—	—	0.20–2.60		—	—	—	—
	(*Carcinus maenas* + gull)	(33; 42)	(33; 42)		(33; 42)						(33; 42)					

E	*Neoechinorhynchus rutili* (*Cypria ophthalmica* + fish)	—	180 (44)	—	0.16–0.24 (8; 44)	—	—	—	—	20–57 (35; 44)	> 5 (44)	—	—	—	—
	Neoechinorhynchus emydis (*Cypria maculata* + mollusc + turtle)	—	—	—	0.48 (22)	—	—	—	—	21 (22)	> 5 (22)	—	—	—	—

[a] In many cases, the values given represent order of magnitude estimates made by us from information given in the references quoted

[b] All parameters measuring birth or death are given as instantaneous rates: π = rate of mortality per adult worm per day; T_1 = prepatent period of adult worm (days); δ = coefficient measuring the degree of density dependence in adult worm mortality; k = parameter of negative binomial distribution measuring the degree of overdispersion in number of adult worms per definitive host; λ = rate of production of eggs per female worm per day; γ = rate of mortality per egg per day; η = rate of transmission per egg per intermediate host per day; μ = rate of mortality per larval parasite per day; T_2 = prepatent period of larval parasite (days); j = parameter of negative binomial distribution measuring the degree of overdispersion in number of cystacanths per intermediate host; α = proportional increase in intermediate host susceptibility to predation (per parasite); ρ = rate of predation per uninfected intermediate host per definitive host per day; a = rate of birth per intermediate host per day; b = rate of mortality per intermediate host per day; r = natural intrinsic growth rate of intermediate host population $(a-b)$

[c] A = Archiacanthocephala; P = Palaeacanthocephala; E = Eoacanthocephala

[d] In cases where two or more intermediate host species may be used, the one in which the parasite has most often been studied is given; no attempt to identify the species of definitive hosts has been made

[e] The figures in brackets correspond to the following references, given in full in the bibliography: (1) Amin (1975*c*); (2) Amin (1981); (3) Amin, Burns & Redlin (1980); (4) Awachie (1965); (5) Awachie (1966); (6) Brattey (1980); (7) Camp & Huizinga (1980); (8) Chappell (1969*a*); (9) Cooper (1965); (10) Crofton (1971*a*); (11) Crompton, Arnold & Barnard (1972); (12) Crompton & Harrison (1965); (13) Crompton *et al.* (1984); (14) Crompton & Walters (1972); (15) Crompton & Whitfield (1968*a*); (16) DeGiusti (1949*a*); (17) Esch *et al.* (1976); (18) Harwood & James (1979); (19) Hine & Kennedy (1974*a*, *b*); (20) Hirsch (1980); (21) Holland (1983*a*); (22) Hopp (1954); (23) Hynes (1955); (24) Hynes & Nicholas (1957); (25) Hynes & Nicholas (1963); (26) Kates (1943); (27) Kennedy & Rumpus (1977); (28) Kennedy, Broughton & Hine (1978); (29) King & Robinson (1967); (30) Lackie (1972*a*, *b*); (31) Lee (1981); (32) Larkin (1948); (33) Liat & Pike (1980); (34) Moore (1946*a*); (35) Merritt & Pratt (1964); (36) Meyer (1932); (37) Nicholas & Hynes (1958); (38) Nickol (1977); (39) Kates (1942); (40) Steel (1961); (41) Tedla & Fernando (1970); (42) Thompson (personal communication); (43) Uznanski & Nickol (1982); (44) Walkey (1967); (45) Bendell (1955); (46) Moore (1983); (47) Moore & Bell (1983); (48) Nickol & Dappen (1982); (49) Paris & Pitelka (1962); (50) Schmidt & Olsen (1964)

parasite persistence to occur, the numerical value of R_0 must be greater than or equal to unity.

From the model described in §10.2, the basic reproductive rate of the acanthocephalan life history shown in Fig. 10.1 may be derived:

$$R_0 = \frac{\Phi T_1 T_2 D_1 D_2}{M_1 M_2 M_3} \quad (10.14)$$

where $T_1 = \lambda\eta H$, $T_2 = \alpha\rho L$, $M_1 = (\pi+\delta)$, $M_2 = (b+\mu+\alpha\rho L)$ and $M_3 = (\eta H+\gamma)$ (see Appendix 1).

The basic reproductive rate is thus the product of the net rate of transmission and reproduction of the parasite, $\Phi T_1 T_2$ (where T_1 represents transmission to the intermediate host and T_2 represents transmission to the definitive host), and the expected lifespans of the egg ($1/M_3$), cystacanth (D_2/M_2) and adult worm (D_1/M_1). Reproductive success is thus determined by the relative rates of transmission and loss. The effects of changes in the size of some of the life history parameters on the magnitude of R_0 are shown in Fig. 10.6. R_0 increases with the size of both host populations, but is more critically dependent on the number of intermediate than definitive hosts. Long life expectancies of both adult worms and intermediate hosts, and short developmental time delays in the intermediate host also tend to magnify the value of R_0.

Since the value of R_0 must be greater than or equal to unity, it can be seen from eqn 10.14 that there exist critical host densities below which the parasite is unable to persist. As is the case for other macroparasites with indirect life histories, the transmission rate consists of a product of the final and intermediate host densities (Anderson, in press). Acanthocephalans may thus survive temporary reductions in the abundance of one of their two host species, provided the density of the other host in the life history is sufficiently high to maintain the rate of transmission above its threshold level. This may perhaps have generated one of the major selective pressures for the evolution of life history strategies involving alternate periods of development or reproduction in different species of host (Anderson, in press).

Rearrangement of eqn 10.14 gives the following expression for H_T, the threshold number of intermediate hosts necessary for acanthocephalan persistence in a population of L definitive hosts:

$$H_T = \frac{\gamma M_1 M_2}{\eta(D_1 D_2 \lambda\Phi T_2 - M_1 M_2)} \quad (10.15)$$

For parasite persistence to occur, $\lambda\Phi D_1 D_2 T_2 > M_1 M_2$. This inequality is normally satisfied as a result of the large rates of egg production (λ) observed for acanthocephalan parasites. Persistence of parasites with lower rates of fecundity may still occur, however, if the parasite has short prepatent time delays. As would be expected, the threshold intermediate host population size necessary for parasite persistence is minimized by factors leading to high rates of parasite transmission and low rates of

Fig. 10.6. Predicted influence of the numerical value of various acanthocephalan life history parameters on the magnitude of the basic reproductive rate of the parasite, R_0. In each case the parameter values used are essentially those given for *Polymorphus minutus* in Table 10.2 The dotted vertical line indicates the threshold size of the host population necessary for parasite transmission.

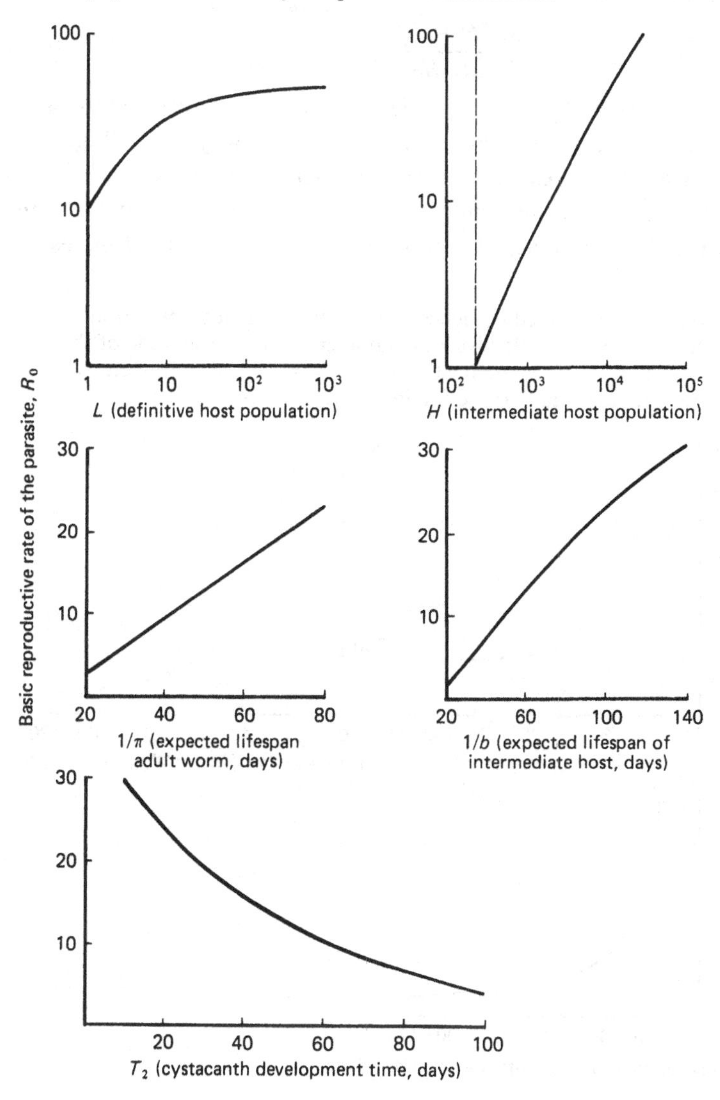

parasite mortality, such as high intermediate host and adult worm longevity (Fig. 10.7). Interestingly, it is predicted that parasite persistence may occur when less than one definitive host is present (Figs 10.6*a* and 10.7*a*). This may be interpreted as showing that under some conditions, the parasite's existence in a population of intermediate hosts may be ensured by the occasional temporary visits of very small numbers of the definitive host species.

A similar expression may be derived for L_T, the threshold number of definitive hosts necessary to allow parasite persistence in a population of H intermediate hosts:

$$L_T = \frac{(b+\mu) M_1(H+H_0)}{\alpha\rho[\lambda D_1 D_2 H - M_1(H+H_0)]} \tag{10.16}$$

where $H_0 = \gamma/\eta$ (see Appendix 1). The threshold, L_T, is affected by changes in the values of life history parameters in the same way as H_T (Fig. 10.8); it varies inversely, for example, with the inherent rate at which definitive hosts attack and consume unparasitised prey. Provided that the transmission rate is maintained at high levels by parasite-induced increase in

Fig. 10.7. Predicted influence of the numerical value of various acanthocephalan life history parameters on the magnitude of the threshold intermediate host density, H_T. Again the parameter values used are essentially those for *Polymorphus minutus*.

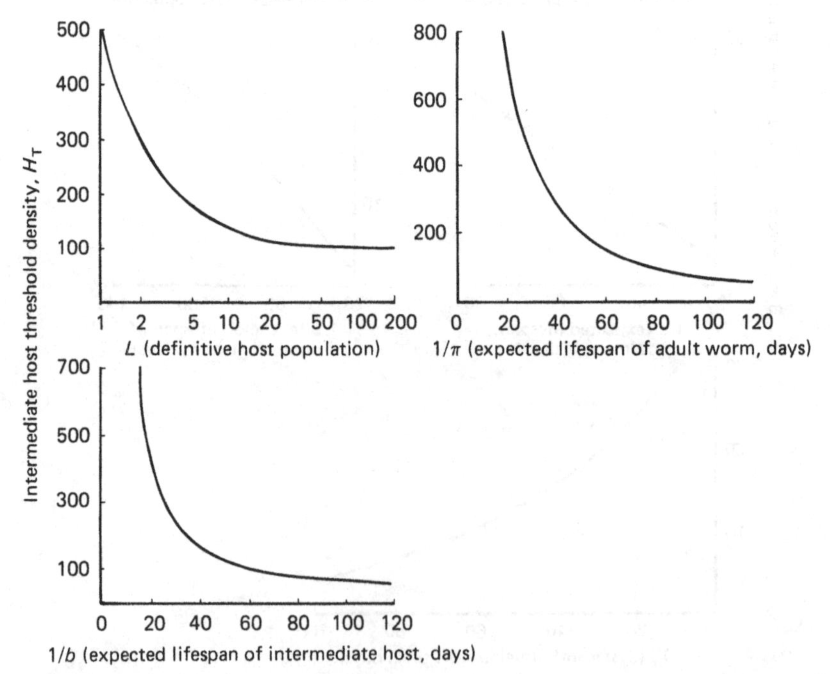

host susceptibility to predation, the threshold number of predators (definitive hosts) necessary for parasite persistence is small in relation to the required number of intermediate hosts (Fig. 10.8).

10.5 Dynamical properties of the model

Insight into the temporal dynamical behaviours of eqns 10.10 to 10.13 may be gained by considering the relative time-scales on which changes in the sizes of the different parasite and host populations operate. Disparities in these time-scales justify the collapse of the four equations to two which describe the life history entirely in terms of the mean intensity of adult parasites (N) and the intermediate host population density (H). This technique retains sufficient detailed information to allow formal local stability analysis of the dynamical properties of the model at equilibrium, whilst also allowing construction of the two-dimensional phase planes of expected patterns of population behaviour. Details of this are given in Appendix 2. The resultant two equations are:

$$\frac{dH}{dt} = H\left[(c_1 - c_2) - \frac{\alpha\lambda N L c_2 D_2}{(H+H_0)(c_3+c_7)}\right] \qquad (10.17)$$

$$\frac{dN}{dt} = N\left[\frac{\lambda H c_7 D_1 D_2}{(H+H_0)(c_3+c_7)} - c_4 - \delta N\left(\frac{k+l}{k}\right)\right] \qquad (10.18)$$

Fig. 10.8. Predicted influence of the numerical value of various acanthocephalan life history parameters on the magnitude of the threshold definitive host density, L_T. The dotted line indicates the threshold intermediate host density, H_T.

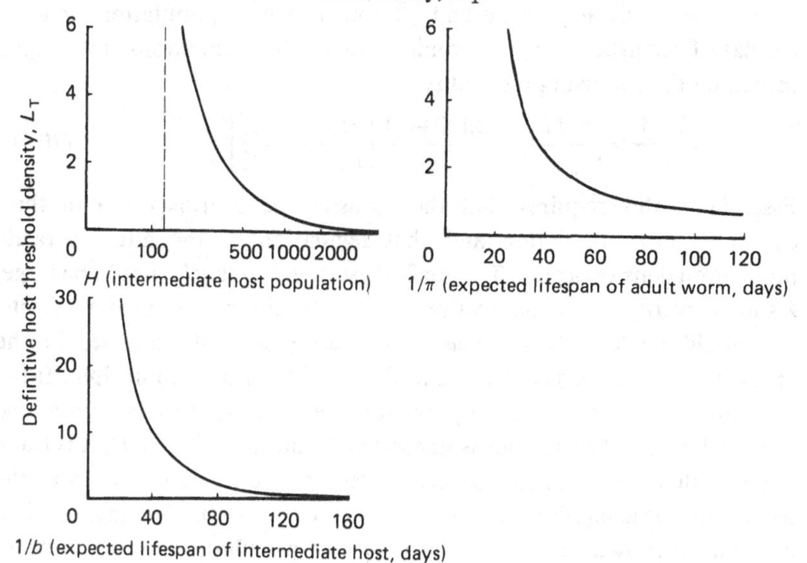

(where $c_1 = a-b = r$; $c_2 = \rho L$; $c_3 = b+\mu$; $c_4 = \pi+\delta$; $H_0 = \gamma/\eta$ and $c_7 = [\alpha c_2 + (c_1 - c_2)(j+1/j)]$). The analysis produces a relatively simple expression for the mean number of cystacanths per intermediate host at equilibrium, M^*:

$$M^* = P_1^*/H^* = (r-\rho L)/\alpha\rho L \tag{10.19}$$

Here, r is the natural intrinsic growth rate of the intermediate host population and the other parameters are as defined in Table 10.2. Eqn 10.19 suggests that the mean equilibrium burden of parasites per intermediate host is inversely associated with the total predation pressure. In addition, it indicates that if the expected growth rate of the prey population, r, is less than the inherent net rate of attack by predators, ρL, no stable equilibrium can exist and the predators eradicate the prey.

The quadratic nature of eqns 10.17 and 10.18 suggests that more than one combination of intermediate host and parasite population densities will satisfy the zero-growth equilibrium conditions ($dN/dt = dH/dt = 0$). These are illustrated in the phase-plane analysis of population behaviour for the basic model shown in Fig. 10.9. Unlike models for cestode parasites with similar life histories to those of acanthocephalans (Keymer, 1982*b*), there exist two positive equilibrium points (intersections) for biologically realistic combinations of parameter values. These occur as a direct consequence of extending the cestode model, which assumes both host populations to be constant, to include the dynamics of the intermediate host. The parasites and predators interact at low intermediate host densities to create a stable equilibrium at which the parasite persists at a low endemic level. Both parasite and intermediate host populations tend to oscillate if perturbed from this equilibrium, but the oscillations are damped providing the following inequality is met:

$$\delta\frac{k+1}{k} > \frac{\alpha\lambda L c_2}{(c_3+c_7)} > \frac{\delta j(c_1+c_2)}{\lambda c_7}\left[\frac{H}{H_0}(c_3+c_7)\right] \tag{10.20}$$

Essentially, this requires that the parasites be overdispersed in their definitive host population and that regulation of the adult parasite population density occurs. To satisfy these conditions, H_0 should be large, λ should be of intermediate value and the density of intermediate hosts, H, should be low. Low levels of overdispersion of parasites in the intermediate host also contribute to the stability of the equilibrium. If the inequality is transgressed, the populations enter a limit cycle, and as the overall dynamical behaviour is dependent upon the value of H, it is quite possible that these cycles may cross the breakpoint that runs from the second unstable equilibrium to the H axis (Fig. 10.9). This may lead to unbounded growth of the intermediate host population and an increase in parasite numbers to a level determined by the density-dependent constraints on establishment in the definitive host population. The model can thus show two different patterns of behaviour and any interaction is dependent upon the initial number of intermediate hosts. When this exceeds a threshold value, the parasites and predators are unable to control the growth of the intermediate host population which then increases exponentially, to be eventually constrained by factors not yet included in the model, such as nutrient resource limitation.

A third pattern of dynamical behaviour is possible if some form of additional regulation is included in the equation for intermediate host population growth (Appendix 2). This produces a further intersection of the two zero population growth isoclines at an upper intermediate host population density (Fig. 10.10). This equilibrium is asymptotically stable, with both parasites and intermediate hosts expected to remain at a fixed population density for prolonged periods of time. The other two intersections correspond directly to the two equilibria of our basic model: at the

Fig. 10.9. The behaviour of the isoclines $dH/dt = 0$ and $dN/dt = 0$ in the $H-N$ phase-plane (from eqns 10.10 to 10.13). The two intersections (A and B) represent alternative possible equilibrium states for H^* (intermediate host population size) and N^* (mean adult worm burden per host). Equilibrium A is a stable oscillatory equilibrium; equilibrium B is always unstable. The dotted line from B to the H axis, corresponds to the intermediate host densities where the parasite fails to regulate their population growth. At intermediate host densities greater than this the parasite fails to regulate these hosts and their population grows exponentially. The arrows indicate the direction of movement of the populations in the different regions of this phase-plane when perturbed from equilibrium.

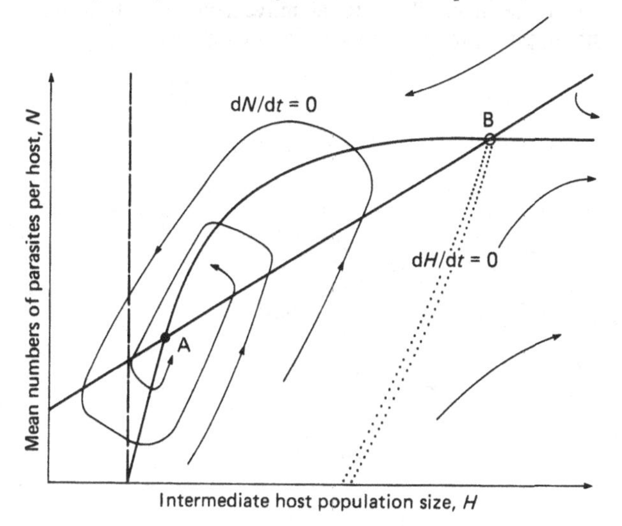

lower point both parasite and intermediate host populations show a tendency to oscillate, whilst the middle intersection corresponds to a breakpoint in intermediate host population density above which both populations grow asymptotically towards the third stable equilibrium.

The effects of numerical variation in parameter values on these equilibrium densities may be investigated by sensitivity analysis; some results of this exercise are qualitatively displayed in Table 10.3. These results suggest that the upper stable intersection is analogous to the single equilibrium of the Anderson & May (1978) direct life history models used as the framework for this chapter. The lower oscillatory equilibrium also has some analogous properties, but the influence of the predator–prey relationship required for the transmission of cystacanths to the definitive host tends to produce the oscillatory behaviour characteristic of many predator–prey models (Hassell, 1978). The position of the breakpoint between these two stable equilibria is determined by the position of the unstable intersection point. Increases in the values of parameters which constrain the growth rate of the intermediate host population (e.g. number of

Fig. 10.10. The behaviour of the isoclines $dH/dt = 0$ and $dN/dt = 0$ in the $H-N$ phase-plane in the more realistic model which includes additional forms of regulation in the intermediate host population (Appendix 2). The axes are as for Fig. 10.9 with the two intersections A and B corresponding to the equilibria A and B in this previous figure; A is stable but oscillatory, whilst B is an unstable equilibrium. The third intersection C, corresponds to an asymptotically stable upper equilibrium. This is attained by both populations when the initial intermediate host population size exceeds the critical densities indicated by the dotted line passing through B and intersecting the H axis. The arrows again illustrate the movement of the two populations in the phase-plane, when perturbed from equilibrium.

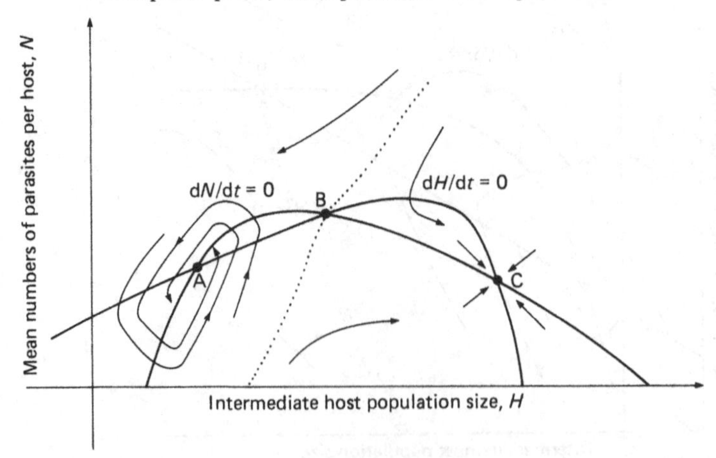

predators or rate of intermediate host mortality) tend to increase the critical intermediate host population size at which this outbreak to a higher population level occurs. As these factors are mainly determined by various aspects of the biology of the intermediate host, they further emphasize the importance of the collection of quantitative population information on these species. In particular, changes in temperature or the build-up of nutrients leading to increases in the birth rate of the intermediate host will be important in determining the sudden outbreaks of the parasites which may occur when the intermediate host population exceeds the critical threshold level. Such outbreaks are characteristic features of many acanthocephalan infections (Swennen & Van Den Broek, 1960; Garden, Rayski & Thom, 1964; Itämies, Valtonen & Fagerholm, 1980).

The influence of the statistical distribution of cystacanths within the intermediate host population is illustrated in Fig. 10.11. Increases in the degree of overdispersion (low values of j) tend to enhance the transmission of the parasite to the definitive host, thus raising the equilibrium mean adult worm burden ($N^* = P_2^*/L$). However, these increases occur without reductions in the intermediate host population density, and so the impact of definitive host predation on the intermediate host population is diminished, and the threshold density of intermediate hosts reduced. The

Table 10.3. *Synopsis of the sensitivity of the three equilibria to parameter value variation*

	Lower stable equilibrium		Intermediate unstable equilibrium		Upper stable equilibrium	
Parameter	H_A	N_A	H_B	N_B	H_C	N_C
Number of definitive hosts, L	+	−	+	−	−	−
Strength of density-dependent parasite mortalities or degree of overdispersion in definitive host, δk	0	0	−	−	+	−
Intermediate host population growth rate, r	−	0	−	+	+	+
Degree of overdispersion in intermediate hosts, j	0	+	−	+	0	+

The +, −, and 0 signs indicate, respectively, that the equilibrium population size increases, decreases or remains the same as the parameter of interest is increased in magnitude. H and N represent the intermediate host population size and the mean number of parasites per definitive host at the joint equilibrium. The subscripts A, B and C correspond to the intersections thus named in Fig. 10.10

propensity of the parasite and intermediate host populations to break away to their upper equilibrium levels is consequently enhanced by high levels of parasite aggregation in the intermediate host.

This result is of relevance to the debate which has recently arisen over the significance of parasite-induced host mortality and overdispersion. Hirsch (1980) has criticized Crofton's (1971*a*) seminal analyses of the data collected by Hynes & Nicholas (1963) pertaining to the distribution of *Polymorphus minutus* in different populations of *Gammarus pulex*. This reanalysis of the *P. minutus* data is used by Hirsch as a basis for further

Fig. 10.11. The top diagram illustrates the predicted influence of the degree of overdispersion in the distribution of cystacanth numbers per intermediate host (j) on the size of the intermediate host population density at the lower oscillatory stable equilibrium, H^* (corresponding to A in Figs 10.9 and 10.10). The middle diagram depicts the effect of similar variations on the magnitude of N^* at this equilibrium. The bottom diagram illustrates the effect of changes in j on the position of the unstable equilibrium (B in Figs 10.9 and 10.10), which essentially corresponds to the critical intermediate host density at which the parasites fail to regulate the intermediate host population size.

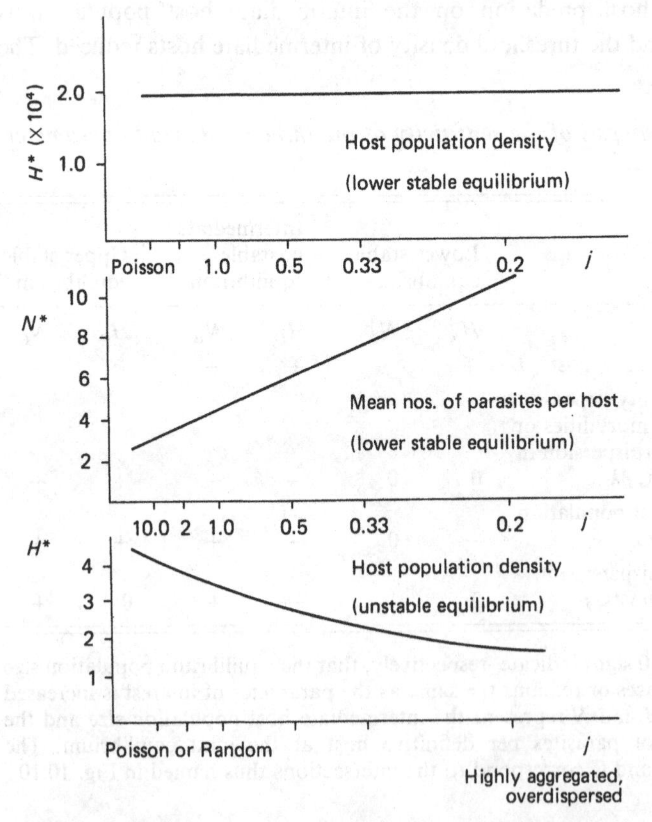

criticisms of Crofton's theoretical work on the population dynamics of parasite life histories (Crofton, 1971*b*). We suggest that these criticisms in no way detract from Crofton's fundamental conclusions concerning the importance of the interaction between overdispersed parasite distributions and regulatory mechanisms such as pathogenicity. The density-dependent constraints on the population growth of adult worms in the definitive host considered in the present model act in an exactly analogous manner to the parasite-induced host mortality considered by Crofton (1971*b*) and by Anderson & May (1978). In all cases, overdispersion in the distribution of parasite numbers per host enhances density-dependent control by concentrating its effects so that a large proportion of the parasite population is affected. The interaction between relatively weak density-dependent constraints and high degrees of parasite aggregation in the definitive host is thought to be the most powerful source of regulation in most parasite populations. In contrast, the influence of acanthocephalan overdispersion in the intermediate host is less obvious. At low parasite population densities, it might be advantageous as it would increase the effective rate of transmission between intermediate and definitive hosts. At higher densities, however, it might tend to destabilize the system by minimizing the impact of the parasite on the population growth of the intermediate host. This may partly explain why overdispersed distributions are characteristic of adult acanthocephalans in their definitive hosts, whilst the distributions of the larval stages in the intermediate hosts tend to be random or only slightly overdispersed (Table 10.2).

10.6 Field studies of acanthocephalan population behaviour

10.6.1 *Longitudinal studies*

Three longitudinal patterns in the prevalence of acanthocephalan populations have been noted. Some species have been found to remain remarkably constant in prevalence over long periods of time, with no marked seasonal changes (Fig. 10.12). We would suggest that these species have dynamics corresponding to those of populations present at the upper stable equilibrium in our phase-plane analysis (intersection C of Fig. 10.10). We would expect this pattern of dynamic behaviour to be exhibited in habitats where both the definitive and intermediate hosts were present throughout the year and where the intermediate host population density remained relatively constant at a level determined by factors external to the parasite–predator–prey relationship. This seems to be likely for the two examples given.

More commonly recorded patterns of behaviour amongst acanthocephalan species are the oscillations in prevalence of varying magnitude illustrated in Fig. 10.13. These oscillations are invariably assumed to be a consequence of seasonal fluctuation in the temperature of the habitat, which, in turn, leads to variations in the rates at which the physiological mechanisms of transmission operate. Although the temperature dependence of several developmental processes, such as intermediate host population growth rate and cystacanth development time, is of importance in determining the magnitude of the cycles observed, the above analysis suggests that life histories of this type would be expected *a priori* to show oscillatory behaviour under certain conditions and combinations of parameter values. Thus, although temperature-related or seasonal changes in host behaviour may be mechanistically important in initiating the outbreak of the parasites at the beginning of any annual cycle, the actual tendency to oscillate may be inherently dependent on the fact that transmission proceeds *via* a predator–prey relationship. The more complex seasonal patterns observed in some of the studies illustrated in Fig. 10.13 may thus reflect more subtle patterns of behaviour than the straightforward temperature dependence of the underlying physiological processes. Unfortunately, the intermediate host population density has not been monitored in any of the published studies. Further work on this, together with laboratory studies on the temperature characteristics of the physiolo-

Fig. 10.12. Long-term and seasonal changes in the prevalence of (*a*) and (*b*) *Pomphorhynchus laevis* in *Gammarus pulex* (closed circles) and dace (open circles) (Kennedy & Rumpus, 1977) and (*c*) and (*d*) *Macracanthorhynchus hirudinaceus* in swine (Dunagan & Miller, 1980).

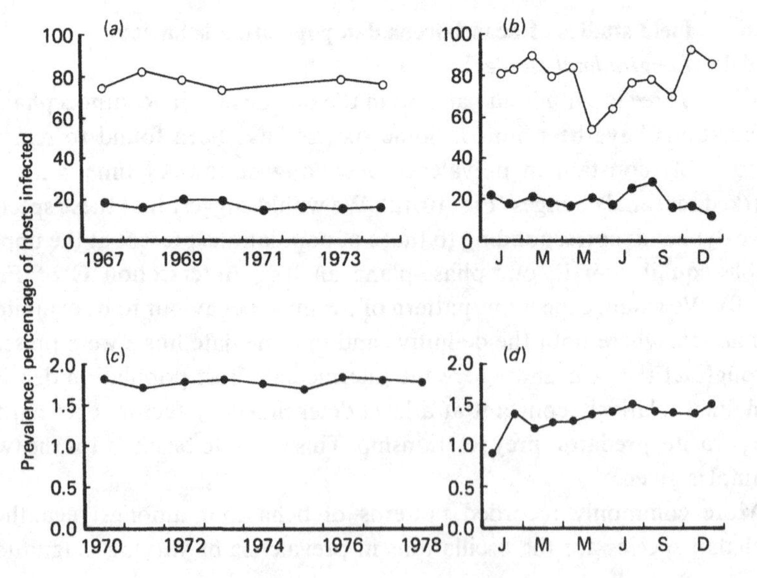

gical host–parasite relationship, is required before the observed patterns of seasonal dynamical behaviour may be fully understood.

A third phenomenon, which has been noted extensively but which proves difficult to monitor with precision, is the occurrence of sudden outbreaks

Fig. 10.13. Seasonal changes in the prevalence of (*a*) *Corynosoma constrictum* in pintail (Buscher, 1965); (*b*) *Echinorhynchus salmonis* in yellow perch (Tedla & Fernando, 1970); (*c*) *Neoechinorhynchus saginatus* in fallfish (Muzzall & Bullock, 1978); (*d*) *Echinorhynchus salmonis* in whitefish (Valtonen, 1980); (*e*) *Acanthocephalus dirus* in *Lepomis cyanellus*, *L. macrochirus*, *Semotilus atromaculatus*, *Cyprinus carpio* and *Carassius auratus* (Muzzall & Rabalais, 1975*a*); (*f*) *Acanthocephalus dirus* in creek chub (Camp & Huizinga, 1980).

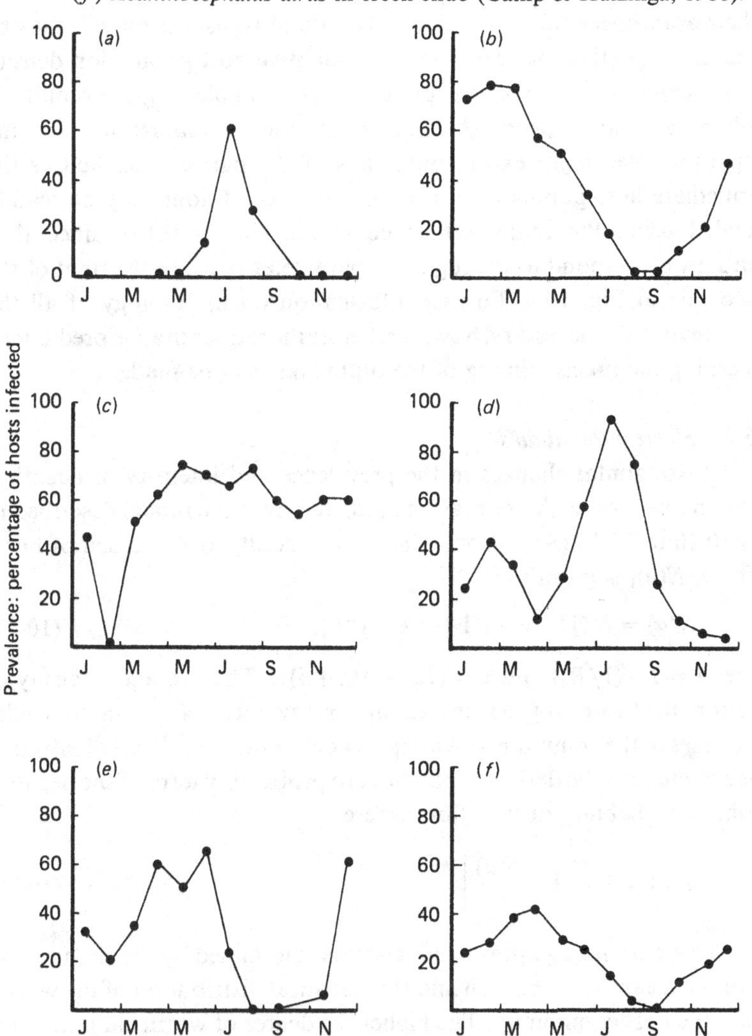

of infection in populations of definitive hosts, often causing extensive morbidity and mortality (Takos & Thomas, 1958; Swennen & Van Den Broek, 1960; Hynes & Nicholas, 1963; Garden *et al.*, 1964; Itämies *et al.*, 1980). These outbreaks usually coincide with rapid increases in the size of intermediate host populations and thus occur at the times of year when intermediate host reproduction is at a maximum and when relatively high temperatures increase the rate of parasite maturation. The occurrence of outbreaks may also be enhanced at these times by changes in the feeding biology of the definitive host. For example, whereas adult ducks are polyphagous predators, newly fledged ducklings feed almost exclusively on the aquatic arthropod species which form the intermediate hosts of many of their acanthocephalan parasites. The rate of transmission will thus be increased at the time of year when the definitive host population density is at a maximum. If the intermediate host is also able to grow rapidly in numbers, as may happen when excess nutrients minimize intraspecific competition, we might expect outbreaks of the parasite (as well as the intermediate host population) to occur. Such conditions may be readily modelled using the initial set of equations (10.10–10.13), since they essentially correspond to the region of parameter space to the right of the dotted line in Fig. 10.9. Further information on the biology of all the species involved is necessary, however, before more quantitative predictions concerning the precise timing of the outbreaks may be made.

10.6.2 *Horizontal studies*

Horizontal changes in the prevalence and intensity of acanthocephalan infection with age may be predicted from the model described in eqns 10.10 to 10.13 (Anderson, 1982*a*). The adult worm burden of hosts of age a, $N(a)$, is given by:

$$N(a) = N^*[1-e^{-\gamma}][1+A\,e^{-\gamma}]^{-1} \tag{10.21}$$

where $A = 1-(1/R_0)$ and $\gamma = (R_0-1)(\pi+\delta)a$. The average intensity of infection therefore approaches an upper asymptote (N^*) in the older age-classes of the community. An expression for the prevalence of infection at age a, $P(a)$, can be derived from the zero probability term of the negative binomial probability distribution, where

$$P(a) = 1-\left[1+\frac{N(a)}{k}\right]^{-k} \tag{10.22}$$

The shape of this age-prevalence curve is determined by the change with age of the mean worm burden and the statistical distribution of the worms within the host community. The higher the degree of worm clumping (the lower the value of k), the lower the asymptotic value of the prevalence of infection in the older age-classes of hosts (Anderson, 1982*a*). Observed patterns of the size-prevalence of acanthocephalan infections are shown in Fig. 10.14, but unfortunately no data for prevalence or intensity of infection with host age are available. Studies to determine the age characteristics of infection would be of considerable significance as a means of the estimation of R_0.

Fig. 10.14. The prevalence and intensity of acanthocephalan infections in relation to host size. (*a*) and (*b*) *Octospinifer macilentus* and *Pamphorhynchus bulbocolli* in whitesuckers (Muzzall, 1980); (*c*) and (*d*) *Neoechinorhynchus saginatus* in fallfish (Muzzall & Bullock, 1978); (*e*) *Neoechinorhynchus rutili* in sticklebacks (Walkey, 1967).

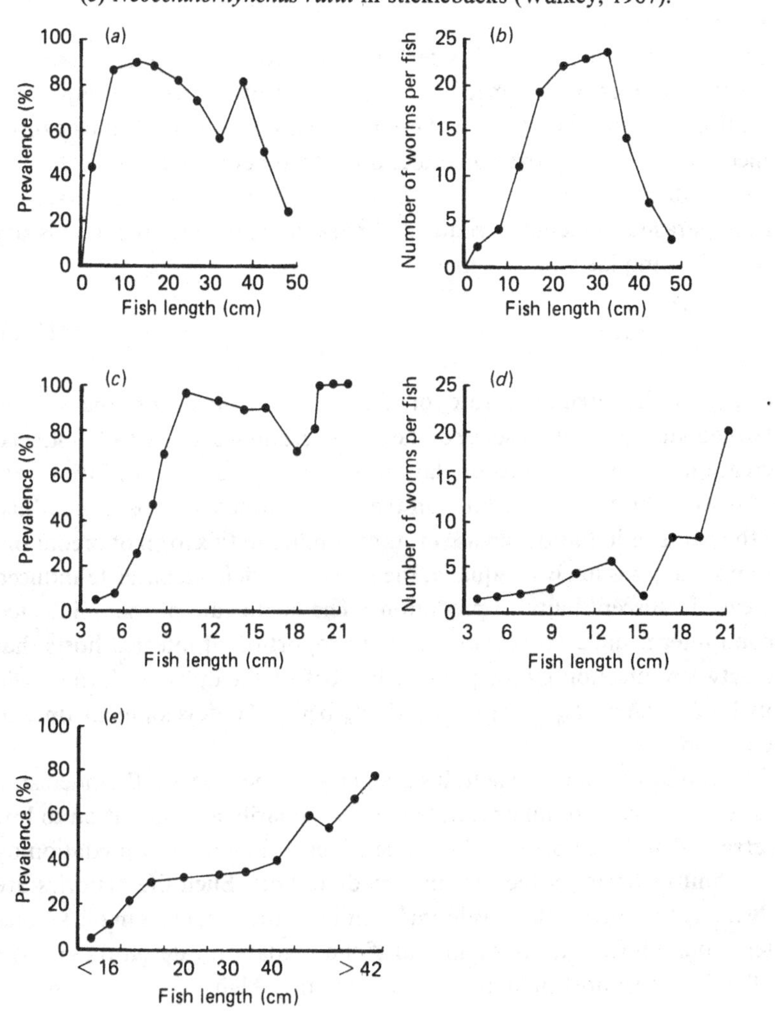

10.7 More complex models for three-host life cycles

10.7.1 *Incorporation of a second intermediate host into the life history*

The basic model described in §10.2 is constructed to mimic the simplest form of acanthocephalan life history. Some species utilize more than one intermediate host; for example *Neoechinorhynchus emydis*, a parasite of *Graptemys geographica*, undergoes sequential stages of development in ostracods (where the acanthellae and cystacanths develop) and then in paratenic snails, where the cysts become infective to the turtle (Hopp, 1954). An obvious prerequisite for the development of this type of life history is that during the time of evolutionary process, the first intermediate host should be preyed upon by the second intermediate host, and that both should be potential prey items in the diet of the definitive host predator.

The basic model can be fairly readily modified to incorporate a second sequential species of intermediate host (Appendix 3). Several important conditions for the evolution of these more complex relationships can be gained by considering the numerical difference between the R_0 values of each possible life history strategy. Essentially, it should be advantageous for the parasite to include a sequential paratenic host into the life history if the following inequality is met:

$$\frac{\nu\zeta \cdot \varepsilon\sigma D_3}{(\omega+\nu\zeta L)} \gg \alpha\rho \tag{10.23}$$

Here, ζ is the predation rate of the definitive hosts on the second intermediate host in the absence of the parasite, and ν is the parasite-induced increase in susceptibility to predation of these hosts. The predation rate of the second intermediate host on the first intermediate host is σ, whilst ε is the parasite-induced increase in susceptibility to this form of predation. (We note that ε is likely to equal α, the previously defined parasite-induced increase in susceptibility to predation.) The death rate of the uninfected second intermediate hosts is ω, and the proportion of infected hosts that die between infection by the parasite but before the cysts are formed will thus be D_3, where $D_3 = \exp[-Y_3\omega]$, Y_3 being the development time of the cyst in days.

The conditions under which this inequality is met suggest that inclusion of a second species of intermediate host is primarily an adaptation to low or erratic definitive host population densities or low rates of predation by the definitive hosts on the first intermediate host. Such life histories are only likely to evolve if the development time of infective cysts in the second intermediate host is fairly rapid and if the predation rate on these hosts by the definitive host predators is already fairly high.

Although analysis of the dynamics of life histories of this type are likely to prove less transparent than those for the simpler two-host model, it seems probable that in most respects they will be essentially similar. The presence of a further time delay in the life history is likely to reduce the stability of the lower equilibrium in the previous analysis of dynamical behaviour. The whole host–parasite relationship will also now be dependent on the density and carrying capacity of the species used as a second intermediate host. As fluctuations in the density of this species will determine the effective transmission rate to the definitive host, it seems sensible to suggest that these species are likely to be either conspicuously numerous or dynamically very stable.

10.7.2 *Utilization of more than one species of definitive host by the parasite*

A second life history variation seen in several species of Acanthocephala is the use of two or more alternative species of definitive host. *Leptorhynchoides thecatus*, for example, has been recovered from 83 species of definitive host from three vertebrate classes (Lincicome & Van Cleave, 1949). Rather than describe in detail the algebraic modifications to the basic equations needed to incorporate these additional species of definitive hosts into the model, we simply point out that the method used proceeds in a manner similar to that described in Appendix 3 for the addition of a second sequential species of intermediate host. The only prerequisite for this type of life history to evolve within the framework of an existing food-web would seem to be a certain degree of physiological elasticity on the part of the parasite. Relaxation of the constraints of host specificity presumably evolves as a means of increasing the rate of transmission and its primary advantage would seem to be as a hedge against the temporary or permanent disappearance of one or more of the potential species of definitive host (Anderson, in press). It would be predicted that population regulation of the parasites should be achieved by the operation of density-dependent constraints on parasite populations in the major definitive host species. Regulatory processes occurring in the minor definitive hosts (i.e., species which account for only small proportions of the total parasite transmission) would not be expected to exert effective control on overall parasite population size (see, for example, Leong & Holmes, 1981). Variations in the pathogenicity of the parasite in each species of definitive host would, however, be likely to contribute to overall parasite population regulation.

10.8 Discussion

The analyses described in this chapter are based on methods which have proved extremely influential in studies of the population behaviour of macroparasites with direct life history (Anderson & May, 1978, 1982; May & Anderson, 1978; Anderson, 1980). The application of these methods to acanthocephalan biology is complicated by the complexity of the life histories involved and also by the present scarcity of empirical data. The areas in which more laboratory studies could profitably be undertaken are shown in Table 10.2. Most notably these include the possible density-dependent constraints that may be acting on adult worm establishment and fecundity, and the survival rates of each of the different life history stages as well as that of the intermediate host species. (We note that, as a result of the differences in the time-scales of various host and parasite population changes, the most valuable information is that relating to the longest lived stages within each life history.) In addition, future field studies would be of significantly increased value if they attempted to determine the time- and age-dependent nature of acanthocephalan–host population interactions. When combined with data from laboratory experiments on the temperature dependence of the physiological processes, these should eventually allow the development of models whose quantitative predictions could be tested against long-term data sets collected under natural conditions. We also suggest that field studies in which attempts are made to collect quantitative information on the population densities of both the intermediate and definitive hosts (as well as that of the parasite) would be of considerable value.

Despite these reservations, we feel that the methods described facilitate an understanding of acanthocephalan life histories and population biology which would not be readily apparent from a less formal consideration of empirical data. The most important unifying concept, in this respect, is the basic reproductive rate of the parasite (Anderson, 1980). This index may be used not only to compare the reproductive success of different parasite species, both within and between phyla, but also to provide insights into the reasons why the more complex life histories have evolved. We are aware that perhaps the major omission from the present application of these ideas is the subject of genetics. Inclusion of genetic variability would extend the scope of questions concerning the mechanisms of parameter value evolution, but would require the detailed monitoring of laboratory and field systems in order to obtain data on interstrain or interspecies variation in host–parasite interactions. In view of their potential as laboratory and field models, perhaps the Acanthocephala will in the future prove suitable subjects for the study of genetic variability.

Acknowledgements
We are very grateful to Dr D. W. T. Crompton for guiding us through the published literature, and to Professors R. M. Anderson and R. M. May for helpful discussions and advice. We thank Mrs S. Friend for preparation of the typescript.

Appendix 1

This appendix illustrates how eqns 10.10 to 10.13 (see page 358) are collapsed to give a single expression for R_0, the basic reproductive rate of the parasite. The original four equations are:

$$\frac{dP_2}{dt} = \alpha\rho L\left[\frac{P_1}{H}+\frac{P_1^{\,2}}{H^2}\left(\frac{j+1}{j}\right)\right]D_1\,H(t-\Upsilon_1) - \pi P_2 - \delta L\left[\frac{P_2}{L}+\frac{P_2^{\,2}}{L^2}\left(\frac{k+1}{k}\right)\right] \qquad (10.A1)$$

$$\frac{dP_1}{dt} = \eta WHD_2(t-\Upsilon_2)-(b+\mu)P_1 - \alpha\rho LH\left[\frac{P_1}{H}+\frac{P_1^{\,2}}{H^2}\left(\frac{j+1}{j}\right)\right] \qquad (10.A2)$$

$$\frac{dW}{dt} = \lambda P_2\Phi-\gamma W-\eta WH \qquad (10.A3)$$

$$\frac{dH}{dt} = (a-b)\,H-\rho LH-\alpha\rho LP_1 \qquad (10.A4)$$

If we assume that the dynamics of the free-living eggs act on a faster time-scale than the rest of the life cycle stages, then $dW/dt \simeq 0$ and the equilibrium number of free-living eggs at any time, W^*, will be given by solving eqn 10.A3 at this equilibrium:

$$W^* = \lambda P_2\Phi/\eta(H_0+H) \qquad (10.A5)$$

where $H_0 = \gamma/\eta$, a parameter that varies inversely with the efficiency of transmission from definitive to intermediate hosts.

These two expressions may then be substituted into eqn 10.A2, and, making the further approximation that when the parasite is first introduced to a habitat, $[(P_1/H)+(P_1^{\,2}/H^2)(j+1/j)] \simeq P_1/H$,

$$\frac{dP_1}{dt} = \frac{\lambda P_2\Phi D_2\,H(t-\Upsilon_2)}{(H_0+H)} - (b+\mu+\alpha\rho L)\,P_1 \qquad (10.A6)$$

At equilibrium, $dP_1{}^*/dt = 0$

$$P_1{}^* = \frac{\lambda P_2\Phi D_2\,H(t-\Upsilon_2)}{(H_0+H)\,(b+\mu+\alpha\rho L)} \qquad (10.A7)$$

This may then be substituted into eqn 10.A1 to give one expression for the rate of growth of the adult parasite population.

$$\frac{dP_2}{dt} = P_2\left[\frac{\alpha\rho L\lambda\Phi D_1 D_2 H}{(H_0+H)(b+\mu+\alpha\rho L)} - (\pi+\delta) - \frac{P_2}{L}\delta\left(\frac{k+1}{k}\right)\right] \quad (10.A8)$$

The expression for R_0 is obtained by considering the rate of growth of the adult parasite population when the parasite is first introduced, e.g. density-dependent mortalities are effectively zero, $P_2/L(k+1/k)\delta = 0$

$$R_0 = \frac{\alpha\rho L\lambda\Phi H D_1 D_2}{(H_0+H)(b+\mu+\alpha\rho L)(\pi+\delta)} \quad (10.A9)$$

Equation 10.14 (page 362) is obtained by resubstituting $\gamma/\eta = H_0$ into this formula and dividing the terms into those that describe each of the transmission (birth) and death processes at each life cycle stage.

The terms for the two transmission thresholds, H_T and L_T, are obtained by rearranging eqn 10.A10 for the special cases when $R_0 = 1$.

Appendix 2

This appendix describes how the dynamical behaviour of the model may be discussed in terms of N, the mean numbers of parasites per definitive host, and H, the intermediate host population density. As in Appendix 1, we shall assume that the free-living egg stages operate on a faster time-scale than the rest of the life cycle and that $dW/dt = 0$ and $H_0 = \gamma/\eta$. The mean numbers of cystacanths per intermediate host, M^*, may be readily found by setting eqn 10.A4 to its equilibrium value, $dH/dt = 0$, and rearranging:

$$P_1^*/H^* = M^* = \frac{r-\rho L}{\alpha\rho L} \quad (10.A10)$$

where $r = a-b$.

If we assume that the developmental time delays Y_1 and Y_2 are short, then D_1 and D_2 will approximately equal unity and $H_t = H_{t-Y_1} = H_{t-Y_2}$. Substituting eqn 10.A10 into eqn 10.A2 gives

$$\frac{dP_1}{dt} = \frac{\lambda P_2 \Phi D_2 H}{(H+H_0)} - (b+\mu+\alpha\rho L)P_1 - (r-\rho L)\left(\frac{j+1}{j}\right)P_1$$

At equilibrium, $dP_1/dt = 0$ and

$$P_1^* = \lambda P_2 \Phi D_2 H/(H+H_0)[b+\mu+\alpha\rho L+(r-\rho L)(j+1/j)] \quad (10.A11)$$

If we assume that the mating function, Φ, is always satisfied and thus equal to unity at equilibrium, eqn 10.A11 may be substituted into eqn 10.A1 and eqn 10.A4 to give two equations which describe the dynamics of the model in terms of the adult parasite population and the intermediate host population. Division of the former by L gives an expression for the mean parasite burden per definitive host:

$$\frac{dH}{dt} = H\left[(c_1-c_2) - \frac{\alpha\lambda NLc_2 D_2}{(H+H_0)(c_3+c_7)}\right] \tag{10.A12}$$

$$\frac{dN}{dt} = N\left[\frac{\lambda Hc_7 D_1 D_2}{(H+H_0)(c_3+c_7)} - c_4 - \delta N\left(\frac{k+1}{k}\right)\right] \tag{10.A13}$$

Here the equations have been simplified further by combining groups of parameters together: thus, $c_1 = a-b = r$; $c_2 = \rho L$; $c_3 = b+\mu$; $c_4 = \pi+\delta$; and $c_7 = \alpha c_2 + (c_1-c_2)(j+1/j)$.

The equilibrium intermediate host population densities are found by setting $dN^*/dt = dH^*/dt = 0$ and solving to give

$$H^* = \frac{\alpha\lambda N^* Lc_2 D_2}{(c_1-c_2)(c_3+c_7)} - H_0 \tag{10.A14}$$

and
$$N^* = \frac{-B\pm\sqrt{B^2-4AC}}{2A} \tag{10.A15}$$

where $A = \delta(k+1/k)$; $B = -\{[\lambda c_7 D_1 D_2/(c_3+c_7)] - c_4\}$; $C = [c_7 H_0(c_1 - c_2)]/\alpha Lc_2 D_2$. Providing $B^2 > 4AC$, there will be two equilibrium values of N^* and H^*. The dynamical behaviour of these equilibrium population densities may be examined by neighbourhood stability analysis. Writing $H_{(t)} = H^* + x(t)$ and $N_{(t)} = N^* + y(t)$ the results of a small perturbation away from (H^*, N^*) may be considered by Taylor expanding, ignoring higher-order terms and considering the expression for the temporal behaviour of the perturbations dx/dt and dy/dt.

$$\frac{dx}{dt} = \left[(c_1-c_2) - \frac{\alpha Lc_2\lambda D_2 NH_0}{(c_3+c_7)(H+H_0)^2}\right]\cdot x - \left[\frac{\alpha Lc_2\lambda D_2 H}{(H+H_0)}\right]\cdot y \tag{10.A16}$$

$$\frac{dy}{dt} = \left[\frac{\lambda c_7 D_1 D_2 NH_0}{(c_3+c_7)(H+H_0)^2}\right]\cdot x - \delta N\left(\frac{k+1}{k}\right)\cdot y \tag{10.A17}$$

The stability-determining damping rates (or eigenvalues), λ, are given by substitution of the above into $\lambda^2 + A\lambda + B = 0$. The Routh Hurvitz criteria can then be used to determine whether (H^*, N^*) is stable. The resultant expressions for A and B must both be greater than zero.

$$A = \delta N^*\left(\frac{k+1}{k}\right) + \left(\frac{H_0}{H^*+H_0} - 1\right)(c_1-c_2) \tag{10.A18}$$

$$B = (c_1-c_2)\left\{\left(\frac{H_0}{H^*+H_0}\right)\left[\delta N^*\left(\frac{k+1}{k}\right) + H_0 c_7\right] - \delta N^*\left(\frac{k+1}{k}\right)\right\} \tag{10.A19}$$

Thus the equilibrium will be stable when N^* and H^* are small, but increasingly unstable as H^* gets larger. The phase-plane analysis of this relation is depicted in Fig. 10.9. The tendency of the lower stable equilibrium to oscillate is controlled by the stabilizing influence of the over-dispersed distribution of the parasites in their definitive hosts combined with the density-dependent reduction in parasite survival in these hosts. The dampened oscillations become limit cycles when the following inequality is transgressed:

$$\delta\left(\frac{k+1}{k}\right) > \frac{\alpha\lambda L c_2}{(c_3+c_7)} > \frac{\delta\left(\frac{k+1}{k}\right)(c_1-c_2)}{\lambda c_7}\left[\frac{H}{H_0}(c_3+c_7)\right] \quad (10.A20)$$

Further regulation may be incorporated into the intermediate host population by the addition of an additional parameter, β, which corresponds to the density-dependent increase in mortality of the intermediate hosts due to factors other than the parasite. Equations 10.A2 and 10.A4 now become

$$\frac{dP_1}{dt} = \eta W D_2 H(t-Y_2) - (b+\mu+\beta H) P_1 - \alpha\rho L H\left[\frac{P_1}{H} + \frac{P_1^2}{H^2}\left(\frac{j+1}{j}\right)\right] \quad (10.A21)$$

$$\frac{dH}{dt} = (a-b-\beta H) H - \rho L H - \alpha\rho L P_1 \quad (10.A22)$$

Stability analysis of this more realistic model follows the recipe outlined above and described in detail by Dobson (unpublished). The dynamical consequences of this modification to the basic model are discussed in the main text.

Appendix 3

This appendix describes how R_0 is derived for more complex acanthocephalan life cycles. Initially we will consider life cycles that contain a second sequential species of intermediate host. It is a prerequisite of the evolution of such a relationship that H and S (the second intermediate host species) are both potential prey items in the predator L's diet; also H should be a potential prey of species S. If we assume that the mating function, Φ, is unity and that the population densities of all potential hosts are constant, then the dynamics of the three parasite populations will be given by the following equations:

$$\frac{dP_1}{dt} = \frac{\lambda P_2 H D_2}{(H+H_0)} - (b+\mu) P_1 - \alpha\rho L P_1 \left[1+\frac{P_1}{H}\left(\frac{j+1}{j}\right)\right] - \varepsilon\sigma P_1 S\left[1+\frac{P_1}{H}\left(\frac{j+1}{j}\right)\right] \quad (10.A23)$$

$$\frac{dP_T}{dt} = \varepsilon\sigma P_1 S\left[1+\frac{P_1}{H}\left(\frac{j+1}{j}\right)\right] D_3 - \omega P_T - V\xi L P_T\left[1+\frac{P_T}{S}\left(\frac{l+1}{l}\right)\right] \quad (10.A24)$$

$$\frac{dP_2}{dt} = \alpha\rho P_1 L D_1\left[1+\frac{P_1}{H}\left(\frac{j+1}{j}\right)\right] + v\xi L P_T D_1\left[1+\frac{P_T}{S}\left(\frac{l+1}{l}\right)\right] - (\pi+\delta) P_2 - \delta N\left(\frac{j+1}{j}\right) P_2 \quad (10.A25)$$

Here, σ is the predation rate of S on H, ε is the parasite-induced increase in susceptibility to predation [realistically $\varepsilon \approx \alpha$], l is the degree of overdispersion of the parasites P_T in their second intermediate host, S, ξ is the predation rate of the definitive hosts on these second intermediate hosts, v is their parasite-induced increased susceptibility to predation and ω their natural death rate. There are now three developmental time delays: D_1 and D_2 correspond to the same delays in eqns 10.A1 and 10.A2 (and in eqns 10.1 and 10.2); D_3 is the proportion of cystacanths ingested by the second intermediate host that survive to be cysts infective to the definitive host, thus $D_3 = \exp[-Y_3\omega]$ where Y_3 is the cyst development time in days.

When the parasite is first introduced into the host populations, it grows at a rate R_0, note that as the equations are written, both the two-host and the sequential three-host life cycles are likely to occur, neither is obligatory. When P_1, P_T and P_2 are very small,

$$\left[1+\frac{P_1}{H}\left(\frac{k+1}{k}\right)\right] \simeq \left[1+\frac{P_T}{S}\left(\frac{l+1}{l}\right)\right] \simeq 1$$

If $dP_1{}^*/dt \simeq 0$,

$$P_1{}^* = \frac{\lambda P_2 H}{(b+\mu+\alpha\rho L+\varepsilon\sigma S)(H+H_0)}$$

Similarly, at $dP_T{}^*/dt \simeq 0$,

$$P_T{}^* = \frac{\varepsilon\sigma P_1 S D_3}{(\omega+v\xi L)} \simeq \frac{\lambda P_2 H D_2 D_3 \varepsilon\sigma}{(\omega+v\xi L)(b+\mu+\alpha\rho L+\varepsilon\sigma S)(H+H_0)}$$

Substitution into eqn 10.A25 thus gives a single equation for the growth of the parasite population:

$$\frac{dP_2}{dt} = P_2\left[\frac{\alpha\rho L D_1 D_2 \lambda H}{(b+\mu+\alpha\rho L+\varepsilon\sigma S)(H+H_0)} + \frac{v\xi L D_1 D_2 D_3 \lambda H \varepsilon\sigma}{(\omega+v\xi L)(b+\mu+\alpha\rho L+\varepsilon\sigma S)(H+H_0)} - (\pi+\delta) - \delta N\left(\frac{k+1}{k}\right)\right]$$

This may be simplified to give a compound expression for R_0 when both two-host and three-host life cycles are being pursued:

$$R_0 = \frac{LH\lambda D_1 D_2}{(b+\mu+\alpha\rho L+\varepsilon\sigma S)(\pi+\delta)(H+H_0)}\left[\alpha\rho + \frac{V\xi\cdot\varepsilon\sigma D_3}{(\omega+v\xi L)}\right] \tag{10.A26}$$

When $\alpha\rho \gg V\xi\cdot\varepsilon\sigma D_3/(\omega+v\xi L)$, the equation is identical to the equation for a basic two-host life cycle, eqn 10.A9. As the second term in the brackets becomes increasingly large, however, it becomes more advantageous to pursue the more complex sequential three-host life cycle.

The inclusion of two or more species of definitive host can be explored in a similar way, the algebraic details are essentially similar (Dobson, unpublished).

11

Regulation and dynamics of acanthocephalan populations

C. R. Kennedy

11.1 Introduction

Acanthocephalans are in many respects particularly suitable for studies of population dynamics, since the life history generally only comprises three infrapopulations: the egg, the larval stages in the arthropod intermediate hosts and the adult in the principal and auxiliary definitive hosts. Transfer of the egg to the arthropod and from the intermediate to the definitive host is always passive, by ingestion, and only the egg stage is free-living. Furthermore, the parasites only reproduce sexually, so that infrapopulations within a host can only increase by immigration.

Despite the suitability of the group and the number of investigations into life cycles and seasonality of occurrence, the amount of useful data in relation to regulation and dynamics of acanthocephalan populations is rather limited. Many host–acanthocephalan interactions have been studied in depth, but regulatory mechanisms have been considered relatively infrequently. This is due (1) to the problems of sampling all the stages in a life cycle simultaneously, (2) to the majority of ecological studies having concentrated primarily on a single host species, (3) to a shortage of data on transmission dynamics from one infrapopulation to another, and (4) to the majority of field investigations having been conducted in freshwater habitats. The full implications of these limitations will become apparent, but because of them the approach adopted in the following account has been to review the data on population dynamics in each host population separately, and then to consider regulation in suprapopulations in the light of current ideas on regulation and stability of parasite populations.

11.2 Theoretical background to studies on regulation and stability of parasites

The major recent advances in understanding the regulation and stability of parasite populations have originated not only from new information as such but also, and primarily, from the development of a conceptual framework into which new data can be fitted and in the light of which existing data can be reinterpreted.

Stability of a parasite population can only be achieved by the operation of density-dependent factors functioning as regulatory processes. The pioneer and classic study of Crofton (1971*a*, *b*) on regulation of an acanthocephalan population showed how the host and parasite populations could interact and achieve dynamic stability provided that the parasite was overdispersed throughout its host population and attained a level of infection at which it was lethal to its host. By viewing stability in a wider context, Bradley (1974) demonstrated that regulation could occur at the level of the host individual or the host population and he recognized three ways in which parasite population levels could be determined and/or regulated: by transmission (Type I); at the host population level, by parasite pathogenesis and subsequent density-dependent parasite-induced host mortality in association with an overdispersed frequency distribution and by complete host immunity to reinfection (Type II); and at the host individual level, by immunity to superinfection or by intraspecific competition resulting in density-dependent parasite fecundity or mortality (Type III). Transmission rates (Type I) contribute to establishing population levels but do not regulate them: the population thus determined will be unstable unless Type II or Type III mechanisms are operating on it in addition. Only these mechanisms can regulate populations and produce stability.

The most comprehensive and recent studies on regulation and stability are those of Anderson & May (1978, 1979) and May & Anderson (1978, 1979). They recognized that stabilizing and destabilizing processes generally operate simultaneously on parasite populations. They identified the principal stabilizing processes as being overdispersion, parasite-induced host mortality when there was a non-linear relation between parasite burden and host death rate, and density-dependent constraints on parasite population growth, specifically density-dependent parasite reproduction and mortality achieved through intraspecific competition or host immune responses. The principal destabilizing factors are parasite-induced reduction in the host's reproductive potential, parasite reproduction within the host (which can be ignored with reference to the exclusively sexual Acanthocephala), and time lags in the development of transmission stages. Parasite-induced host mortality is thus viewed as a series of graded responses, and not in the light of a determined lethal level. Anderson & May considered that there is normally a degree of tension between stabilizing and destabilizing processes in populations, and that destabilizing influences are often counteracted by equally strong stabilizing processes. Thus, where overdispersion is low and distribution close to random, compensation may occur in the form of strong density-dependent parasite survival. Similarly, parasite inhibition of host reproduction or long time lags in the system may be compensated by overdispersion.

On theoretical grounds, Anderson (1976) had previously shown that density-dependent regulation at any stage in the life cycle or within any single infrapopulation was capable of regulating the whole suprapopulation. The terms infrapopulation and suprapopulation are used here as originally defined by Esch *et al.* (1975): all individuals of a single parasite species within an individual host are regarded as an infrapopulation, whereas all individuals of the same parasite species in all stages of development within all hosts in an ecosystem constitute a suprapopulation. Holmes *et al.* (1977) also demonstrated that a single regulatory mechanism in a single infrapopulation in one host species was sufficient to regulate the whole suprapopulation, no matter how many other species harboured infections, provided the flow through the single host species was adequate. In other words, only a single regulatory mechanism is needed to regulate the whole suprapopulation, and this can operate in any host. These considerations are particularly relevant to acanthocephalan populations because they are frequently found in several species of vertebrate (Holmes *et al.*, 1977; Hine & Kennedy, 1974*a*; Amin & Burrows, 1977) in addition to their principal (or required *sensu* Holmes, 1979) definitive hosts.

11.3 Regulation and dynamics of acanthocephalan infrapopulations

11.3.1 *Free-living stages*

The only free-living stage in the life cycle of acanthocephalans is the egg or shelled acanthor, but it is probably the least studied of all the stages. Virtually nothing is known about the pattern of egg dispersal in natural habitats, and very little about the dynamics of transmission to the intermediate host.

Fully developed eggs are infective to arthropods immediately upon leaving the parent without needing a period of further development (Parshad & Crompton, 1981). The general situation appears to be that egg survival is influenced primarily by the abiotic factors of temperature and moisture. From the limited data available, and by analogy with cestodes and other parasite groups (Keymer & Anderson, 1979; Keymer, 1980), it seems likely that acanthocephalan egg survival and infectivity are age-dependent processes, as they in turn depend upon the endogenous and non-renewable food reserves in the egg, and that natural mortality will increase exponentially with age. The best strategy would seem to be that of maximizing survival in the early part of the lifespan of the egg when infectivity is greatest. Mortality rates and age-dependent infectivity will thus probably prove to be important determinants of the efficiency of transmission of eggs to the intermediate host, but they are both density-independent processes and so are not involved in regulation of population sizes. No field study appears to have been carried out on arthropod feeding patterns in relation to the ingestion of acanthocephalan eggs, so nothing is known about the natural dynamics of transmission to the arthropods.

Even the data available on survival are very limited. Survival of the eggs of *Macracanthorhynchus hirudinaceus*, which is a parasite of terrestrial organisms, is primarily dependent upon the abiotic factors of temperature and moisture (Kates, 1942). Freezing temperatures, high temperatures (above 45 °C) and drying all reduce survival: at 37 °C eggs can survive for up to 368 days, at 5 °C for up to 551 days, and in soil for up to 3.5 years.

Table 11.1. *Longevity of some acanthocephalan life cycle stages*

Species	Survival time (months)		Survival time (days)		
	Egg	Cyst-acanth	Adult	Host	Reference
Echinorhynchus truttae	9	7.5	85	Fish	Awachie (1966)
Acanthocephalus clavula	?	3	48	Fish	Rojanapaibul (1977)
Acanthocephalus dirus	9	9	?	Fish	Seidenberg (1973)
Polymorphus minutus	2	15	50	Bird	Hynes & Nicholas (1963); Crompton & Whitfield (1968*b*)
Mediorhynchus centurorum	?	6	120	Bird	Nickol (1977)
Moniliformis moniliformis	⩾ 3	2–5	144	Mammal	Crompton *et al.* (1972); Lackie (1972*a*, *b*)
Macracanthorhynchus hirudinaceus	18	⩾ 2	c. 365	Mammal	Kates (1942, 1944)

Data on eggs and larval stages are average times at normal habitat temperatures, but will vary according to temperature (Parshad & Crompton, 1981). Data on adults are average times for females; male lifespans are generally shorter

However, both high and low temperatures also reduce infectivity, which also declines with age even under favourable conditions. In contrast, eggs of *Polymorphus minutus*, which is a parasite of aquatic organisms, are not as resistant to changes in physical conditions, but are very susceptible to desiccation and are much shorter lived than those of *M. hirudinaceus*. Infectivity declines particularly between 9 and 16 days at 17 °C, and is lost after 3–4 weeks (Hynes & Nicholas, 1957, 1963). The eggs survive longer (2.5 months) at lower temperatures (−22 °C), but infectivity is very low. Crompton & Whitfield (1968*a*) reported eggs as surviving rather longer, for about 2 months, but confirmed that infectivity declines with age and less favourable conditions. Crompton (1970) concluded that acanthocephalan eggs usually survive for about 6–9 months and further information is given in Table 11.1.

Extended survival times would introduce time lags into the parasite system, especially in species such as *Macracanthorhynchus hirudinaceus*, but if most infections of the intermediate host took place in the early part of the lifespan of the egg, such time lags would be very short and thus would not be important destabilizing factors. Extended survival appears to be of more significance in relation to transmission (see Chapter 9) by enabling the population to overwinter or survive other unfavourable conditions.

11.3.2 *Acanthocephalans in intermediate hosts*

General pattern of population dynamics. Successful infection of an intermediate host requires not only compatible host and parasite species but also compatibility between parasite strain and host species. Acanthocephalans in one location are very specific for their intermediate host, although they may use different species of intermediate host in different parts of their range (Kennedy *et al.*, 1978). Furthermore, different strains of parasite may be specific to different species of host. Larvae of *Polymorphus minutus* originating from adults derived from *Gammarus pulex* can only infect *G. pulex*, not *G. lacustris*, and *vice versa* (Hynes & Nicholas, 1958). Parasites in the wrong species of host are killed in a short time (2–3 weeks) by a host response, but in the right species the host response has no obvious effect upon them and may even protect them (Crompton, 1975; Uznanski & Nickol, 1980*a*, *b*). No data are available on the proportion of eggs lost by being ingested by the wrong host in natural conditions.

In general terms, infection levels of acanthocephalan larval stages in their intermediate hosts will be dependent upon the rates of ingestion of eggs, upon the host effects upon the parasites and upon the parasite effects upon the hosts. It can be assumed that the efficiency and rate of infection of arthropods will depend primarily upon the density of eggs and intermediate hosts and upon the feeding behaviour of the latter. The data of Lackie (1972*a*) tend to support this assumption. His experimental investigation of the relation between exposure density of eggs of *Moniliformis moniliformis* and resultant parasite burden in the cockroach, *Periplaneta americana*, revealed a curvilinear relation between the two variables, with parasite burden reaching a plateau rather than increasing indefinitely (Fig. 11.1). The data are indicative of a functional response on the part of the host to changes in the density of eggs, similar to that observed in infections of intermediate hosts by cestode eggs (Keymer & Anderson, 1979; Keymer, 1980). The plateau is thought to result either from host saturation, or from constraints imposed by the time required by the host to handle and consume the eggs (Keymer, 1982).

Where maturation of the adult acanthocephalan and/or egg production and release are seasonal, transmission to the intermediate host is likely to take place only at restricted times of the year (Kennedy, 1977; Muzzall, 1978; Camp & Huizinga, 1980; Liat & Pike, 1980), due to the limited survival time and infectivity of the eggs. Seasonal population cycles in larval acanthocephalans may also be due to seasonal changes in the population composition of the intermediate host itself, and in particular to seasonal appearance of young arthropods and to seasonal mortality of adult arthropods (Awachie, 1965; Seidenberg, 1973; Muzzall, 1978; Amin *et al.*, 1980). If, however, conditions are such that eggs are long lived and are produced and shed throughout the year (Rumpus, 1973; Hine & Kennedy, 1974*b*), transmission can potentially take place at all times of the year and there should be no seasonal cycle in population levels in the intermediate host.

Fig. 11.1. Functional response in the relation between infective stage exposure density and the resultant parasite burden of the exposed host population. Transmission of *Moniliformis moniliformis* to the intermediate host *Periplaneta americana*. The line indicates the predictions of a functional response model. (Based on data from Lackie, 1972*a*, and reproduced from Keymer, 1982, by permission of the author.)

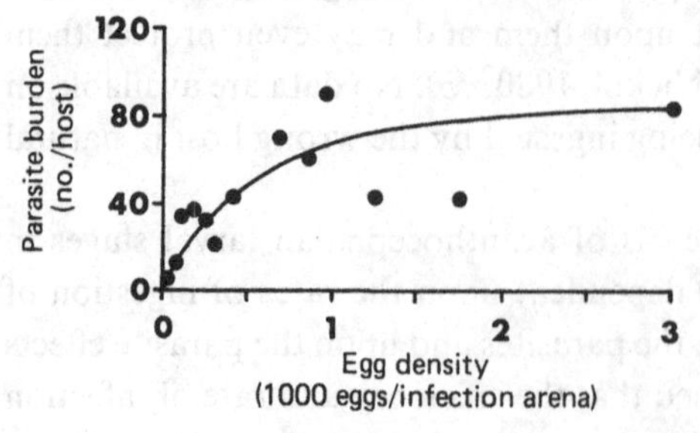

In some cases hosts of all ages are susceptible to infection (Hynes & Nicholas, 1963), and so infection levels increase with size and age of host as exposure time to infection increases (e.g. Liat & Pike, 1980). Infection levels, however, often decline in the largest or oldest hosts, as a result of parasite-induced reduction in host growth rate and/or death of older infected hosts, so that the highest infection levels are frequently found in hosts of medium size and age (Rumpus, 1973; Muzzall, 1978) (Fig. 11.2). Rates of growth and development of the parasite (and of the intermediate host) are temperature dependent and vary between species (Parshad & Crompton, 1981; Tokeson & Holmes, 1982) and so the parasite's larval span is to a large extent determined by the climatic conditions in the habitat. In some species, e.g. *Polymorphus marilis*, low temperatures in winter induce a facultative diapause in the cycle; larval stages overwinter in a dormant condition in the intermediate host, thus introducing time lags into the system (Tokeson & Holmes, 1982).

Once fully developed, the cystacanth serves as a resting stage, seldom dying within the host but surviving as long as its host, generally for a period

Fig. 11.2. The distribution of *Pomphorhynchus laevis* in *Gammarus pulex*. (*a*) Incidence of *P. laevis* in *G. pulex* of different sizes. (*b*) Changes in intensity of infection in relation to size of host. (Reproduced from Hine & Kennedy, 1974*b*, by permission of the authors and the Fisheries Society of the British Isles.)

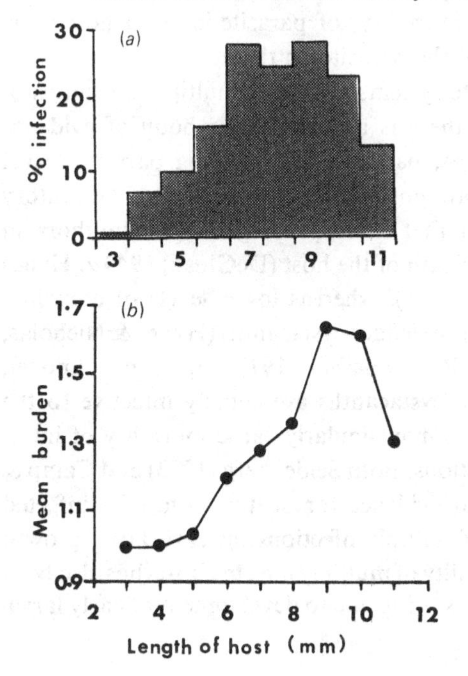

of several months to a year depending on the host species (Table 11.1). Multiple infections by a single species (e.g. Hynes & Nicholas, 1958; Rojanapaibul, 1977) or mixed species of parasite (Awachie, 1967) are possible, and the parasites are nearly always overdispersed throughout the host population (Crofton, 1971*a*; Rumpus, 1973; Hirsch, 1980). Prevalence levels in intermediate hosts in natural populations are very variable, ranging from 60% (Seidenberg, 1973) to less than 1% (Amin, 1978*b*) depending on parasite species and locality, and even between 80% and 3% for the same species at different sites within a single locality (Hine & Kennedy, 1974*a*).

Operation of regulatory processes. Although Lackie (1972*a*) demonstrated a functional response on the part of the intermediate host *Periplaneta americana* to changes in density of eggs of *Moniliformis moniliformis* under experimental infection conditions, there is no indication that such a response would play any significant role in the regulation of a naturally occurring population (Keymer, 1982). Lackie (1972*a*) found no evidence of further regulatory processes acting after ingestion, and indeed although many aspects of this host–parasite system have been studied in depth both by him and by others, relatively little is known about the possible existence of any regulatory mechanisms. Superinfections of cockroaches are possible, and there is no evidence of competition between parasites for resources, of density-dependent parasite mortality, of parasite-induced host mortality, or of any other effects of the parasite on the host.

In several other host–parasite systems, although multiple infections of individual hosts are possible, there is a considerable body of evidence suggesting that heavy infections, particularly by young parasite larval stages, are detrimental to both host and parasite. Under laboratory conditions it has been shown that mass penetration of acanthors in arthropods often results in the death of the host (DeGiusti, 1949*a*; Hynes & Nicholas, 1958; Crompton, 1975), whereas lower levels of crowding result in smaller, stunted and more slender cystacanths (Hynes & Nicholas, 1957, 1963; Awachie, 1966; Rojanapaibul, 1977). It is not known, however, whether such stunted cystacanths are equally infective to the definitive host. Heavy infections may similarly cause mortality of intermediate hosts under field conditions. Both Seidenberg (1973) and Camp & Huizinga (1980) have provided evidence for deaths of heavily infected *Asellus intermedius* as a result of multiple infections and early larval growth of *Acanthocephalus dirus*. Mortality of multiply infected hosts has also been reported by Amin *et al.* (1980) as being due to development of early larval stages of *A. dirus*. Such parasite-induced host mortality is due to overcrowding in the early developmental stages of the parasite and is density dependent, but it must be distinguished from host mortality associated with the presence of fully developed cystacanths. Since the host death occurs when the acanthocephalans are in the early stages of development and so are unable to infect the definitive host, the dead hosts and their contained parasites are lost from the system completely.

As the acanthocephalan larvae grow and develop, they may also affect intermediate host growth and reproduction. Reductions in growth rates of intermediate hosts have been reported for *Polymorphus minutus* by Hynes & Nicholas (1958, 1963), for *Pomphorhynchus laevis* by Rumpus (1973), for *Echinorhynchus truttae* by Awachie (1966) and for *Acanthocephalus dirus* by Oetinger & Nickol (1981). This will in turn lead to a loss in intermediate host production, and in one locality where prevalences of *P. laevis* in *Gammarus pulex* were around 12%, this reduction was calculated as being of the order of 11.7% (Rumpus, 1973). Respiration of arthropods may also be affected by the presence of cystacanths (Rumpus & Kennedy, 1974).

Many, though by no means all, larval acanthocephalans are also capable of reducing the reproductive capacity of their intermediate hosts. The extent of this reduction varies from species to species, ranging from rendering females completely sterile, through reduced fecundity, to having no detectable effect. *Acanthocephalus lucii* and *A. dirus* render female isopods completely sterile (Seidenberg, 1973; Muzzall & Rabalais, 1975*b*; Brattey, 1980; Oetinger & Nickol, 1981). Infected females contain no ovarian tissue, are never found in amplexus and never carry eggs, whereas the reproduction of male isopods is completely unaffected by the presence of the parasites. Larvae of *Polymorphus minutus* also interfere with the reproductive activity of female, but not male, *Gammarus pulex* and *G. lacustris*: they prevent immature females from becoming mature, and mature females from developing ovaries and mating (Hynes, 1955; Hynes & Nicholas, 1958, 1963; Crompton & Harrison, 1965). Sterilization is not, however, always complete, and Hynes (1955) reported that 10% of infected *G. lacustris* females contained eggs. *Pomphorhynchus laevis* also reduces the reproductive capacity of some females, yet has no effect (Rumpus, 1973) upon male *G. pulex*. It has a detrimental effect upon host reproduction but does not render all females completely sterile: in a locality where parasite prevalence was 26%, Rumpus (1973) calculated that only 2.3% of the total number of female *G. pulex* were actually affected by sterility. By contrast, some species of Acanthocephala, including

Echinorhynchus truttae and *Leptorhynchoides thecatus*, appear to have no effect upon the fertility and reproductive capacity of their intermediate hosts (Awachie, 1966; Uznanski & Nickol, 1980).

Cystacanths of many acanthocephalan species also affect either the pigmentation and/or the behaviour of their intermediate hosts in such a way as to render the infected individuals more vulnerable to predation (see Chapter 9). Change in pigmentation of infected arthropods has been reported as being caused by, amongst other species, *Acanthocephalus lucii* and *A. dirus* by Seidenberg (1973), Camp & Huizinga (1979, 1980), Brattey (1980) and Oetinger & Nickol (1981), and the whole topic has been reviewed by Oetinger & Nickol (1981). They concluded that the acanthocephalan infection is the only factor responsible for the altered pigmentation, and that it increases the vulnerability of infected individuals to predation. Some acanthocephalans, including *Polymorphus minutus*, *P. marilis*, *P. paradoxus*, *Pomphorhynchus laevis* and *A. dirus* alter the behaviour, especially the escape responses, of their intermediate hosts (Bethel & Holmes, 1973, 1977; Holmes & Bethel, 1972; Hindsbo, 1972; Muzzall & Rabalais, 1975*c*; Kennedy *et al.*, 1978; Camp & Huizinga, 1979). Other species, however, show no effects on host behaviour or pigmentation (Awachie, 1966; Bethel & Holmes, 1973). Induced behavioural changes (1) appear in all cases to be produced by the presence of cystacanths, not earlier stages, (2) appear to be due to the presence of a single cystacanth not to multiple infections, and (3) appear to render infected individuals more vulnerable to predation. Increased vulnerability is due primarily to behavioural changes induced by the parasite rather than to increased conspicuousness resulting from altered pigmentation. The changed behaviour facilitates transmission to the principal definitive host but may also expose the infected intermediate host to predation by other species of vertebrate, in some of which the parasite will fail to develop. Such losses will be inevitable and may be tolerable, and may well represent a trade-off against the increased probability of transmission to the principal definitive host. However, if habitat conditions are such that the proportion of parasites failing to infect the definitive host(s) becomes too high, there may well be a significant loss of parasites from the system.

The regulatory consequences of acanthocephalan-induced intermediate host mortality were first examined in detail by Crofton (1971*a*, *b*). Using the field data collected by Hynes & Nicholas (1963) on the distribution of *Polymorphus minutus* in *Gammarus pulex* at several sites along a small stream, he found that the parasite population was overdispersed (variance greater than the mean); at the stations closest to the source of the infection, where the parasite's population levels were highest, the distribution could best be described by the truncated negative binomial model (Fig. 11.3). This was interpreted as indicating that there was a lethal level of the parasite, equivalent to the truncation point, above which the parasites induced host mortality. Host death was thus related to the intensity of the infection and this, in Crofton's view, prevented the contained parasites from completing their life cycle. Since some hosts normally acquired a lethal level of infection, and the proportion that did so was related to overall infection levels, Crofton believed that the parasite population could be regulated by this mechanism. In his view, the higher the infection level, the greater the degree of overdispersion and so the greater the proportion of parasites removed from the system by the death of the most heavily infected hosts.

This interpretation, however, was challenged by Hirsch (1980), who used the same data as Crofton and additional data from two other stations on the same stream. He attempted to fit five theoretical frequency distributions to the data, and then tested the goodness of fit to each model, which Crofton had not done. Hirsch found (Table 11.2) that the negative binomial model provided the best description of the frequency distributions at all stations except two, and that the data from only two stations could be best described by the *truncated* negative binomial. He concluded that the intensity of truncation was not related to the intensity of exposure and thus to the infection level, and that therefore Crofton's assumption that truncated distributions arose from parasite-induced host mortality seemed

Fig. 11.3. The frequency distribution of *Polymorphus minutus* in a population of *Gammarus pulex*. Solid circles indicate predicted values assuming a negative binomial distribution. (Based on data from Crofton, 1971*a*.)

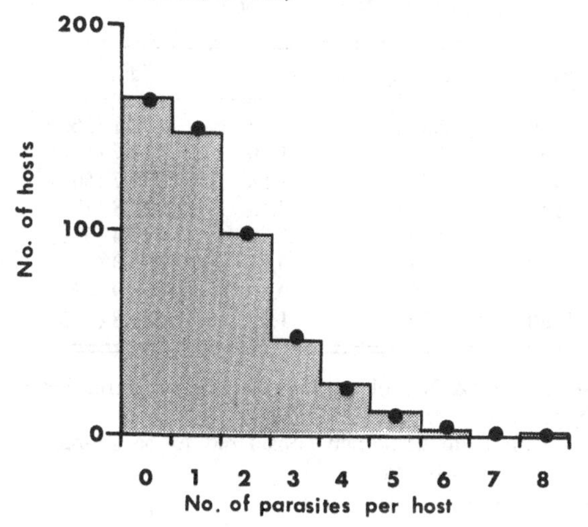

unlikely. In Hirsch's view, alternative explanations were possible: for example, if the probability of *Gammarus pulex* containing a heavy infection of *Polymorphus minutus* increased with the age of the arthropod, and if the probability of death of *G. pulex* increased with its age, then a truncated distribution would result but host mortality would not have been parasite-induced.

Other investigators have also observed that the overdispersion of cystacanths in their intermediate hosts can be described by the truncated negative binomial model, but there is no consensus of opinion as to whether it results from parasite-induced host mortality nor on the role of this as a regulatory agent. Rumpus (1973) found that the distribution of *Pomphorhynchus laevis* in a population of *Gammarus pulex* could best be described by a truncated negative binomial model but considered that, in view of the low prevalence of the parasite and the paucity of multiple infections, this mechanism was unlikely to have any regulatory effects on the parasite population. By contrast, Amin *et al.* (1980) found that the distribution of *Acanthocephalus dirus* in its intermediate host could best be described by a truncated negative binomial model in spring only, when the cystacanth population was at its peak and mortality of arthropods greatest. They suggested therefore that parasite-induced host mortality amongst heavily infected arthropods could be regulating the parasite population. However, the explanation advanced by Hirsch (1980) to

Table 11.2. *Characteristics of negative binomial distributions generated for data from Hynes & Nicholas (1963) on* Polymorphus minutus *in* Gammarus pulex. *Chi-square values are from tests of overdispersion using the variance/mean ratio. Data from Hirsch (1980)*

Station[a]	Location[b]	Year	Mean	k	χ^2 (df)
4	−50	1959	0.38	0.33	699.1 (360)[c]
5 (1)	10	1959	2.43	1.10	1719.7 (554)[c]
6 (2)	250	1959	1.42	1.58	931.7 (508)[c]
7 (4)	800	1959	1.67	32.48	425.8 (386)
8	2000	1959	0.76	−23.98	527.4 (532)
9 (6)	2500	1959	0.27	0.61	274.8 (190)[c]
6 (3)	250	1960	0.60	0.30	1628.9 (632)[c]
7 (5)	800	1960	0.89	1.27	460.8 (275)[c]

[a] Station numbers are those of Hynes & Nicholas (1963). Numbers in parentheses are station numbers used by Crofton (1971*a*)
[b] Location in yards downstream from duck pen considered to be source of infection. Negative value is distance upstream
[c] Significantly overdispersed ($p < 0.05$)

account for the *P. minutus–G. pulex* situation would appear to be equally applicable to this case.

The whole field of parasite-induced host mortality and its regulatory consequences has recently been examined in great detail by Uznanski & Nickol (1980), both in general terms and with reference to a particular system. In a thorough experimental laboratory study of *Leptorhynchoides thecatus* in *Hyalella azteca* they found no evidence at all of lethal or sublethal effects of the parasite on its host (Fig. 11.4), and concluded that the parasite population was not being regulated by mortality of heavily infected intermediate hosts. There was no negative effect of infection intensity on host or parasite growth and no significant difference in survival between infected and uninfected arthropods: mortality of arthropods was not directly related to the infection process nor to parasite intensity levels. The distribution of the parasite, however, could be described by the negative binomial model, even though it did not induce host mortality. Furthermore, Uznanski & Nickol were unable to detect any host response that was effective against the parasite. They concluded that whereas death of an abnormally heavily infected host individual did occasionally occur, this could not be taken as evidence of regular parasite-induced host mortality. They suggested, indeed, that much of the evidence invoked in other studies to support parasite-induced host mortality was inadequate. Truncated negative binomials could be generated by other

Fig. 11.4. Survival of *Hyalella azteca* after exposure to *Leptorhynchoides thecatus* eggs in the laboratory. Triangles indicate unexposed amphipods: solid circles indicate exposed amphipods and common points. (Reproduced from Uznanski and Nickol, 1980, by permission of the authors and the American Society of Parasitologists.)

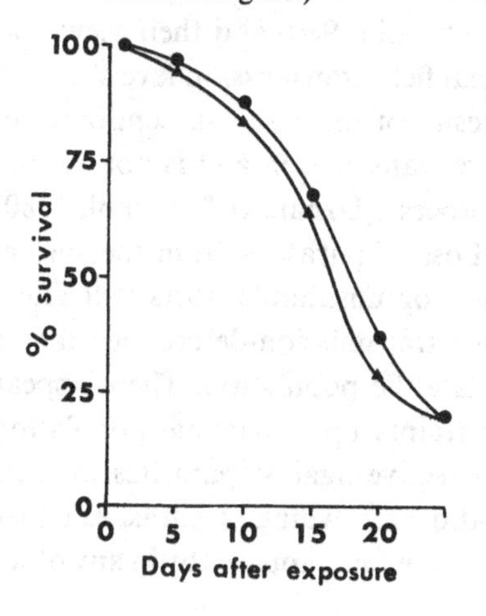

means, and the link between parasites and host death was frequently unproven. As they also emphasized, predation upon and death through predation of infected intermediate hosts is necessary for continuation of the parasite's life cycle. If parasite-induced host mortality is regulating population levels, then it is necessary to show that death of the infected hosts terminates the cycle by death of the parasite before it infects the definitive host. Neither Crofton nor other workers have shown this to be the case. Uznanski & Nickol also believed that many of the laboratory studies that demonstrated parasite-induced host mortality had involved selected hosts and abnormally high and hence lethal parasite levels. They concluded that death of infected host individuals could and did occur, but that the mechanism of this was often selective predation: the effect of such predation was to facilitate transfer and assist transmission to the definitive host. There was, therefore, little, if any, concrete evidence for regulation of acanthocephalan infrapopulations in their intermediate hosts by parasite-induced host mortality at transmission levels normally prevailing in the field.

Conclusions. When viewed in the context of regulatory processes, acanthocephalan infrapopulations in their intermediate hosts exhibit a variety of stabilizing and destabilizing factors such as were identified and predicted by Anderson & May (1978) and May & Anderson (1978) from their studies on theoretical models. Considering the stabilizing factors first, all populations appear to be overdispersed and this can be described by the simple or truncated negative binomial models. There is some evidence of parasite-induced host mortality due to heavy infections of early developmental stages resulting in host death and consequent loss of contained parasites from the system, but this must be evaluated with reference to the reservations expressed by Uznanski & Nickol (1980) and their view that such mortality may be unusual at normal field transmission levels. Death of hosts containing cystacanths as a result of increased susceptibility to predation appears primarily to facilitate transmission and is not density dependent, and so is not a regulatory process (Uznanski & Nickol, 1980) despite earlier views to the contrary. Loss of parasites from the system through selective predation to auxiliary or unsuitable hosts will affect population levels, but it appears to be a transmission-determined factor (Bradley Type I) and so may not regulate the population. There appear also to be few density-dependent constraints upon parasite population growth, in that host responses are ineffective against parasites in their required hosts and heavy infections and overcrowding of parasites either kill the host or result in smaller cystacanths but not apparently in any other effects on the parasite. Destabilizing factors are also present in many populations. The principal one appears to be parasite-induced reduction in host reproductive potential. In addition, because the cystacanth serves as a resting stage it can introduce time lags into the system, although these, including winter diapause, appear to be relatively short compared to the time lags in some cestode and digenean cycles which can extend over several years.

In none of the populations studied is there any real evidence of compensation such as May & Anderson (1978) suggested might occur, in that, for example, inhibition of host reproduction is not counteracted by marked overdispersion. Both stabilizing and destabilizing processes can be demonstrated in many acanthocephalan infrapopulations, but there is not as yet sufficient evidence to indicate whether the populations are actually regulated. Results from laboratory studies cannot indicate what actually happens in the field, and the majority of field studies have continued for only a relatively short term, generally 1 or 2 years at the most, which is not a sufficiently long period to detect major changes in population levels. In some populations there is no evidence for the operation of regulatory or destabilizing processes, other than overdispersion. In these populations especially, although it may well also be the case in all the populations studied, it can be presumed that transmission factors, which appear to be density-independent and so non-regulatory, must play a major role in determining infrapopulation levels.

11.3.3 *Acanthocephalans in definitive hosts*

General pattern of population dynamics in fish hosts. Although there have been a large number of studies on acanthocephalan infrapopulation dynamics in fish hosts, the great majority were concerned primarily with seasonal dynamics, particularly seasonal cycles in infection incidence and intensity and their causal factors. Few of these investigations provided data that are relevant in the context of regulatory processes. Seasonal dynamics have also been the subject of recent reviews (Kennedy, 1977; Chubb, 1982) and because they are discussed in the context of transmission in Chapter 9 they will not be considered in detail here. Generally the situation appears to be that the majority of acanthocephalan infrapopulations in fish are in a state of dynamic equilibrium between recruitment and loss of parasites, and there is consensus that the major determinants of population levels are water temperature (acting directly or indirectly), availability of infected hosts and infective larvae, and host diet and feeding behaviour. When any or all of these factors vary seasonally, seasonal cycles in infection levels may result. The dynamics of transmission between intermediate host and fish have never been studied in any real quantitative detail but it would seem likely that the efficiency and rates of transmission will depend primarily upon the densities of the cystacanths, intermediate and definitive hosts and the dietary behaviour of the latter. Although many parasite species are known to be able to employ paratenic hosts in their life cycles, there appear to be no quantitative ecological data available on the dynamics of acanthocephalan populations in such hosts.

Once in their fish hosts, acanthocephalans are seldom long lived (Table 11.1) and indeed the whole life cycle (all hosts included) of the majority of species infecting fish appears generally to be less than 1 year (Awachie, 1966; Rojanapaibul, 1977; Kulachkova & Timofeeva, 1977; Camp & Huizinga, 1980; Amin *et al.*, 1980); development and, frequently, survival rates appear to be temperature dependent. The majority of parasites that successfully establish in fish probably survive for their normal lifespan, that of males being less than that of females (Parshad & Crompton, 1981). Information on acanthocephalan fecundity and the factors affecting it in fish is very sparse. There is no actual evidence that acanthocephalan fecundity declines in heavy infections, as is commonly observed with cestodes, and indeed there is every possibility that it may rise since the probability of mating will increase with population density. All acanthocephalans appear to exhibit habitat specificity within the alimentary tract to some degree. The site range is extended at high population densities

Fig. 11.5. The distribution of *Pomphorhynchus laevis* in the alimentary tract of goldfish one week after experimental infections with different parasite burdens. (Reproduced from Kennedy, Broughton & Hine, 1976, by permission of the authors.)

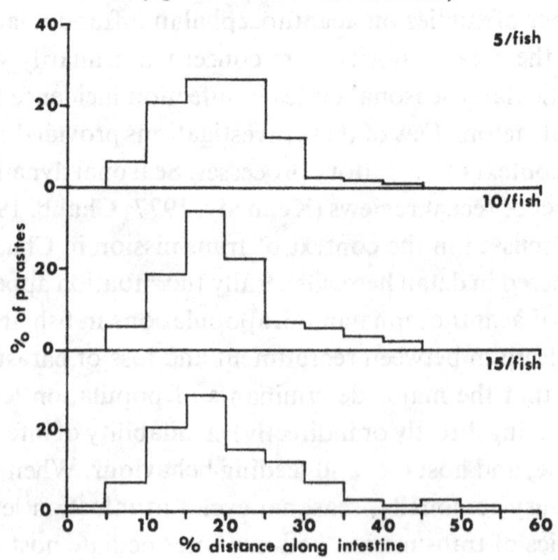

(Fig. 11.5) (Uglem & Beck, 1972; Crompton, 1973; Kennedy *et al.*, 1976; Kennedy & Lord, 1982), but acanthocephalans are not displaced from their optimum site by parasites of other phyla: rather the reverse (Holmes, 1961, 1962*a*; Chappell, 1969*a*; Amin, 1975*c*; Grey & Hayunga, 1980). Many species may cause considerable local damage to the host intestine (e.g. Bullock, 1963; Chaicharn & Bullock, 1967; Hine & Kennedy, 1974*a*), but they seldom cause host mortality, although occasional deaths of heavily infected fish have been recorded (Hentschel, 1979; Pennycuick, 1971*a*). Apart from local inflammatory and tissue responses, host reactions appear to be ineffective against the parasites (Harris, 1972).

Although the majority of studies have been concerned with seasonal dynamics, a few authors have investigated the establishment and course of infection in fish by acanthocephalans under field and laboratory situations specifically in the context of the operation of regulatory processes. The following account will therefore concentrate only on these investigations as other studies contribute little data of relevance to the present topic.

Operation of regulatory processes in fish hosts. Acanthocephalans of fish are generally overdispersed throughout their host population (Pennycuick, 1971*c*; Hine & Kennedy, 1974*b*) and the negative binomial often provides a suitable descriptive model.

Awachie (1965, 1966, 1972*a*, *b*) investigated in considerable detail the population dynamics of *Echinorhynchus truttae* in *Salmo trutta*, in both the field and laboratory. He demonstrated that the infrapopulation in fish was in a state of dynamic equilibrium, with a seasonal cycle in intensity of infection but not in incidence. Changes in the infrapopulations in intermediate and definitive hosts were complementary, and populations in the fish were related to seasonal variation in fish feeding activity, composition of diet and water temperature. The parasite was able to establish in fish at all temperatures examined, but the rate of establishment was higher at cold temperatures and rate of loss greater at warm temperatures. In the course of an infection, parasites moved down the intestine and males were lost first. The parasite infrapopulation was overdispersed and superimposed infections were possible. Doubling the experimental infection levels from 15 to 30 parasites per fish had no effect upon parasite establishment or survival. At higher levels of infection there was no evidence of a reduction in the rate of parasite growth but the probability of copulation was increased. Starvation of the fish did not produce any loss of parasites and the only factor that appeared to influence parasite expulsion was elevated water temperature. There was no apparent immunity to reinfection; fish were unable to mount any effective response against the acanthocephalan and the parasite caused no damage, either local or general, to its host. When a secondary infection was superimposed upon a primary one, there was again no decline in parasite establishment or evidence of stunted growth. There thus appear to be no mechanisms capable of preventing superinfections, and in this species population levels appear to be limited by availability of infective cystacanths, host feeding behaviour and transmission factors. It is an example of a Bradley Type I situation, where parasite density is determined by density-independent factors and the population is unregulated.

Infrapopulations of *Acanthocephalus clavula* in fish also appear to be in a state of dynamic equilibrium. In localities in which there are seasonal changes in fish diet and feeding behaviour and in the availability of the intermediate host it may exhibit seasonal cycles in incidence and intensity of infection (Pennycuick, 1971*b*), but elsewhere, and especially when it is in its principal ('required' *sensu* Holmes, 1979) definitive host, *Anguilla anguilla*, recruitment and maturation take place throughout the year and there is no seasonal cycle in population levels (Chubb, 1964; Kennedy, unpublished). The parasite is overdispersed throughout the eel population but appears to have no effect upon condition or mortality of eels; the eel cannot mount an effective response against the parasite (Kennedy, unpublished), and neither host starvation nor water temperature affect the rate of loss of parasites (Kennedy & Lord, 1982). In eels, therefore, there appear to be no mechanisms operative that are capable of regulating the size of the parasite population. Thus, population size appears to be determined solely by density-independent transmission factors. In three-spined sticklebacks, *Gasterosteus aculeatus*, however, which are not the principal but are auxiliary definitive hosts and which are much smaller than eels, heavy infections can reduce the condition factor of the fish and may have the potential to kill them (Pennycuick, 1971*a, c*). Consequently Pennycuick suggested that the parasite population could be regulated by density-dependent parasite-induced host mortality. The evidence for this, however, was not unequivocal, as interpretation of the data was complicated by the presence of the cestode *Schistocephalus solidus* in the same stickleback population, and this species is known to cause host mortality by rendering the fish more vulnerable to avian predation. It appears possible, however, that some *A. clavula*–host systems may be regulated by parasite-induced host mortality in a Bradley Type II manner.

Uznanski & Nickol (1982) investigated the population dynamics of *Leptorhynchoides thecatus* in its natural definitive host *Lepomis cyanellus* by feeding the fish different densities of cystacanths in the laboratory.

There was no host response to parasites in their normal site in the intestine, but parenteral individuals were encapsulated and destroyed. The distribution of parasites in the intestine was similar at all exposure levels. At low dosage levels, 80% of the parasites established and there was no decline in parasite numbers over the next 2 weeks. At high levels (40 cystacanths per fish), however, both percentage recovery and number of parasites recovered declined between 3 and 14 days postinfection to levels similar to those in fish fed 20 cystacanths (Fig. 11.6). Uznanski & Nickol concluded that activation of the parasite and site selection were density-independent processes, but that establishment and survival were density dependent. *Lepomis* appeared to provide resources for only 10 to 15 parasites: below this limit all parasites survived, but above it excesses were lost. This thus appears to be an example of regulation by spatial or other resource limitation in individual hosts (Bradley Type III) around a ceiling level, restricting the maximum numbers of established adults. It is therefore the first demonstrable experimental example of such density-dependent regulation of acanthocephalan infrapopulations in fish.

The population dynamics of *Echinorhynchus salmonis* in all its hosts has been studied in considerable detail by Holmes *et al.* (1977) in a single lake, Cold Lake, in Alberta, Canada. This parasite dominates the lake parasite fauna, comprising 50% of all the parasites, and occurs in 10 species of host. It only develops fully, however, in salmonids (Chapter 7), and the whitefish, *Coregonus clupeaformis*, was identified as the principal host (Table 11.3) since it harboured 80% of all the gravid females and the flow

Fig. 11.6. Mean number of *Leptorhynchoides thecatus* recovered from *Lepomis cyanellus* fed 10, 20 and 40 cystacanths. (Reproduced from Uznanski & Nickol, 1982, by permission of the authors and the American Society of Parasitologists.)

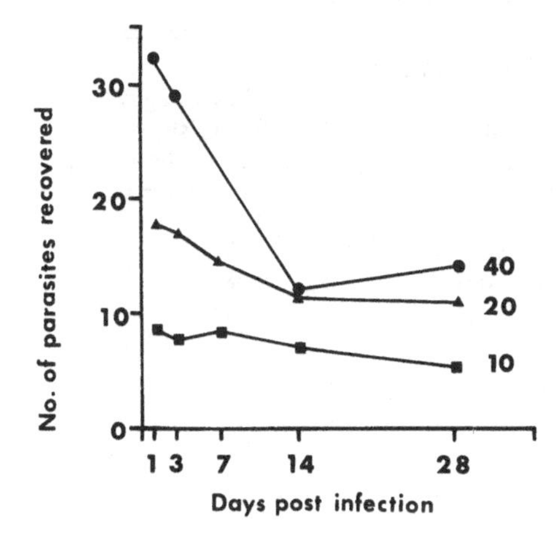

through this species was sufficient to support the suprapopulation (Holmes *et al.*, 1977; Leong & Holmes, 1981). The authors identified several regulatory mechanisms operating on the acanthocephalan infrapopulations in the different species of fish. In whitefish they found that the mean number of gravid female parasites per fish remained constant, despite an increase in the number of parasites per fish at different seasons and as the fish aged, and that there was a significant negative regression of the percentage of gravid females on the number of acanthocephalans in individual fish. They concluded therefore that infrapopulations of *E. salmonis* in whitefish were regulated by density-dependent parasite maturation (a Bradley Type III mechanism). In cisco, *C. artedii*, however, parasite populations were unregulated, and levels of infection were determined by transmission density-independent factors (Type I). In lake trout, *Salvelinus namaycush*, they found a negative regression of the percentage of gravid female acanthocephalans on the total number of cestodes of the genus *Eubothrium* present, and, to a lesser degree of significance, on the total number of acanthocephalans. Again this was considered to be evidence of Type III regulation. In coho, *Oncorhynchus kisutch*, population levels appeared to be determined primarily by transmission (Type I) factors, although there was some evidence for the death of heavily infected hosts and Type II regulation. Infection levels in all other hosts were unregulated and determined by transmission. The authors did not attempt to verify their conclusions experimentally, but the field data certainly suggested that different regulatory mechanisms were operating in the four different species of salmonid.

Table 11.3. *Simplified scheme of derived relative flow rates for* Echinorhynchus salmonis *in a community of fishes in Cold Lake, Alberta, Canada. Data from Holmes, Hobbs & Leong* (1977)

Fish host	Flow to fish by ingestion of infected intermediate hosts	Flow of eggs from fish to intermediate host
Whitefish (*Coregonus clupeaformis*)	0.702	0.518
Cisco (*Coregonus artedii*)	0.11	0.37
Lake trout (*Salvelinus namaycush*)	0.051	0.084
Coho (*Oncorhynchus kisutch*)	0.005	0.018
Pike (*Esox lucius*) Burbot (*Lota lota*) Suckers (*Catostomus* spp.) Walleye (*Stizostedion vitreum*) Stickleback (*Pungitius pungitius*)	0.131	0.011

The flow of parasites, however, was undoubtedly greatest through whitefish (Table 11.3), so the authors constructed a model to test whether regulation of the infrapopulation in this host alone would be sufficient to regulate the whole suprapopulation. They demonstrated that this was in fact the case, but only if egg production from the non-regulated part of the suprapopulation was low and less than production from the regulated infrapopulation, i.e. the total flow through the unregulated infrapopulations had to be less than the flow through the regulated one. Adding regulation in other species had little effect upon the parasite suprapopulation size, but increased the range of densities within which the population could be regulated. If the flow through the required host was small, the regulatory mechanisms had to be very efficient and *vice versa*, but the precise mechanism of regulation was not very important. Thus, stability of the parasite suprapopulation was possible by a continuing, as opposed to a ceiling, regulatory process. They further emphasized that their results demonstrated that parasite infrapopulations could exhibit different population dynamics and regulatory mechanisms in different species of host, and so great care should be taken in interpreting field and laboratory data from studies that may not reveal the mechanisms operating in nature and in practice.

The warning is a salutary one, as in Lake Michigan the flow pattern of *Echinorhynchus salmonis* is rather different (Amin & Burrows, 1977). In this lake it infects 14 species of fish, maturing in all of them, including the non-salmonid species: in contrast to the situation in Cold Lake, the flow of parasites through the different host species is more even, and the principal hosts appear to be coho, *Oncorhynchus kisutch*, and smelt, *Osmerus mordax*. In Lake Ontario (Tedla & Fernando, 1970), perch, *Perca fluviatilis*, and in the Baltic Sea (Valtonen, 1980), whitefish, *Coregonus nasus*, appear to be principal hosts. In none of these localities, however, have the population dynamics of the parasite been studied in all the species of host and in sufficient detail to determine whether regulatory processes are operating, and to enable a comparison to be made with Cold Lake.

In the River Avon, England, *Pomphorhynchus laevis* also occurred in all the species of fish in the locality, but in no species could any effect on host growth or mortality be demonstrated (Hine & Kennedy, 1974*a*, *b*). The parasite infrapopulations were in a state of dynamic equilibrium and did not exhibit any obvious seasonal cycle in incidence or intensity of infection (Hine & Kennedy, 1974*b*) (Fig. 11.7). Field data suggested that the principal determinants of infection levels were host diet and feeding behaviour, and that infrapopulation sizes were dependent primarily upon transmission processes. Laboratory studies showed that superimposed infections were possible and that at high densities of infection there was no reduction of parasite establishment in dace, *Leuciscus leuciscus*, an auxiliary host, (Kennedy, 1972) and only a very slight reduction in rainbow trout, *Salmo gairdneri*, a principal host (Broughton, 1974). The rate of parasite establishment in dace was higher at low temperatures: at higher temperatures establishment rates were lower, but in the field this was compensated for by increased intensity of fish feeding and so a greater rate of ingestion (Kennedy, 1972). Thus, recruitment and loss of parasites balanced out over the year and resulted in a non-seasonal pattern of prevalence (Hine & Kennedy, 1974*b*). Establishment and survival of the parasites in both dace and rainbow trout in the laboratory appeared to be independent of the size of the infection and the number of parasites already present and therefore to be density independent (Kennedy, 1972, 1974). No effect of parasite density upon the rate of parasite growth could be demonstrated at the densities employed in the experimental infections, nor was there any significant relation between the proportion of gravid females and the density of infection in the principal host species in the field. Harris (1972) showed that the principal hosts produced antibodies against the parasite but that these antibodies were ineffective. It was therefore concluded that there was no density-dependent mortality of parasites in auxiliary host species and in one species of principal host (Broughton, 1974; Kennedy, 1974). Long-term field data, however, indicated that infection levels had remained stable over a period of 9 years (Table 11.4),

Fig. 11.7. Seasonal changes in the incidence of *Pomphorhynchus laevis* in the River Avon. (*a*) Adult parasites in dace. (*b*) Cystacanths in *Gammarus pulex*. (Reproduced from Hine & Kennedy, 1974*b*, by permission of the authors and the Fisheries Society of the British Isles.)

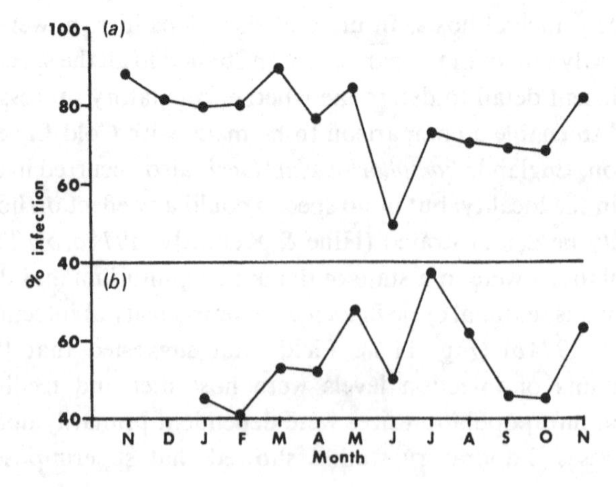

and thus strongly suggested that regulatory processes were operating (Kennedy & Rumpus, 1977). A possible mechanism for such regulation was suggested by Kennedy (1977), based on habitat specificity of the parasite within the host intestine. *P. laevis* can grow outside the optimum site but attains a smaller size (Kennedy *et al.*, 1976). It was suggested that in crowded infections more of the parasites inhabit less-favourable sites in the intestine, these parasites are smaller and so the average size of the parasite and, it is therefore presumed, its fecundity decline. The only supporting evidence is that in heavy infections of the principal hosts in the field there is a higher ratio of small: large parasites than in lighter infections and the average parasite size is reduced. The suggestion is therefore that regulation could occur through competition for favoured space (a Type III mechanism); experimental verification of this hypothesis has still to be undertaken.

General pattern of population dynamics in bird and mammal hosts. There are so few studies of acanthocephalan infrapopulations in natural popula-

Table 11.4. *Long-term changes in the size of the population of* Pomphorhynchus laevis *in the River Avon. Data from Kennedy & Rumpus (1977)*

	Levels in *Gammarus pulex*		Levels in *Leuciscus leuciscus*	
	Mean prevalence	Mean intensity	Mean prevalence	Mean intensity
Annual changes				
1966	No sample taken		74%	7.06
1967	19.2%	1.5	81%	9.6
1968	18.7%	1.4	79%	9.3
1969	20.1%	1.6	72%	6.4
1970	19.8%	1.3	No sample taken	
1971	16.7%	1.5	No sample taken	
1972	17.6%	1.5	No sample taken	
1973	17.1%	1.4	78%	5.1
1974	18.0%	1.4	75%	4.3
Monthly changes				
1966	No sample taken		max. 100.0%	12.7
			min. 53.2%	3.5
1967	—	—	max. 90.4%	18.0
			min. 40.9%	3.3
1968	max. 36.9%	1.8	—	—
	min. 14.1%	1.2		
1969	max. 24.8%	2.1	—	—
	min. 4.8%	1.0		

tions of birds and mammals that it is difficult to produce any general statements about the patterns of population dynamics. It would seem very likely that general considerations applicable to acanthocephalans of fish would also apply to these natural populations: these may again be in a state of dynamic equilibrium, and seasonal cycles could be determined by similar factors, although the parasites are often longer lived in homeothermic hosts (Table 11.1) and temperature will correspondingly have less effect.

Operation of regulatory processes in bird and mammal hosts. Both a dynamic equilibrium and seasonality of occurrence have been reported for *Polymorphus botulus* in eider ducks by Liat & Pike (1980). The parasite occurred more frequently and in larger numbers in juvenile ducks. Infection levels showed seasonal peaks, due primarily to a seasonal peak in availability of cystacanths which coincided with the developing infections in young eiders. The time of minimum availability of intermediate hosts and the lowest level of infection coincided with the drop in infection rates in the adult definitive host. The parasite is believed to be short lived and there was a natural regular loss of parasites from the birds, accentuated at times of stress and host starvation. The authors considered that a regular low level of parasite infection was sufficient to maintain the observed infrapopulation levels and that this continual low intake and continual natural loss of a parasite with a short lifespan would result in a dynamic equilibrium. The decline in levels in older ducks was attributed to a change in diet or increased host resistance. Population levels were thus considered to be determined largely by transmission factors.

Natural populations of domestic ducks can apparently acquire infections of *Polymorphus minutus* in all seasons, and any seasonality of occurrence seems to be due to the effects of temperature on larval growth and to the seasonal dietary and migratory behaviour of the ducks (Nicholas & Hynes, 1958; Hynes & Nicholas, 1963). Ducks can occasionally be found harbouring high levels of infection, but the numbers of parasites per host have to be very heavy to cause host disease or death. In such heavy infections, parasites may be smaller but there was no evidence of reduced fecundity, and in experimental infections there was no decrease in the rate of establishment of primary infections at high parasite densities. However, establishment of secondary infections was partially inhibited by the presence of a primary infection. It appears that the parasite exhibits a strong site preference within the intestine (Crompton & Whitfield, 1968*a*) and that parasites establishing outside the optimum site fail to survive or are more likely to be expelled. Thus parasites in secondary infections frequently find the optimum site occupied by acanthocephalans of the primary infection, and so competition for a favoured site causes a decline in the numbers establishing. Those individuals of a secondary infection that are able to establish in the optimum site develop normally (Nicholas & Hynes, 1958). If ducks are exposed to continual infections, this reduction in establishment keeps the infrapopulation low, and only when a bird is exposed to a new, massive infection is it damaged or killed. Natural loss of parasites may be accentuated by starvation of the hosts (Crompton, 1975), but the natural earlier loss of males and short lifespan of the females (Table 11.1) ensures a regular turnover of parasites and results in a dynamic equilibrium. Egg production does not take place at a constant rate but reaches a peak: the time of copulation may be delayed by the distribution of male and female parasites in the intestine and the period of egg release is variable and depends on the number of females present (Crompton & Whitfield, 1968*a*). It has not yet been demonstrated, however, that fecundity is density dependent, and the only regulatory process so far identified is competition for favoured space (a Type III mechanism).

The only species of acanthocephalan with a mammalian host whose population dynamics have been studied in any real detail, and then only in the laboratory, is *Moniliformis moniliformis*. There is, even so, an absence of basic quantitative data on many aspects of its biology such as parasite-induced host mortality and pattern of distribution in its host population, and density-dependent regulatory factors have seldom been the subject of specific investigation. Despite individual host differences in susceptibility, possibly genetically based, the majority of parasites in primary infections establish and, after an initial period of loss, numbers remain constant until about the seventh week (Burlingame & Chandler, 1941). At this time most males are lost, the loss of females commencing after 10 weeks. Some of this loss is due to the failure of parasites to establish in their optimum site, but within the optimum site growth of individuals appears to be unaffected by parasite density. Establishment of parasites is unaffected by host diet but, in general terms, parasite survival, growth and reproduction appear to be proportional to the amount of available glucose and other carbohydrates in the host diet (Nesheim *et al.*, 1977, 1978). In rats fed on a normal diet, most parasite deaths appear to be due to natural senescence following reproduction (Crompton & Walters, 1972; Crompton *et al.*, 1972; Crompton *et al.*, 1976). The time and duration of egg release depend on a number of factors (Parshad & Crompton, 1981) but fecundity has not yet been shown to be density-dependent (frequently demonstrated in cestode infections). In concurrent infections, acanthocephalans appear to be superior competitors. They are smaller but no effect on their fecundity was observed (Holmes, 1962). The acanthocephalans do not retard the growth or affect the state of nutrition of their host, nor do they compete for nutrients with their host except under special circumstances (Nesheim *et al.*, 1978). Thus, although overcrowding and poor host nutrition could lead to a decline in acanthocephalan fecundity, such a combination of circumstances appears to be rare. In superimposed infections, parasites from the primary infection are not affected and there is initially no decline in the rate of establishment of the secondary infection. However, only a proportion of the parasites, those for which there is space, establish in the optimum site. The majority of secondary parasites, especially if the primary parasites are large and mature, establish outside the optimum site and are soon lost (Burlingame & Chandler, 1941). Those in the optimum site survive and grow at the normal rate. The number of parasites in an individual rat is thus related to the size of the optimum site, is determined by competition for attachment sites (space) and varies between host individuals. Infection levels in individual hosts are therefore regulated in a density-dependent manner around a ceiling level by spatial or other resource limitation in a fashion similar to regulation in infrapopulations of *Leptorhynchoides thecatus*.

Andreassen (1975*a, b*) repeated the experiments of Burlingame & Chandler (1941) but using greater levels of infection with *Moniliformis moniliformis*. At levels above 100 cystacanths per rat, the rate of recovery of parasites was very low (10–20%) at 8 weeks. The loss of parasites occurred between weeks 4 and 8 when establishment fell from 85% to 15%. At lower infection levels expulsion of parasites occurred later. Expulsion was prevented by administration of cortisone to the rats, when 85% of the parasites survived to eight weeks and the total mass of the parasite burden was higher than in untreated controls. Rats given 100 cystacanths in a primary infection, dosed with anthelmintic drugs and then challenged with a secondary infection, showed a lower recovery and retarded growth of secondary parasites. Parasites in secondary infections were always smaller and lighter and in a more posterior site than in primary infections. Andreassen (1975*a, b*) was also able to demonstrate the presence of reaginic antibodies in association with the host rejection, and so concluded that there was some evidence for an effective host immune response to the parasite. The situation is thus very similar to that for the cestode *Hymenolepis diminuta* in rats, where both intraspecific competition and host immune responses have been suggested as explanations of the crowding effect. In infections of *M. moniliformis*, the immune response may occur in addition to the competition for space, or may be a more correct explanation for the data of Burlingame & Chandler (1941), but in either case there is clear evidence of Type III density-dependent regulation of the parasite.

Conclusions. Under field conditions, infrapopulation sizes of acanthocephalans in their definitive hosts, whether fish, bird or mammal, appear to be influenced primarily by density-independent transmission factors. The most important of these appear to be availability of cystacanths, host diet and feeding behaviour, densities of intermediate and definitive hosts, and, especially in fish, water temperature.

It is clear, however, particularly from laboratory studies, that regulatory mechanisms do exist in many definitive host–parasite infrapopulations. Most of the parasite populations studied are overdispersed throughout their host population. Parasite-induced host mortality occasionally happens and when it does it is regulatory in that the parasites are removed from the system, but it appears in general to be of very rare occurrence as the great majority of acanthocephalans are not pathogenic at the levels of infection that occur in nature. However, density-dependent constraints on parasite population growth occur far more frequently than they do in intermediate host–parasite systems. Such constraints may take the form of host immune responses, as in *Moniliformis moniliformis*, or competition for limited resources of space around a ceiling level, as in *Leptorhynchoides thecatus*, *M. moniliformis*, *Polymorphus minutus* and possibly *Pomphorhynchus laevis*, or of density-dependent parasite maturation and fecundity, as in *Echinorhynchus salmonis* in whitefish. All these are Type III mechanisms, and only for *Acanthocephalus clavula* in its auxiliary host and, possibly, for *E. salmonis* in coho, has it been suggested that regulation may involve a Type II response. Destabilizing processes, in contrast, are far less common in definitive host–parasite systems when compared with intermediate host–parasite systems. There is no evidence that the reproduction of any species of definitive host is affected by its acanthocephalan parasites, and the short lifespans of the parasites ensure that time lags are of little importance. It appears therefore that regulatory processes are fairly common in acanthocephalan–definitive host systems.

It is not clear, however, to what extent these regulatory processes actually operate in natural, as opposed to laboratory, situations. The most convincing evidence of their operation and effectiveness has generally come from laboratory investigations, where infection levels are usually higher than in the field. It may therefore be that at transmission rates normally encountered in the field, regulatory factors do not operate, but when transmission efficiency increases for any of a variety of reasons the mechanisms for regulation are in existence and able to come into operation.

This may explain why some investigations, such as those of *Echinorhynchus truttae* and *Acanthocephalus clavula* in eels, provide no evidence of the existence or functioning of any regulatory processes: at the time of the investigations the infection levels were just too low.

11.4 Regulation and dynamics of acanthocephalan suprapopulations

Investigations into the regulation and dynamics of acanthocephalan infrapopulations have revealed both the existence and method of operation of regulatory processes in many systems, but it is both relevant and essential to enquire whether in fact suprapopulations of acanthocephalans are stable and regulated in natural situations. Mathematical models have revealed that a single regulatory process operating at any point in the life cycle and in any host infrapopulation may be sufficient to produce stability (Anderson, 1976; Holmes *et al.*, 1977). The process is likely to be most effective if it operates on the infrapopulation in the principal definitive or intermediate host, but it could operate instead, or in addition, in auxiliary hosts. The existence of more than one mechanism in the same or a different species of host will increase the effectiveness of the regulation (Holmes *et al.*, 1977). These, however, are theoretical considerations. There is actually very little evidence of acanthocephalan suprapopulation stability, largely because of a dearth of information since few field studies have extended over more than a year or 2 years at the most.

Positive evidence of suprapopulation stability has been obtained for only one species of acanthocephalan: *Pomphorhynchus laevis* in the River Avon. Over a period of 9 years, infrapopulation levels remained stable in both the intermediate host, *Gammarus pulex*, and in an auxiliary definitive host species (Kennedy & Rumpus, 1977), and variation in infection levels between years was less than the variation within any single year (Table 11.4). Over the same period, the pattern of changes along the length of the river also remained constant, as did the pattern of frequency distribution in the auxiliary host population. Small annual changes in population levels were noted, but they did not fit into any pattern. The authors considered that it was valid to accept these criteria as a reasonable indication of the size and stability of the *P. laevis* suprapopulation in the river since: (1) *G. pulex* was the only intermediate host at this locality, and any changes in infrapopulation size in any species of definitive host or in suprapopulation size would have been reflected in changes in infection levels in the intermediate host; and (2) although only the auxiliary definitive host species, *Leuciscus leuciscus*, was examined, there was no reason to assume that the proportion of the parasite population located in dace had changed or that the dace infrapopulation had undergone changes in size independent of changes in other species of fish. Kennedy & Rumpus therefore concluded that the suprapopulation of *P. laevis* at this locality was stable, and that such stability implied the existence of regulatory mechanisms: the paradox is that no regulatory process has actually yet been demonstrated to be operating on any of the infrapopulations in the river, although the existence of a Type III mechanism has been postulated.

Dunagan & Miller (1980) have also provided evidence for stability of *Macracanthorhynchus hirudinaceus* over a period of 18 years, although their study was not confined to a single locality but related to a restricted geographical area. They found that some local populations had clearly declined in some localities, but over the wider area studied population levels appeared to have remained stable over the period. It is not clear, however, whether there were several suprapopulations or only a single suprapopulation within their area of study and whether they were therefore really considering global or suprapopulation stability: in either case, they provided no information on the processes responsible for the observed stability.

There is, on the other hand, some clear evidence that suprapopulation levels may change considerably from year to year. Several authors have noted that epidemics of acanthocephalans can occur (Liat & Pike, 1980; Hynes & Nicholas, 1963; Chapter 9), but although this suggests that transmission rates have risen and/or regulatory mechanisms have broken down, the precise details, circumstances and reasons are seldom, if ever, known. Hynes & Nicholas (1963), for example, suggested that the major population changes in *Polymorphus minutus* are due to movements of ducks between localities, which results in the decline of existing populations and the establishment of new ones, sometimes to epidemic levels if a population of ducks encounters massive parasite infections for the first time. There is, however, no real evidence in support of this suggestion. Hynes (1955) also concluded that *P. minutus* could be responsible for the disappearance of populations of *Gammarus lacustris*, and so also of the parasite, from some localities. Again, however, no such disappearance was documented in detail.

Annual changes in acanthocephalan infrapopulations in fish, however, are rather better documented. Pennycuick (1971*b*) recorded a steady, though inexplicable, decline in the population of *Acanthocephalus clavula* in one locality over a period of 2 years. Large changes in suprapopulations of *A. clavula* over short periods have also been recorded. In one locality, Slapton Ley, Devon, infection levels in perch, *Perca fluviatilis*, declined from an incidence of 42.1% and a mean of 1.89 parasites per fish to values of 7.3% and 0.146, respectively, over a period of only 12 months. In a separate locality, the River Clyst, Devon, the population has been studied continuously over a period of 3 years. At the start of the investigation in 1979, incidence and intensity of infection in eels, *Anguilla anguilla* were 62% and 2.8 parasites per fish, respectively. By 1980, incidence had fallen to 10% and intensity to 0.15, and by 1982 the respective figures were 4% and 0.06 (Fig. 11.8). Over the same period the distribution of the parasite had changed from being overdispersed to being nearly random. The change was due to a decline in recruitment in the fish, but no regulatory factor has yet been identified. There was no competition between parasites, the parasites had no effect upon the host, and there was no effective host response against the parasite. Kennedy & Lord (1982) have demonstrated that there is no competition for limited resources of space as the parasite survives and reproduces throughout the whole length of the intestinal tract (Kennedy & Lord, 1982). The pattern of decline and the apparent absence of any regulatory processes would suggest that this population may be unregulated and unstable.

Thus, on the basis of the very limited data available, it would appear that in some localities some acanthocephalan suprapopulations are stable and regulated but that in others they may not be. It must be emphasized, however, that there is still far too little evidence on which to base any generalizations: only two examples of long-term acanthocephalan suprapopulation survival have actually been documented, and there is as yet no recorded evidence of local suprapopulation extinction.

11.5 General conclusions

In many respects the major conclusion to be drawn from this review is that there are insufficient appropriate data to generate general

Fig. 11.8. Annual changes in the incidence of infection of *Acanthocephalus clavula* in eels in the River Clyst.

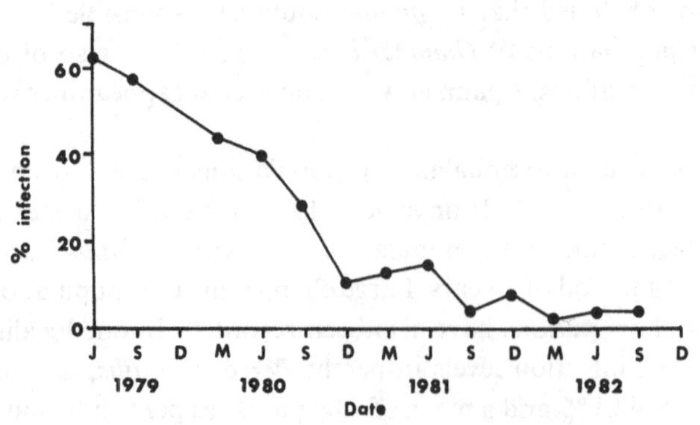

principles. The review has shown that several different regulatory processes have been identified in acanthocephalan populations, in both intermediate and definitive host infrapopulations. These processes are all potentially capable of stabilizing suprapopulations, but there is in fact little or no concrete evidence that they do so. No regulatory process has yet been identified as operating upon *Echinorhynchus truttae* in either host or upon *Acanthocephalus clavula* in eels or its intermediate host, yet the former population appears to be stable and the latter unstable. Regulatory processes have been identified as operating on *E. salmonis* in its fish hosts, both principal and auxiliary, but there is no concomitant evidence of stability of the natural population since this was not the aim of the investigation and the system was not studied over a sufficiently long period. Regulatory processes have similarly been identified in infrapopulations of *Leptorhynchoides thecatus*, *Moniliformis moniliformis* and *Polymorphus minutus*, but the mechanisms were identified in laboratory populations and there are no contemporary field investigations that demonstrate population stability. Field data, in fact, indicate that despite the theoretical and experimental demonstrations of regulatory processes, infection levels are probably determined primarily by density-independent transmission rates, although regulatory mechanisms may come into operation at high parasite densities. It is probably true, for example, that any species of acanthocephalan is potentially able to induce host mortality at sufficiently high density and under certain conditions, but the extent to which this ever occurs naturally and whether it acts in a density-dependent manner in natural infections are unknown. The data available suggest that if it occurs it is relatively infrequent. In many of the acanthocephalan populations destabilizing processes can also be identified, but again there are not sufficient data to determine whether these dominate and result in unstable populations or whether there is compensation for them in the form of regulatory mechanisms. On the other hand, there exists the paradox of stability being shown to exist in suprapopulations of *Pomphorhynchus laevis* and *Macracanthorhynchus hirudinaceus*, but with no experimental evidence of the stabilizing processes responsible.

Such a situation is clearly unsatisfactory, but the present review does provide some explanations, and justifications, for this, and indications for future work. In the first place, many studies have been undertaken with other objectives in mind, and data pertinent to population regulation have been provided almost incidentally. Secondly, there are far too few studies on acanthocephalan infrapopulations in all their hosts at the same time: too many studies have been restricted to a single species of host, and the parasite in its other hosts has been ignored. As Holmes *et al.* (1977) pointed out, interpretation of such studies must be viewed with caution because both population dynamics and regulatory mechanisms may vary in different hosts. When the parasite infects hosts of several species, it is essential to know the flow through each species. Thirdly, the most obvious explanations of situations may not be the correct ones. Hirsch (1980) and Uznanski & Nickol (1980) have shown that a truncated negative binomial model may not always be the *best* fit to the data, and that even when it is, it does not necessarily imply parasite-induced host mortality since other explanations are generally possible. Fourthly, there are no studies on egg populations, and few detailed quantitative studies on the dynamics of transmission from egg to intermediate host, or from intermediate to definitive host. Fifthly, there is a paucity of information on many aspects of acanthocephalan biology crucial to an understanding of regulatory processes, in particular on acanthocephalan fecundity and its relation to parasite infection levels. Finally, there are far too few long-term studies on any species in a single habitat. What are clearly needed now are investigations directed specifically towards identifying and quantifying potential density-dependent factors, to solving particular problems and to obtaining the right quality and quantity of information for this. Until the results of studies directed and integrated towards these ends are available, the situation is likely to remain unclear.

References

Abele, L.G. & Gilchrist, S. (1977). Homosexual rape and sexual selection in acanthocephalan worms. *Science*, **197**, 81–3.

Acholonu, A.D. (1976). Helminth fauna of saurians from Puerto Rico with observations on the life cycle of *Lueheia inscripta* (Westrumb, 1821) and description of *Allopharynx puertoricensis* sp. n. *Proceedings of the Helminthological Society of Washington*, **43**, 106–16.

Acholonu, A.D. & Finn, O.J. (1974). *Moniliformis moniliformis* (Acanthocephala: Moniliformidae) in the cockroach *Periplaneta americana* in Puerto Rico. *Transactions of the American Microscopical Society*, **93**, 141–2.

Ahearn, G.A. (1982). Water and solute transport by crustacean gastrointestinal tract. In *Membrane Physiology of Invertebrates*, ed. R.B. Podesta, pp. 261–337. New York: Marcel Dekker, Inc.

Al-Rawas, A.Y., Mirza, M.Y., Shafig, M.A. & Al-Kindy, L. (1977). First finding of *Moniliformis moniliformis* (Bremser, 1811) Travassos, 1915 (Acanthocephala: Oligacanthorhynchidae) in Iraq from a human child. *Journal of Parasitology*, **63**, 396–7.

Amin, O.M. (1975*a*). *Acanthocephalus parksidei* sp. n. (Acanthocephala: Echinorhynchidae) from Wisconsin fishes. *Journal of Parasitology*, **61**, 301–6.

Amin, O.M. (1975*b*). Variability in *Acanthocephalus parksidei*, Amin, 1974 (Acanthocephala: Echinorhynchidae). *Journal of Parasitology*, **61**, 307–17.

Amin, O.M. (1975*c*). Host and seasonal associations of *Acanthocephalus parksidei* Amin, 1974 (Acanthocephala: Echinorhynchidae) in Wisconsin fishes. *Journal of Parasitology*, **61**, 318–29.

Amin, O.M. (1978*a*). Effect of host spawning on *Echinorhynchus salmonis* Müller, 1784 (Acanthocephala: Echinorhynchidae) maturation and localization. *Journal of Fish Diseases*, **1**, 195–7.

Amin, O.M. (1978*b*). On the crustacean hosts of larval acanthocephalan and cestode parasites in south-western Lake Michigan, U.S.A. *Journal of Parasitology*, **64**, 842–5.

Amin, O.M. (1980). *Fessisentis tichiganensis* sp. nov. (Acanthocephala: Fessisentidae) from Wisconsin fishes, with a key to species. *Journal of Parasitology*, **66**, 1039–45.

Amin, O.M. (1981). The seasonal distribution of *Echinorhynchus salmonis* (Acanthocephala: Echinorhynchidae) among rainbow smelt, *Osmerus mordax* Mitchell, in Lake Michigan. *Journal of Fish Biology*, **19**, 467–74.

Amin, O.M. (1982). Description of larval *Acanthocephalus parksidei* Amin, 1975 (Acanthocephala: Echinorhynchidae) from its isopod intermediate host. *Proceedings of the Helminthological Society of Washington*, **49**, 235–45.

Amin, O.M. & Burrows, J.M. (1977). Host and seasonal associations of *Echinorhynchus salmonis* (Acanthocephala: Echinorhynchidae) in Lake Michigan fishes. *Journal of the Fisheries Research Board of Canada*, **34**, 325–31.

Amin, O.M., Burns, L.A. & Redlin, M.J. (1980). The ecology of *Acanthocephalus parksidei* (Acanthocephala: Echinorhynchidae) in its isopod intermediate host. *Proceedings of the Helminthological Society of Washington*, **47**, 37–46.

Anantaraman, S. & Ravindranath, M.H. (1973). Chemical nature of the hooks of the cystacanth of *Moniliformis moniliformis*. *Acta Histochemica*, **47**, 124–31.

Anantaraman, S. & Ravindranath, M.H. (1976). Histochemical characteristics of the egg envelopes of *Acanthosentis* sp. (Acanthocephala). *Zeitschrift für Parasitenkunde*, **48**, 227–38.

Anantaraman, S. & Subramoniam, T. (1975). Oogenesis in *Acanthosentis oligospinus* n. sp., an acanthocephalan parasite of the fish, *Macrones gulio*. *Proceedings of the Indian Academy of Sciences*, **82B**, 139–45.

Andersen, K. (1978). The helminths in the gut of perch (*Perca fluviatilis* L.) in a small oligotrophic lake in southern Norway. *Zeitschrift für Parasitenkunde*, **56**, 17–27.

Anderson, R.M. (1976). Dynamic aspects of parasite population ecology. In *Ecological Aspects of Parasitology*, ed. C.R. Kennedy, pp. 431–62. Amsterdam: North Holland Publishing Company.

Anderson, R.M. (1978). The regulation of host population growth by parasitic species. *Parasitology*, **76**, 119–57.

Anderson, R.M. (1980). The dynamics and control of direct life-cycle helminth parasites. *Lecture Notes in Biomathematics*, **39**, 278–322.

Anderson, R.M. (1982*a*). Epidemiology. In *Modern Parasitology*, ed. F.E.G. Cox, pp. 204–51. Oxford & London: Blackwell Scientific Publications.

Anderson, R.M. (1982*b*). Hookworm and roundworm infections. In *Population Dynamics of Infectious Diseases*, ed. R.M. Anderson, pp. 67–108. London: Chapman and Hall.

Anderson, R.M. Trematoda. In *Reproductive Biology of Invertebrates*, vol. 6, *Asexual Propagation and Reproductive Strategies*, ed. K.G. Adiyodi and R.G. Adiyodi. Chichester, New York, Brisbane, Toronto, Singapore: John Wiley. (In press.)

Anderson, R.M. & May, R.M. (1978). Regulation and stability of host–parasite population interactions. I. Regulatory processes. *Journal of Animal Ecology*, **47**, 219–47.

Anderson, R.M. & May, R.M. (1979). Population biology of infectious diseases: I. *Nature, London*, **280**, 361–7.

Anderson, R.M. & May, R.M. (1982). Population dynamics of human helminth infections: control by chemotherapy. *Nature, London*, **297**, 557–63.

Andreassen, J. (1975*a*). Immunity to the acanthocephalan *Moniliformis dubius* infections in rats. *Norwegian Journal of Zoology*, **23**, part 3, 195–6.

Andreassen, J. (1975*b*). Reaginic antibodies in response to *Moniliformis dubius* infections in rats. *Norwegian Journal of Zoology*, **23**, part 3, 196.

Andres, A. (1878). Uber den Weiblichen Geschelechtsapparat d. *Echinorhynchus gigas* Rudolphi – Ein Beitrag zur Anatomie der Acanthocephalen. *Morphologisches Jahrbuch*, **4**, 584–91.

Andryuk, L.V. (1974). The development of *Acanthocephalus lucii* (Muller, 1776), Lühe, 1911 (Echinorhynchidae) in the intermediate host. *Byulleten Vsesoyuznogo Instituta Gelmintologiya im K.K. Skryabina*, **13**, 9–13.

Andryuk, L.V. (1976). The infection of *Asellus aquaticus* with *Acanthocephalus* larvae in the Upper Dniepr Basin (USSR). (Abstract) In *Vsesoyuznyi Simpozium po bolenzyam i parazitam vodnykh bespozvonochnykh*, Leningrad, 28–30 January 1976. Tezisy doklador. Leningrad: Izdatel'stvo 'Nauka'.

Andryuk, L.V. (1979*a*). The cycle of development of *Acanthocephalus anguillae*. *Zoologichesky Zhurnal*, **63**, 168–74.

Andryuk, L.V. (1979*b*). The life-cycle of *Acanthocephalus lucii* (Echinorhynchidae). *Parazitologiya*, **13**, 530–9.

Andryuk, L.V. (1979*c*). On the life-cycle of acanthocephalans from the genus *Acanthocephalus* Koelreuther. *Doklady Timiryazevskoi Sel'skokhozyaistvennoi Akademii*, **225**, 101–4.

Anvar, A.K. & Paran, T.P. (1976). *Periplaneta americana* (L.) as intermediate host of *Moniliformis moniliformis* (Bremser) in Penang, Malaysia. *Southeast Asian Journal of Tropical Medicine and Public Health*, **7**, 415–16.

Anya, A.O. (1976). Physiological aspects of reproduction in nematodes. *Advances in Parasitology*, **14**, 267–351.

Arai, H.P. (1980). Migratory activity and related phenomena in *Hymenolepis diminuta*. In *Biology of the Tapeworm* Hymenolepis diminuta, ed. H.P. Arai, pp. 615–37. London and New York: Academic Press.

Asaolu, S.O. (1977). *Studies on the reproductive biology of Moniliformis (Acanthocephala)*. Ph D dissertation, University of Cambridge.

Asaolu, S.O. (1980). Morphology of the reproductive system of female *Moniliformis dubius* (Acanthocephala). *Parasitology*, **81**, 433–46.

Asaolu, S.O. (1981). Morphology of the reproductive system of male *Moniliformis dubius* (Acanthocephala). *Parasitology*, **82**, 297–309.

Asaolu, S.O., Whitfield, P.J., Crompton, D.W.T. & Maxwell, L. (1981). Observations on the development of ovarian balls of *Moniliformis* (Acanthocephala). *Parasitology*, **83**, 23–32.

Atkinson, K.H. & Byram, J.E. (1976). The structure of the ovarian ball and oogenesis in *Moniliformis dubius* (Acanthocephala). *Journal of Morphology*, **148**, 391–426.

Atrashkevich, G.I. (1975*a*). A contribution to the study of the life history of *Arhythmorhynchus comptus* Van Cleave et Rausch, 1950 (Acanthocephala: Polymorphidae). In *Paraziticheskie organizmy severo-vostoka Azii*, pp. 241–6. Vladivostok, USSR: Akademiya Nauk SSSR, Dal'nevostochnyi Nauchnyi Tsentr.

Atrashkevich, G.I. (1975*b*). The finding of cystacanths of *Polymorphus strumosoides* Lundstrom, 1942 (Acanthocephala: Polymorphidae) in *Gammarus pulex* Linne, 1758 in the delta of the River Chaun (north-western Chukotka). In *Paraziticheskie organizmy severo-vostoka Azii*, pp. 247–53. Vladivostok, USSR: Akademiya Nauk SSSR, Dal'nevostochnyi Nauchnyi Tsentr.

Atrashkevich, G.I. (1979*a*). Post-embryonic development of *Arhythmorhynchus petrochenkoi* (Schmidt, 1969) (Acanthocephala: Polymorphidae). In *Ekologiya i morfologiya gel'mintov pozvonochnykh Chukotki*, pp. 73–80. Moscow, USSR: 'Nauka'.

Atrashkevich, G.I. (1979*b*). Ecological characteristics and specificity of the dominant species of bird acanthocephalans in the Chaun lowlands. In *Ekologiya i morfologiya gel'mintov pozvonochnykh Chukotki*, pp. 81–92. Moscow, USSR: 'Nauka'.

Atrashkevich, G.I. (1982). *Filicollis trophimenkoi* n. sp. (Acanthocephala: Polymorphidae) from Anatidae in north-western Chukotka. *Parazitologiya*, **16**, 102–6.

Austin, C.R. (1965). *Fertilization*. Prentice-Hall: Englewood Cliffs, New Jersey.

Austin, C.R. (1982). The egg. In *Reproduction in Mammals. 1. Germ Cells and Fertilization*, eds C.R. Austin and R.V. Short, pp. 46–62. Cambridge University Press.

Avery, R.A. (1971). Helminth parasite populations in newts and their tadpoles. *Freshwater Biology*, **1**, 113–19.

Awachie, J.B.E. (1963). On the development and life history of *Metechinorhynchus truttae* (Schrank, 1788) Petrochenko, 1956 (Acanthocephala). *Parasitology*, **53**, 3P.

Awachie, J.B.E. (1965). The ecology of *Echinorhynchus truttae* Schrank, 1788 (Acanthocephala) in a trout stream in North Wales. *Parasitology*, **55**, 747–62.

Awachie, J.B.E. (1966). The development and life-history of *Echinorhynchus truttae* Schrank 1788 (Acanthocephala). *Journal of Helminthology*, **40**, 11–32.

Awachie, J.B.E. (1967). Experimental studies on some host–parasite relationships of Acanthocephala. Co-invasion of *Gammarus pulex* L. by *Echinorhynchus truttae* Schrank, 1788 and *Polymorphus minutus* (Goeze, 1782). *Acta Parasitologica Polonica*, **15**, 69–74.

Awachie, J.B.E. (1972*a*). Experimental studies on some host–parasite relationships of the Acanthocephala. Effects of primary heavy infection and superimposed infection of *Salmo trutta* L. by *Echinorhynchus truttae* Schrank, 1788. *Acta Parasitologica Polonica*, **20**, 375–82.

Awachie, J.B.E. (1972*b*). Experimental studies on some host–parasite relationships of the Acanthocephala. Effect of host size and starvation on *Echinorhynchus truttae* Schrank, 1788, in its definitive host. *Acta Parasitologia Polonica*, **20**, 383–8.

Baer, J.C. (1951). *Ecology of Animal Parasites*. Urbana: University of Illinois Press.

Baer, J.C. (1961). Platyhelminthes, Mesozoaires, Acanthocephales, Nemertiens. In *Traité de Zoologie*, vol. 4, ed. P.P. Grassé, pp. 733–82. Paris: Masson et Cie.

Balesdent, M.L. (1965). Recherches sur la sexualité et le déterminisme des caractères sexuels d'*Asellus aquaticus* Linné (Crustacé: Isopode). *Bulletin de l'Academie et de la Société Lorraines des Sciences*, **5**, 1–273.

Baltzer, C. (1880). Aur Kenntnis der Echinorhynchen. *Archiv für Naturgeschichte*, **46**, 1–40.

Bangham, R.V. (1955). Studies on fish parasites of Lake Huron and Manitoulin Island. *American Midland Naturalist*, **53**, 184–94.

Barabashova, V.N. (1971). The structure of the integument and its role in the vital activity of Acanthocephala. *Parazitologiya*, **5**, 446–54.

Barrett, J. (1976*a*). Bioenergetics in helminths. In *Biochemistry of Parasites and Host–Parasite Relationships*, ed. H. Van den Bossche, pp. 67–80. Amsterdam, New York and Oxford: Elsevier/North Holland.

Barrett, J. (1976*b*). Energy metabolism and infection in helminths. *Symposia of the British Society for Parasitology*, **15**, 121–44.

Barrett, J. (1981). *Biochemistry of Parasitic Helminths*. Baltimore: University Park Press.

Barrett, J. & Butterworth, P.E. (1968). The carotenoids of *Polymorphus minutus* (Acanthocephala) and its intermediate host, *Gammarus pulex*. *Comparative Biochemistry and Physiology*, **27**, 575–81.

Barrett, J. & Butterworth, P.E. (1971). Wax esters in the cystacanths of *Polymorphus minutus* (Acanthocephala). *Lipids*, **6**, 763–7.

Barrett, J. & Butterworth, P.E. (1973). The carotenoid pigments of six species of adult Acanthocephala. *Experientia*, **29**, 651–3.

Barrett, J. & Körting, W. (1976). Studies on beta-oxidation in the adult liver fluke *Fasciola hepatica*. *International Journal for Parasitology*, **6**, 155–7.

Barrett, J. & Körting, W. (1977). Lipid catabolism in the plerocercoids of *Schistocephalus solidus* (Cestoda: Pseudophyllidea). *International Journal for Parasitology*, **7**, 419–22.
Barrett, J., Cain, G.D. & Fairbairn, D. (1970). Sterols in *Ascaris lumbricoides* (Nematoda), *Macracanthorhynchus hirudinaceus* and *Moniliformis dubius* (Acanthocephala), and *Echinostoma revolutum* (Trematoda). *Journal of Parasitology*, **56**, 1004–8.
Barrett, J., Coles, G.C. & Simpkin, K.G. (1978). Pathways of acetate and propionate production in adult *Fasciola hepatica*. *International Journal for Parasitology*, **8**, 117–23.
Barus, V. & Tenora, F. (1976). New data on parasitic nematodes and acanthocephalans recovered from Amphibia and Reptilia from Afghanistan. *Acta Universitatis Agriculturae, Sbornick Vysoke skoly zemedelske v Brne*, **24**, 339–50.
Barus, V., Kullmann, E. & Tenora, F. (1970). Neue erkenntnisse über Nematoden und Acanthocephalen aus Nagetieren Afghanistans. *Vestnik Ceskoslovenske Spolecnosti Zoologicke*, **34**, 263–76.
Batchvarov, G. (1974). Method of transmission and experimental specificity of *Acanthocephalus anthuris* (Acanthocephala). *Comptes Rendus Hebdomadaires des Séances de l'Académie des Sciences, Paris*, **278**, 3339–41.
Beames, C.G. & Fisher, F.M. (1964). A study on the neutral lipids and phospholipids of the Acanthocephala *Macracanthorhynchus hirudinaceus* and *Moniliformis dubius*. *Comparative Biochemistry and Physiology*, **13**, 401–12.
Beaver, P.C. (1969). The nature of visceral larval migrans. *Journal of Parasitology*, **55**, 3–12.
Beermann, I., Arai, H.P. & Costerton, J.W. (1974). The ultrastructure of the lemnisci and body wall of *Octospinifer macilentus* (Acanthocephala). *Canadian Journal of Zoology*, **52**, 553–5.
Befus, A.D. & Podesta, R.B. (1976). Intestine. In *Ecological Aspects of Parasitology*, ed. C.R. Kennedy, pp. 303–25. Amsterdam, New York and Oxford: North Holland.
Behm, C.A. & Bryant, C. (1975). Studies of regulatory metabolism in *Moniezia expansa*: General considerations. *International Journal for Parasitology*, **5**, 209–17.
Behm, C.A. & Bryant, C. (1982). Phosphoenolpyruvate carboxykinase from *Fasciola hepatica*. *International Journal for Parasitology*, **12**, 271–8.
Beitinger, T.L. & Hammen, C.S. (1971). Utilization of oxygen, glucose, and carbon dioxide by *Echinorhynchus gadi*. *Experimental Parasitology*, **30**, 224–32.
Bendell, J.F. (1955). Disease as a control of a population of Blue Grouse, *Dendrogapus obscurus fuliginosus* (Ridgway). *Canadian Journal of Zoology*, **33**, 195–223.
Bergin, T.J. (1976). Stranded marine mammals. *Australian Veterinary Practitioner*, **6**, 41–5.
Berman, W.F., Bautista, J.O., Rogers, S. & Segal, S. (1976). Metabolism and transport of galactose by rat intestine. *Biochimica et Biophysica Acta*, **455**, 90–101.
Bethel, W.M. & Holmes, J.C. (1973). Altered evasive behavior and responses to light in amphipods harboring acanthocephalan cystacanths. *Journal of Parasitology*, **59**, 945–56.
Bethel, W.M. & Holmes, J.C. (1974). Correlation of development of altered evasive behavior in *Gammarus lacustris* (Amphipoda) harboring cystacanths of *Polymorphus paradoxus* (Acanthocephala) with the infectivity to the definitive host. *Journal of Parasitology*, **60**, 272–4.
Bethel, W.M. & Holmes, J.C. (1977). Increased vulnerability of amphipods to predation

owing to altered behavior induced by larval acanthocephalans. *Canadian Journal of Zoology*, **55**, 110–15.

Bhamburkar, B.L., Garde, V.R. & Shastri, U.V. (1970). First record of incidence of larval phases of acanthocephalan parasites in cockroaches (*Periplaneta americana* L.) in India. *Indian Journal of Entomology*, **32**, 89.

Bibby, M.C. (1972). Population biology of the helminth parasites of *Phoxinus phoxinus* (L.), the minnow, in a Cardiganshire lake. *Journal of Fish Biology*, **4**, 289–300.

Bieler, W. (1913). Uber den Kittapparat von *Neorhynchus*. *Zoologischer Annalen*, **41**, 234–6.

Bird, A.F. & Bird, J. (1969). Skeletal structures and integument of Acanthocephala and Nematoda. In *Chemical Zoology*, vol. 3, ed. M. Florkin and B.T. Sheer, pp. 253–88. London and New York: Academic Press.

Bloom, W. & Fawcett, D.W. (1968). *A Textbook of Histology*. Philadelphia and London: W.B. Saunders & Co.

Bogitsh, B.J. (1961). Histological and histochemical observations on the nature of the cyst of *Neoechinorhynchus cylindratus* in *Lepomis* sp. *Proceedings of the Helminthological Society of Washington*, **28**, 75–81.

Bone, L.W. (1974*a*). The chromosomes of *Leptorhynchoides thecatus* (Acanthocephala). *Journal of Parasitology*, **60**, 818.

Bone, L.W. (1974*b*). The chromosomes of *Neoechinorhynchus cylindratus* (Acanthocephala). *Journal of Parasitology*, **60**, 731–2.

Bone, L.W. (1976). *Anterior Neuromorphology and Neurosecretion of* Moniliformis dubius *Meyer, 1932* (*Acanthocephala*). Ph D dissertation, University of Arkansas.

Bonfonte, R., Faust, E.C. & Giraldo, L.E. (1961). Parasitologic surveys in Cali, Departamento del Valle, Colombia. IX. Endoparasites of rodents and cockroaches in Ward Siloe, Cali, Colombia. *Journal of Parasitology*, **47**, 843–6.

Borgarenko, L.F. (1975). Larvae of the genus *Mediorhynchus* in *Gammarus*. In *Zoologicheskii, sbornik. Chast' II*, pp. 74–6. Dushanbe, USSR: 'Donish'.

Borgarenko, L.F., Bronshpits, G.M. & Solodenko, T.E. (1975). The prevalence of infection in the invertebrates of the Nadoksai ravine (northern Tadzhikistan). In *Zoologicheskii sbornic. Chast' II*, pp. 81–90. Dushanbe, USSR: 'Donish'.

Bowen, R.C. (1967). Defence reactions of certain spirobolid millipedes to larval *Macracanthorhynchus ingens*. *Journal of Parasitology*, **53**, 1092–5.

Boxrucker, J.C. (1979). Effects of a thermal effluent on the incidence and abundance of the gill and intestinal metazoan parasites of the Black Bullhead. *Parasitology*, **78**, 195–206.

Bozkov, D. (1976). On the postcycle parasitism in helminths and its biological significance. *Angewandte Parasitologie*, **17**, 85–8.

Bozkov, D. (1980). Experimental studies on the passage of mature helminths from *Rana ridibunda* Pall. to *Bufo viridis* Laur. *Khelmintologiya*, **10**, 24–8.

Bradley, D.J. (1974). Stability in host–parasite systems. In *Ecological Stability*, ed. M.B. Usher and M.H. Williamson, pp. 71–87. London: Chapman & Hall.

Branch, S.I. (1970*a*). *Moniliformis dubius* and *Macracanthorhynchus hirudinaceus*: Na, K, Ca and Mg content, and Na and K active transport. *Experimental Parasitology*, **27**, 33–43.

Branch, S.I. (1970*b*). *Moniliformis dubius*: osmotic responses. *Experimental Parasitology*, **27**, 44–52.

Branch, S.I. (1970*c*). Accumulation of amino acids by *Moniliformis dubius* (Acanthocephala). *Experimental Parasitology*, **27**, 95–9.

Brand, T. von (1939*a*). Chemical and morphological observations upon the composition of *Macracanthorhynchus hirudinaceus* (Acanthocephala). *Journal of Parasitology*, **25**, 329–42.

Brand, T. von (1939*b*). The glycogen distribution in the body of Acanthocephala. *Journal of Parasitology*, **25**, 22S.

Brand, T. von (1940). Further observations upon the composition of Acanthocephala. *Journal of Parasitology*, **26**, 301–7.

Brand, T. von (1973). *Biochemistry of Parasites*, 2nd edn. London and New York: Academic Press.

Brand, T. von (1979). *Biochemistry and Physiology of Endoparasites*. Amsterdam, New York and Oxford: Elsevier/North Holland.

Brand, T. von & Saurwein, J. (1942). Further studies upon the chemistry of *Macracanthorhynchus hirudinaceus. Journal of Parasitology*, **28**, 315–18.

Brandes, G. (1899). Das Nervensystem des als Nemathelminthen zusammengefassen Wurmtypen. *Abhandlungen Deutsches Naturforschendes Gesellschaft* (Halle), **21**, 271–99.

Brattey, J. (1980). Preliminary observations on larval *Acanthocephalus lucii* (Müller, 1776) (Acanthocephala: Echinorhynchidae) in the isopod *Asellus aquaticus* (L.). *Parasitology*, **81**, xlix–l.

Bremser, J.G. (1811). *Notitia/insignis vermium intestinalium collections vindobonesis*. Viennae. 31 pp.

Brennan, B.M. & Cheng, T.C. (1975). Resistance of *Moniliformis dubius* to the defense reactions of the American cockroach, *Periplaneta americana. Journal of Invertebrate Pathology*, **26**, 65–73.

Broughton, P.F. (1974). *Specificity of* Pomphorhynchus laevis (*Acanthocephala*) *with Special Reference to Growth in Trout and* in vitro. M.Sc. dissertation, University of Exeter.

Brownell, W.N. (1970). Comparison of *Mysis relicta* and *Pontoporeia affinis* as possible intermediate hosts for the acanthocephalan *Echinorhynchus salmonis. Journal of the Fisheries Research Board of Canada*, **27**, 1864–6.

Brumpt, E. & Desportes, C. (1938). Hôtes intermédiaires expérimentaux de deux éspèces d'acanthocéphales (*Prosthenorchis spirula* et *P. elegans*) parasites des lemuriens et des singes. *Annales de Parasitologie Humaine et Comparée*, **16**, 301–4.

Brumpt, E. & Urbain, A. (1938). Epizootie vermineuse par acanthocéphales (*Prosthenorchis*) ayant sévi à la singerie du Museum de Paris. *Annales de Parasitologie Humaine et Comparée*, **16**, 289–300.

Bryant, C. (1970). Electron transport in parasitic helminths and protozoa. *Advances in Parasitology*, **8**, 139–72.

Bryant, C. (1975). Carbon dioxide utilization, and the regulation of respiratory metabolic pathways in parasitic helminths. *Advances in Parasitology*, **13**, 35–69.

Bryant, C. (1978). The regulation of respiratory metabolism in parasitic helminths. *Advances in Parasitology*, **16**, 311–31.

Bryant, C. & Behm, C.A. (1976). Regulation of respiratory metabolism in *Moniezia expansa* under aerobic and anaerobic conditions. In *Biochemistry of Parasites and Host–Parasite Relationships*, ed. H. Van den Bossche, pp. 89–94. Amsterdam, New York and Oxford: Elsevier/North Holland.

Bryant, C. & Nicholas, W.L. (1965). Intermediary metabolism in *Moniliformis dubius* (Acanthocephala). *Comparative Biochemistry and Physiology*, **15**, 103–12.

Bryant, C. & Nicholas, W.L. (1966). Studies on the oxidative metabolism of *Moniliformis dubius* (Acanthocephala). *Comparative Biochemistry and Physiology*, **17**, 825–40.

Buckner, R.L. (1977). Evaluation of variation in species of *Fessisentis* (Acanthocephala), with observations on the life cycle of *Fessisentis fessus*. *Dissertation Abstracts International*, **37**B, 3291–2.

Buckner, S.C. & Nickol, B.B. (1975). Host specificity and lack of hybridization of *Moniliformis clarki* (Ward, 1917) Chandler, 1921, and *Moniliformis moniliformis* (Bremser, 1811) Travassos, 1915. *Journal of Parasitology*, **61**, 991–5.

Buckner, R.L. & Nickol, B.B. (1978). Redescription of *Fessisentis vancleavei* (Hughes and Moore, 1943) Nickol, 1972 (Acanthocephala: Fessisentidae). *Journal of Parasitology*, **64**, 635–7.

Buckner, R.L. & Nickol, B.B. (1979). Geographic and host-related variation among species of *Fessisentis* (Acanthocephala) and confirmation of the *Fessisentis fessus* life cycle. *Journal of Parasitology*, **65**, 161–6.

Buckner, R.L., Overstreet, R.M. & Heard, R.W. (1978). Intermediate hosts for *Tegorhynchus furcatus* and *Dollfusentis chandleri* (Acanthocephala). *Proceedings of the Helminthological Society of Washington*, **45**, 195–201.

Bullock, T.H. & Horridge, G.A. (1965). *Structure and Function in the Nervous System of Invertebrates*, vol. 1. San Francisco: Freeman Press.

Bullock, W.L. (1949*a*). Histochemical studies on the Acanthocephala I. The distribution of lipase and phosphatase. *Journal of Morphology*, **84**, 185–200.

Bullock, W.L. (1949*b*). Histochemical studies on the Acanthocephala II. The distribution of glycogen and fatty substances. *Journal of Morphology*, **84**, 201–26.

Bullock, W.L. (1957*a*). The acanthocephalan parasites of the fishes of the Texas coast. *Publications of the Institute of Marine Science, University of Texas*, **4**, 278–83.

Bullock, W.L. (1957*b*). *Octospiniferoides chandleri* n. gen., n. sp., a neoechinorhynchid acanthocephalan from *Fundulus grandis* Baird and Girard on the Texas coast. *Journal of Parasitology*, **43**, 97–100.

Bullock, W.L. (1958). Histochemical studies on the Acanthocephala III. Comparative histochemistry of alkaline glycerophosphatase. *Experimental Parasitology*, **7**, 51–68.

Bullock, W.L. (1960). Some acanthocephalan parasites of Florida fishes. *Bulletin of Marine Science of the Gulf and Caribbean*, **10**, 481–4.

Bullock, W.L. (1962). A new species of *Acanthocephalus* from New England fishes with observations on variability. *Journal of Parasitology*, **48**, 442–51.

Bullock, W.L. (1963). Intestinal histology of some salmonid fishes with particular reference to the histopathology of acanthocephalan infections. *Journal of Morphology*, **112**, 23–44.

Bullock, W.L. (1969). Morphological features as tools and pitfalls in acanthocephalan systematics. In *Problems in Systematics of Parasites*, ed. G.D. Schmidt, pp. 9–24. Baltimore: University Park Press.

Burlingame, P.L. & Chandler, A.C. (1941). Host–parasite relations of *Moniliformis dubius* (Acanthocephala) in albino rats, and the environmental nature of resistance to single and superimposed infections with this parasite. *American Journal of Hygiene*, **33**, 1–21.

Burow, C.H.A. (1836). *Echinorhynchi strumosi* anatome. Dissert Zootomica. In Albertina Literarum Univ. veniam legendi capessiturus die XIV. M. Julii A. MDCCCXXXVI. 28 pp. 21 pl. Regiomonti Prussorum.

Buscher, H.N. (1965). Dynamics of the intestinal helminth fauna in three species of ducks. *Journal of Wildlife Management*, **29**, 772–81.

Butterworth, P.E. (1967). Aspects of the nutrition of *Polymorphus minutus* during its development. *Parasitology*, **58**, 3P.

Butterworth, P.E. (1969*a*). The development of the body wall of *Polymorphus minutus* (Acanthocephala) in its intermediate host *Gammarus pulex*. *Parasitology*, **59**, 373–88.
Butterworth, P.E. (1969*b*). *Studies on the physiology and development of* Polymorphus minutus (*Acanthocephala*). Ph D dissertation, University of Cambridge.
Byram, J.E. (1971). *Studies on the morphology, function, and phylogenetic implications of the acanthocephalan absorptive surface*. Ph D dissertation, Rice University. (*Dissertation Abstracts International*, **32B**, 2006.)
Byram, J.E. & Fisher, F.M. (1973). The absorptive surface of *Moniliformis dubius* (Acanthocephala) I. Fine Structure. *Tissue and Cell*, **5**, 553–79.
Byram, J.E. & Fisher, F.M. (1974). The absorptive surface of *Moniliformis dubius* (Acanthocephala) II. Functional aspects. *Tissue and Cell*, **6**, 21–42.
Byrd, E.E. & Denton, F. (1949). The helminth parasites of birds. II. A new species of Acanthocephala from North American birds. *Journal of Parasitology*, **35**, 391–410.
Cabib, E. & Leloir, L.F. (1958). The biosynthesis of trehalose phosphate. *Journal of Biological Chemistry*, **231**, 259–75.
Cable, R.M. & Dill, W.T. (1967). The morphology and life history of *Paulisentis fractus* Van Cleave and Bangham, 1949 (Acanthocephala: Neoechinorhynchidae). *Journal of Parasitology*, **53**, 810–17.
Cain, G.D. (1970). Collagen from the giant acanthocephalan *Macracanthorhynchus hirudinaceus*. *Archives of Biochemistry and Biophysics*, **141**, 264–70.
Calow, P. (1978). *Life Cycles*. London: Chapman and Hall.
Camp, J.W. & Huizinga, H.W. (1979). Altered color, behavior and predation susceptibility of the isopod, *Asellus intermedius*, infected with *Acanthocephalus dirus*. *Journal of Parasitology*, **65**, 667–9.
Camp, J.W. & Huizinga, H.W. (1980). Seasonal population interactions of *Acanthocephalus dirus* (Van Cleave, 1931) in the creek chub, *Semotilus atromaculatus*, and isopod, *Asellus intermedius*. *Journal of Parasitology*, **66**, 299–304.
Carafoli, E. & Scarpa, A. (ed.) (1982). Transport ATPases. *Annals of the New York Academy of Sciences*, **402**.
Carey, M.C., Small, D.M. & Bliss, C.M. (1983). Lipid digestion and absorption. *Annual Review of Physiology*, **45**, 651–77.
Castro, G.A., Johnson, L.R., Copeland, E.M. & Dudrick, S.J. (1976). Course of infection with enteric parasites in hosts shifted from enteral to total parenteral nutrition. *Journal of Parasitology*, **62**, 353–9.
Chaicharn, A. & Bullock, W.L. (1967). The histopathology of acanthocephalan infections in suckers with observations on the intestinal histology of two species of catastomid fishes. *Acta Zoologica, Stockholm*, **48**, 19–42.
Chandler, A.C. (1935). Parasites of fishes in Galveston Bay. *Proceedings of the United States National Museum*, **83**, 123–58.
Chandler, A.C. (1946*a*). Helminths of armadillos, *Dasypus novemcinctus*, in eastern Texas. *Journal of Parasitology*, **32**, 237–41.
Chandler, A.C. (1946*b*). Some observations on the anatomy of certain male Acanthocephala. *Transactions of the American Microscopical Society*, **65**, 304–10.
Chandler, A.C. (1949). *Introduction to Parasitology*, 8th edn. New York: John Wiley.
Chandler, A.C. (1953). An outbreak of *Prosthenorchis* (Acanthocephala) infection in primates in the Houston Zoological Garden, and a report of this parasite in *Nasua narica* in Mexico. *Journal of Parasitology*, **39**, 226.
Chandler, A.C. & Melvin, D.M. (1951). A new cestode, *Oochoristica pennsylvanica*, and

some new or rare helminth host records from Pennsylvania mammals. *Journal of Parasitology*, **37**, 106–9.

Chappell, L.H. (1969*a*). Competitive exclusion between two intestinal parasites of three-spined stickleback *Gasterosteus aculeatus* L. *Journal of Parasitology*, **55**, 775–8.

Chappell, L.H. (1969*b*). The parasites of the three-spined stickleback *Gasterosteus aculeatus* L. from a Yorkshire pond. II. Variation of the parasite fauna with sex and size of fish. *Journal of Fish Biology*, **1**, 339–47.

Chebotarev, R.S. (1954). New data on the biology of *Macracanthorhynchus* which causes a disease of pigs. *Zoologicheskii Zhurnal*, USSR, **33**, 1206–9.

Chibichenko, N.T. & Mamytova, S. (1978). Gammarids as intermediate hosts of helminths in Kirgizia (USSR). In *Biologicheskie osnovy rybnogo kozyaistava vodoemov Srednei Asii i Kazakstan*, pp. 498–9. Frunze, USSR; 'Ilim'.

Cholodkovsky, N.A. (1897). O Sistematichenskom polozhenii skrebnei (Classification of the Acanthocephala) – Trudy SPb. *Obshchestva Estestvoispytatelei*, **28**, 14–20.

Christensen, H.N. (1975). *Biological Transport*, 2nd edn. Reading, Massachusetts: W.A. Benjamin, Inc.

Christensen, H.N., de Cespedes, C., Handlogten, M.E. & Ronquist, G. (1973). Energization of amino acid transport, studied for the Ehrlich ascites tumor cell. *Biochimica et Biophysica Acta*, **300**, 487–522.

Christiansen, M. (1948). Epidemiagtigt Sygdomsudbrund blandt Ederfugle (*Somateria mollissima* L.) ved Bornholm, foraarsaget af dyriske Snyltere. *Dansk Ornitologisk Forenings Tidsskrift*, **42**, 41–7.

Christie, J.R. (1929). Proceedings of the Helminthological Society of Washington, One hundred sixteenth meeting. *Journal of Parasitology*, **15**, 281–93.

Chubb, J.C. (1964). Occurrence of *Echinorhynchus clavula* (Dujardin, 1845) nec Hamann, 1892 (Acanthocephala) in the fish of Llyn Tegid (Bala Lake), Merionethshire. *Journal of Parasitology*, **50**, 52–9.

Chubb, J.C. (1982). Seasonal occurrence of helminths in freshwater fishes Part IV. Adult Cestoda, Nematoda and Acanthocephala. *Advances in Parasitology*, **20**, 1–292.

Chubb, J.C., Awachie, J.B.E. & Kennedy, C.R. (1964). Evidence for a dynamic equilibrium in the incidence of Cestoda and Acanthocephala in the intestines of freshwater fish. *Nature, London*, **203**, 986–7.

Clark, G.M., O'Meara, D. & Van Weelden, J.W. (1958). An epizootic among eider ducks involving an acanthocephalid worm. *Journal of Wildlife Management*, **22**, 204–5.

Clausen, T. (1975). The effect of insulin on glucose transport in muscle cells. In *Current Topics in Membranes and Transport*, vol. 6, ed. F. Bronner and A. Kleinzeller, pp. 169–226. London and New York: Academic Press.

Clegg, J.S. (1965). The origin of trehalose and its significance during the formation of encysted dormant embryos of *Artemia salina*. *Comparative Biochemistry and Physiology*, **14**, 135–43.

Cloquet, J. (1824). *Anatomie de vers intestinaux Ascaride lumbricoide et Echinorhynque géant*. Paris.

Cohen, J. (1977). *Reproduction*. London: Butterworths.

Collins, G.D. (1971). *Mediorhynchus grandis* in a short-tailed shrew, *Blarina brevicauda*, from South Dakota. *Journal of Parasitology*, **57**, 1038.

Colowick, S.P. (1973). The hexokinases. In *The Enzymes*, 3rd edn, vol. 9, ed. P.D. Boyer, pp. 1–48. London and New York: Academic Press.

Conway Morris, S. & Crompton, D.W.T. (1982). The origins and evolution of the Acanthocephala. *Biological Reviews of the Cambridge Philosophical Society*, **57**, 85–115.

Cooper, W.E. (1965). Dynamics and production of a natural population of a fresh-water amphipod, *Hyalella azteca*. *Ecological Monographs*, **35**, 377–94.

Copland, W.O. (1956). Notes on the food and parasites of pike (*Esox lucius*) in Loch Lomond. *Glasgow Naturalist*, **18**, 230–5.

Cornford, E.M. & Oldendorf, W.H. (1979). Transintegumental uptake of metabolic substrates in male and female *Schistosoma mansoni*. *Journal of Parasitology*, **65**, 357–63.

Cornish, R.A. & Bryant, C. (1975). Studies of regulatory metabolism in *Moniezia expansa*: Glutamate, and the absence of the γ-aminobutyrate pathway. *International Journal for Parasitology*, **5**, 355–62.

Cornish, R.A., Wilkes, J. & Mettrick, D.F. (1981*a*). A study of phosphoenolpyruvate carboxykinase from *Moniliformis dubius* (Acanthocephala). *Molecular and Biochemical Parasitology*, **2**, 151–66.

Cornish, R.A., Wilkes, J. & Mettrick, D.F. (1981*b*). The levels of some metabolites in *Moniliformis dubius* (Acanthocephala). *Journal of Parasitology*, **67**, 754–6.

Coronel Guevara, M.L. (1953). *Observations sobre el ciclo biologico de Moniliformis moniliformis (Bremser, 1811)*. Thesis, Mexico. 27 pp. (*Helminthological Abstracts*, **22**, 144.)

Cram, E.B. (1931). Recent findings in connection with parasites of game birds. *Transactions of the American Game Conference*, **18**, 243–7.

Crane, R.K. (1960). Intestinal absorption of sugars. *Physiological Reviews*, **40**, 789–825.

Crane, R.K. (1965). Na^+-dependent transport in the intestine and other animal tissues. *Federation Proceedings*, **24**, 1000–6.

Crane, R.K. (1968). Absorption of sugars. In *Handbook of Physiology, Section 6: Alimentary Canal*, vol. 3, ed. C.F. Code, pp. 1323–51. Washington, D.C.: American Physiological Society.

Crane, R.K. (1975). A digestive-absorptive surface as illustrated by the intestinal cell brush border. *Transactions of the American Microscopical Society*, **94**, 529–44.

Crites, J.L. (1964). A millipede, *Narceus americanus*, as a natural intermediate host of an acanthocephalan. *Journal of Parasitology*, **50**, 293.

Crofton, H.D. (1971*a*). A quantitative approach to parasitism. *Parasitology*, **62**, 179–94.

Crofton, H.D. (1971*b*). A model of host–parasite relationships. *Parasitology*, **63**, 343–64.

Croll, N.A. (1966). *Ecology of Parasites*. Cambridge, Massachusetts: Harvard University Press.

Crompton, D.W.T. (1963). Morphological and histochemical observations on *Polymorphus minutus* (Goeze, 1782) with special reference to the body wall. *Parasitology*, **53**, 663–85.

Crompton, D.W.T. (1964). The envelope surrounding *Polymorphus minutus* (Goeze, 1782) (Acanthocephala) during its development in the intermediate host, *Gammarus pulex*. *Parasitology*, **54**, 721–35.

Crompton, D.W.T. (1965). A histochemical study of the distribution of glycogen and oxidoreductase activity in *Polymorphus minutus* (Goeze, 1782) (Acanthocephala). *Parasitology*, **55**, 503–14.

Crompton, D.W.T. (1967). Studies on the haemocytic reaction of *Gammarus* spp., and its relationship to *Polymorphus minutus* (Acanthocephala). *Parasitology*, **57**, 389–401.

Crompton, D.W.T. (1969). On the environment of *Polymorphus minutus* (Acanthocephala) in ducks. *Parasitology*, **59**, 19–28.

Crompton, D.W.T. (1970). *An Ecological Approach to Acanthocephalan Physiology*. Cambridge University Press.

Crompton, D.W.T. (1972*a*). The growth of *Moniliformis dubius* (Acanthocephala) in the intestine of male rats. *Journal of Experimental Biology*, **56**, 19–29.

Crompton, D.W.T. (1972*b*). Monorchic *Moniliformis dubius* (Acanthocephala). *International Journal for Parasitology*, **2**, 483.

Crompton, D.W.T. (1973). The sites occupied by some parasitic helminths in the alimentary tract of vertebrates. *Biological Reviews of the Cambridge Philosophical Society*, **48**, 27–83.

Crompton, D.W.T. (1974). Experiments on insemination in *Moniliformis dubius* (Acanthocephala). *Parasitology*, **68**, 229–38.

Crompton, D.W.T. (1975). Relationships between Acanthocephala and their hosts. In *Symbiosis*, ed. D.H. Jennings and D.L. Lee, pp. 467–504. *Symposia of the Society for Experimental Biology*, 29. London: Cambridge University Press.

Crompton, D.W.T. (1983*a*). Acanthocephala. In *Reproductive Biology of Invertebrates. I. Oogenesis, Oviposition, and Oosorption*, ed. K.G. Adiyodi and R.G. Adiyodi, pp. 257–68. Chichester, New York, Brisbane, Toronto, Singapore: John Wiley & Sons.

Crompton, D.W.T. (1983*b*). Acanthocephala. In *Reproductive Biology of Invertebrates. II. Spermatogenesis and Sperm Function*, ed. K.G. Adiyodi and R.G. Adiyodi, pp. 257–67. Chichester, New York, Brisbane, Toronto, Singapore: John Wiley & Sons.

Crompton, D.W.T. Acanthocephala. In *Reproductive Biology of Invertebrates*, vol. 3, *Accessory Sex Glands*, ed. K.G. Adiyodi and R.G. Adiyodi. Chichester, New York, Brisbane, Toronto, Singapore: Wiley. (In press.)

Crompton, D.W.T. Acanthocephala. In *Reproductive Biology of Invertebrates*, vol. 4, *Fertilization, Development and Parental Care*, ed. K.G. Adiyodi and R.G. Adiyodi. Chichester, New York, Brisbane, Toronto, Singapore: Wiley. (In press.)

Crompton, D.W.T. Acanthocephala. In *Reproductive Biology of Invertebrates*, vol. 5, *Sexual Differentiation and Behaviour*, ed. K.G. Adiyodi and R.G. Adiyodi. Chichester, New York, Brisbane, Toronto, Singapore: Wiley. (In press.)

Crompton, D.W.T. & Edmonds, S.J. (1969). Measurements of the osmotic pressure in the habitat of *Polymorphus minutus* (Acanthocephala) in the intestine of domestic ducks. *Journal of Experimental Biology*, **50**, 69–77.

Crompton, D.W.T. & Harrison, J.G. (1965). Observations on *Polymorphus minutus* (Goeze, 1782) (Acanthocephala) from a wildfowl reserve in Kent. *Parasitology*, **55**, 345–55.

Crompton, D.W.T. & Lee, D.L. (1963). Structural and metabolic components of acanthocephalan body wall. *Parasitology*, **53**, 3–4P.

Crompton, D.W.T. & Lee, D.L. (1965). The fine structure of the body wall of *Polymorphus minutus* (Goeze, 1782) (Acanthocephala). *Parasitology*, **55**, 357–64.

Crompton, D.W.T. & Lockwood, A.P.M. (1968). Studies on the absorption and metabolism of D-(U-^{14}C) glucose by *Polymorphus minutus* (Acanthocephala) *in vitro*. *Journal of Experimental Biology*, **48**, 411–25.

Crompton, D.W.T. & Walters, D.E. (1972). An analysis of the course of infection of *Moniliformis dubius* (Acanthocephala) in rats. *Parasitology*, **64**, 517–23.

Crompton, D.W.T. & Ward, P.F.V. (1967*a*). Lactic and succinic acids as excretory products of *Polymorphus minutus* (Acanthocephala) *in vitro*. *Journal of Experimental Biology*, **46**, 423–30.

Crompton, D.W.T. & Ward, P.F.V. (1967*b*). Production of ethanol and succinate by *Moniliformis dubius* (Acanthocephala). *Nature, London*, **215**, 964–5.

Crompton, D.W.T. & Ward, P.F.V. (1984). Selective metabolism of L-serine by *Moniliformis* (Acanthocephala) *in vitro*. *Parasitology*, **89**, 133–44.

Crompton, D.W.T. & Whitfield, P.J. (1968*a*). The course of infection and egg production of *Polymorphus minutus* (Acanthocephala) in domestic ducks. *Parasitology*, **58**, 231–46.

Crompton, D.W.T. & Whitfield, P.J. (1968*b*). A hypothesis to account for the anterior migrations of adult *Hymenolepis diminuta* (Cestoda) and *Moniliformis dubius* (Acanthocephala) in the intestine of rats. *Parasitology*, **58**, 227–9.

Crompton, D.W.T. & Whitfield, P.J. (1974). Observations on the functional organization of the ovarian balls of *Moniliformis* and *Polymorphus* (Acanthocephala). *Parasitology*, **69**, 429–43.

Crompton, D.W.T., Arnold, S. & Barnard, D. (1972). The patent period and production of eggs of *Moniliformis dubius* (Acanthocephala) in the small intestine of male rats. *International Journal for Parasitology*, **2**, 319–26.

Crompton, D.W.T., Arnold, S. & Walters, D.E. (1976). The number and size of ovarian balls of *Moniliformis* (Acanthocephala) from laboratory rats. *Parasitology*, **73**, 65–72.

Crompton, D.W.T., Keymer, A.E. & Arnold, S.E. (1984). Investigating overdispersion: *Moniliformis* (Acanthocephala) and rats. *Parasitology*, **88**, 317–31.

Crompton, D.W.T., Singhvi, A. & Keymer, A. (1982). Effects of host dietary fructose on experimentally stunted *Moniliformis* (Acanthocephala). *International Journal for Parasitology*, **12**, 117–21.

Crompton, D.W.T., Keymer, A., Singhvi, A. & Nesheim, M.C. (1983). Rat dietary fructose and the intestinal distribution and growth of *Moniliformis* (Acanthocephala). *Parasitology*, **86**, 57–81.

Crompton, D.W.T., Singhvi, A., Nesheim, M.C. & Walters, D.E. (1981). Competition for dietary fructose between *Moniliformis dubius* (Acanthocephala) and growing rats. *International Journal for Parasitology*, **11**, 457–61.

Crook, J.R. (1964). The role of intermediate hosts in the ecological distribution of some endoparasites in the Bonneville Basin, Utah. *Dissertation Abstracts*, **25**, 2109.

Crook, J.R. & Grundmann, A.W. (1964). The life history and larval development of *Moniliformis clarki* (Ward, 1917). *Journal of Parasitology*, **50**, 689–93.

Cross, S.X. (1934). A probable case of non-specific immunity between two parasites of ciscoes of the Trout Lake region of northern Wisconsin. *Journal of Parasitology*, **20**, 244–5.

Crowe, J.H., Crowe, L.M. & Mouradian, R. (1983). Stabilization of biological membranes at low water activities. *Cryobiology*, **20**, 346–56.

Dappen, G.E. & Nickol, B.B. (1981). Unaltered hematocrit values and differential hemocyte counts in acanthocephalan-infected *Armadillidium vulgare*. *Journal of Invertebrate Pathology*, **38**, 209–12.

Das, E.N. (1950). On some juvenile forms of Acanthocephala of the genus *Centrorhynchus* from India. *Indian Journal of Helminthology*, **2**, 49–56.

Das, E.N. (1952). On some interesting larval stages of an acanthocephalan, *Centrorhynchus batrachus* sp. nov., from *Rana tigrina* (Daud) from India. *Records of The Indian Museum*, **50**, 147–56.

Das, E.N. (1957*a*). Les stades larvaires de *Centrorhynchus falconis* E.N. Das, 1950. *Annales de Parasitologie Humaine et Comparée*, **32**, 71–82.

Das, E.N. (1957*b*). On juvenile and adult forms of *Pseudoporrorchis indicus*, a new species of Acanthocephala from India. *Journal of Parasitology*, **43**, 659–63.

Davenport, R. (1979). *An Outline of Animal Development*. Reading, Massachusetts, and London: Addison-Wesley Publishing Co.

Daynes, P. (1966). Note sur le cycle biologique de *Macracanthorhynchus hirudinaceus* (Pallas, 1781) à Madagascar. *Revue d'Elevage et de Médecien Veterinaire des Pays Tropicaus*, **19**, 277–82.

DeGiusti, D.L. (1939). Preliminary note on the life cycle of *Leptorhynchoides thecatus*, an acanthocephalan parasite of fish. *Journal of Parasitology*, **25**, 180.

DeGiusti, D.L. (1949*a*). The life cycle of *Leptorhynchoides thecatus* (Linton), an acanthocephalan of fish. *Journal of Parasitology*, **35**, 437–60.

DeGiusti, D.L. (1949*b*). Partial development of *Echinorhynchus coregoni* in *Hyalella azteca* and the cellular reaction of the amphipod to the parasite. *Journal of Parasitology*, **35** (6, Sec. 2), 31.

DeMont, D.J. & Corkum, K.C. (1982). The life cycle of *Octospiniferoides chandleri* Bullock, 1957 (Acanthocephala: Neoechinorhynchidae) with some observations on parasite-induced, photophilic behavior in ostracods. *Journal of Parasitology*, **68**, 125–30.

Denbo, J. (1971). *Osmotic and ionic regulation in* Macracanthorhynchus hirudinaceus (*Acanthocephala*). Masters thesis: Southern Illinois University.

Dendinger, J.E. & Roberts, L.S. (1977). Glycogen synthase in the rat tapeworm, *Hymenolepis diminuta*. II. Control of enzyme activity by glucose and glycogen. *Comparative Biochemistry and Physiology*, **58B**, 231–6.

Denny, M. (1968). The life-cycle and ecology of *Polymorphus marilis* Van Cleave, 1939. *Parasitology*, **58**, 23P.

Denny, M. (1969). Life-cycles of helminth parasites using *Gammarus lacustris* as an intermediate host in a Canadian lake. *Parasitology*, **59**, 795–827.

Devine, M.C. (1975). Copulatory plugs in snakes: enforced chastity. *Science*, **187**, 844–5.

Diamond, J.M. (1971). Standing-gradient model of fluid transport in epithelia. *Federation Proceedings*, **30**, 6–13.

Diesing, K.M. (1851). *Systema Helminthum*. Vindabonae, Vol. 2. 588 pp.

Dill, W.T. (1975). The life-cycle and ontogeny of *Atactorhynchus verecundus* Chandler, 1935 (Acanthocephala: Neoechinorhynchidae). *Dissertation Abstracts International*, **36B**(1), 41.

Dimitrova, E. (1963). A contribution to the topographic distribution of lipids in *Macracanthorhynchus hirudinaceus* (Pallas, 1781) (Travassos, 1917). *Izvestiya Tsentralnata Khelmintologichna Laboratoriya, Bulgarska akademiya na naukite*, **8**, 69–78.

DiPolo, R. & Beaugé, L. (1983). The calcium pump and sodium–calcium exchange in squid axons. *Annual Review of Physiology*, **45**, 313–24.

Dollfus, R.P. (1938). Etude morphologique et systematique de deux éspèces d'Acanthocephales, parasites de lemuriens et de singes. Revue critique du genère *Prosthenorchis* Travassos. *Annales de Parasitologie Humaine et Comparée*, **16**, 385–422.

Dollfus, R.P. (1957). Miscellanea helminthologica Maroccana XXII. Annototiens au sujet de divers acanthocephales dont et a ete question dans Miscellanea Helminthologica Maroccana I (1951) et XI (1933). *Archives de l'Institut Pasteur du Maroc*, **5**, 403–7.

Dollfus, R.P. & Dalens, H. (1960). *Prosthorhynchus cylindraceus* (Goeze, 1782) au stade juvenile, chez un isopode terrestre. Acanthocephala–Polymorphidae. *Annales de Parasitologie Humaine et Comparée*, **35**, 347–9.

Dollfus, R.P. & Golvan, Y. (1956). Mission M. Blanc-F. d'Aubenton (1954). V.

Acanthocephales de poissons du Niger. *Bulletin de l'Institut Français d'Afrique Noire*, Series A, **18**, 1086–109.

Donahue, M.J., Yacoub, N.J., Kaeini, M.R., Tu, S., Hodzi, R.A. & Harris, B.G. (1981). Studies on potential carbohydrate regulatory enzymes and metabolite levels in *Macracanthorhynchus hirudinaceus*. *Journal of Parasitology*, **67**, 756–8.

Dubinin, V.B. (1948). Developmental cycle of *Macracanthorhynchus catulinus* Kostylew, 1927. *Doklady Akademii Nauk SSSR*, **60**, 1109–11.

Dubnitski, A.A. (1957). The infestation of fur-bearing animals by acanthocephalan parasites. *Karakulevodstvo i Zverovodstvo*, **10**, 52–4.

Dunagan, T.T. (1962). Studies on in vitro survival of Acanthocephala. *Proceedings of the Helminthological Society of Washington*, **29**, 131–5.

Dunagan, T.T. (1963). Glycogen depletion in *Neoechinorhynchus* spp. (Acanthocephala) from turtles. *Journal of Parasitology*, **49**, Suppl., 18–19.

Dunagan, T.T. (1964). Studies on the carbohydrate metabolism of *Neoechinorhynchus* spp. (Acanthocephala). *Proceedings of the Helminthological Society of Washington*, **31**, 166–72.

Dunagan, T.T. & de Luque, O. (1966). Isozyme patterns for lactic and malic dehydrogenases in *Macracanthorhynchus hirudinaceus* (Acanthocephala). *Journal of Parasitology*, **52**, 727–9.

Dunagan, T.T. & Miller, D.M. (1970). Major nerves in the anterior nervous system of *Macracanthorhynchus hirudinaceus* (Acanthocephala). *Comparative Biochemistry and Physiology*, **37**, 235–42.

Dunagan, T.T. & Miller, D.M. (1973). Some morphological and functional observations on *Fessisentis fessus* Van Cleave (Acanthocephala) from the dwarf salamander, *Siren intermedia* Le Conte. *Proceedings of the Helminthological Society of Washington*, **40**, 209–16.

Dunagan, T.T. & Miller, D.M. (1974). Muscular anatomy of the praesoma of *Macracanthorhynchus hirudinaceus* (Acanthocephala). *Proceedings of the Helminthological Society of Washington*, **41**, 199–208.

Dunagan, T.T. & Miller, D.M. (1975). Anatomy of the cerebral ganglion of the male acanthocephalan *Moniliformis dubius*. *Journal of Comparative Neurology*, **164**, 483–94.

Dunagan, T.T. & Miller, D.M. (1976). Nerves originating from the cerebral ganglion of *Moniliformis moniliformis* (Acanthocephala). *Journal of Parasitology*, **62**, 442–50.

Dunagan, T.T. & Miller, D.M. (1977). A new ganglion in male *Moniliformis moniliformis* (Acanthocephala). *Journal of Morphology*, **152**, 171–6.

Dunagan, T.T. & Miller, D.M. (1978*a*). Anatomy of the genital ganglion of the male acanthocephalan, *Moniliformis moniliformis*. *Journal of Parasitology*, **64**, 431–5.

Dunagan, T.T. & Miller, D.M. (1978*b*). Determination of dorsal and ventral surfaces in histological preparations of Archiacanthocephala. *Proceedings of the Helminthological Society of Washington*, **45**, 256–7.

Dunagan, T.T. & Miller, D.M. (1978*c*). Muscles of the reproductive system of male *Moniliformis moniliformis* (Acanthocephala). *Proceedings of the Helminthological Society of Washington*, **45**, 69–76.

Dunagan, T.T. & Miller, D.M. (1979). Genital ganglion and associated nerves in male *Macracanthorhynchus hirudinaceus* (Acanthocephala). *Proceedings of the Helminthological Society of Washington*, **46**, 106–14.

Dunagan, T.T. & Miller, D.M. (1980). *Macracanthorhynchus hirudinaceus* from swine: an eighteen-year record of acanthocephala from Southern Illinois. *Proceedings of the Helminthological Society of Washington*, **47**, 33–6.

Dunagan, T.T. & Miller, D.M. (1981). Anatomy of the cerebral ganglion in *Oligacanthorhynchus tortuosa* (Acanthocephala) from the opossum (*Didelphis virginiana*). *Journal of Parasitology*, **67**, 881–5.

Dunagan, T.T. & Miller, D.M. (1983). Apical sense organ of *Macracanthorhynchus hirudinaceus* (Acanthocephala). *Journal of Parasitology*, **69**, 897–902.

Dunagan, T.T. & Scheifinger, C.C. (1966*a*). Studies on the TCA cycle of *Macracanthorhynchus hirudinaceus* (Acanthocephala). *Comparative Biochemistry and Physiology*, **18**, 663–7.

Dunagan, T.T. & Scheifinger, C.C. (1966*b*). Studies of glycolytic enzymes from *Macracanthorhynchus hirudinaceus* (Acanthocephala). *Journal of Parasitology*, **52**, 730–4.

Dunagan, T.T. & Yau, T.M. (1968). Oligosaccharidases from *Macracanthorhynchus hirudinaceus* (Acanthocephala) from swine. *Comparative Biochemistry and Physiology*, **26**, 281–9.

Eckert, J. (1961). Bemerkenswirte Falle von Helminthenbefall bei Zootieren. *Monatshefte für Veterinarmedizin*, **16**, 851–6.

Edmonds, S.J. (1955). Acanthocephala collected by the Australian National Antarctic Research Expedition on Heard Island and Macquarie Island during 1948–50. *Transactions of the Royal Society of South Australia*, **78**, 141–4.

Edmonds, E.J. (1957*a*). Acanthocephala. *British, Australian, New Zealand Antarctic Research Expedition 1929–1931. Reports – Series B (Zoology and Botany)*, **6**, 93–8.

Edmonds, S.J. (1957*b*). Australian Acanthocephala No. 10. *Transactions of the Royal Society of South Australia*, **80**, 76–80.

Edmonds, S.J. (1965). Some experiments on the nutrition of *Moniliformis dubius* Meyer (Acanthocephala). *Parasitology*, **55**, 337–44.

Edmonds, S.J. (1966). Hatching of the eggs of *Moniliformis dubius*. *Experimental Parasitology*, **19**, 216–26.

Edmonds, S.J. & Dixon, B.R. (1966). Uptake of small particles by *Moniliformis dubius* (Acanthocephala). *Nature, London*, **209**, 99.

Elbein, A.D. (1974). The metabolism of α,α-trehalose. *Advances in Carbohydrate Chemistry and Biochemistry*, **30**, 227–56.

Elkins, C.A. & Nickol, B.B. (1983). The epizootiology of *Macracanthorhynchus ingens*. *Journal of Parasitology*, **69**, 951–6.

El Mofty, M.M. & Smyth, J.D. (1964). Endocrine control of encystation in *Opalina ranarum* parasitic in *Rana temporaria*. *Experimental Parasitology*, **15**, 185–99.

Engelbrecht, H. (1957). Einige Bemerkungen zu *Pomphorhynchus laevis* (Zoega, Müller, 1776) als Parasit in *Pleuronectes flexus* and *Pleuronectes platessa*. *Zentralblatt für Bakteriologie, Parasitenkunde, Infektionskrankheiten und Hygiene*, Abteilung 1, **168**, 474–9.

Esch, G.W. (1977). Parasitism and r- and K- selection. In *Regulation of Parasite Populations*, ed. G.W. Esch, pp. 9–62. New York and London: Academic Press.

Esch, G.W., Gibbons, J.W. & Bourque, J.E. (1975). Analysis of the relationship between stress and parasitism. *American Midland Naturalist*, **93**, 339–53.

Esch, G.W., Campbell, G.C., Conners, R.E. & Coggins, J.R. (1976). Recruitment of helminth parasites by Bluegills (*Lepomis macrochiras*) using a modified live-box technique. *Transactions of the American Fisheries Society*, **105**, 486–90.

Eure, H. (1976). Seasonal abundance of *Neoechinorhynchus cylindratus* taken from largemouth bass (*Micropterus salmoides*) in a heated reservoir. *Parasitology*, **73**, 355–70.

Fairbairn, D. (1958). Trehalose and glucose in helminths and other invertebrates. *Canadian Journal of Zoology*, **36**, 787–95.

Fairbairn, D. (1970). Biochemical adaptation and loss of genetic capacity in helminth parasites. *Biological Reviews of the Cambridge Philosophical Society*, **45**, 29–72.

Farland, W.H. & MacInnis, A.J. (1978). In vitro thymidine kinase activity: present in *Hymenolepis diminuta* (Cestoda) and *Moniliformis dubius* (Acanthocephala), but apparently lacking in *Ascaris lumbricoides* (Nematoda). *Journal of Parasitology*, **64**, 564–5.

Farzaliev, A.M. & Petrochenko, V.I. (1980). New data on the life cycle of the acanthocephalan *Macracanthorhynchus catulinus* Kostylew, 1927 (Acanthocephala), a parasite of carnivores. *Trudy Vsesoyuznogo Instituta Gel'mintologii im K. I. Skryabina (Ekologiya gel'mintov, epizoologiya gel'mintozov i mery bor'by s nimi)*, **25**, 140–4.

Faust, E.C. (1929). *Human Helminthology. A Manual for Clinicians, Sanitarians and Medical Zoologists*. Philadelphia: Lea & Febiger.

Fioravanti, C.F. & Saz, H.J. (1980). Energy metabolism of adult *Hymenolepis diminuta*. In *Biology of the Tapeworm* Hymenolepis diminuta, ed. H.P. Arai, pp. 463–504. London and New York: Academic Press.

Fish, P.G. (1972). Notes on *Moniliformis clarki* (Ward) (Acanthocephala: Moniliformidae) in West Central Indiana. *Journal of Parasitology*, **58**, 147.

Fisher, F.M. (1960). On Acanthocephala of turtles, with a description of *Neoechinorhynchus emyditoides* n. sp. *Journal of Parasitology*, **46**, 257–66.

Fisher, F.M. (1964). Synthesis of trehalose in Acanthocephala. *Journal of Parasitology*, **50**, 803–4.

Fisher, R.A. (1930). *The Genetical Theory of Natural Selection*. Oxford: Clarendon Press.

Florescu, B. (1936). Le cycle évolutif de *Polymorphus minutus* Goeze en Roumanie (Acanthocephala). *Musée d'Histoire Naturelle de Chisinau–Roumanie*, **7**, 61–9.

Fotedar, D.N. (1968). New species of *Neoechinorhynchidae*, Hamann, 1892 from *Oreinus sinuatus*, freshwater fish in Kashmir. *Kashmir Science*, **5**, 147–52.

Fotedar, D.N., Raina, M.K. & Dhar, R.L. (1977). Life cycle stages and experimental transmission of *Polymorphus magnus* Skrjabin, 1913 from *Gammarus pulex* to *Anas platyrhynchos domesticus* in Kashmir, India. *Abstracts of the 1st National Congress of Parasitology, Baroda, February, 1977, Indian Society for Parasitology*, **1**, 24–5.

Franzen, A. (1956). On spermiogenesis, morphology of the spermatozoon, and biology of fertilization among invertebrates. *Zoologiska Bidrag från Uppsala*, **31**, 355–482.

Fresi, E. (1967*a*). Ricerche sulla sessualita di *Asellus* (*Proasellus*) *coxalis* Dollfus (Crust. Isop.). I. Effeti del parassitismo da larve di Acanthocefalo. *Archivio Zoologica Italiano*, **52**, 277–86.

Fresi, E. (1967*b*). Ricerche sulla sessualita di *Asellus* (*Proasellus*) *coxalis* Dollfus (Crust. Isop.). II. Pseudormafroditismo. *Archivio Zoologica Italiano*, **52**, 287–303.

Friedman, S. (1978). Trehalose regulation, one aspect of metabolic homeostasis. *Annual Review of Entomology*, **23**, 389–407.

Furtado, J.I. (1963). On *Acanthogyrus partispinus* nov. sp. (Quadrigyridae, Acanthocephala) from a Malayan cyprinid *Hampala macrolepidota* Van Hasselt. *Zeitschrift für Parasitenkunde*, **23**, 219–25.

Gaevskaya, A.V. (1977). The helminth fauna of the Atlantic squid *Stenoteuthis pteropus* (Steenstrup). *Nauchnye Doklady Vysshei Shkoly, Biologicheskie Nauki*, 1977, **8**, 47–52.

Gaevskaya, A.V. & Nigmatullin, C.M. (1981). Some ecological aspects of parasitic relations of *Sthenoteuthis pteropus* in the tropical Atlantic Ocean. *Nauchnye Doklady Vysshei Shkoly, Biologicheskie Nauki*, 1981, **1**, 52–7.

Gafurov, A.K. (1970). Acanthocephalan and cestode larvae found in tenebrionid beetles in Tadzhikstan. *Izvestiya Akademii Fanhoi RSS Tocikiston, Biologiya*, **3**(40), 36–9.

Gafurov, A.K. (1975). New intermediate hosts of the acanthocephalan *Mediorhynchus papillosus* Van Cleave, 1916. In *Zoologicheskii sbornik, chast' II*, pp. 103–4. Dushanbe, USSR: 'Donish'.

Gamj, P., Murer, H. & Kinne, R. (1979). Calcium ion transport across plasma membranes isolated from rat kidney cortex. *Biochemical Journal*, **178**, 549–57.

Garden, E.A., Rayski, C. & Thom, V.M. (1964). A parasitic disease in eider ducks. *Bird Study*, **11**, 280–7.

Gegenbauer, C. (1859). *Grundzuge der Vergleichende Anatomie*. Leipzig.

George, P.V. & Nadakal, A.M. (1973). Studies on the life cycle of *Pallisentis nagpurensis* Bhalerao, 1931 (Pallisentidae; Acanthocephala) parasitic in the fish *Ophiocephalus striatus* (Bloch). *Hydrobiologia*, **42**, 31–43.

George, P.V. & Nadakal, A.M. (1984). Studies on encapsulation of immature juvenile of the acanthocephalid worm, *Pallisentis nagpurensis* Bhalerao, 1931 on the liver of definitive host, *Ophicephalus striatus* (Bloch.). *Japanese Journal of Parasitology*, **32**, 387–91.

Gettier, D.A. (1942). Studies on the saline requirements of *Neoechinorhynchus emydis*. *Proceedings of the Helminthological Society of Washington*, **9**, 75–8.

Ginger, C.D. & Fairbairn, D. (1966). Lipid metabolism in helminth parasites I. The lipids of *Hymenolepis diminuta* (Cestoda). *Journal of Parasitology*, **52**, 1086–96.

Glynn, I.M. & Karlish, S.J.D. (1975). The sodium pump. *Annual Review of Physiology*, **37**, 13–55.

Gmelin, J.F. (1788–1793). *Caroli a Linné-Systema naturae per regna tria naturae, secundum classes, ordines, genera, species cum characteribus differentiis, synonymis, locis. Editio decima tertia aucta, reformata.* Vol. 1 (7 pts) 4120 pp.; Vol 2 (2 pts) 1661 pp.; Vol. 3, 476 pp.

Goeze, J.A.E. (1782). *Versuch einer Naturgeschichte der Eingeweidewurmer thierischer Körper*. Blankenburg. xi+471 pp.

Goldstein, J.L., Anderson, R.G.W. & Brown, M.S. (1979). Coated pits, coated vesicles and receptor-mediated endocytosis. *Nature, London*, **279**, 679–85.

Golubev, A.I. & Sal'nikov, V.V. (1979). The ultrastructure of the specialized junctions between the neurone and the intracellular substance in the cerebral ganglion of *Echinorhynchus gadi*. *Parasitologiya*, **21**, 1100–2.

Golvan, Y.J. (1956*a*). Acanthocéphales d'oiseaux. Première note. Description d'*Arhythmorhynchus longicollis* (Villot, 1875) et revision du genre *Arhythmorhynchus* Lühe, 1911 (Acanthocephala). *Annales de Parasitologie Humaine et Comparée*, **31**, 199–224.

Golvan, Y.J. (1956*b*). Le genre *Centrorhynchus* Lühe, 1911 (Acanthocephala – Polymorphidae). Revision des espèces européennes et description d'une nouvelle espèce africaine parasite de Rapace diurne. *Bulletin de l'Institut Français d'Afrique Noire, Series A*, **18**, 732–91.

Golvan, Y.J. (1956*c*). Acanthocéphales d'oiseaux. Note additionelle. *Pseudoporrorchis rotundatus* (O.V. Linstow, 1897) (Palaeacanthocephala – Polymorphidae), parasite d'un Cucullidae, *Centropus madagascariensis* (Briss.). *Bulletin de la Société Zoologique de France*, **81**, 339–44.

Golvan, Y.J. (1956*d*). Acanthocephales d'Amazonie. Redescription d'*Oligacanthorhynchus iheringi* Travassos, 1916 et description de *Neoechinorhynchus buttnerae*, n. sp. (Neoacanthocephala – Neoechinorhynchidae). *Annales de Parasitologie Humaine et Comparée*, **31**, 500–24.

Golvan, Y.J. (1958). Le Phylum des Acanthocephala. Premier note. Sa place dans l'échelle zoologique. *Annales de Parasitologie Humaine et Comparée*, **33**, 538–602.

Golvan, Y.J. (1959*a*). Le Phylum des Acanthocephala. Deuxième note. La Classe de Eoacanthocephala (Van Cleave, 1936). *Annales de Parasitologie Humaine et Comparée*, **34**, 5–52.

Golvan, Y.J. (1959*b*). Protonephridies et taxonomie des acanthocephales. *Proceedings of the XVth International Congress of Zoology*, 960.

Golvan, Y.J. (1960–1961). Le Phylum des Acanthocephala. Troisième note. La Classe des Palaeacanthocephala (Meyer, 1931). *Annales de Parasitologie Humaine et Comparée*, **35**, 76–91, 138–65, 350–86, 573–93, 713–23; **36**, 76–91, 612–47, 717–36.

Golvan, Y.J. (1962). Le Phylum Acanthocephala (Quatrième note). La Classe des Archiacanthocephala (A. Meyer, 1931). *Annales de Parasitologie Humaine et Comparée*, **37**, 1–72.

Golvan, Y.J. (1969). Systematique des Acanthocephales (Acanthocephala Rudolphi, 1801), L'ordre des Palaecanthocephala Meyer, 1931, La superfamille des Echinorhynchidea (Cobbold, 1876) Golvan et Houin 1973. *Mémoires du Muséum Nationale d'Histoire Naturelle*, **47**, 1–373.

Golvan, Y.J. & Ormières, R. (1957). Presence d'un juvenile du genère *Gordiorhynchus* A. Meyer, 1931 (Acanthocephala, Polymorphidae) chez un acridien du Congo Belge. *Parc National Albert*, Deuxième Série, **5**, 85–9.

Golvan Y.J. & Theodorides, J. (1960). Cycle évolutif d'un acanthocephale parasite de gerbillides du genre *Meriones* en Iran. *Comptes Rendus Hebdomadaires des Séances de l'Académie des Sciences*, **250**, 224–5.

Golvan, Y.J., Gracia-Rodrigo, A. & Diaz-Ungria, C. (1964). *Megapriapus ungriai* (Gracia-Rodrigo, 1960) n. gen. (Palaeacanthocephala) parasite d'une Pasternague d'eau douce du Vénézuéla (*Potamotrygon bystrix*). *Annales de Parasitologie Humaine et Comparée*, **39**, 53–9.

Gonzalez, J.P. & Mishra, G.S. (1976). Natural infection of cockroaches (Insecta: Dictyoptera) in Tunis; their role as intermediate hosts of *Mastophorus muris* and of *Moniliformis moniliformis*. *Archives de l'Institut Pasteur de Tunis*, **53**, 211–38.

Grabda, J. & Slosarczyk, W. (1981). Parasites of marine fishes from New Zealand. *Acta Ichthyologica et Piscatoria*, **11**, 85–103.

Grabda-Kazubska, B. (1964). Observations on the armature of embryos of acanthocephalans. *Acta Parasitologica Polonica*, **12**, 215–31.

Graeber, K. & Storch, V. (1978). Elektronmikroskopische und morphometrische Untersuchungen am Integument der Acanthocephala (Aschelminthes). *Zeitschrift für Parasitenkunde*, **57**, 121–35.

Graff, D.J. (1964). Metabolism of C^{14}-glucose by *Moniliformis dubius* (Acanthocephala). *Journal of Parasitology*, **50**, 230–4.

Graff, D.J. (1965). The utilization of $C^{14}O_2$ in the production of acid metabolites by *Moniliformis dubius* (Acanthocephala). *Journal of Parasitology*, **51**, 72–5.

Graff, D. & Allen, K. (1963). Glycogen content in *Moniliformis dubius* (Acanthocephala). *Journal of Parasitology*, **49**, 204–8.

Graff, D.J. & Kitzman, W.B. (1965). Factors influencing the activation of acanthocephalan cystacanths. *Journal of Parasitology*, **51**, 424–9.

Grassi, B. & Calandruccio, S. (1888). Ueber einen *Echinorhynchus*, welcher im Menschen parasitirt und dessen Zwischenwirte ein *Blaps* ist. *Centralblatt für Bakteriologie und Parasitenkunde*, **3**, 521–5.

Graybill, W.H. (1902). Some points in the structure of the Acanthocephala. *Transactions of the American Microscopical Society*, **23**, 191–200.

Greeff, R. (1864). Untersuchen über den Bau and die Naturgeschichte von *Echinorhynchus miliarius* Zenker (*Echinorhynchus polymorphus*). *Archiv für Naturgeschichte (Berlin)*, **20**, 98–140.

Grenquist, P. (1970). Väkäkärsämatojen aiheuttanasta haahkojen kuolleisuudesta. *Suomen Riista*, **22**, 24–34.

Grey, A.J. & Hayunga, E.G. (1980). Evidence for alternative site selection by *Glaridacris laruei* (Cestodea: Caryophyllidea) as a result of interspecific competition. *Journal of Parasitology*, **66**, 371–2.

Gunn, R.B. (1980). Co- and counter-transport mechanisms in cell membranes. *Annual Review of Physiology*, **42**, 249–59.

Gupta, N.K. & Jain, M. (1975). On two already known species of Acanthocephala of the genus *Porrorchis* Fukui, 1929 (Gigantorhynchidea: Prosthorhynchidae) from Chandigarh, India. *Acta Parasitologica Polonica*, **23**, 381–7.

Gupta, P.V. (1950). On some stages in the development of the acanthocephalan genus *Centrorhynchus*. *Indian Journal of Helminthology*, **2**, 41–8.

Gupta, V. & Fatma, S. (1983). On five new species of the genus *Centrorhynchus* Lühe, 1911 (Acanthocephala: Centrorhynchidae) from avian and amphibian hosts of Lucknow, Uttar Pradesh. *Indian Journal of Helminthology*, **33**, 105–20.

Guraya, S.S. (1969). Histochemical observations on the developing acanthocephalan oocyte. *Acta Embryologiae Experimentalis*, **1**, 147–55.

Guraya, S.S. (1971). Morphological and histochemical observations on the acanthocephalan spermatogenesis. *Acta Morphologica Neerlando-Scandinavica*, **9**, 75–83.

Haffner, K. von (1942*a*). Untersuchungen über das Urogenitalsystem der Acanthocephalen. I. Teil. Das Urogenitalsystem von *Oligacanthorhynchus thumbi* forma juv. *Zeitschrift für Morphologie und Oekologie der Tiere*, **38**, 251–94.

Haffner, K. von (1942*b*). Untersuchungen über das Urogenitalsystem der Acanthocephalen. II. Teil. Das Urogenitalsystem von *Gigantorhynchus echinodiscus* Diesing. *Zeitschrift für Morphologie und Oekologie der Tiere*, **38**, 295–316.

Haffner, K. von (1942*c*). Untersuchungen über das Urogenitalsystem der Acanthocephalen. III. Teil. Theoretische Betrachtungen auf Grund eigener Ergebnisse. *Zeitschrift für Morphologie und Oekologie der Tiere*, **38**, 317–33.

Haffner, K. von (1943). Die Sinnesorgan an der Russelspitze von Acanthocephalen. *Zeitschrift für Morphologie und Oekologie der Tiere*, **40**, 80–92.

Hair, J.D. (1969). Geographical distribution and seasonal incidence in the helminth parasites of the starling, *Sturnus vulgaris* L. *Program and Abstracts of the 44th Annual Meeting of the American Society of Parasitologists*, p. 42.

Haley, A.J. & Bullock, W.L. (1952). Comparative histochemical studies on the cement glands of certain Acanthocephala. *Journal of Parasitology*, **38**, 25–6.

Halvorsen, O. (1972). Studies on the helminth fauna of Norway XX. Seasonal cycles of fish parasites in the River Glomma. *Norwegian Journal of Zoology*, **20**, 9–18.

Hamann, O. (1889). Vorlaufige Mitteilungen zur Morphologie der Echinorhynchen. *Nachrichten K. Ges. Wiss. Georg-Augusts. Univ. Gottingen*, **6**, 85–9.

Hamann, O. (1891*a*). Monographie der Acanthocephalen (Echinorhynchen). Ihre Entwicklungsgeschichte, Histogenie und Anatomie nebst Beitragen zur Systematik und Biologie. *Jenaische Zeitschrift für Naturwissenschaft*, **25**, 113–231.

Hamann, O. (1891*b*). Die kleineren Süsswasserfische als Haupt- und Zwischenwirte des *Echinorhynchus proteus* Westr. *Centralblatt für Bakteriologie und Parasitenkunde*, **10**, 791–2.

Hamann, O. (1892). Das System der Acanthocephalen. *Zoologischer Anzieger*, **15**, 195–7.

Hamann, O. (1895). *Die Nemathelminthen. Beiträge zur Kenntnis ihrer Entwicklung, ihres Baues und ihrer Lebensgeschichte. Zweites Heft. 1. Monographie der Acanthocephalen* (*Echinorhynchen*). Jena. 120 pp.

Hammond, R.A. (1966*a*). The proboscis mechanisms of *Acanthocephalus ranae*. *Journal of Experimental Biology*, **45**, 203–13.

Hammond, R.A. (1966*b*). Changes of internal hydrostatic pressure and body shape in *Acanthocephalus ranae*. *Journal of Experimental Biology*, **45**, 197–202.

Hammond, R.A. (1967). The fine structure of the trunk and praesoma wall of *Acanthocephalus ranae* (Schrank, 1788), Lühe, 1911. *Parasitology*, **57**, 475–86.

Hammond, R.A. (1968*a*). Observations on the body surface of some acanthocephalans. *Nature, London*, **218**, 872–3.

Hammond, R.A. (1968*b*). Some observations on the role of the body wall of *Acanthocephalus ranae* in lipid uptake. *Journal of Experimental Biology*, **48**, 217–25.

Harada, I. (1931). Das Nervensystem von *Bolbosoma turbinella* (Dies). *Japanese Journal of Zoology*, **3**, 161–99.

Harms, C.E. (1963). The development and cultivation of the acanthocephalan, *Octospinifer macilentis* (macilentis) Van Cleave, 1919. *Dissertation Abstracts*, **23**, 2632–3.

Harms, C.E. (1965*a*). The life cycle and larval development of *Octospinifer macilentis* (Acanthocephala: Neoechinorhynchidae). *Journal of Parasitology*, **51**, 286–93.

Harms, C.E. (1965*b*). *In vitro* cultivation of an acanthocephalan, *Octospinifer macilentis*. *Experimental Parasitology*, **17**, 41–5.

Harpur, R.P. (1963). Maintenance of *Ascaris lumbricoides in vitro*. II. Changes in muscle and ovary carbohydrates. *Canadian Journal of Biochemistry and Physiology*, **41**, 1673–89.

Harris, J.E. (1970). Precipitin production by chub (*Leuciscus cephalus*) to an intestinal helminth. *Journal of Parasitology*, **56**, 1035.

Harris, J.E. (1972). The immune response of a cyprinid fish to infections of the acanthocephalan *Pomphorhynchus laevis*. *International Journal for Parasitology*, **2**, 459–69.

Harwood, R.F. & James, M.T. (1979). *Entomology in Human and Animal Health*. London: Macmillan Inc.

Hasan, R. & Qasim, S.Z. (1960). Of *Pallisentis basiri* Farooqi (Acanthocephala) in the liver of *Trichogaster chuna* (Ham.). *Zeitschrift für Parasitenkunde*, **20**, 152–6.

Hassell, M. (1978). *The Dynamics of Arthropod Predator–Prey Systems*. Princeton, New Jersey: Princeton University Press.

Helle, E. & Valtonen, E.T. (1980). On the occurrence of *Corynosoma* spp. (Acanthocephala) in ringed seals (*Pusa hispida*) in the Bothnian Bay, Finland. *Canadian Journal of Zoology*, **58**, 298–303.

Helle, E. & Valtonen, E.T. (1981). Comparison between spring and autumn infection by *Corynosoma* (Acanthocephala) in the ringed seal *Pusa hispida* in the Bothnian Bay of the Baltic Sea. *Parasitology*, **82**, 287–96.

Henle, F.G.J. (1840). Uber verschiedene Lebensperioden der Eingeweidewurmer. *N. Notiz. Geb. Nat.-u Heilk*, (270), **13**, 86–7.

Hentschel, H. (1979). Occurrence of *Echinorhynchus truttae* (Acanthocephala) in the epirhythron of a brook in the Mittelbirge, West Germany. *Bericht der Naturhistorischen Gesellschaft zu Hannover*, **22**, 109–24.

Hers, H.G. (1976). The control of glycogen metabolism in the liver. *Annual Review of Biochemistry*, **45**, 167–89.

Hibbard, K.M. & Cable, R.M. (1968). The uptake and metabolism of tritiated glucose, tyrosine, and thymidine by adult *Paulisentis fractus* Van Cleave and Bangham, 1949 (Acanthocephala: Neoechinorhynchidae). *Journal of Parasitology*, **54**, 517–23.

Hightower, K., Miller, D.M. & Dunagan, T.T. (1975). Physiology of the body wall muscles in an acanthocephalan. *Proceedings of the Helminthological Society of Washington*, **42**, 71–9.

Hindsbo, O. (1972). Effects of *Polymorphus* (Acanthocephala) on colour and behaviour of *Gammarus lacustris*. *Nature, London*, **238**, 333.

Hine, P.M. (1977). *Acanthocephalus galaxii* n. sp. parasitic in *Galaxias maculatus* (Jenyns, 1842) in the Lema Stream, New Zealand. *Journal of the Royal Society of New Zealand*, **7**, 51–7.

Hine, P.M. & Kennedy, C.R. (1974*a*). Observations on the distribution, specificity and pathogenicity of the acanthocephalan *Pomphorhynchus laevis* (Muller). *Journal of Fish Biology*, **6**, 521–35.

Hine, P.M. & Kennedy, C.R. (1974*b*). The population biology of the acanthocephalan *Pomphorhynchus laevis* (Müller) in the River Avon. *Journal of Fish Biology*, **6**, 665–79.

Hirsch, R.P. (1980). Distribution of *Polymorphus minutus* among its intermediate hosts. *International Journal for Parasitology*, **10**, 243–8.

Hnath, J.G. (1969). Transfer of an adult acanthocephalan from one fish host to another. *Transactions of the American Fisheries Society*, **98**, 332.

Holcman-Spector, B., Mañe-Garzón, F. & Dei-Cas, E. (1977). Une larva cystacantha (Acanthocephala) de la cavidad general de *Chasmagnathus granulata* Dana, 1851. *Revista de Biologia del Uruguay*, **5**, 67–76.

Holland, C.V. (1983*a*). *Interactions between three species of helminth parasite in the rat small intestine*. Ph D thesis, University of Cambridge.

Holland, C.V. (1983*b*). Interactions between *Moniliformis* (Acanthocephala) and *Nippostrongylus* (Nematoda) in the small intestine of laboratory rats. *Parasitology*, **88**, 303–15.

Holloway, H.L. (1966). *Prosthorhynchus formosum* (Van Cleave, 1918) in songbirds, with notes on acanthocephalans as potential parasites of poultry in Virginia. *Virginia Journal of Science*, **17** (new series), 149–54.

Holloway, H.L. & Bier, J.W. (1967). Notes on the host specificity of *Corynosoma hamanni* (Linstow, 1892). *Bulletin of the Wildlife Disease Association*, **3**, 76–7.

Holloway, H.L. & Nickol, B.B. (1970). Morphology of the trunk of *Corynosoma hamanni* (Acanthocephala: Polymorphidae). *Journal of Morphology*, **130**, 151–2.

Holmes, J.C. (1961). Effects of concurrent infections on *Hymenolepis diminuta* (Cestoda) and *Moniliformis dubius* (Acanthocephala). I. General effects and comparison with crowding. *Journal of Parasitology*, **47**, 209–16.

Holmes, J.C. (1962*a*). Effects of concurrent infections on *Hymenolepis diminuta* (Cestoda) and *Moniliformis dubius* (Acanthocephala). II. Effects on growth. *Journal of Parasitology*, **48**, 87–96.

Holmes, J.C. (1962*b*). Effects of concurrent infections on *Hymenolepis diminuta* (Cestoda) and *Moniliformis dubius* (Acanthocephala). III. Effects in hamsters. *Journal of Parasitology*, **48**, 97–100.

Holmes, J.C. (1973). Site selection by parasitic helminths: interspecific interactions, site segregation, and their importance to the development of helminth communities. *Canadian Journal of Zoology*, **51**, 333–47.

Holmes, J.C. (1979). Parasite populations and host community structure. In *Host–Parasite Interfaces*, ed. B.B. Nickol, pp. 27–46. New York and London: Academic Press.

Holmes, J.C. & Bethel, W.M. (1972). Modification of intermediate host behaviour by parasites. In *Behavioural Aspects of Parasite Transmission*, ed. E.U. Canning and C.A. Wright, pp. 123–49. *Journal of the Linnean Society*, **51**, Suppl. 1.

Holmes, J.C., Hobbs, R.P. & Leong, T.S. (1977). Populations in perspective: community organisation and regulation of parasite populations. In *Regulation of Parasite Populations*, ed. G.W. Esch, pp. 209–45. New York: Academic Press.

Holwerda, D.A. & de Zwaan, A. (1980). On the role of fumarate reductase in anaerobic carbohydrate catabolism of *Mytilus edulis* L. *Comparative Biochemistry and Physiology*, **67B**, 447–53.

Honegger, P. & Semenza, G. (1973). Multiplicity of carriers for free glucalogues in hamster small intestine. *Biochimica et Biophysica Acta*, **318**, 390–410.

Hopp, W.B. (1946). Notes on the life history of *Neoechinorhynchus emydis* (Leidy), an acanthocephalan parasite of turtles. *Proceedings of the Indiana Academy of Science*, **55**, 183.

Hopp, W.B. (1954). Studies on the morphology and life cycle of *Neoechinorhynchus emydis* (Leidy), an acanthocephalan parasite of the map turtle, *Graptemys geographica* (Le Sueur). *Journal of Parasitology*, **40**, 284–99.

Hopper, A.F. & Hart, N.H. (1980). *Foundations of Animal Development*. Oxford: Oxford University Press.

Horvath, K. (1971). Glycogen metabolism in larval *Moniliformis dubius* (Acanthocephala). *Journal of Parasitology*, **57**, 132–6.

Horvath, K. (1972). Glycolytic enzymes in larval *Moniliformis dubius*. *Journal of Parasitology*, **58**, 1219–20.

Horvath, K. & Fisher, F.M. (1971). Enzymes of CO_2 fixation in larval and adult *Moniliformis dubius* (Acanthocephala). *Journal of Parasitology*, **57**, 440–2.

Houslay, M.D. & Stanley, K.K. (1982). *Dynamics of Biological Membranes*. New York: Wiley-Interscience.

Huffman, D.G. & Bullock, W.L. (1975). Meristograms: graphical analysis of serial variation of proboscis hooks of *Echinorhynchus* (Acanthocephala). *Systematic Zoology*, **24**, 333–45.

Huffman, D.G. & Nickol, B.B. (1979). Meristogram analysis of the acanthocephalan genus *Pomphorhynchus* in North America. *Journal of Parasitology*, **64**, 851–9.

Hutton, T.L. & Oetinger, D.F. (1980). Morphogenesis of the proboscis hook of an archiacanthocephalan, *Moniliformis moniliformis* (Bremser, 1811) Travassos, 1915. *Journal of Parasitology*, **66**, 965–72.

Hyman, L.H. (1951). *The Invertebrates*, vol. 3, *Acanthocephala, Aschelminthes, and Entoprocta*, 72 pp. New York: McGraw Hill.

Hynes, H.B.N. (1955). The reproductive cycle of some British freshwater Gammaridae. *Journal of Animal Ecology*, **24**, 352–87.

Hynes, H.B.N. & Nicholas, W.L. (1957). The development of *Polymorphus minutus* (Goeze, 1782) (Acanthocephala) in the intermediate host. *Annals of Tropical Medicine and Parasitology*, **51**, 380–91.

Hynes, H.B.N. & Nicholas, W.L. (1958). The resistance of *Gammarus* spp. to infection by *Polymorphus minutus* (Goeze, 1782) (Acanthocephala). *Annals of Tropical Medicine and Parasitology*, **52**, 376–83.

Hynes, H.B.N. & Nicholas, W.L. (1963). The importance of the acanthocephalan *Polymorphus minutus* as a parasite of domestic ducks in the United Kingdom. *Journal of Helminthology*, **37**, 185–98.

Insler, G.D. & Roberts, L.S. (1980). Developmental physiology of cestodes. XVI. Effects of certain excretory products on incorporation of ^{3}H-thymidine into DNA of *Hymenolepis diminuta*. *Journal of Experimental Zoology*, **211**, 55–61.

Itämies, J., Valtonen, E.T. & Fagerholm, H.P. (1980). *Polymorphus minutus* infestation in eiders and its role as a possible cause of death. *Annales Zoologici Fennici*, **17**, 285–9.

Ivanova, G.V. & Makhanbetov, S.H. (1975). Innervation of the genital system of the male *Polymorphus phippsi* Kostylew, 1922. *Trudy Gel'mintologicheskoi Laboratorii Akademii Nauk SSR* (*Issedovani po sistematike, zhiznemyn tsiklam i biokimii gel'mintov*), **25**, 33–7.

Ivashkin, V.M. (1956). On the life cycle of the acanthocephalan *Moniliformis moniliformis* (Bremser, 1811) Travassos, 1915. *Trudy Gel'mintologicheskoi Laboratorii. Trudii Gel'mintologicheskoi Akademii Nauk SSR*, **8**, 31–2.

Ivashkin, V.M. (1972). Infections of Amphipoda from streams in the Carpathian mountains by acanthocephalan larvae. In *Parazity vodnykh bespozvonochnykh zhivotnykh. I Vsesoyuznyi Simpozium. L'vov, USSR: Izdatel'stvo L'vovskogo Universiteta*, 31–3.

Ivashkin, V.M. & Shmitova, G. Ya. (1969). The life-cycle of *Mediorhynchus papillosus*. *Trudy Gel'mintologicheskoi Laboratorii, Akademii Nauk SSR*, **20**, 62–3.

Jackson, D.P., Chan, A.H. & Cotter, D.A. (1982). Utilization of trehalose during *Dictyostelium discoideum* spore germination. *Developmental Biology*, **90**, 369–74.

James, P.M. (1954). On some helminths from British small mammals with a redescription of *Echinorhynchus rosai* Porta, 1910. *Journal of Helminthology*, **28**, 183–8.

Janssens, P.A. & Bryant, C. (1969). The ornithine-urea cycle in some parasitic helminths. *Comparative Biochemistry and Physiology*, **30**, 261–72.

Jarling, C. (1983). The helminth fauna of the smelt (*Osmerus eperlanus*) in the Elbe estuary. *Archiv für Hydrobiologie Supplementband* **61**, 377–95.

Jennings, J.B. & Calow, P. (1975). The relationship between high fecundity and the evolution of endoparasitism. *Oecologia*, **21**, 109–15.

Jensen, T. (1952). The life cycle of the fish acanthocephalan, *Pomphorhynchus bulbocolli* (Linkins) Van Cleave, 1919, with some observations on larval development *in vitro*. Ph D thesis, University of Minnesota. *Dissertation Abstracts International*, **12**, 607.

Jewell, P.A. (1976). Selection for reproductive success. In *Reproduction in Mammals*, vol. 6, ed. C.R. Austin and R.V. Short, pp. 71–109. Cambridge University Press.

Jilek, R. (1978). Seasonal occurrence and host specificity of *Gracilisentis gracilisentis* and *Tanaorhamphus longirostris* (Acanthocephala: Neoechinorhynchidae) in Crab Orchard Lake, Illinois. *Journal of Parasitology*, **64**, 951–2.

Jilek, R. & Crites, J.L. (1979). Morphological studies on the proboscis of *Gracilisentis gracilisentis* (Acanthocephala: Neoechinorhynchidae). *Transactions of the American Microscopical Society*, **98**, 243–7.

John, B. (1957). The chromosomes of zooparasites. I. *Acanthocephalus ranae* (Acanthocephala: Echinorhynchidae). *Chromosoma*, **8**, 730–8.

Johnson, C.A. & Harkema, R. (1971). The life history of *Pomphorhynchus rocci* Cordonnier and Ward, 1967 (Acanthocephala) in the striped bass, *Morone saxatilis*. *Bulletin of the Association of Southeastern Biologists*, **18**, 40.

Johnston, J.M. (1968). Mechanism of fat absorption. In *Handbook of Physiology, Section 6: Alimentary Canal*, vol. 3, ed. C.F. Code, pp. 1353–75. Washington, D.C.: American Physiological Society.

Johnston, T.H. (1914). Some new Queensland endoparasites. *Proceedings of the Royal Society of Queensland*, **26**, 76–84.

Johnston, T.H. & Deland, E.W. (1929). Australian Acanthocephala, No. 2. *Transactions of the Royal Society of South Australia*, **53**, 155–66.

Johnston, T.H. & Edmonds, S.J. (1948). Australian Acanthocephala, No. 7. *Transactions of the Royal Society of South Australia*, **72**, 69–76.

Johnston, T.H. & Edmonds, S.J. (1952). Australian Acanthocephala, No. 9. *Transactions of the Royal Society of South Australia*, **75**, 16–21.

Johnstone, R.M. (1972). Glycine accumulation in absence of Na^+ and K^+ gradients in Ehrlich ascites cells: shortfall of the potential energy from the ion gradients for glycine accumulation. *Biochimica et Biophysica Acta*, **282**, 366–73.

Johnstone, R.M. (1974). Role of ATP on the initial rate of amino acid uptake in Ehrlich ascites cells. *Biochimica et Biophysica Acta*, **356**, 319–30.

Johnstone, R.M. & Laris, P.C. (1980). Bimodal effects of cellular amino acids on Na^+-dependent amino acid transport in Ehrlich cells. *Biochimica et Biophysica Acta*, **599**, 715–30.

Jones, A.W. & Ward, H.L. (1950). The chromosomes of *Macracanthorhynchus hirudinaceus* (Pallas). *Journal of Parasitology*, **36**, 86.

Jones, M. (1928). An acanthocephalid, *Plagiorhynchus formosus*, from the chicken and the robin. *Journal of Agricultural Research*, **36**, 773–5.

Kabilov, T. (1969). The occurrence of the acanthocephalan *Mediorhynchus papillosus* Van Cleave, 1916 in Coleoptera. *Materialy k Nauchnoi Konferentsii Vsesoiuznogo Obschchestva Gel'mintologov*, 105–7.

Kaiser, J. (1887). Uber die Entwicklung des *Echinorhynchus gigas*. *Zoologischer Anzeiger*, **10**, 414–19.

Kaiser, J. (1891). Die Acanthocephalen und ihre Entwicklung. *Bibliotheca Zoologica Kassel*, **7**, 136–48.

Kaiser, J. (1892). Die Nephridien der Acanthocephalen. *Zentralblatt für Bakteriologie, Parasitenkunde, Infektionskrankheiten und Hygiene*, **11**, 44–9.

Kaiser, J. (1893). Die Acanthocephalen und ihre Entwicklung. *Bibliotheca Zoologica*, **11**, Heft 7. 148 pp.

Kaiser, J. (1913). *Die Acanthocephalen und ihre Entwicklung. Beitrage zur Kenntnis der Histologie, Ontogenie und Biologie einiger einheimischen Echinorhynchen.* Leipzig: Jachner & Fischer. 66 pp.

Kashnikov, A.A. (1972). On the biology of *Macracanthorhynchus hirudinaceus* (Pallas, 1781) in Belorussia. *Trudy (Nauchnye trudy) Nauchno-Issledovatel'skogo Veterinarnogo Instituta Belorusskoi SSR*, **10**, 144–7.

Kates, K.C. (1942). Viability of eggs of the swine thorn-headed worm (*Macracanthorhynchus hirudinaceus*). *Journal of Agricultural Research*, **64**, 93–100.

Kates, K.C. (1943). Development of the swine thorn-headed worm *Macracanthorhynchus hirudinaceus* in its intermediate host. *American Journal of Veterinary Research*, **4**, 173–81.

Kates, K.C. (1944). Some observations on experimental infections of pigs with the thorn-headed worm *Macracanthorhynchus hirudinaceus*. *American Journal of Veterinary Research*, **5**, 166–72.

Kaw, B.L. (1941). Studies on the helminth parasites of Kashmir; Part 1. Description of some new species of the genus *Pomphorhynchus* Monticelli (1905). *Proceedings of the Indian Academy of Sciences, Section B*, **13**, 369–78.

Keithly, J.S. & Ulmer, M.J. (1965). Experimental development of cystacanths of *Polymorphus* sp. in the amphipod, *Hyalella azteca*. *Journal of Parasitology*, **51**, 60.

Kennedy, C.R. (1970). The population biology of helminths of British freshwater fish. *Symposia of the British Society for Parasitology*, **9**, 145–59.

Kennedy, C.R. (1972). The effects of temperature and other factors upon the establishment and survival of *Pomphorhynchus laevis* (Acanthocephala) in goldfish, *Carassius auratus*. *Parasitology*, **65**, 283–94.

Kennedy, C.R. (1974). The importance of parasite mortality in regulating the population size of the acanthocephalan *Pomphorhynchus laevis* in goldfish. *Parasitology*, **68**, 93–101.

Kennedy, C.R. (1977). The regulation of fish parasite populations. In *Regulation of Parasite Populations*, ed. G.W. Esch, pp. 63–109. New York: Academic Press.

Kennedy, C.R. Acanthocephala. In *Reproductive Biology of Invertebrates*, vol. 6, *Asexual Propagation and Reproductive Strategies*, ed. R.G. Adiyodi and K.G. Adiyodi. Chichester, New York, Brisbane, Toronto, Singapore: Wiley. (In press.)

Kennedy, C.R. & Lord, D. (1982). Habitat specificity of the acanthocephalan *Acanthocephalus clavula* (Dujardin, 1845) in eels *Anguilla anguilla* (L.). *Journal of Helminthology*, **56**, 121–9.

Kennedy, C.R. & Rumpus, A. (1977). Long term changes in the size of the *Pomphorhynchus laevis* (Acanthocephala) population in the River Avon. *Journal of Fish Biology*, **10**, 35–42.

Kennedy, C.R., Broughton, P.F. & Hine, P.M. (1976). The sites occupied by the acanthocephalan *Pomphorhynchus laevis* in the alimentary canal of fish. *Parasitology*, **72**, 195–206.

Kennedy, C.R., Broughton, P.F. & Hine, P.M. (1978). The status of brown and rainbow trout, *Salmo trutta* and *S. gairdneri* as hosts of the acanthocephalan, *Pomphorhynchus laevis*. *Journal of Fish Biology*, **13**, 265–75.

Kennedy, M. (1982). A redescription of *Acanthocephalus bufonis* (Shipley, 1903) Southwell and MacFie (Acanthocephala: Echinorhynchidae) from the black-spotted toad, *Bufo melanostictus*, from Bogor, Indonesia. *Canadian Journal of Zoology*, **60**, 356–60.

Keppner, E.J. (1974). The life history of *Paulisentis missouriensis* n. sp. (Acanthocephala: Neoechinorhynchidae) from the creek chub *Semotilus atromaculatus*. *Transactions of the American Microscopical Society*, **93**, 89–100.

Keymer, A.E. (1980). The influence of *Hymenolepis diminuta* on the survival and fecundity of the intermediate host, *Tribolium confusum*. *Parasitology*, **81**, 405–21.

Keymer, A.E. (1982*a*). Density-dependent mechanisms in the regulation of intestinal helminth populations. *Parasitology*, **84**, 573–87.

Keymer, A.E. (1982*b*). Tapeworm infections. In *Population Dynamics of Infectious Diseases*, ed. R.M. Anderson, pp. 109–38. London: Chapman and Hall.

Keymer, A.E. & Anderson, R.M. (1979). The dynamics of infection of *Tribolium confusum* by *Hymenolepis diminuta*: the influence of infective stage density and spatial distribution. *Parasitology*, **79**, 195–207.

Keymer, A.E., Crompton, D.W.T. & Walters, D.E. (1983). Parasite population biology and host nutrition: dietary fructose and *Moniliformis* (Acanthocephala). *Parasitology*, **87**, 265–78.

Khatkevich, L.M. (1975). The histological structure of the uterine bell, uterus and vagina of the acanthocephalan *Echinorhynchus gadi* Mueller, 1776. *Trudy* (1), *Moskovoskogo Meditsinskogo Instituts* (*Aktual'nye*) *Voprosy Sovremennoy Parazitologii*, **84**, 107–11.

Khlebovich, V.V. & Mikhailova, O. Yu. (1976). Osmotic tolerance and adaptation to a hypotonic medium in *Echinorhynchus gadi* (Acanthocephala: Echinorhynchidae). *Parazitologiya*, **10**, 444–8.

Kilejian, A. (1963). The effect of carbon dioxide on glycogenesis in *Moniliformis dubius* (Acanthocephala). *Journal of Parasitology*, **49**, 862–3.

Kilian, R. (1932). Zur Morphologie und Systematik der Gigantorhynchidae (Acanthocephala). *Zeitschrift für Wissenschaftliche Zoologie, Abteilung A*, **141**, 246–345.

Killick, K.A. & Wright, B.E. (1975). Trehalose synthesis during differentiation in *Dictyostelium discoideum*. *Archives of Biochemistry and Biophysics*, **170**, 634–43.

King, D. & Robinson, E.S. (1967). Aspects of the development of *Moniliformis dubius*. *Journal of Parasitology*, **53**, 142–9.

Kinne, R. (1976). Properties of the glucose transport system in the renal brush border membrane. In *Current Topics in Membranes and Transport*, vol. 8, ed. F. Bronner and A. Kleinzeller, pp. 209–67. London and New York: Academic Press.

Klesov, M.D. & Kovalenko, I.I. (1967). The biology, epizootiology and prophylaxis of helminths of ducks on the Azov coast. *Veterinariya*, **44**, 3–7.

Knupffer, P. (1888). Beitrag zur anatomie des ausfuhrungsganges der Weiblichen geschlechtsprodakte einiger acanthocephalen. *Mémoires de l'Academie Impériale des Sciences de St Petersbourg*, **7**, 36(12), 17 pp.

Kobayashi, M. (1959). Studies on the Acanthocephala (3). Studies on the anomaly of testis of *Centrorhynchus elongatus* Yamaguti. *Japanese Journal of Parasitology*, **8**, 423.

Koelreuther, J.T. (1771). Descriptio cyprini rutili quem halawel russi vocant historicoanatomica. *Nov. Comment. Ac. Sci. Petropol.*, **15**, 494–503.

Köhler, P. & Bachmann, R. (1980). Mechanisms of respiration and phosphorylation in *Ascaris* muscle mitochondria. *Molecular and Biochemical Parasitology*, **1**, 75–90.

Köhler, P., Bryant, C. & Behm, C.A. (1978). ATP synthesis in a succinate decarboxylase system from *Fasciola hepatica* mitochondria. *International Journal for Parasitology*, **8**, 399–404.

Kohn, P., Dawes, E.D. & Duke, J.W. (1965). Absorption of carbohydrates from the intestine of the rat. *Biochimica et Biophysica Acta*, **107**, 358–62.

Komarova, M.S. (1950). On the question of the life cycle of proboscis worm *Acanthocephalus lucii*. *Doklady Akademii Nauk SSR*, **70**, 359–60.

Komarova, T.I. (1969). Acanthocephalan larvae, parasites of benthic Crustacea in water-reservoirs in the Danube delta. *Problemy Parazitologiya*, 1969, 122–3.

Komuniecki, R., Komuniecki, P.R. & Saz, H.J. (1981*a*). Relationships between pyruvate decarboxylation and branched-chain volatile acid synthesis in *Ascaris* mitochondria. *Journal of Parasitology*, **67**, 601–8.

Komuniecki, R., Komuniecki, P.R. & Saz, H.J. (1981*b*). Pathway of formation of branched-chain volatile fatty acids in *Ascaris* mitochondria. *Journal of Parasitology*, **67**, 841–6.

Körting, W. & Barrett, J. (1977). Carbohydrate catabolism in the plerocercoids of *Schistocephalus solidus* (Cestoda: Pseudophyllidea). *International Journal for Parasitology*, **7**, 411–17.

Körting, W. & Fairbairn, D. (1972). Anaerobic energy metabolism in *Moniliformis dubius* (Acanthocephala). *Journal of Parasitology*, **58**, 45–50.

Kotelnikov, G.A. (1954). *Filicollis* in domestic ducks. *Veterinariya*, **31**, 30–2.

Kovalenko, I.I. (1960). Study of the life-cycles of some helminths of domestic ducks from farms on the Azov coast. *Doklady Akademii Nauk SSSR*, **133**, 1259–61.

Krebs, J.R. & Davies, N.B. (1981). *An Introduction to Behavioural Ecology*. Oxford and London: Blackwell Scientific Publications.

Krotov, A.I. (1969). Physiology of helminths (nutrition, osmoregulation, excretion). *Meditsinskaya Parazitologiya i Parazitarnye Bolezni*, **38**, 226–33.

Kulachkova, V.G. & Timofeeva, T.A. (1977). The acanthocephalan *Echinorhynchus gadi* (Zoega) from relict cod in Lake Mogil'noe. *Parazitologiya*, **11**, 316–20.

Kurandina, D.P. (1975). The parasite fauna of Amphipoda in the Kremenchug Reservoir. In *Problemy Parazitologii. Materialy VIII Nauchnoi Konferentsii Parazitologov UkSSR. Chast' 1*, pp. 296–7. Kiev: Izdatel'stvo 'Nauka Dumka'.

Kurbanov, M.N. (1978*a*). The development of *Acanthocephalus ranae* (Schrank, 1788) Lühe, 1911 in the intermediate host. *Izvestiya Akademii Nauk Azerbaidzhanskoi SSR, Biologicheske Nauki*, **1**, 114–19.

Kurbanov, M.N. (1978*b*). The development of *Sphaerirostris teres* (Rud., 1819) in the final host. *Izvestiya Akademii Nauk Azerbaidzhanskoi SSR, Biologicheske Nauki*, **5**, 7477.

Lackie, A.M. (1974). The activation of cystacanths of *Polymorphus minutus* (Acanthocephala) *in vitro*. *Parasitology*, **68**, 135–46.

Lackie, A.M. (1975). The activation of infective stages of endoparasites of vertebrates. *Biological Reviews of the Cambridge Philosophical Society*, **50**, 285–323.

Lackie, A.M. (1978). Acanthocephala. In *Methods of Culturing Parasites in vitro*, ed. A.E.R. Taylor and J.R. Baker, pp. 279–85. London and New York: Academic Press.

Lackie, A.M. & Lackie, J.M. (1979). Evasion of the insect immune response by *Moniliformis dubius* (Acanthocephala): further observations on the origin of the envelope. *Parasitology*, **79**, 297–301.

Lackie, J.M. (1972*a*). The course of infection and growth of *Moniliformis dubius* (Acanthocephala) in the intermediate host *Periplaneta americana*. *Parasitology*, **64**, 95–106.

Lackie, J.M. (1972*b*). The effect of temperature on the development of *Moniliformis dubius* (Acanthocephala) in the intermediate host, *Periplaneta americana*. *Parasitology*, **65**, 371–7.

Lackie, J.M. (1975). The host specificity of *Moniliformis dubius* (Acanthocephala), a parasite of cockroaches. *International Journal for Parasitology*, **5**, 301–7.

Lackie, J.M. & Rotheram, S. (1972). Observations on the envelope surrounding *Moniliformis dubius* (Acanthocephala) in the intermediate host, *Periplaneta americana*. *Parasitology*, **65**, 303–8.

Lange, H. (1970). Uber Struktur und Histochemie des Integumentes von *Echinorhynchus gadi* Muller (Acanthocephala). *Zeitschrift für Zellforschung und Mikroscopische Anatomie*, **104**, 149–64.

LaNoue, K.F. & Schoolwerth, A.C. (1979). Metabolite transport in mitochondria. *Annual Review of Biochemistry*, **48**, 871–922.

Lantz, K.E. (1974). Acanthocephalan occurrence in cultured red crawfish. *Proceedings of the 27th Annual Conference of the Southeastern Association of Game and Fish Commissioners, 1973*, 735–8.

Larkin, R.A. (1948). *Pontoporeia* and *Mysis* in Athabasca, Great Bear and Great Slave Lakes. *Bulletin of the Fisheries Research Board Canada*, **78**, 1–23.

Laurie, J.S. (1957). The *in vitro* fermentation of carbohydrates by two species of cestodes and one species of Acanthocephala. *Experimental Parasitology*, **6**, 245–60.

Laurie, J.S. (1959). Aerobic metabolism of *Moniliformis dubius* (Acanthocephala). *Experimental Parasitology*, **8**, 188–97.

Le Clerc, D. (1715). Historia naturalis et medica latorum lumbricorum intra Hominem et alia animalia, mascentium, ex variis auctoribus et propriis observationibus, accessit, horum occasione, de ceteris quoque hominum vermibus, tum de omnium origine, tandemque de remediis quibus pelli possint, disquisito. Genevae. 16+456 pp., 13 pls.

Lebedev, B.I. (1967). *Australorhynchus tetramorphacanthus* n.g., n.sp. (Acanthocephala: Rhadinorhynchidae) from fish found in Australian and New Zealand Seas. *Zoologicheskii Zhurnal*, **46**, 279–82.

Lee, D.L. (1966). The structure and composition of the helminth cuticle. *Advances in Parasitology*, **4**, 187–254.

Lee, D.L. (1972). The structure of the helminth cuticle. *Advances in Parasitology*, **10**, 347–79.

Lee, R.L.G. (1981). Ecology of *Acanthocephalus lucii* (Muller, 1776) in perch, *Perca fluviatilis*, in the Serpentine, London, U.K. *Journal of Helminthology*, **55**, 149–54.

Leeuwenhoek, A. von (1695). Derde Vervolg der Brieven, geschreven aan de Koniglijke Societiet tet London. *Tot Delft*. 1693–1695, *St Missive*, 518–20.

Leng, Y.-J., Huang, W.-O. & Liang, P.-N. (1981). Human infection with *Macracanthorhynchus hirudinaceus* Travassos, 1916 in Guangdong Province, with notes on its prevalence in China. *Annals of Tropical Medicine and Parasitology*, **77**, 107–9.

Leong, T.S. & Holmes, J.C. (1981). Communities of metazoan parasites in open water fishes of Cold Lake, Alberta. *Journal of Fish Biology*, **18**, 693–713.

LeRoux, M.L. (1931). Castration parasitaire et caractères sexuels secondaires chez les Gammariens. *Comptes Rendus Hebdomadaires des Séances de l'Académie des Sciences*, **192**, 889–91.

Leuckart, R. (1848). Uber die morphologie und die Verwandschaftverhaltnisse der wirbellosen Thiere. Ein Beitrag zur Charakteristik und Classification der Thierischen Formen. Braunschweig., 180 pp.

Leuckart, R. (1862). Helminthologische Experimentaluntersuchungen III. Uber *Echinorhynchus. Nachricht. Gross. August. Univer. Gen. Wiss. Gottingen*, **22**, 433–47.

Leuckart, R. (1873). Commentatio de statu et embryonali et larvali echinorhynchorum eorumque metamorphosi. Decanatsprogramm 1873. 33 pp. Lipsiae. Abstract in *Ztschr. Ges Naturw.*, 42n. f., v. 8, 208–9.

Leuckart, T. (1876). *Die menschlichen Parasiten und die von ihnen herrührenden Krankheiten*, Band 2. Leipzig and Heidelberg: C.F. Winter'sche Verlagshandlung.

Levitt, D.G. (1980). The mechanism of the sodium pump. *Biochimica et Biophysica Acta*, **604**, 321–45.

Lewis, K.R. & John, B. (1968). The chromosomal basis of sex determination. *International Review of Cytology*, **23**, 277–379.

Li, H.C. (1982). Phosphoprotein phosphatases. In *Current Topics in Cellular Regulation*, vol. 21, ed. B.L. Horecker and E.R. Stadtman, pp. 129–74. London and New York: Academic Press.

Liat, L.B. & Pike, A.W. (1980). The incidence and distribution of *Profilicollis botulus* (Acanthocephala), in the eider duck, *Somateria mollissima*, and in its intermediate host the shore crab, *Carcinus maenas*, in north east Scotland. *Journal of Zoology*, **190**, 39–51.

Lincicome, D.R. & Van Cleave, H.J. (1949). Distribution of *Leptorhynchoides thecatus*, a common acanthocephalan parasitic in fishes. *American Midland Naturalist*, **41**, 421–31.

Lincicome, D.R. & Whitt, A. (1947). Occurrence of *Neoechinorhynchus emydis* (Acanthocephala) in snails. *Transactions of the Kentucky Academy of Science*, **12**, 19.

Lingard, A.M. & Crompton, D.W.T. (1972). Observations on the establishment of *Polymorphus minutus* (Acanthocephala) in the intestine of domestic ducks. *Parasitology*, **65**, 159–65.

Linstow, O. von (1872). Zur Anatomie und Entwicklungsgeschichte des *Echinorhynchus angustus* Rudolphi. *Archiv für Naturgeschichte*, **38**, 6–16.

Lisowski, E.A. (1979). Variations in body color and eye pigmentation of *Asellus brevicauda* Forbes (Isopoda: Asellidae) in a southern Illinois cave stream. *Bulletin of the National Speleological Society*, **41**, 11–14.

Logachev, E.D., Bruskin, B.R. & Kesten, L.A. (1961). Reaction of tissues of *Gammarus lacustris* to parasitization by *Polymorphus magnus* acanthellae. *Helminthologia*, **3**, 226–33.

Loomis, S.H., Madin, K.A.C. & Crowe, J.H. (1980). Anhydrobiosis in nematodes: biosynthesis of trehalose. *Journal of Experimental Zoology*, **211**, 311–20.

Lühe, M. (1904–1905). Geschichte und Ergebnisse der Echinorhychen-Forschung bis auf Westrumb (1821). *Zoologische Ann. Zeitschrift*, **1**, 139–250, 251–353.

Lühe, M. (1911). Acanthocephalen. Register der Acanthocephalen und parasitischen Plattwurmer geordnet nach ihren Wirten. *Susswasserfauna Deutschlands Exkursionfauna*, **16**, 114.

Lukacsovics, F. (1959). *Polymorphus minutus* Goeze (Acanthocephala) Larva hatasa a *Gammarus roeseli* Gerv. (Amphipoda) fajra. *Annales Instituti Biologici (Tihany). Hungaricae Academiae Scientiarum*, **26**, 31–9.

Lumsden, R.D. (1975). Surface ultrastructure and cytochemistry of parasitic helminths. *Experimental Parasitology*, **37**, 267–339.

McAlister, R.O. & Fisher, F.M. (1972). The biosynthesis of trehalose in *Moniliformis dubius* (Acanthocephala). *Journal of Parasitology*, **58**, 51–62.

Macdonald, G. (1965). The dynamics of helminth infections, with special reference to schistosomes. *Transactions of the Royal Society of Tropical Medicine and Hygiene*, **59**, 489–506.

Machado, D.A. (1950). Revisão do género *Prosthenorchis* Travassos 1915 (Acanthocephala). *Memoirs, Institute Oswaldo Cruz*, **48**, 495–544.

M'ackness, B. (1977). Infestation of the pheasant coucal by the acanthocephalan *Porrorchis hylae* (Edmonds). *Emu*, **77**, 39.

Makhanbetov, S.H. (1972). The structure of the nerve endings of the acanthocephalan *Polymorphus phippsi* Kostylov, 1922. *Gel'minty pischevynkh produktov, Tezisy dokladov mezhrespublikananskoy, Nauchnoi konferentsii, Samarkand*, 1972, 22–5.

Makhanbetov, S.H. (1974). Paired lateral nerve in the body wall of the acanthocephalan *Polymorphus phippsi* (Kostylov, 1922). *Vestnik Sel'skokhozyaistyennoi Nauk, Kazastana (Kazastant Auyh. Sharuashylyk Gylymynyn Habarshysy)*, **1**, 104–6.
Makhanbetov, S.H. (1975*a*). The ultrastructure and localization of commissures in the anterior end of the acanthocephalan *Polymorphus phippsi* (Kostylew, 1922). *Trudy (1) Moskovskogo Instituta (Aktual'nye) voprosy sovremennoy parazitologii*, **84**, 96–7.
Makhanbetov, S.H. (1975*b*). The topography and histological structure of the nervous system of acanthocephalans. *Nektorye voprosy biologii i razuitiya zhivotnykh kh. vyp. 1 Tbilisi, USSR: Izdatel'stvo Tbilisskogo Universiteta*, 202–5.
Maksimova, A.P. (1972). Death of young swans due to helminths in the Kurgal'dzhin reserve. *Trudy VII Vsesoyuznoi Konferentsii po Prirodnoi Ochagovosti Boleznei i Obshchim Voprosam Parazitologii Zhivotnykh*, 1972, 125–9.
Malathi, P. & Crane, R.K. (1969). Phlorizin hydrolase: A β-glucosidase of hamster intestinal brush border membrane. *Biochimica et Biophysica Acta*, **173**, 245–56.
Malathi, P., Ramaswamy, K., Caspary, W.F. & Crane, R.K. (1973). Studies on the transport of glucose from disaccharides by hamster small intestine *in vitro*. I. Evidence for a disaccharidase-related transport system. *Biochimica et Biophysica Acta*, **307**, 613–26.
Manter, H.W. (1928). Notes on the eggs and larvae of the thorny-headed worm of hogs. *Transactions of the American Microscopical Society*, **47**, 342–7.
Marchand, B. & Mattei, X. (1976*a*). La spermatogenèse des Acanthocéphales. 1. L'appareil centriolaire et flagellaire au cours de la spermatogenèse d'*Illiosentis furcatus* var. *africana* Golvan, 1956 (Palaeacanthocephala: Rhadinorhynchidae). *Journal of Ultrastructure Research*, **54**, 347–58.
Marchand, B. & Mattei, X. (1976*b*). La spermatogenèse des Acanthocéphales. 2. Variation du nombre des fibres centrales dan la flagellum spermatique d'*Acanthosentis tilapiae* Baylis, 1947 (Eoacanthocephala: Quadrigyridae). *Journal of Ultrastructure Research*, **55**, 391–9.
Marchand, B. & Mattei, X. (1976*c*). Ultrastructure du spermatozoide de *Centrorhynchus milvus* Ward, 1956 (Palaeacanthocephala: Polymorphidae). *Comptes Rendus de Séances de la Société Biologique*, **170**, 237–40.
Marchand, B. & Mattei, X. (1976*d*). Présence de flagelles spermatiques dans les sphères ovariennes des éoacanthocéphales. *Journal of Ultrastructure Research*, **56**, 331–8.
Marchand, B. & Mattei, X. (1977*a*). La spermatogenèse des Acanthocéphales. 3. Formation du dérivé centriolaire au cours de la spermiogenèse de *Serrasentis socialis* Van Cleave, 1924 (Palaeacanthocephala: Gorgorhynchidae). *Journal of Ultrastructure Research*, **59**, 263–71.
Marchand, B. & Mattei, X. (1977*b*). Un type nouveau de structure flagellaire. Type 9+n. *Journal of Cell Biology*, **72**, 707–13.
Marchand, B. & Mattei, X. (1978*a*). La spermatogenèse des Acanthocéphales. 4. Le dérivé nucléocytoplasmique. *Biologie Cellulaire*, **31**, 79–90.
Marchand, B. & Mattei, X. (1978*b*). La spermatogenèse des Acanthocéphales. 5. Flagellogenèse chez un éoacanthocephala: mise en place et désorganisation de l'axoneme spermatique. *Journal of Ultrastructure Research*, **63**, 41–50.
Marchand, B. & Mattei, X. (1979). La fécondation chez les Acanthocéphales. I. Modifications ultrastructurales des sphères ovariennes et des Fémelles de l'Acanthocéphale *Neoechinorhynchus agilis*. *Journal of Ultrastructure Research*, **66**, 32–9.

Marchand, B. & Mattei, X. (1980). Fertilization in Acanthocephala. II. Spermatozoon penetration of oocyte, transformation of gametes and elaboration of the fertilization membrane. *Journal of Submicroscopic Cytology*, **12**, 95–105.

Margolis, L. (1958). The occurrence of juvenile *Corynosoma* (Acanthocephala) in Pacific salmon (*Oncorhynchus* spp.). *Journal of the Fisheries Research Board of Canada*, **15**, 983–90.

Markov, G.S., Zinyakova, M.P. & Lutta, A.S. (1967). New data in the parasitology of central Asiatic *Vipera lebetina. Sbornik Nauchnykh Rabot Volgogradski Pedagogicheski Institut*, 1967, 98–105.

Markov, G.S., Bogdanov, O.P., Makeev, V.M. & Khutoryanski, A.A. (1968). New data on the helminth fauna of *Naja oxiana*. In *Papers on Helminthology Presented to Academician K.I. Skryabin on His 90th Birthday*, pp. 244–8. Moscow: Izdatelstvo Akademii Nauk SSSR.

Markowski, S. (1971). On some species of parasitic worms in the 'Discovery' collections obtained in the years 1925–1936. *Bulletin of the British Museum (Natural History), Zoology*, **21**, 51–65.

Marshall, J.P. (1976). Observations on the envelope surrounding *Pomphorhynchus laevis* (Acanthocephala) in its intermediate host *Gammarus pulex*. *Parasitology*, **73**, xxix.

May, R.M. (1977). Togetherness among schistosomes: its effects on the dynamics of the infection. *Mathematical Biosciences*, **35**, 301–45.

May, R.M. & Anderson, R.M. (1978). Regulation and stability of host–parasite population interactions. II. Destabilizing processes. *Journal of Animal Ecology*, **47**, 249–68.

May, R.M. & Anderson, R.M. (1979). Population biology of infectious diseases: II. *Nature, London*, **280**, 455–61.

Mercer, E.H. & Nicholas, W.L. (1967). The ultrastructure of the capsule of the larval stages of *Moniliformis dubius* (Acanthocephala) in the cockroach *Periplaneta americana*. *Parasitology*, **57**, 169–74.

Merritt, S.V. & Pratt, I. (1964). The life history of *Neoechinorhynchus rutili* and its development in the intermediate host (Acanthocephala: Neoechinorhynchidae). *Journal of Parasitology*, **50**, 394–400.

Mettrick, D.F. (1980). The intestine as an environment for *Hymenolepis diminuta*. In *Biology of the Tapeworm* Hymenolepis diminuta, ed. H.P. Arai, pp. 281–356. London and New York: Academic Press.

Mettrick, D.F. & Podesta, R.B. (1974). Ecological and physiological aspects of helminth–host interactions in the mammalian gastrointestinal canal. *Advances in Parasitology*, **12**, 183–278.

Mettrick, D.F., Budziakowski, M.E. & Podesta, R.B. (1979). Net fluxes of electrolytes in the rat intestine infected with *Moniliformis dubius* (Acanthocephala). *Canadian Journal of Physiology and Pharmacology*, **57**, 882–6.

Metzler, D.E. (1977). *Biochemistry: The Chemical Reactions of Living Cells*. London and New York: Academic Press.

Meyer, A. (1928). Die Furchung nebst Eibildung, Reifung und Befruchtung des *Gigantorhynchus gigas* (Ein Beitrag zur Morphologie der Acanthocephalen). *Zoologisches Jahrbücher. Abteilung für Anatomie und Ontogenie der Tiere*, **50**, 117–218.

Meyer, A. (1931*a*). Urhautzelle, Hautbahn und plasmodiale Entwicklung der Larve von *Neoechinorhynchus rutili* (Acanthocephala). *Zoologische Jahrbücher. Abteilung für Anatomie und Ontogenie der Tiere*, **53**, 103–26.

Meyer, A. (1931*b*). Neue Acanthocephalen aus dem Berliner Museum. Bergründung eines neuen Acanthocephalensystems auf Grund einer Untersuchung der Berliner Sammlung. *Zoologische Jahrbücher. Abteilung für Systematik, Okologie und Geographie der Tiere*, **62**, 53–108.

Meyer, A. (1931*c*). Das urogenitale Organ von *Oligacanthorhynchus taenioides* (Diesing) ein neuer Nephridialtypus bein den Acanthocephalen. *Zeitschrift für Wissenschaftliche Zoologies, Abteilung A*, **138**, 88–98.

Meyer, A. (1932). Acanthocephala. In *Dr H.G. Bronn's Klassen und Ordnungen des Tier-Reichs*, vol. **4**, pp. 1–332. Leipzig: Akademische Verlagsgesellschaft MBH.

Meyer, A. (1933). Acanthocephala. In *Dr H.G. Bronn's Klassen und Ordnungen des Tierreichs*, vol. **4**, pp. 333–582. Leipzig: Akademische Verlagsgesellschaft MBH.

Meyer, A. (1936). Die plasmodiale Entwicklung und Formbildung des Reisenkratzers *Macracanthorhynchus hirudinaceus* (Pallas). I Teil. *Zoologische Jahrbücher. Abteilung für Anatomie und Ontogenie der Tiere*, **62**, 111–72.

Meyer, A. (1937). Die plasmodiale Entwicklung und Formbildung des Riesenkratzers (*Macracanthorhynchus hirudinaceus*). II Teil. *Zoologische Jahrbücher. Abteilung für Anatomie und Ontogenie der Tiere*, **63**, 1–36.

Meyer, A. (1938*a*). Die plasmodiale Entwicklung und Formbildung des Riesenkratzers *Macracanthorhynchus hirudinaceus* (Pallas). III Teil. *Zoologische Jahrbücher. Abteilung für Anatomie und Ontogenie der Tiere*, **64**, 131–97.

Meyer, A. (1938*b*). Die plasmodiale Entwicklung und Formbildung des Riesenkratzers (*Macracanthorhynchus hirudinaceus* (Pallas)). IV. Allgemeiner Teil. *Zoologische Jahrbücher. Abteilung für Anatomie und Ontogenie der Tiere*, **64**, 199–242.

Mied, P.A. & Bueding, E. (1979). Glycogen synthase of *Hymenolepis diminuta*. I. Allosteric activation and inhibition. *Journal of Parasitology*, **65**, 14–24.

Miller, D.M. & Dunagan, T.T. (1971). Studies on the rostellar hooks of *Macracanthorhynchus hirudinaceus* (Acanthocephala) from swine. *Transactions of the American Microscopical Society*, **90**, 329–35.

Miller, D.M. & Dunagan, T.T. (1976). Body wall organization of the acanthocephalan, *Macracanthorhynchus hirudinaceus*: a reexamination of the lacunar system. *Proceedings of the Helminthological Society of Washington*, **43**, 99–106.

Miller, D.M. & Dunagan, T.T. (1977). The lacunar system and tubular muscles in Acanthocephala. *Proceedings of the Helminthological Society of Washington*, **44**, 201–5.

Miller, D.M. & Dunagan, T.T. (1978). Organization of the lacunar system in the acanthocephalan *Oligacanthorhynchus tortuosa*. *Journal of Parasitology*, **64**, 436–9.

Miller, D.M. & Dunagan, T.T. (1983). A support cell to the apical and lateral sensory organs in *Macracanthorhynchus hirudinaceus* (Acanthocephala). *Journal of Parasitology*, **69**, 534–8.

Miller, D.M., Dunagan, T.T. & Richardson, J. (1973). Anatomy of the cerebral ganglion of the female acanthocephalan, *Macracanthorhynchus hirudinaceus*. *Journal of Comparative Neurology*, **152**, 403–15.

Miller, D.M., Wong, B.S. & Dunagan, T.T. (1981). Membrane potentials in an acanthocephalan worm (*Macracanthorhynchus hirudinaceus*). *Parasitology*, **83**, 33–41.

Miller, M.A. (1943). Studies on the developmental stages and glycogen metabolism of *Macracanthorhynchus hirudinaceus* in the Japanese beetle larva. *Journal of Morphology*, **73**, 19–41.

Mingazzini, P. (1896). Nuove ricerche sul parassitismo. *Ricerca Laboratorio di Anatomia Normale, Roma*, **5**, 169–87.

Miremad-Gassmann, M. (1981*a*). Contribution à la connaissance de la biologie de *Moniliformis moniliformis* Bremser, 1811 (Acanthocephala). Influence de la résistance de l'hôte, *Rattus norvegicus* Berkenhout, 1769, sur le parasite. *Annales de Parasitologie Humaine et Comparée*, **56**, 407–21.

Miremad-Gassmann, M. (1981*b*). Contribution à la connaissance des relations immunologiques entre *Moniliformis moniliformis* Bremser, 1811 (Acanthocephala) et *Rattus norvegicus* Berkenhout, 1769, var. *albinos*. *Acta Tropica*, **38**, 137–47.

Mishra, F.A.Z. (1978). The occurrence of *Acanthocephalus lucii* (Müller, 1777) in the fish of the Shropshire Union Canal, Cheshire, England. *Annals of Zoology*, **14**, 181–8.

Moeller, H. (1978). The effects of salinity and temperature on the development and survival of fish parasites. *Journal of Fish Biology*, **12**, 311–23.

Möller, H. von (1974). Untersuchungen über die Parasiten der Flunder (*Platichthys flesus* L.) in der Kieler Förde. *Berichte der Deutschen Wissenschaftlichen Kommission für Meeresforschung*, **23**, 136–49.

Möller, H. von (1975). Parasitological investigations on the European eelpout (*Zoarces viviparus* L.) in the Kiel-Fjord (Western Baltic). *Berichte der Deutschen Wissenschaftlichen Kommission für Meeresforschung*, **24**, 63–70.

Monné, L. (1959). On the external cuticles of various helminths and their role in the host parasite relationships, a histochemical study. *Archives of Zoology*, **12**, 343–58.

Monné, L. (1964). Chemie und Bildung der Embryophoren von *Polymorphus botulus* Van Cleave (Acanthocephala). *Zeitschrift für Parasitenkunde*, **25**, 148–56.

Monné, L. & Hönig, G. (1954). On the embryonic envelopes of *Polymorphus botulus* and *P. minutus* (Acanthocephala). *Archiv für Zoologie*, **7**, 257–60.

Moore, D.V. (1942). An improved technique for the study of the acanthor stage in certain acanthocephalan life histories. *Journal of Parasitology*, **28**, 495.

Moore, D.V. (1946*a*). Studies on the life history and development of *Moniliformis dubius* Meyer, 1933. *Journal of Parasitology*, **32**, 257–71.

Moore, D.V. (1946*b*). Studies on the life history and development of *Macracanthorhynchus ingens* Meyer, 1933, with a redescription of the adult worm. *Journal of Parasitology*, **32**, 387–99.

Moore, D.V. (1962). Morphology, life history and development of the acanthocephalan *Mediorhynchus grandis* Van Cleave, 1916. *Journal of Parasitology*, **48**, 76–86.

Moore, J. (1983). Responses of an avian predator and its isopod prey to an acanthocephalan parasite. *Ecology*, **64**, 1000–15.

Moore, J. & Bell, D.H. (1983). Pathology (?) of *Plagiorhynchus cylindraceus* in the starling, *Sturnus vulgaris*. *Journal of Parasitology*, **69**, 387–90.

Moore, J.G. (1970). Epizootic of acanthocephaliasis among primates. *Journal of the American Veterinary Medical Association*, **157**, 699–705.

Moravec, F. & Barus, V. (1971). Studies on parasitic worms from Cuban fishes. *Vestnik Ceskoslovenske Spolecnosti Zoologicke*, **35**, 56–74.

Mordvinova, T.N. & Parukhin, A.M. (1978). *Golvanacanthus problematicus* n. sp.; a new acanthocephalan from *Gammarus olivii* in the Black Sea. *Biologiya Morya, Kiev* (Parazitofauna zhivotnykh Yuzhnykh morei), Issue 45, 42–4.

Morin, M.L., Renquist, D.M., Johnson, D.K. & Strumpf, I.J. (1980). Flexible fiberoptic proctoscopy compared with fecal examination techniques for diagnosis of *Prosthenorchis* infection in squirrel monkeys (*Saimiri sciureus*). *Laboratory Animal Science*, **30**, 1009–11.

Morozov, Y.F. (1959). The occurrence of *M. hirudinaceus* (Pallas) in geotropes in the Białowieza forest. *Doklady Akademii Nauk SSSR*, **3**, 430–1.

Moskalev, V.A. (1976). On the infestation of the Fuligulinae with intestinal helminths as related to their population density. *Zoologicheskii Zhurnal*, **55**, 1612–16.

Müller, O.F. (1776). Zoologiae Danicae prodromus, seu animalium Daniae et norvegiae indigenarum characters, nomina, et synonyma imprimis popularium. Havniae, XXXII+282 pp.

Muller, R. (1975). *Worms and Disease*. London: Heinemann.

Munro, W.R. (1953). Intersexuality in *Asellus aquaticus* L. parasitized by a larval acanthocephalan. *Nature, London*, **172**, 313.

Murdoch, W.W. & Oaten, A. (1975). Predation and population stability. *Advances in Ecological Research*, **9**, 1–131.

Muzzall, P.M. (1978). The host–parasite relationships and seasonal occurrence of *Fessisentis friedi* (Acanthocephala: Fessisentidae) in the isopod (*Caecidotea communis*). *Proceedings of the Helminthological Society of Washington*, **45**, 77–82.

Muzzall, P.M. (1980). Ecology and seasonal abundance of three acanthocephalan species infecting white suckers in SE New Hampshire. *Journal of Parasitology*, **66**, 127–33.

Muzzall, P.M. (1981). Parasites of the isopod, *Caecidotea azteca*, in New Hampshire. *Proceedings of the Helminthological Society of Washington*, **48**, 91–2.

Muzzall, P.M. & Bullock, W.L. (1978). Seasonal occurrence and host–parasite relationships of *Neoechinorhynchus saginatus* Van Cleave and Bangham, 1949 in the fallfish *Semotilus corporalis* (Mitchell). *Journal of Parasitology*, **64**, 860–5.

Muzzall, P.M. & Rabalais, F.C. (1975*a*). Studies on *Acanthocephalus jacksoni* Bullock, 1962 (Acanthocephala: Echinorhynchidae). I. Seasonal periodicity and new host records. *Proceedings of the Helminthological Society of Washington*, **42**, 31–4.

Muzzall, P.M. & Rabalais, F.C. (1975*b*). Studies on *Acanthocephalus jacksoni* Bullock, 1962 (Acanthocephala: Echinorhynchidae). II. An analysis of the host–parasite relationship of larval *Acanthocephalus jacksoni* in *Lirceus lineatus* (Say.). *Proceedings of the Helminthological Society of Washington*, **42**, 35–8.

Muzzall, P.M. & Rabalais, F.C. (1975*c*). Studies on *Acanthocephalus jacksoni* Bullock, 1962 (Acanthocephala: Echinorhynchidae). III. The altered behavior of *Lirceus lineatus* (Say.) infected with cystacanths of *Acanthocephalus jacksoni*. *Proceedings of the Helminthological Society of Washington*, **42**, 116–18.

Nagasawa, K., Egusa, S. & Ishino, K. (1982). Occurrence of *Acanthocephalus minor* (Acanthocephala) in two types of the goby *Chaenogobius annularis*. *Japanese Journal of Ichthyology*, **29**, 229–31.

Naidenova, N.N. & Zuev, G.V. (1978). The helminth fauna of *Sthenoteuthis pteropus* (Steenstrup) in the east-central Atlantic. *Biologiya Morya, Kiev* (Parazitofauna zhivothykh Yuzhnykh morei), **45**, 55–64.

Nazarova, N.S. (1959). A new intermediate host of *Moniliformis moniliformis* (Bremser, 1811). *Trudy Gel'mintologicheskoi Laboratorii. Akademiya Nauk SSR*, **9**, 203–5.

Neiland, K.A. (1962). Alaskan species of acanthocephalan genus *Corynosoma* Luehe, 1904. *Journal of Parasitology*, **48**, 69–75.

Nesheim, M.C., Crompton, D.W.T., Arnold, S. & Barnard, D. (1977). Dietary relations between *Moniliformis* (Acanthocephala) and laboratory rats. *Proceedings of the Royal Society of London (B)*, **197**, 363–83.

Nesheim, M.C., Crompton, D.W.T., Arnold, S. & Barnard, D. (1978). Host dietary starch and *Moniliformis* (Acanthocephala) in growing rats. *Proceedings of the Royal Society of London (B)*, **202**, 399–408.

Newton, A.A. (1982). Viruses – exploiters or dependents of the host? *Parasitology*, **85**, 189–216.

Nicholas, W.L. (1967). The biology of the Acanthocephala. *Advances in Parasitology*, **5**, 205–46.

Nicholas, W.L. (1973). The biology of the Acanthocephala. *Advances in Parasitology*, **11**, 671–706.

Nicholas, W.L. & Grigg, H. (1965). The *in vitro* culture of *Moniliformis dubius* (Acanthocephala). *Experimental Parasitology*, **16**, 332–40.

Nicholas, W.L. & Hynes, H.B.N. (1958). Studies on *Polymorphus minutus* (Goeze, 1782) (Acanthocephala) as a parasite of the domestic duck. *Annals of Tropical Medicine and Parasitology*, **52**, 36–47.

Nicholas, W.L. & Hynes, H.B.N. (1963*a*). Embryology, postembryonic development and phylogeny of the Acanthocephala. In *The Lower Metazoa*, ed. E.C. Daugherty, pp. 385–402. Berkeley: University of California Press.

Nicholas, W.L. & Hynes, H.B.N. (1963*b*). The embryology of *Polymorphus minutus* (Goeze, 1782) (Acanthocephala). *Proceedings of the Zoological Society of London*, **141**, 791–801.

Nicholas, W.L. & Mercer, E.H. (1965). The ultrastructure of the tegument of *Moniliformis dubius* (Acanthocephala). *Quarterly Journal of Microscopical Science*, **106**, 137–46.

Nickol, B.B. (1967). *Fessisentis necturorum* sp. n. (Acanthocephala: Fessisentidae), a parasite of the Gulf Coast waterdog, *Necturus beyeri*. *Journal of Parasitology*, **53**, 1292–4.

Nickol, B.B. (1969). Acanthocephala of Louisiana Piscidae with description of a new species of *Mediorhynchus*. *Journal of Parasitology*, **55**, 324–8.

Nickol, B.B. (1972). *Fessisentis*, a genus of acanthocephalans parasitic in North American poikilotherms. *Journal of Parasitology*, **58**, 282–9.

Nickol, B.B. (1977). Life history and host specificity of *Mediorhynchus centurorum* Nickol, 1969 (Acanthocephala: Gigantorhynchidae). *Journal of Parasitology*, **63**, 104–11.

Nickol, B.B. & Dappen, G.E. (1982). *Armadillidium vulgare* (Isopoda) as an intermediate host of *Plagiorhynchus cylindraceus* (Acanthocephala) and isopod response to infection. *Journal of Parasitology*, **68**, 570–5.

Nickol, B.B. & Heard, R.W. (1973). Host–parasite relationships of *Fessisentis necturorum* (Acanthocephala: Fessisentidae). *Proceedings of the Helminthological Society of Washington*, **40**, 204–8.

Nickol, B.B. & Holloway, H.L. (1968). Morphology of the presoma of *Corynosoma hamanni* (Acanthocephala: Polymorphidae). *Journal of Morphology*, **124**, 217–26.

Nickol, B.B. & Kocan, A. (1982). *Andracantha mergi* (Acanthocephala: Polymorphidae) from American bald eagles, *Haliaeetus leucocephalus*. *Journal of Parasitology*, **68**, 168–9.

Nickol, B.B. & Oetinger, D.F. (1968). *Prosthorhynchus formosus* from the short-tailed shrew (*Blarina brevicauda*) in New York state. *Journal of Parasitology*, **54**, 456.

Noll, W. (1950). Eine Masseninfektion von *Gammarus pulex fossarum* koch mit *Polymorphus minutus* Goeze. Nachrichten der Sammelstelle für Schmarotzerbestimmung. *Nachrichten des Naturwissenshaftlichen Museums der Stadt Aschaffenburg*, **29**, 13–15.

Oetinger, D.F. & Nickol, B.B. (1974). A possible function of the fibrillar coat in *Acanthocephalus jacksoni* eggs. *Journal of Parasitology*, **60**, 1055–6.

Oetinger, D.F. & Nickol, B.B. (1981). Effects of acanthocephalans on pigmentation of freshwater isopods. *Journal of Parasitology*, **67**, 672–84.

Oetinger, D.F. & Nickol, B.B. (1982*a*). Spectrophotometric characterization of integumental pigments from uninfected and *Acanthocephalus dirus*-infected *Asellus intermedius*. *Journal of Parasitology*, **68**, 270–5.

Oetinger, D.F. & Nickol, B.B. (1982*b*). Developmental relationships between acanthocephalans and altered pigmentation in freshwater isopods. *Journal of Parasitology*, **68**, 463–9.

Okorokov, V.I. (1953). Acanthocephalates of wild and domestic birds of Chelyabinsk Region. In *Contributions to Helminthology Published to Commemorate the 75th Birthday of K.I. Skrjabin*, pp. 461–3.

Olson, R.E. (1970). The life cycle and larval development of *Echinorhynchus lageniformis* Ekbaum, 1983 (Acanthocephala: Echinorhynchidae). *Journal of Parasitology*, **56**, 253–4.

Olson, R.E. & Pratt, I. (1971). The life cycle and larval development of *Echinorhynchus lageniformis* Ekbaum, 1938 (Acanthocephala: Echinorhynchidae). *Journal of Parasitology*, **57**, 143–9.

Ono, S. (1933). Studies on the life-history of *Echinorhynchus gigas* in Manchuria. I. On the encysted larvae of *Echinorhynchus* found in *Harpalus tridens*, *Gymnopleurus mopsus* and *Phyllodromia germanica*, and the natural infestation of suslik with the parasite under discussion. *Journal of the Japanese Society of Veterinary Science*, **12**, 61–8.

Oparin, P.G. (1962). Biology of *Macracanthorhynchus hirudinaceus* of pigs in the Primorsk territory. *Trudi Dalnevostochnogo Nauchno-Issledovatelskogo Veterinarynogo Instituta*, **4**, 93–104.

Orecchia, P., Paggi, L. & Hannuna, S. (1970). Su alcuni nuovi reperti parassitologici in *Pagellus erythrinus*. *Parassitologia*, **12**, 135–40.

Orecchia, P., Paggi, L., Manilla, G. & Rossi, G. (1978). Parasite fauna of *Salmo trutta* in the River Tirino. Note IV. Investigation of the intermediate hosts of *Cyathocephalus truncatus* and *Dentitruncus truttae*. *Parassitologia*, **20**, 175–81.

Overstreet, R.M. (1983). Aspects of the biology of the red drum, *Sciaenops ocellatus*, in Mississippi. *Gulf Coast Research Reports*, Suppl. 1, 45–68.

Ovington, K.S. & Bryant, C. (1981). The role of carbon dioxide in the formation of end-products by *Hymenolepis diminuta*. *International Journal for Parasitology*, **11**, 221–8.

Oxoid (1977). *Laboratory Animal Diets*. Basingstoke: Oxoid Ltd.

Padmavathi, P. (1967). Likely paratenic avian hosts for the thorny-headed worm *Oncicola* in and around Madras. *Indian Veterinary Journal*, **44**, 727–9.

Page, C.R. & MacInnis, A.J. (1971). Membrane transport of nucleosides by cestodes and acanthocephalans. *Program of the 46th Annual Meeting of the American Society of Parasitologists*, Abstract 152.

Pagenstecher, H.A. (1859). Uber einige Organisationsverhalnisse besonders die weiblichen Geschlechsorgane von *Echinorhynchus proteus*. *Versam. deutsch. Naturforsch. Artze Carlsruhe. Bericht*, **14**, 133–4.

Pagenstecher, H.A. (1863). Zur anatomie von *Echinorhynchus proteus*. *Zeitschrift für Wissenschaftliche Zoologie*, **13**, 413–21.

Pallas, P.S. (1760). Dissertatio-medica inauguralis de infectis viventibus intr viventia, Lugduni, Batavorum, St Petersburg. In *Thesaurus diss. progr. aliorumque opus select.* (Sandifort, Eduardus), **1**, 247–96.

Pallas, P.S. (1766). *Elenchus zoophytorum sistens generum adumbrationes generaliores et specierum cognitarum succinctas descriptiones cum selectis auctorum synonymis.* Hagae Comitum. 451 pp.

Pallas, P.S. (1781). Einige Erinnerungen die Bandwurmer betreffend; in Beziehung auf das zuolfte und vierzehnte Stuck des Naturforschers. *Neue nordl. Beytrage Physik Geogr. Erd und Volkerbeschr., Naturg. Okonomie, St Petersburg and Leipzig*, **2**, 58–82.

Paperna, I. & Zwerner, D.E. (1976). Parasites and diseases of striped bass, *Morone saxatilis* (Walbaum), from the lower Chesapeake bay. *Journal of Fish Biology*, **9**, 267–87.

Pappas, P.W. & Read, C.P. (1972*a*). Trypsin inactivation by intact *Hymenolepis diminuta. Journal of Parasitology*, **58**, 864–71.

Pappas, P.W. & Read, C.P. (1972*b*). Inactivation of α- and β-chymotrypsin by intact *Hymenolepis diminuta* (Cestoda). *Biological Bulletin*, **143**, 605–16.

Pappas, P.W. & Read, C.P. (1974). Relation of nucleoside transport and surface phosphohydrolase activity in the tapeworm, *Hymenolepis diminuta. Journal of Parasitology*, **60**, 447–52.

Pappas, P.W. & Read, C.P. (1975). Membrane transport in helminth parasites: a review. *Experimental Parasitology*, **37**, 469–530.

Pappas, P.W., Uglem, G.L. & Read, C.P. (1974). Anion and cation requirements for glucose and methionine accumulation by *Hymenolepis diminuta* (Cestoda). *Biological Bulletin*, **146**, 56–66.

Parenti, U. & Antoniotti, M.L. (1967). Oogenesis in *Echinorhynchus truttae* (Schrank). *Nature, Milano*, **58**, 89–91.

Parenti, U., Antoniotti, M.L. & Beccio, C. (1965). Sex ratio and sex digamety in *Echinorhynchus truttae. Experientia*, **21**, 657–8.

Paris, O.P. & Pitelka, F.A. (1962). Population characteristics of the terrestrial isopod *Armadillidium vulgare* in California grassland. *Ecology*, **43**, 229–48.

Parker, J.W. (1909). *Echinorhynchus canis. American Veterinary Review*, **35**, 702–4.

Parshad, V.R. & Crompton, D.W.T. (1981). Aspects of acanthocephalan reproduction. *Advances in Parasitology*, **19**, 73–138.

Parshad, V.R. & Guraya, S.S. (1977*a*). Morphological and histochemical observations on the ovarian balls of *Centrorhynchus corvi* (Acanthocephala). *Parasitology*, **74**, 243–53.

Parshad, V.R. & Guraya, S.S. (1977*b*). Comparative histochemical observations on the excretory system of helminth parasites. *Zeitschrift für Parasitenkunde*, **52**, 81–9.

Parshad, V.R. & Guraya, S.S. (1977*c*). Correlative biochemical and histochemical observations on the lipids of *Centrorhynchus corvi* (Acanthocephala). *Annales de Biologie Animale, Biochemie, Biophysique*, **17**, 953–9.

Parshad, V.R. & Guraya, S.S. (1978*a*). Morphological and histochemical observations on oocyte atresia in *Centrorhynchus corvi* (Acanthocephala). *Parasitology*, **77**, 133–8.

Parshad, V.R. & Guraya, S.S. (1978*b*). Phosphatases in helminths: effects of pH and various chemicals and anthelmintics on the enzyme activities. *Veterinary Parasitology*, **4**, 111–20.

Parshad, V.R. & Guraya, S.S. (1979). Some observations on the testicular changes in an acanthocephalan, *Centrorhynchus corvi* in natural infections of the crow, *Corvus splendens. International Journal of Invertebrate Reproduction*, **1**, 262–6.

Parshad, V.R., Crompton, D.W.T. & Martin, J. (1980). Observations on the surface morphology of the ovarian balls of *Moniliformis* (Acanthocephala). *Parasitology*, **81**, 423–31.

Parshad, V.R., Crompton, D.W.T. & Nesheim, M.C. (1980). The growth of *Moniliformis* (Acanthocephala) in rats fed on various monosaccharides and disaccharides. *Proceedings of the Royal Society of London (B)*, **209**, 299–315.

Pearse, B.M.F. & Bretscher, M.S. (1981). Membrane recycling by coated vesicles. *Annual Review of Biochemistry*, **50**, 85–101.

Penny, C. & Knapton, R.W. (1977). Record of an American robin killing a shrew. *Canadian Field Naturalist*, **91**, 393.

Pennycuick, L. (1971 *a*). Quantitative effects of three species of parasites on a population of three-spined sticklebacks, *Gasterosteus aculeatus* L. *Journal of Zoology*, **165**, 143–62.

Pennycuick, L. (1971 *b*). Seasonal variations in the parasite infections in a population of three-spined sticklebacks, *Gasterosteus aculeatus*. L. *Parasitology*, **63**, 373–88.

Pennycuick, L. (1971 *c*). Frequency distributions of parasites in a population of three-spined sticklebacks, *Gasterosteus aculeatus* L., with particular reference to the negative binomial distribution. *Parasitology*, **63**, 389–406.

Pereira, C. (1936). Nota sobre a acanthella de *Oncicola oncicola* (V. Ihering, 1892). *Archivos do Instituto Biologico, Sao Paulo*, **7**, 245–52.

Persson, L., Borg, K. & Fält, H. (1974). On the occurrence of endoparasites in eider ducks in Sweden. *Viltrevy Swedish Wildlife*, **9**, 1–24.

Petrochenko, V.A. (1949). Elucidation of the life cycle of the acanthocephalan *Polymorphus magnus* Skrjabin, 1913, parasite of domestic and wild ducks. *Dokladi Akademii Nauk SSSR*, **66**, 137–40.

Petrochenko, V.I. (1952). On the position of the Acanthocephala in the zoological system. (Phylogenetic connection of the Acanthocephala with other groups of invertebrates.) *Zoologicheskii Zhurnal*, **31**, 288–327.

Petrochenko, V.I. (1953). Thorny-headed worms (Acanthocephala) of U.S.S.R. amphibians. In *Contributions to Helminthology Published to Commemorate the 75th Birthday of K.I. Skrjabin*, pp. 508–17.

Petrochenko, V.I. (1956). *Acanthocephala of Domestic and Wild Animals*, vol. 1. Moscow: Izdatel'stvo Akademii Nauk SSSR.

Petrochenko, V.I. (1958). *Acanthocephala of Domestic and Wild Animals*, vol. 2. Moscow: Izdatel'stvo Akademii Nauk SSSR.

Petrov, Yu. F. (1973). The survival of the ova of some helminths of aquatic birds in the water bodies of northern and central Kazakhstan. *Sbornik Nauchnykh trudov Vsesoyuznyi Nauchno-Issledovatel'skii po Boleznyam Ptits*, 1973, 231–4.

Petrov, Yu. F. & Nikitin, V.M. (1975). Changes in protein and protein fractions in the serum of ducks during *Tetrameres*, *Streptocara*, *Polymorphus* and *Hymenolepis* infections. *Materialy Nauchnykh konferentsii Vsesoyuznogo Obshchestva Gel'mintologov*, 1975, 88–92.

Peura, R., Valtonen, E.T. & Crompton, D.W.T. (1982). The structure of the mature oocyte and zygote of *Corynosoma semerme* (Acanthocephala). *Parasitology*, **84**, 475–9.

Pflugfelder, O. (1949). Histophysiologische Untersuchungen über die Fettresorption darmloser Parasiten: Die Funktion der Lemnsiken der Acanthocephalen. *Zeitschrift für Parasitenkunde*, **14**, 274–80.

Pflugfelder, O. (1956). Abwehrreaktion der Wirtstiere von *Polymorphus boschadis* Schr. (Acanthocephala). *Zeitschrift für Parasitenkunde*, **17**, 371–82.

Pietrzak, S.M. & Saz, H.J. (1981). Succinate decarboxylation to propionate and the associated phosphorylation in *Fasciola hepatica* and *Spirometra mansonoides*. *Molecular and Biochemical Parasitology*, **3**, 61–70.

Plagemann, P.G.W., Wohlhueter, R.M., Graff, J., Erbe, J. & Wilkie, P. (1981). Broad specificity hexose transport system with differential mobility of loaded and empty carrier, but with directional symmetry, is a common property of mammalian cell lines. *Journal of Biological Chemistry*, **256**, 2835–42.

Podder, T.N. (1938). A new species of Acanthocephala, *Acanthosentis dattai* n. sp. from a freshwater fish of Bengal, *Barbus ticto* (Ham. and Buch.) and *B. stigma* (Cuv. and Val.). *Parasitology*, **30**, 171–5.

Podesta, R.B. (1977). *Hymenolepis diminuta*: unstirred layer thickness and effects on active and passive transport kinetics. *Experimental Parasitology*, **43**, 12–21.

Podesta, R.B. (1980). Concepts of membrane biology in *Hymenolepis diminuta*. In *Biology of the Tapeworm* Hymenolepis diminuta, ed. H.P. Arai, pp. 505–49. London and New York: Academic Press.

Podesta, R.B. (1982). Membrane physiology of helminths. In *Membrane Physiology of Invertebrates*, ed. R.B. Podesta, pp. 121–77. New York: Marcel Dekker, Inc.

Podesta, R.B. & Holmes, J.C. (1970). The life cycles of three polymorphids (Acanthocephala) occurring as juveniles in *Hyalella azteca* (Amphipoda) at Cooking Lake, Alberta. *Journal of Parasitology*, **56**, 1118–23.

Podesta, R.B., Stallard, H.E., Evans, W.S., Lussier, P.E., Jackson, D.J. & Mettrick, D.F. (1977). *Hymenolepis diminuta*: determination of unidirectional uptake rates for nonelectrolytes across the surface 'epithelial' membrane. *Experimental Parasitology*, **42**, 300–17.

Poinar, G.O. (1969). Arthropod immunity to worms. In *Immunity to Parasitic Animals*, vol. 1, ed. G.J. Jackson, R. Herman and I. Singer, pp. 173–210. New York: Appleton-Century-Crofts.

Potashner, S.J. & Johnstone, R.M. (1971). Cation gradients, ATP and amino acid accumulation in Ehrlich ascites cells. *Biochimica et Biophysica Acta*, **233**, 91–103.

Prychitko, S.B. & Nero, R.W. (1983). Occurrence of the acanthocephalan *Echinorhynchus leidyi* (Van Cleave, 1924) in *Mysis relicta*. *Canadian Journal of Zoology*, **61**, 460–2.

Pujatti, D. (1952). Capiti intermedi di *Centrorhynchus spinosus* Kaiser (1893) (Acanthocephala). *Doriana*, **1**, 1–7.

Purich, D.L., Fromm, H.J. & Rudolph, F.B. (1973). The hexokinases: kinetic, physical and regulatory properties. In *Advances in Enzymology*, vol. 39, ed. A. Meister, pp. 249–326. New York: John Wiley & Sons.

Rahman, M.S. & Mettrick, D.F. (1982). Carbohydrate intermediary metabolism in *Hymenolepis diminuta* (Cestoda). *International Journal for Parasitology*, **12**, 155–62.

Ramaswamy, K., Malathi, P. Caspary, W.F. & Crane, R.K. (1974). Studies on the transport of glucose from disaccharides by hamster small intestine *in vitro*. II. Characteristics of the disaccharidase-related transport system. *Biochimica et Biophysica Acta*, **345**, 39–48.

Rasero, F.S., Monteoliva, M. & Mayor, F. (1968). Enzymes related to 4-aminobutyrate metabolism in intestinal parasites. *Comparative Biochemistry and Physiology*. **25**, 639–701.

Rasin, K. (1949). Postembryonalni vyvoj vrtejse *Leptorhynchoides plagicephalus* (Westrumb, 1821). *Vestnik Ceskoslovenske Spolecnosti Zoologicke*, **3**, 289–94.
Rauther, M. (1930). Sechste Klasse des Cladus Nemathelminthes. Acanthocephala Kratzwurmer. In *Handbuch der Zoologie*, Band 2, ed. W. Kükenthal, pp. 449–82. Berlin and Leipzig: Walter De Gruyter & Co.
Ravindranath, M.H. & Anantaraman, S. (1977). The cystacanth of *Moniliformis moniliformis* (Bremser, 1811) and its relationship with the haemocytes of the intermediate host (*Periplaneta americana*). *Zeitschrift für Parasitenkunde*, **53**, 225–37.
Ravindranathan, R. & Nadakal, A.M. (1971). Carotenoids in an acanthocephalid worm *Pallisentis nagpurensis*. *Japanese Journal of Parasitology*, **20**, 1–5.
Rayski, C. & Garden, E.A. (1961). Life cycle of an acanthocephalan parasite of the eider duck. *Nature, London*, **192**, 185–6.
Read, C.P. (1950*a*). The vertebrate small intestine as an environment for parasitic helminths. *Rice Institute Pamphlet*, **37 (2)**, 1–94.
Read, C.P. (1950*b*). The rat as an experimental host of *Centrorhynchus spinosus* (Kaiser, 1893), with remarks on host specificity of the Acanthocephala. *Transactions of the American Microscopical Society*, **69**, 179–82.
Read, C.P. (1961). The carbohydrate metabolism of worms. In *Comparative Physiology of Carbohydrate Metabolism in Heterothermic Animals*, ed. A.W. Martin, pp. 3–34. Washington: Washington University Press.
Read, C.P. (1971). The microcosm of intestinal helminths. In *Ecology and Physiology of Parasites*, ed. A.M. Fallis, pp. 188–97. Ontario: University of Toronto Press.
Read, C.P. (1973). Contact digestion in tapeworms. *Journal of Parasitology*, **59**, 672–7.
Read, C.P. & Rothman, A.H. (1958). The carbohydrate requirement of *Moniliformis* (Acanthocephala). *Experimental Parasitology*, **7**, 191–7.
Read, C.P., Rothman, A.H. & Simmons, J.E. (1963). Studies on membrane transport, with special reference to parasite–host integration. *Annals of the New York Academy of Sciences*, **113**, 154–205.
Redi, F. (1684). *Osservazioni Intorno Agli Animali Viventi che si Trovano Regli Animali Viventi*. Firenze, 253 pp.
Reinhard, E.G. (1944). A hermit crab as intermediate host of *Polymorphus* (Acanthocephala). *Journal of Parasitology*, **30**, 201.
Reinhard, E.G. (1956). Parasitic castration of crustacea. *Experimental Parasitology*, **5**, 79–107.
Reish, D.J. (1950). Preliminary note on the life cycle of the acanthocephalan *Polymorphus kenti* Van Cleave, 1947. *Journal of Parasitology*, **36**, 496.
Richart, R. & Benirschke, K. (1963). Causes of death in a colony of marmoset monkeys. *Journal of Pathology and Bacteriology*, **86**, 221–3.
Riquier, J.K. (1909). Die Larve von *Pomphorhynchus laevis* Zoega (=*Echinorhynchus proteus* Westr.) in der *Tinca vulgaris* und dessen experimentell erzielte Entwicklung in *Esox lucius*. *Centralblatt für Bakteriologie und Parasitenkunde*, Abteilung I, **52**, 248–52.
Rizhikov, K.M. & Dizer, Y.B. (1954). Biology of *Macracanthorhynchus catulinus* and *Mediorhynchus micracanthus*. *Dokladi Akademii Nauk SSSR*, **95**, 1367–9.
Roach, P.J. (1981). Glycogen synthase and glycogen synthase kinases. In *Current Topics in Cellular Regulation*, vol. 20, ed. B.L. Horecker and E.R. Stadtman, pp. 45–105. London and New York: Academic Press.

Roberts, L.S. (1980). Development of *Hymenolepis diminuta* in its definitive host. In *Biology of the Tapeworm* Hymenolepis diminuta, ed. H.P. Arai, pp. 357–423. London and New York: Academic Press.

Roberts, L.S. & Fairbairn, D. (1965). Metabolic studies on adult *Nippostrongylus brasiliensis* (Nematoda: Trichostrongyloidea). *Journal of Parasitology*, **51**, 129–38.

Robinson, E.S. (1964). Chromosome morphology and behaviour in *Macracanthorhynchus hirudinaceus*. *Journal of Parasitology*, **50**, 694–7.

Robinson, E.S. (1965). The chromosomes of *Moniliformis dubius* (Acanthocephala). *Journal of Parasitology*, **51**, 430–2.

Robinson, E.S. & Strickland, B.C. (1969). Cellular responses of *Periplaneta americana* to acanthocephalan larvae. *Experimental Parasitology*, **26**, 384–92.

Rojanapaibul, A. (1976). The life cycle of *Acanthocephalus clavula* (Dujardin, 1845), an acanthocephalan of fish of Llyn Tegid, North Wales. *Parasitology*, **73**, xi. (Abstract.)

Rojanapaibul, A. (1977). *The Biology and Life History of* Acanthocephalus clavula *Dujardin 1845 in Llyn Tegid (Bala Lake), North Wales*. Ph D thesis: University of Liverpool.

Rolleston, F.S. (1972). A theoretical background to the use of measured concentrations of intermediates in the study of the control of intermediary metabolism. In *Current Topics in Cellular Regulation*, vol. 5, ed. B.L. Horecker and E.R. Stadtman, pp. 47–75. London and New York: Academic Press.

Romanovski, A.B. (1964). Life-cycle of *Polymorphus minutus*. *Veterinariya*, **40**, 40–1.

Rotheram, S. & Crompton, D.W.T. (1972). Observations on the early relationship between *Moniliformis dubius* (Acanthocephala) and the haemocytes of the intermediate host, *Periplaneta americana*. *Parasitology*, **64**, 15–21.

Rothman, A.H. (1967). Ultrastructural enzyme localization in the surface of *Moniliformis dubius* (Acanthocephala). *Experimental Parasitology*, **21**, 42–6.

Rothman, A.H. & Elder, J.E. (1970). Histochemical nature of an acanthocephalan, a cestode and a trematode absorbing surface. *Comparative Biochemistry and Physiology*, **33**, 745–62.

Rothman, A.H. & Fisher, F.M. (1964). Permeation of amino acids in *Moniliformis* and *Macracanthorhynchus* (Acanthocephala). *Journal of Parasitology*, **50**, 410–14.

Rothman, A.H. & Rosario, B. (1961). The structure of the surface of *Macracanthorhynchus hirudinaceus* as seen with the electron microscope. *Journal of Parasitology*, **47**, 25.

Rudolphi, C.A. (1793). *Observationes circa vermes intestinales quas consensu ampliss. facul. philos. in meroriam concilii puriorum sacrorum in Suecia stabiliendorum gratia ante ducentos annos Upsaliae habiti praeside...Domino Ioanne Quistorp...pro summis in philosophia honoribus publice defendet...Mart. mens.* 1793. 46 pp. Gryphiswaldiae.

Rudolphi, C.A. (1801). Biobachtungen über die Eingeweidwurmer. *Wiedemanns Arch. Zool. Zootom.*, II, Bd. I, 1–65. Braunschweig.

Rudolphi, C.A. (1802). Fortsetzung der Biobachtungen über die Eingeweidwurmer. *Wiedemanns Arch. Zool. Zootom.*, II, Bd. II, 1–67. Braunschweig.

Rudolphi, C.A. (1804–1805). *Bemerkungen aus dem Gebiete der Naturgeschichte, Medizin und Tierarzneikunde*, I. Teil., VIII+296 pp.; II. Teil., XVI+ 222 pp. Berlin.

Rudolphi, C.A. (1808–1809). *Entozoorum sive vermium intestinalium historia naturalis.* Vol. I, XXVI+527 pp.; Vol. 2, 257 pp. Amstelaedami.

Rudolphi, C.A. (1814). Erster Nachtrag zu meiner Naturgeschichte der Eingeweidwurmer. *Magaz. Gesellsch. Naturf. Freunde Berlin*, **6**, 83–113.

Rudolphi, C.A. (1819). *Entozoorum synopsis, cui accedunt mantissa duplex et indices locupletissimi.* X+811 pp. Berolini.

Ruff, M.D. & Read, C.P. (1973). Inhibition of pancreatic lipase by *Hymenolepis diminuta. Journal of Parasitology*, **59**, 105–11.

Ruff, M.D., Uglem, G.L. & Read, C.P. (1973). Interactions of *Moniliformis dubius* with pancreatic enzymes. *Journal of Parasitology*, **59**, 839–43.

Rumpus, A. (1973). *The parasites of* Gammarus pulex *in the River Avon, Hampshire.* PhD thesis: University of Exeter.

Rumpus, A.E. (1975). The helminth parasites of the bullhead *Cottus gobio* (L) and the stone loach *Neomacheilus barbatulus* (L) from the River Avon, Hampshire. *Journal of Fish Biology*, **7**, 469–83.

Rumpus, A.E. & Kennedy, C.R. (1974). The effect of the acanthocephalan *Pomphorhynchus laevis* upon the respiration of its intermediate host, *Gammarus pulex. Parasitology*, **68**, 271–84.

Sadaterashvili, Yu. F. (1970). Development of *Macracanthorhynchus hirudinaceus* in intermediate hosts in the Georgian SSR. *Trudy Vsesoyuznogo Instituta Gel'mintologii im K.I. Skryabina*, **16**, 201–8.

Sadaterashvili, Yu. F. (1977). Comparative data on the survival and development of *Macracanthorhynchus hirudinaceus* in piglets infected in summer and in autumn. *Sbornik Trudov Gruzinskogo Zootekhnichesko-Veterinarnogo Uchebno-Issledovatel'skogo Instituta*, **40**, 147–9.

Sadaterashvili, Yu. F. (1978*a*). On the prevalence of *Macracanthorhynchus* larvae in the intermediate hosts in East Georgia. *Nauchnye Trudy, Gruzinskii Zootekhinichesko-Veterinarnyi Nauchno-Issledovatel'skii Institut*, **41**, 105–9.

Sadaterashvili, Yu. F. (1978*b*). Prevalence of *Macracanthorhynchus* larvae in the intermediate host and the migration of beetle larvae in soil related to the season. In *Materialy pervoi (I) Zakavkazskoi Konferentsii po obshchei parasitologii 4–6 Maya 1977, Tbilisi*, pp. 196–203. *Tbilisi: 'Metsniereba'*.

Saefftigen, A. (1884). Zur Organisation der Echynorhynchen. *Morphologisches Jahrbuch*, **10**, 120–71.

Sallee, V.L., Wilson, F.A. & Dietschy, J.M. (1972). Determination of unidirectional uptake rates for lipids across the intestinal brush border. *Journal of Lipid Research*, **13**, 184–92.

Salt, G. (1963). The defence reactions of insects to metazoan parasites. *Parasitology*, **53**, 527–642.

Samuel, G. & Bullock, W.L. (1981). Life cycle of *Paratenuisentis ambiguus* (Van Cleave, 1921) Bullock and Samuel, 1975 (Acanthocephala: Tenuisentidae). *Journal of Parasitology*, **67**, 214–17.

Samuel, N., Nickol, B.B. & Mayes, M.A. (1976). Acanthocephala of Nebraska fishes. *American Midland Naturalist*, **96**, 391–406.

Sandground, J.H. (1926). On an unusual occurrence of *Moniliformis moniliformis* (Acanthocephala) as a parasite of toads and lizards in Central America. *Transactions of the American Microscopical Society*, **45**, 289–97.

Sanford, S.E. (1978). Mortality in mute swans in southern Ontario associated with infestation with the thorny-headed worm, *Polymorphus boschadis. Canadian Veterinary Journal*, **19**, 234–6.

Saxon, D.J. & Dunagan, T.T. (1975). Enzymes of the pentose phosphate pathway in *Macracanthorhynchus hirudinaceus* (Acanthocephala). *Comparative Biochemistry and Physiology*, **50B**, 299–303.

Saxon, D.J. & Dunagan, T.T. (1976). Comparative study of pentose phosphate pathway enzymes in acanthocephalans from homothermic and poikilothermic hosts. *Comparative Biochemistry and Physiology*, **55B**, 377–80.

Saz, H.J. (1969). Carbohydrate and energy metabolism of nematodes and Acanthocephala. In *Chemical Zoology*, vol. 3, ed. M. Florkin and B.T. Scheer, pp. 329–60. London and New York: Academic Press.

Saz, H.J. (1971). Facultative anaerobiosis in the invertebrates: pathways and control systems. *American Zoologist*, **11**, 125–35.

Saz, H.J. (1981). Energy metabolisms of parasitic helminths: adaptations to parasitism. *Annual Review of Physiology*, **43**, 323–41.

Saz, H.J. & Pietrzak, S.M. (1980). Phosphorylation associated with succinate decarboxylation to propionate in *Ascaris* mitochondria. *Archives of Biochemistry and Biophysics*, **202**, 388–95.

Schaefer, P.W. (1970). *Periplaneta americana* (L) as intermediate host of *Moniliformis moniliformis* (Bremser) in Honolulu, Hawaii. *Proceedings of the Helminthological Society of Washington*, **37**, 204–7.

Scheer, D. (1934). *Gammarus pulex* und *Carinogammarus roeselii* als Zwischenwirte von *Polymorphus minutus* (Acanth.). *Zeitschrift für Parasitenkunde*, **7**, 268–72.

Schmidt, G.D. (1964*a*). A note on the Acanthocephala parasitizing amphipod crustacea in a spring-fed pond in Montana. *Canadian Journal of Zoology*, **42**, 718.

Schmidt, G.D. (1964*b*). Life cycle and development of *Prosthorhynchus formosus* (Van Cleave, 1918) Travassos, 1926, an acanthocephalan parasite of birds. *Dissertation Abstracts*, **25**, 2677.

Schmidt, G.D. (1965). *Corynosoma bipapillum* sp. n. from Bonaparte's gull *Larus philadelphia* in Alaska, with a note on *C. constrictum* Van Cleave, 1918. *Journal of Parasitology*, **51**, 814–16.

Schmidt, G.D. (1972*a*). Acanthocephala of captive primates. In *Pathology of Simian Primates*, ed. R.N. T-W-Fiennes, pp. 144–56. Basel: S. Karger.

Schmidt, G.D. (1972*b*). Revision of the class Archiacanthocephala Meyer, 1931 (Phylum Acanthocephala), with emphasis on Oligacanthorhynchidae Southwell et Macfie, 1925. *Journal of Parasitology*, **58**, 290–7.

Schmidt, G.D. (1972*c*). *Oncicola schacheri* sp. n. and other Acanthocephala of Lebanese mammals. *Journal of Parasitology*, **58**, 279–81.

Schmidt, G.D. (1973*a*). Resurrection of *Southwellina* Witenberg, 1932, with a description of *Southwellina dimorpha* sp. n. and a key to genera in Polymorphidae (Acanthocephala). *Journal of Parasitology*, **59**, 299–305.

Schmidt, G.D. (1973*b*). Early embryology of the acanthocephalan *Mediorhynchus grandis* Van Cleave, 1916. *Transactions of the American Microscopical Society*, **92**, 512–16.

Schmidt, G.D. (1975). *Andracantha*, a new genus of Acanthocephala (Polymorphidae) from fish-eating birds, with description of three species. *Journal of Parasitology*, **61**, 615–20.

Schmidt, G.D. (1983). What is *Echinorhynchus pomatostomi* Johnston & Cleland, 1912? *Journal of Parasitology*, **69**, 397–9.

Schmidt, G.D. & Huggins, E.J. (1973). Acanthocephala of South American Fishes. Part I. Eoacanthocephala. *Journal of Parasitology*, **59**, 829–35.

Schmidt, G.D. & Kuntz, R.E. (1967*a*). Notes on the life cycle of *Polymorphus* (*Profilicollis*) *formosus* sp. n. and records of *Arhythmorhynchus hispidus* Van Cleave, 1925 (Acanthocephala) from Taiwan. *Journal of Parasitology*, **53**, 805–9.

Schmidt, G.D. & Kuntz, R.E. (1967*b*). Revision of the Porrorchinae (Acanthocephala: Plagiorhynchidae) with descriptions of two new genera and three new species. *Journal of Parasitology*, **53**, 130–41.

Schmidt, G.D. & Kuntz, R.E. (1969). *Centrorhynchus spilornae* sp. n. (Acanthocephala) and other Centrorhynchidae from the Far East. *Journal of Parasitology*, **55**, 329–34.

Schmidt, G.D. & MacLean, S.A. (1978). *Polymorphus* (*Profilicollis*) *major* Lundstrom, 1942 juveniles in Rock crabs, *Cancer irroratus*, from Maine. *Journal of Parasitology*, **64**, 953–4.

Schmidt, G.D. & Olsen, O.W. (1964). Life cycle and development of *Prosthorhynchus formosus* (Van Cleave, 1918) Travassos, 1926, an acanthocephalan parasite of birds. *Journal of Parasitology*, **50**, 721–30.

Schmidt, G.D., Walley, H.D. & Wijek, D.S. (1974). Unusual pathology in a fish due to the acanthocephalan *Acanthocephalus jacksoni* Bullock, 1962. *Journal of Parasitology*, **60**, 730–1.

Schmidt-Nielsen, K. (1975). *Animal Physiology*. Cambridge University Press.

Schneider, A.F. (1868). Uber den Bau der Acanthocephalen. *Archiv für Anatomie, Physiologie und Wissenschaftliche Medizin*, **1**, 584–97.

Schneider, A.F. (1871). On the development of *Echinorhynchus gigas*. *Annals and Magazine of Natural History*, **4**, 441–3.

Schrank, F.V.P. (1782). Zoologische Biobachtungen. *Natur. 18. Stuck Holle*, 66–85.

Schultz, S.G. & Curran, P.F. (1970). Coupled transport of sodium and organic solutes. *Physiological Reviews*, **50**, 637–718.

Schütze, H.R. & Ankel, W.E. (1976). Population dynamics of *Gammarus pulex fossarum* infected with *Echinorhynchus truttae* in a small stream of Oberhessen. *Zeitschrift für Parasitenkunde*, **50**, 197–8.

Segel, I.H. (1975). *Enzyme Kinetics*. New York: Wiley.

Seidenberg, A.J. (1973). Ecology of the acanthocephalan *Acanthocephalus dirus* (Van Cleave, 1931), in its intermediate host *Asellus intermedius* Forbes (Crustacea: Isopoda). *Journal of Parasitology*, **59**, 957–62.

Semenza, G. (1968). Intestinal oligosaccharidases and disaccharidases. In *Handbook of Physiology, Section 6: Alimentary Canal*, vol. 5, ed. C.F. Code, pp. 2543–66. Washington, D.C.: American Physiological Society.

Sen, P. (1938). On a new species of Acanthocephala, *Acanthosentis holospinus* sp. nov. from the fish *Barbus stigma* (Cuv. and Val.). *Proceedings of the Indian Academy of Sciences*, **7**, 41–6.

Setchell, B.P. (1982). Spermatogenesis and spermatozoa. In *Reproduction in Mammals*, vol. 1, *Germ Cells and Fertilization*, ed. C.R. Austin and R.V. Short, pp. 63–101. Cambridge University Press.

Sharma, S.K. & Wattal, B.L. (1976). First record of a cyclopoid host – *Mesocyclops leuckharti* (Claus) for an acanthocephalous worm – *Acanthosentis dattai* Podder from Delhi (India). *Folia Parasitologica*, **23**, 169–73.

Sharpilo, V.P. & Sharpilo, L.D. (1969). Experimental identification of three larval forms of helminth parasites of reptiles with their mature forms. *Zbirnik Prats' Zoologichnogo Muzeyu*, **33**, 30–5.

Shcherbovich, I.A. (1950). *Macracanthorhynchus* infection of swine. *Veterinariya*, **27**, 6–11.

Shiau, Y.-F. (1981). Mechanisms of intestinal fat absorption. *American Journal of Physiology*, **240**, G1–G9.

Shtein, G.A. (1959). The parasite fauna of aquatic arthopods from some lakes in Karelia. III. Larval Acanthocephala from crustaceans. In *Ekologisheskaya Parazitologiya*. Leningrad.

Siebold, C.T.E. von (1836). Fernere Biobachtungen über die Spermatozoen der wirbellosen Tiere: Die Spermatozoen der Helminthen. *Archiv für Anatomie, Physiologie und Wissenschaftliche Medizin*, 232–3.

Silverstein, S.C., Steinman, R.M. & Cohn, Z.A. (1977). Endocytosis. *Annual Review of Biochemistry*, **46**, 669–722.

Simmonds, F.J. (1960). *Report on a Tour of Commonwealth Countries in Africa, March–June, 1960*. Farnham Royal: Commonwealth Agricultural Bureaux.

Sinnott, E.W., Dunn, L.C. & Dobzhanksky, T. (1958). *Principles of Genetics*. New York and London: McGraw-Hill.

Sita, E. (1949). The life-cycle of *Moniliformis moniliformis* (Bremser, 1811), Acanthocephala. *Current Science*, **6**, 216–18.

Skrjabin, K.I. & Shults, R.E.S. (1931). Gelmintozy cheloveka (Arcanthocephala). In *The Helminthiases of Man (Elements of Medical Helminthology) for Physicians, Veterinarians, Naturalists, and Students*, part 2, pp. 242–50. Moscow.

Smith, M.H. (1969). The pigments of Nematoda and Acanthocephala. In *Chemical Zoology*, vol. 3, ed. M. Florkin and B.T. Scheer, pp. 501–20. London and New York: Academic Press.

Smith, W., Dunagan, T.T. & Miller, D.M. (1979). Fatty acids in female *Macracanthorhynchus hirudinaceus* (Acanthocephala). *Lipids*, **14**, 253–6.

Smyth, J.D. & Smyth, M.M. (1968). Some aspects of host specificity in *Echinococcus granulosus*. *Helminthologia*, **9**, 519–29.

Southwell, T. & MacFie, J.W.S. (1925). On a collection of Acanthocephala in the Liverpool School of Tropical Medicine. *Annals of Tropical Medicine and Parasitology*, **19**, 141–84.

Spencer, L.T. (1974). Parasitism of *Gammarus lacustris* (Crustacea: Amphipoda) by *Polymorphus minutus* (Acanthocephala) in Colorado. *American Midland Naturalist*, **91**, 505–9.

Stalmans, W. & Hers, H.G. (1973). Glycogen synthesis from UDPG. In *The Enzymes*, 3rd edn, vol. 9, ed. P.D. Boyer, pp. 309–61. London and New York: Academic Press.

Starling, J.A. (1972). *Carbohydrate transport and accumulation in the intestinal parasite* Moniliformis dubius *(Acanthocephala)*. Ph D thesis, Rice University. (*Dissertation Abstracts International*, **33B**, 1753–4.)

Starling, J.A. (1975). Tegumental carbohydrate transport in intestinal helminths: correlation between mechanisms of membrane transport and the biochemical environment of absorptive surfaces. *Transactions of the American Microscopical Society*, **94**, 508–23.

Starling, J.A. & Fisher, F.M. (1975). Carbohydrate transport in *Moniliformis dubius* (Acanthocephala). I. The kinetics and specificity of hexose absorption. *Journal of Parasitology*, **61**, 977–90.

Starling, J.A. & Fisher, F.M. (1978). Carbohydrate transport in *Moniliformis dubius* (Acanthocephala). II. Post-absorptive phosphorylation of glucose and the role of trehalose in the accumulation of endogenous glucose reserves. *Journal of Comparative Physiology*, **126**, 223–31.

Starling, J.A. & Fisher, F.M. (1979). Carbohydrate transport in *Moniliformis dubius* (Acanthocephala). III. Post-absorptive fate of fructose, mannose and galactose. *Journal of Parasitology*, **65**, 8–13.

Steel, E.A. (1961). Some observations on the life history of *Asellus aquaticus* (L.) and *Asellus meridianus* Racovitza (Crustacea: Isopoda). *Proceedings of Zoological Society of London*, **137**, 71–87.

Stein, T.A. (1971). Partial purification and properties of glucose-6-phosphate dehydrogenase from *Macracanthorhynchus hirudinaceus* (Acanthocephala). *Comparative Biochemistry and Physiology*, **39B**, 541–9.

Stein, W.D. (1967). *The Movement of Molecules Across Cell Membranes*. London and New York: Academic Press.

Stock, R.D. (1978). Effects of concurrent infections on *Moniliformis dubius* and *Nippostrongylus brasiliensis*. *Proceedings of the Nebraska Academy of Sciences*, **88**, 62.

Stranack, F.R. (1972). The fine structure of the acanthor shell of *Pomphorhynchus laevis* (Acanthocephala). *Parasitology*, **64**, 187–90.

Stranack, F.R., Woodhouse, M.A. & Griffin, R.L. (1966). Preliminary observations on the ultrastructure of the body wall of *Pomphorhynchus laevis* (Acanthocephala). *Journal of Helminthology*, **40**, 395–402.

Stunkard, H.W. (1965). New intermediate hosts in the life cycle of *Prosthenorchis elegans* (Diesing, 1851), an acanthocephalan parasite of primates. *Journal of Parasitology*, **51**, 645–9.

Styczynska, E. (1958). Some observations on the development and bionomics of larvae of *Filicollis anatis* Schrank. (Parasitofauna of the biocoenosis of Druzno Lake. Part VII.) *Acta Parasitologica Polonica*, **6**, 213–24.

Sultanov, M.A., Kabilov, T.K. & Davlatov, N. (1974). Intermediate hosts of the acanthocephalan *Moniliformis moniliformis* (Bremser, 1811) in the Fergana Valley. *Uzbekskii Biologicheskii Zhurnal*, **2**, 55–7.

Sultanov, M.A., Kabilov, T.K. & Siddikov, B.Kh. (1980). Infection of Isopoda with helminth larvae. *Uzbekskii Biologicheskii Zhurnal*, 45–7.

Supowit, S.C. & Harris, B.G. (1976). *Ascaris suum* hexokinase: purification and possible function in compartmentation of glucose 6-phosphate in muscle. *Biochimica et Biophysica Acta*, **422**, 48–59.

Sutherland, D.R. & Holloway, H.L., Jr (1979). Parasites of fish from the Missouri, James, Sheyenne, and Wild Rice rivers in North Dakota. *Proceedings of the Helminthological Society of Washington*, **46**, 128–34.

Swales, W.E. & Gwatkin, R.G. (1948). Experiments to determine the role of the thorny-headed worm, *Macracanthorhynchus hirudinaceus*, in the occurrence of disease of pigs in Canada. *Canadian Journal of Comparative Medicine*, **12**, 297–9.

Swennen, C. & Van Den Broek, E. (1960). *Polymorphus botulus* als parasiet bij de eidereenden in de Waddenzee. *Ardea*, **48**, 90–7.

Takos, M.J. & Thomas, L.J. (1958). The pathology and pathogenesis of fatal infections due to an acanthocephalan parasite of marmoset monkeys. *American Journal of Tropical Medicine and Hygiene*, **7**, 90–4.

Tanaka, R.D. & MacInnis, A.J. (1980). Analyses of the pseudocoelomic fluid from *Moniliformis dubius*. *Journal of Parasitology*, **66**, 354–5.

Taylor, E.W. & Thomas, J.N. (1968). Membrane (contact) digestion in the three species of tapeworm *Hymenolepis diminuta*, *Hymenolepis microstoma* and *Moniezia expansa*. *Parasitology*, **58**, 535–46.

Tedla, S. & Fernando, C.H. (1969). Observations on the seasonal changes of the parasite fauna of yellow perch (*Perca flavescens*) from the Bay of Quinte, Lake Ontario. *Journal of the Fisheries Research Board of Canada*, **26**, 833–43.

Tedla, S. & Fernando, C.H. (1970). Some remarks on the ecology of *Echinorhynchus salmonis* (Muller, 1784). *Canadian Journal of Zoology*, **48**, 317–21.

Tesarcik, J. (1970). Egg production by *Neoechinorhynchus rutili* in fish ponds in southern Bohemia. *Zivocisna Vyroba*, **15**, 785–8.

Tesarcik, J. (1972). Localization of *Neoechinorhynchus rutili* in the intestine of carp. *Parazitologiya*, **6**, 190–1.

Thapar, G.S. (1927). On *Acanthogyrus* n. g. from the intestine of the Indian fish, *Labeo rohita*, with a note on the classification of the Acanthocephala. *Journal of Helminthology*, **5**, 109–20.

Thompson, W.R. (1934). The tachinid parasites of woodlice. *Parasitology*, **26**, 378–400.

Thompson-Cowley, L.L., Helfer, D.H., Schmidt, G.D. & Eltzroth, E.K. (1979). Acanthocephalan parasitism in the western bluebird (*Sialia mexicana*). *Avian Diseases*, **23**, 768–71.

Tilesius (1810). Piscium camtschaticorum Terpuk et Vakhnia. Descriptiones et econes. *Mémoires de l'Academie Impériale des Sciences de St Petersbourg*, **2**, 335–75.

Tinsley, R.C. (1983). Ovoviviparity in platyhelminth life-cycles. *Parasitology*, **86**, 161–96.

Tobias, R.C. & Schmidt, G.D. (1977). In vitro cultivation of *Moniliformis moniliformis* juveniles. *Journal of Parasitology*, **63**, 588–9.

Tokeson, J.P.E. & Holmes, J.C. (1982). The effects of temperature and oxygen on the development of *Polymorphus marilis* (Acanthocephala) in *Gammarus lacustris* (Amphipoda). *Journal of Parasitology*, **68**, 112–19.

Travassos, L. (1917). Contribuicoes para o conhecimento da fauna helmintolojica brazileira. VI. Revisaeo dos acantocefalos brazileiros. Parte 1. Fam. Gigantorhynchidae Hamann, 1892. *Memorias do Instituto Oswaldo Cruz*, **9**, 5–62.

Travassos, L. (1926). Contribuicoes para o conhecimento da fauna helmintologica brasileira. XX. Revisaeo dos acantocefalos brasileiros. Parte II. Fam. Echinorhynchidae. Sf. Centrorhynchinae Travassos, 1919. *Memorias do Instituto Oswaldo Cruz*, **19**, 31–125.

Trifonov, T. (1961). New intermediate host of *Macracanthorhynchus hirudinaceus. Izvestiya na Veterinarniya Institut za Zarazni i Parazitni Bolesti*, **2**, 301–6.

Trifonov, T. (1963). The intermediate host of *Macracanthorhynchus hirudinaceus. Izvestiya na Veterinarniya Institut za Zarazni i Parazitni Bolesti*, **7**, 187–9.

Tsimbalyuk, A.K., Kulikov, U.V., Ardasheva, N.V & Tsimbalyuk, E.M. (1978). Helminths of invertebrates in the littoral zone of Itupur Island. In *Zhivotnyi i Rastitel'nyi mir Shel'fovykh zon Kuril'skikh Ostrovov*, pp. 64–126. Moscow: 'Nauka'.

Tubangui, M.A. (1933). Notes on Acanthocephala in the Philippines. *Philippine Journal of Science*, **50**, 115–28.

T-W-Fiennes, R.N. (1966). The Zoological Society of London report of the society's pathologist for the year 1964. *Journal of Zoology*, **148**, 363–80.

Uglem, G.L. (1972*a*). *A physiological basis for habitat specificity in the acanthocephalan parasites* Neoechinorhynchus cristatus *Lynch, 1936, and* N. crassus *Van Cleave, 1919.* Ph D thesis, University of Idaho. (*Dissertation Abstracts International*, **33B**, 2417.)

Uglem, G.L. (1972*b*). The life cycle of *Neoechinorhynchus cristatus* Lynch, 1936 (Acanthocephala) with notes on the hatching of eggs. *Journal of Parasitology*, **58**, 1071–4.

Uglem, G.L. & Beck, S.M. (1972). Habitat specificity and correlated aminopeptidase activity in the acanthocephalans *Neoechinorhynchus cristatus* and *N. crassus. Journal of Parasitology*, **58**, 911–20.

Uglem, G.L. & Larson, O.R. (1969). The life history and larval development of *Neoechinorhynchus saginatus* Van Cleave and Bangham, 1945 (Acanthocephala: Neoechinorhynchidae). *Journal of Parasitology*, **55**, 1212–17.

Uglem, G.L. & Read, C.P. (1973). *Moniliformis dubius*: uptake of leucine and alanine by adults. *Experimental Parasitology*, **34**, 148–53.

Uglem, G.L., Pappas, P.W. & Read, C.P. (1973). Surface aminopeptidase in *Moniliformis dubius* and its relations to amino acid uptake. *Parasitology*, **67**, 185–95.

Uspenskaja, A. (1960). Parasitofaune der crustaces benthiques de la Mer de Barents (Exposition Preliminaire). *Annales de Parasitologie Humaine et Comparée*, **35**, 221–41.

Uzmann, J.R. (1970). Use of parasites in identifying lobster stocks. *Journal of Parasitology*, **56**, 349.

Uznanski, R.L. & Nickol, B.B. (1976). Structure and function of the fibrillar coat of *Leptorhynchoides thecatus* eggs. *Journal of Parasitology*, **62**, 569–73.

Uznanski, R.L. & Nickol, B.B. (1980*a*). Parasite population regulation: lethal and sublethal effects of *Leptorhynchoides thecatus* (Acanthocephala: Rhadinorhynchidae) on *Hyalella azteca* (Amphipoda). *Journal of Parasitology*, **66**, 121–6.

Uznanski, R.L. & Nickol, B.B. (1980*b*). A sequential ranking system for developmental stages of an acanthocephalan, *Leptorhynchoides thecatus*, in its intermediate host, *Hyalella azteca*. *Journal of Parasitology*, **66**, 506–12.

Uznanski, R.L. & Nickol, B.B. (1982). Site selection, growth, and survival of *Leptorhynchoides thecatus* (Acanthocephala) during the prepatent period in *Lepomis cyanellus*. *Journal of Parasitology*, **68**, 686–90.

Val'ter, E.D. (1976). Helminths of *Caprella septentrionalis* in the Rugozweks bay of the White Sea (USSR). In *Kratkie tezisy dokladov II Vsesoyuznogo simposiums po parazitam i boleznyam morskikh zhivotnykh*, pp. 11–12. Kaliningrad: Ministerstvo Rybnogo Khozyaistva SSSR, Atlant NIRO.

Val'ter, E.D., Kondrashkova, A.N. & Popova, T.I. (1980). *Caprella septentrionalis* Kroyer, an intermediate host of the acanthocephalan *Echinorhynchus gadi* Muller, 1776). In *Voprosy parasitologii vodnykh bespozvonochnyjh zhivotnykh*. (*Tematicheskii Sbornik*.), pp. 19–20. Vilnyus, USSR: Akademiya Nauk Kitovskoi SSR, Institut Zoologii i Parasitologii.

Valtonen, E.T. (1980). *Metechinorhynchus salmonis* (Muller, 1780) (Acanthocephala) as a parasite of the whitefish in the Bothnian Bay. I. Seasonal relationships between infection and fish size. *Acta Parasitologica Polonica*, **27**, 293–300.

Van Assche, J.A., Van Laere, A.J. & Carlier, A.R. (1978). Trehalose metabolism in dormant and activated spores of *Phycomyces blakesleeanus* Burgeff. *Planta*, **139**, 171–6.

Van Cleave, H.J. (1914). Studies on cell constancy in the genus *Eorhynchus*. *Journal of Morphology*, **25**, 253–99.

Van Cleave, H.J. (1916). Seasonal distribution of some Acanthocephala from fresh-water hosts. *Journal of Parasitology*, **2**, 106–10.

Van Cleave, H.J. (1919). Acanthocephala from the Illinois River, with descriptions of species and synopsis of the family Neoechinorhynchidae. *Bulletin of the Illinois State Natural History Survey*, **13**, 225–57.

Van Cleave, H.J. (1920*a*). Notes on the life cycle of two species of Acanthocephala from freshwater fishes. *Journal of Parasitology*, **6**, 167–72.

Van Cleave, H.J. (1920*b*). Sexual dimorphism in the Acanthocephala. *Transactions of the Illinois State Academy of Sciences*, **13**, 280–92.

Van Cleave, H.J. (1921). Acanthocephala parasitic in the dog. *Journal of Parasitology*, **7**, 91–4.
Van Cleave, H.J. (1924). A critical study of the Acanthocephala described and identified by Joseph Leidy. *Proceedings of the Academy of Natural Sciences of Philadelphia*, **76**, 279–334.
Van Cleave, H.J. (1925). Acanthocephala from Japan. *Parasitology*, **17**, 149–56.
Van Cleave, H.J. (1928). Nuclei of the subcuticula in the Acanthocephala. *Zeitschrift für Zellforschung und Mikroscopische Anatomie*, **7**, 109–13.
Van Cleave, H.J. (1936). The recognition of a new order in the Acanthocephala. *Journal of Parasitology*, **22**, 202–6.
Van Cleave, H.J.(1937). Developmental stages in acanthocephalan life histories. In *Papers in Helminthology Published in Commemoration of the 30 Year Jubileum of the Scientific, Educational, and Social Activities of the Honoured Worker of Science, K.I. Skrjabin, M. Ac. Sci. and of the Fifteenth Anniversary of the All-Union Institute of Helminthology*, pp. 739–743. Moscow: The All-Union Lenin Academy of Agricultural Science.
Van Cleave, H.J. (1940). The Acanthocephala collected by the Allan Hancock Pacific Expedition, 1934. *Allan Hancock Foundation Publications*, series 1, **2**, 501–27.
Van Cleave, H.J. (1941*a*). Hook patterns on the acanthocephalan proboscis. *Quarterly Review of Biology*, **16**, 157–72.
Van Cleave, H.J. (1941*b*). Relationships of the Acanthocephala. *American Naturalist*, **75**, 31–47.
Van Cleave, H.J. (1947). A critical review of terminology for immature stages in acanthocephalan life histories. *Journal of Parasitology*, **33**, 118–25.
Van Cleave, H.J. (1948). Expanding horizons in the recognition of a phylum. *Journal of Parasitology*, **34**, 1–20.
Van Cleave, H.J. (1949). Morphological and phylogenetic interpretation of the cement glands in the Acanthocephala. *Journal of Morphology*, **84**, 427–57.
Van Cleave, H.J. (1952). Some host–parasite relationships of the Acanthocephala with special reference to the organs of attachment. *Experimental Parasitology*, **1**, 305–30.
Van Cleave, H.J. (1953). Acanthocephala of North American mammals. *Illinois Biological Monographs*, **23**, 1–179.
Van Cleave, H.J. & Bangham, R.V. (1950). Four new species of the acanthocephalan family Neoechinorhynchidae from fresh water fishes of North America, one representing a new genus. *Journal of the Washington Academy of Sciences*, **39**, 398–409.
Van Cleave, H.J. & Bullock, W.L. (1950). Morphology of *Neoechinorhynchus emydis*, a typical representative of the Eoacanthocephala. I. The praesoma. *Transactions of the American Microscopical Society*, **69**, 288–303.
Van Cleave, H.J. & Rausch, R.L. (1950). A new species of the acanthocephalan genus *Arhythmorhynchus* from sandpipers of Alaska. *Journal of Parasitology*, **36**, 278–83.
Van Cleave, H.J. & Ross, E.L. (1944). Physiological responses of *Neoechinorhynchus emydis* (Acanthocephala) to various solutions. *Journal of Parasitology*, **30**, 369–72.
Van Maren, M.J. (1979*a*). Les crustacés amphipodes comme hôtes intermédiaires de vers parasites de poissons. *Bulletin du Centre d'Etudes et de Recherches Scientifiques de Biarritz*, **12**, 597–8.
Van Maren, M.J. (1979*b*). The amphipod *Gammarus fossarum* Koch (Crustacea) as intermediate host for some helminth parasites, with notes on their occurrence in the final host. *Bijdragen tot de Dierkunde*, **48**, 97–110.
Van Thiel, P.H. & Wiegand-Bruss, C.J.E. (1945). Présence de *Prosthenorchis spirula*

chez les chimpanzés. Son rôle pathogène et son developpement dans *Blattella germanica. Annales de Parasitologie Humaine et Comparée*, **20**, 304–20.

Van Vugt, F., Kalaycioglu, L. & Van den Bergh, S.G. (1976). ATP production in *Fasciola hepatica* mitochondria. In *Biochemistry of Parasites and Host–Parasite Relationships*, ed. H. Van den Bossche, pp. 151–8. Amsterdam, New York and Oxford: Elsevier/North Holland.

Van Vugt, F., Van Der Meer, P. & Van den Bergh, S.G. (1979). The formation of propionate and acetate as terminal processes of energy metabolism of the adult liver fluke *Fasciola hepatica. International Journal of Biochemistry*, **10**, 11–18.

Varute, A.T. & Savant, V.A. (1972). Histopathology of alimentary tract of rats infected by *Moniliformis dubius* (Acanthocephala). 2. Comparative studies on lipids of infected and non-infected intestines. *Indian Journal of Experimental Biology*, **10**, 208–12.

Vergara, L.A. & George-Nascimento, M. (1982). Contribucion al estudio del parasitismo en el congrio colorado *Genypterus chilensis* (Guichenot, 1848). *Boletin Chileno de Parasitologia*, **37**, 9–14.

Verma, S.C. & Datta, M.N. (1929). Acanthocephala from Northern India. I. A new genus *Acanthosentis* from a Calcutta fish. *Annals of Tropical Medicine and Parasitology*, **23**, 483–500.

Vogt, C. (1851). *Zoologische Briefe. Vol. 1 Naturgeschichte der Lebenden und Untergegangen Thiere*. Frankfurt.

Vysotskaya, R.U. & Sidorov, V.S. (1973). Lipid content of some helminths from freshwater fish. *Parazitologiya*, **7**, 51–7.

Wagener, G. (1858). Helminthologische Bemerkungen aus einem Sendschreiben an E.Th.V. Siebold. *Zeitschrift für Wissenschaftliche Zoologie*, **9**, 73–90.

Walkey, M. (1962). Observations on the life history of *Neoechinorhynchus rutili* (Müller, 1776). *Parasitology*, **52**, 18.

Walkey, M. (1967). The ecology of *Neoechinorhynchus rutili* (Müller). *Journal of Parasitology*, **53**, 795–804.

Walton, A.C. (1959). Some parasites and their chromosomes. *Journal of Parasitology*, **45**, 1–20.

Wang, C., Li, D. & Cai, Z. (1981). An investigation on *Dorysthenes paradoxus* infected by the juvenile of *Macracanthorhynchus hirudinaceus* in Zhuanghe County, Liaoning Province. *Acta Zoologica Sinica*, **27**, 371–4.

Wanson, W.W. & Nickol, B.B. (1973). Origin of the envelope surrounding larval acanthocephalans. *Journal of Parasitology*, **59**, 1147.

Wanson, W.W. & Nickol, B.B. (1975). Presomal morphology and development of *Prosthorhynchus formosus*, *Prosthenorchis elegans* and *Moniliformis dubius* (Acanthocephala). *Journal of Morphology*, **145**, 73–83.

Ward, H.L. (1940*a*). Notes on juvenile Acanthocephala. *Journal of Parasitology*, **26**, 191–4.

Ward, H.L. (1940*b*). Studies on the life history of *Neoechinorhynchus cylindratus* (Van Cleave, 1913) (Acanthocephala). *Transactions of the American Microscopical Society*, **59**, 327–47.

Ward, H.L. (1952). Glycogen consumption in Acanthocephala under aerobic and anaerobic conditions. *Journal of Parasitology*, **38**, 493–4.

Ward, H.L. & Nelson, D.R. (1967). Acanthocephala of the genus *Moniliformis* from rodents of Egypt with the description of a new species from the Egyptian Spiny Mouse (*Acomys cahirinus*). *Journal of Parasitology*, **53**, 150–6.

Ward, H.L. & Winter, H.A. (1952). Juvenile Acanthocephala from the yellowfin croaker, *Umbrina roncador*, with description of a new species of the genus *Arhythmorhynchus*. *Transactions of the American Microscopical Society*, **71**, 154–6.

Ward, P.F.V. (1982). Aspects of helminth metabolism. *Parasitology*, **84**, 177–94.

Ward, P.F.V. & Crompton, D.W.T. (1969). The alcoholic fermentation of glucose by *Moniliformis dubius* (Acanthocephala), *in vitro*. *Proceedings of the Royal Society of London* (*B*), **172**, 65–88.

Warden, D.A., Fannin, F.F., Evans, J.O., Hanke, D.W. & Diedrich, D.F. (1980). A hydrolase-related transport system is not required to explain the intestinal uptake of glucose liberated from phlorizin. *Biochimica et Biophysica Acta*, **599**, 664–72.

Warner, F.D. (1974). The fine structure of the ciliary and flagellar axoneme. In *Cilia and Flagella*, ed. M.A. Sleigh, pp. 11–37. New York and London: Academic Press.

Warren, K.S. (1982). Transmission: patterns and dynamics of infectious diseases. Group report. In *Population Biology of Infectious Diseases*, ed. R.M. Anderson and R.M. May, pp. 67–86. (Dahlem Konferenzen. Life Sciences Research Report 25.) Berlin: Springer-Verlag.

Watson, R.A. & Dick, T.A. (1979). Metazoan parasites of whitefish, *Coregonus clupeaformis* (Mitchill) and cisco *C. artedii* Lesueur from Southern Indian Lake, Manitoba. *Journal of Fish Biology*, **15**, 579–87.

Webster, J.D. (1943). Helminths from the robin, with the description of a new nematode, *Porrocaecum brevispiculum*. *Journal of Parasitology*, **29**, 161–3.

Weeden, R.B. & Weeden, J.S. (1973). Record of a robin feeding shrews to its nestlings. *Condor*, **75**, 248.

Weinland, D.F. (1856). On the digestive apparatus of the Acanthocephala. *Proceedings of the American Association for the Advancement of Science, 10th Meet. Albany*, 197–201.

West, A.J. (1963). A preliminary investigation of the embryonic layers surrounding the acanthor of *Acanthocephalus jacksonii* Bullock, 1962 and *Echinorhynchus gadi* (Zoega) Muller. *Journal of Parasitology*, **49** (suppl.), 42–3.

West, A.J. (1964). The acanthor membranes of two species of Acanthocephala. *Journal of Parasitology*, **50**, 731–4.

Westrumb, A.H.L. (1821). *De Helminthibus acanthocephalis. Commentatio historicoanatomic adnexo recensu animalium, in Museo Vindobonensis circa helminthes dissectorum et singularum speciarum harum in illis repertarum. Hanoverae Helwig*. 85 pp., pls. I–III.

Whitfield, P.J. (1968). A histological description of the uterine bell of *Polymorphus minutus* (Acanthocephala). *Parasitology*, **58**, 671–82.

Whitfield, P.J. (1969). *Studies on the Reproduction of Acanthocephala*. Ph D dissertation, University of Cambridge.

Whitfield, P.J. (1970). The egg sorting function of the uterine bell of *Polymorphus minutus* (Acanthocephala). *Parasitology*, **61**, 111–26.

Whitfield, P.J. (1971*a*). Spermiogenesis and spermatozoan ultrastructure in *Polymorphus minutus* (Acanthocephala). *Parasitology*, **62**, 415–30.

Whitfield, P.J. (1971*b*). Phylogenetic affinities of Acanthocephala: an assessment of ultrastructural evidence. *Parasitology*, **63**, 49–58.

Whitfield, P.J. (1973). The egg envelopes of *Polymorphus minutus* (Acanthocephala). *Parasitology*, **66**, 387–403.

Whitlock, S.C. (1939). Snails as intermediate hosts of Acanthocephala. *Journal of Parasitology*, **25**, 443.

Wilkes, J., Cornish, R.A. & Mettrick, D.F. (1981). Purification and properties of phosphoenolpyruvate carboxykinase from *Hymenolepis diminuta* (Cestoda). *Journal of Parasitology*, **67**, 832–40.

Wilkes, J., Cornish, R.A. & Mettrick, D.F. (1982*a*). Fumarase activity in *Moniliformis dubius* (Acanthocephala). *Journal of Parasitology*, **68**, 162–3.

Wilkes, J., Cornish, R.A. & Mettrick, D.F. (1982*b*). Purification and properties of phosphoenolpyruvate carboxykinase from *Ascaris suum*. *International Journal for Parasitology*, **12**, 163–71.

Wilson, D.B. (1978). Cellular transport mechanisms. *Annual Review of Biochemistry*, **47**, 933–65.

Wilson, F.A. & Dietschy, J.M. (1974). The intestinal unstirred layer: its surface area and effect on active transport kinetics. *Biochimica et Biophysica Acta*, **363**, 112–26.

Wilson, F.A. & Treanor, L.L. (1975). Characterization of the passive and active transport mechanisms of bile acid uptake into rat isolated intestinal epithelial cells. *Biochimica et Biophysica Acta*, **406**, 280–93.

Witenberg, G. (1932*a*). Akanthocephalen Studien. I. Uber einige für Systematik der Akanthocephalen wichtige anatomisch Merkmale. *Bollettino di Zoologia, Publicato dall'Unione Zoologica Italiana*, **3**, 243–52.

Witenberg, G. (1932*b*). Akanthocephalen Studien. II. Uber das System der Akanthocephalen. *Bollettino di Zoologia, Publicato dall'Unione Zoologica Italiana*, **3**, 253–66.

Wolffhügel, K. (1924). Versuche mit dem Riesenkratzer (*Macracanthorhynchus hirudinaceus* [Pallas], Syn. *Echinorhynchus gigas* Goeze). *Zeitschr. Infektionskrantheiten, Parasit. Krankh. Hyg. Haustiere*, **26**, 177–207.

Womersley, C. & Smith, L. (1981). Anhydrobiosis in nematodes. I. The role of glycerol, myo-inositol and trehalose during desiccation. *Comparative Biochemistry and Physiology*, **70B**, 579–86.

Wong, B.S., Miller, D.M. & Dunagan, T.T. (1979). Electrophysiology of acanthocephalan body wall muscles. *Journal of Experimental Biology*, **82**, 273–80.

Wright, R.D. (1970). Surface ultrastructure of the acanthocephalan lemnisci. *Proceedings of the Helminthological Society of Washington*, **37**, 52–6.

Wright, R.D. (1971). The egg envelopes of *Moniliformis dubius*. *Journal of Parasitology*, **57**, 122–31.

Wright, R.D. & Lumsden, R.D. (1968). Ultrastructural and histochemical properties of the acanthocephalan epicuticle. *Journal of Parasitology*, **54**, 1111–23.

Wright, R.D. & Lumsden, R.D. (1969). Ultrastructure of the tegumentary pore-canal system of the acanthocephalan *Moniliformis dubius*. *Journal of Parasitology*, **55**, 993–1003.

Wright, R.D. & Lumsden, R.D. (1970). The acanthor tegument of *Moniliformis dubius*. *Journal of Parasitology*, **56**, 727–35.

Wyatt, G.R. (1967). The biochemistry of sugars and polysaccharides in insects. *Advances in Insect Physiology*, **4**, 287–360.

Yalynskaya, N.S. (1980). Protein changes in the body cavity of *Gammarus* (Crustacea: Amphipoda) infected with helminths. In *Voprosy parasitologii vodnykh bespozvonochnykh zhivotnykh*. (*Tematicheskii Sbornik*), pp. 108–10. Vilnyus, USSR: Akademiya Nauk Litovskoi SSR, Institut Zoologii I Parazitologii.

Yamaguti, S. (1935). Studies on the helminth fauna of Japan. Part 8. Acanthocephala, I. *Japanese Journal of Zoology*, **6**, 247–78.

Yamaguti, S. (1939). Studies on the helminth fauna of Japan. Part 29. Acanthocephala, II. *Japanese Journal of Zoology*, **13**, 317–51.

Yamaguti, S. (1963). Acanthocephala. In *Systema Helminthum*, vol. 5, pp. 1–423. New York and London: Wiley Interscience.

Yamaguti, S. & Miyata, I. (1942). *Uber die Entwicklungsgeschichte von Moniliformis dubius* Meyer, *1933* (*Acanthocephala*) *mit besonders Berucksichtgung seiner Entwicklung im Zwischenwirt*. Kyoto: Parasitologisches Laboratorium der Kaiserlichen Universität zu Kyoto.

Yarzhinsky, T. (1868*a*). Material zur kenntnis des Onegasees und der Onegaumgebung hauptsächlich in zoologischer Hinsicht. *Arb. ersten Versamml. Russ. Naturforscher St Petersburg*.

Yarzhinsky, T. (1868*b*). Untersuchungen uber den Bau des Nervensystems der Echinorhynchen. *Arb. ersten Versamml. Russ. Naturforscher St Petersburg*, 298–310.

Yarzhinsky, T. (1870*a*). Der Bau der Echinorhynchen. *Arb. ersten Versamml. Russ. Naturforscher St Petersburg*, 60.

Yarzhinsky, T. (1870*b*). Echinorhynchen, die sich in Fischen des Finnischen Meerbusens aufhalten. *Arb. ersten Versamml. Russ. Naturforscher St Petersburg*, 57.

Young, B.W. & Lewis, P.D. (1977). Growth of an acanthocephalan on the chick chorioallantois. *Proceedings of the Montana Academy of Sciences*, **37**, 88.

Zalenskii, V. (1870). Zametki ob organizatsii *Echinorhynchus angustatus*. *Zapiski Obshchestva Estestvoipytatelei, Kiev*, **1**, 305–18.

Zeder, J.G.H. (1800). *Erster Nachtrag zur Naturgeschichte der Eingeweidewurmer, mit Zusatzen und Anmerkungen*. 320 pp., 6 pls. Leipzig.

Zeder, J.G.H. (1803). *Anleitung zur Naturgeschichte der Eingeweidewurmer*. 432 pp., 4 pls. Bamberg.

Zeledón, R. & Arroyo, G. (1960). Presencia de formas larvarias de *Oncicola oncicola* (Acanthocephala) en una gallina domestica. *Revista de Biologia Tropical*, **8**, 197–9.

Author index

Index to acanthocephalan species

(Page numbers in *italic type* refer to figures)

Acanthogyrus (cont.)
nigeriensis Dollfus & Golvan, 1956, 65
oligospinus (Anantaraman, 1969) comb. n., 65, 229, 236, 253, 261
papilo Troncy & Vassiliades, 1974, 65
partispinus Furtado, 1963, 65, 248
periophthalmi (Wang, 1980) comb. n., 65
scomberomori (Wang, 1980) comb. n., 65
shashiensis (Tso, Chen & Chien, 1974) comb. n., 65
similis (Wang, 1980) comb. n., 65
sircari (Podder, 1941) Dollfus & Golvan, 1956, 65
thapari (Parasad, Sahay & Shambhunath, 1969) comb. n., 66
tilapiae (Baylis, 1948) Dollfus & Golvan, 1956, 66, 227, 229, 236, 242, 245, 253, 254
tripathi Rai, 1967, 64
vittatusi (Verma, 1973) comb. n., 66
Acanthosentis Verma & Datta, 1929 – See *Acanthogyrus* Thapar, 1927
acanthuri Cable & Quick, 1954 – See *Acanthogyrus acanthuri* (Cable & Quick, 1954) Golvan, 1959
anguillae Wang, 1981 – See *Acanthogyrus anguillae* (Wang, 1981) comb. n.
antspinus Verma & Datta, 1929 – See *Acanthogyrus antspinus* (Verma & Datta, 1929) Dollfus & Golvan, 1956
arii Bilqees, 1971 – See *Acanthogyrus arii* (Bilqees, 1971) comb. n.
bacailai Verma, 1973 – See *Acanthogyrus bacailai* (Verma, 1973) comb. n.
betawi Tripathi, 1956 – See *Acanthogyrus antspinus* (Verma & Datta, 1929) Dollfus & Golvan, 1956
cameroni Gupta & Kajaji, 1969 – See *Acanthogyrus cameroni* (Gupta & Kajaji, 1969) comb. n.
dattai Podder, 1938 – See *Acanthogyrus dattai* (Podder, 1938) Dollfus & Golvan, 1956
giuris Soota & Sen, 1956 – See *Acanthogyrus giuris* (Soota & Sen, 1956) comb. n.
golvani Gupta & Jain, 1980 – See *Acanthogyrus golvani* (Gupta & Jain, 1980) comb. n.
hilsai Pal, 1963 – See *Acanthogyrus indica* (Tripathi, 1959) Chubb, 1962
holospinus Sen, 1937 – See *Acanthogyrus holospinus* (Sen, 1937) Dollfus & Golvan, 1956
indicus Tripathi, 1959 – See *Acanthogyrus indica* (Tripathi, 1959) Chubb, 1982
maroccanus Dollfus, 1951 – See *Acanthogyrus maroccanus* (Dollfus, 1951) Dollfus & Golvan, 1956
oligospinus Anantaraman, 1969 – See *Acanthogyrus oligospinus* (Anantaraman, 1969) comb. n.
periophthalmi Wang, 1980 – See *Acanthogyrus periophthalmi* (Wang, 1980) comb. n.
scomberomori Wang, 1980 – See *Acanthogyrus scomberomori* (Wang, 1980) comb. n.
shashiensis Tso, Chen & Chien, 1974 – See *Acanthogyrus shashiensis* (Tso, Chen & Chien, 1974) comb. n.
similis Wang, 1980 – See *Acanthogyrus similis* (Wang, 1980) comb. n.
sircari Podder, 1941 – See *Acanthogyrus sircari* (Podder, 1941) Dollfus & Golvan, 1956
thapari Parasad, Sahay & Shambhunath, 1969 – See *Acanthogyrus thapari* (Parasad, Sahay & Shambhunath, 1969) comb. n.
tilapiae Baylis, 1948 – See *Acanthogyrus tilapiae* (Baylis, 1948) Dollfus & Golvan, 1956
vittatusi Verma, 1973 – See *Acanthogyrus vittatusi* (Verma, 1973) comb. n.
Allorhadinorhynchus Yamaguti, 1959
segmentatus Yamaguti, 1959, 41
Andracantha Schmidt, 1975
gravida (Alegret, 1941) Schmidt, 1975, 59
mergi (Lundström, 1941) Schmidt, 1975, 59
phalacrocoracis (Yamaguti, 1939) Schmidt, 1975, 59
Apororhynchus Shipley, 1899
aculeatum Meyer, 1931, 31
amphistomi Byrd & Denton, 1949, 31
bivoluerus Das, 1952, 31
chauhani Sen, 1975, 31
hemignathi (Shipley, 1896) Shipley, 1899, 31
paulonucleatus Hoklova & Cimbaluk, 1971, 31
silesiacus Okulewicz & Maruszewski, 1980, 31
Arhynchus Shipley, 1896 – See *Apororhynchus* Shipley, 1899
hemignathi Shipley, 1896 – See *Apororhynchus hemignathi* (Shipley, 1896) Shipley, 1899
Arhythmacanthus Yamaguti, 1935 – See *Heterosentis* Van Cleave, 1931
fusiformis Yamaguti, 1935 – See *Heterosentis fusiformis* (Yamaguti, 1935) Tripathi, 1959
overstreeti Schmidt & Paperna,

Arhythmacanthus (cont.)
1978 – See *Heterosentis overstreeti* (Schmidt & Paperna, 1978) comb. n.
paraplagusiarum Nickol, 1972 – See *Heterosentis paraplagusiarum* (Nickol, 1972) comb. n.
plotosi (Yamaguti, 1935) Schmidt & Paperna, 1978 – See *Heterosentis plotosi* Yamaguti, 1935
septacanthus Sita in Golvan, 1969 – See *Heterosentis septacanthus* (Sita in Golvan, 1969) comb. n.
thapari Gupta & Fatma, 1979 – See *Heterosentis thapari* (Gupta & Fatma, 1979) comb. n.
Arhythmorhynchus Lühe, 1911
anser Florescu, 1941 – See *Arhythmorhynchus longicollis* (Villot, 1875) Lühe, 1912
brevis Van Cleave, 1916, 59
capellae Schmidt, 1963 – See *Arhythmorhynchus jeffreyi* Schmidt, 1973
capellae (Yamaguti, 1935) Schmidt, 1973, 59
comptus Van Cleave & Rausch, 1950, 59, 301
distinctus Baer, 1956, 59
duocinctus Chandler, 1935 – See *Southwellina hispida* (Van Cleave, 1925) Witenberg, 1932
eroliae (Yamaguti, 1939) Schmidt, 1973, 59
frassoni (Molin, 1858) Lühe, 1911, 59, 311
frontospinosus (Tubangui, 1935) Yamaguti, 1963, 59
fuscus Harada, 1929 – See *Southwellina hispida* (Van Cleave, 1925) Witenberg, 1932
hispidus Van Cleave, 1925 – See *Southwellina hispida* (Van Cleave, 1925) Witenberg, 1932
jeffreyi Schmidt, 1973, 59
johnstoni Golvan, 1960, 59
limosae Edmonds, 1971, 60
longicollis (Villot, 1875) Lühe, 1912, 60
macracanthus Ward & Winter, 1952 – See *Southwellina macracanthus* (Ward & Winter, 1952) Schmidt, 1973
petrochenkoi (Schmidt, 1969) Atrashkevich, 1979, 60, 301
plicatus (Linstow, 1883) Meyer, 1932, 60
pumilirostris Van Cleave, 1916, 60
sachalinenis Krotov & Petrochenko, 1958 – See *Arhythmorhynchus teres* Van Cleave, 1920
siluricola Dollfus, 1929, 60
teres Van Cleave, 1920, 60
tigrinus Moghe & Das, 1953, 60
trichocephalus (Leuckart, 1876) Lühe, 1912, 60, 75
tringi Gubanov, 1952, 60
uncinatus (Kaiser, 1893) Lühe, 1912, 60, 96, 301, 311
xeni Atrashkevich, 1978, 60
Aspersentis Van Cleave, 1929
austrinus Van Cleave, 1929 – See *Aspersentis megarhynchus* (Linstow, 1892) Golvan, 1960
heteracanthus (Linstow, 1896) Golvan, 1969 – See *Heterosentis heteracanthus* (Linstow, 1896) Van Cleave, 1931
johni (Baylis, 1929) Chandler, 1934, 45
megarhynchus (Linstow, 1892) Golvan, 1960, 45
Atactorhyncus Chandler, 1935
mugilis Machado, 1951 – See *Floridosentis mugilis* (Machado, 1951) Bullock, 1962
verecundus Chandler, 1935, 67, 129, 305
Australorhyncus Lebedev, 1967
tetramorphacanthus Lebedev, 1967, 49, 298

Bolborhynchoides Achmerov, 1959
exiguus (Achmerov & Dombrowskaja-Achmerova, 1941) Achmerov, 1959, 46
Bolborhynchus Achmerov & Dombrowskaja-Achmerova, 1941 – See *Bolborhynchoides* Achmerov, 1959
exiguus Achmerov & Dombrowskaja-Achmerova, 1941 – See *Bolborhynchoides exiguus* (Achmerov & Dombrowskaja-Achmerova, 1941) Achmerov, 1959
Bolborhynchus Porta, 1906 – See *Bolbosoma* Porta, 1908
brevicollis (Malm, 1867) Porta, 1906 – See *Bolbosoma brevicolle* (Malm, 1867) Porta, 1908
turbinella (Diesing, 1851) Porta, 1906 – See *Bolbosoma turbinella* (Diesing, 1851) Porta, 1908
Bolbosentis Belous, 1952 – See *Micracanthorhynchina* Strand, 1936
sajori Belous, 1952 – See *Micracanthorhynchina sajori* (Belous, 1952) Golvan, 1969
Bolbosoma Porta, 1908
balaenae (Gmelin, 1790) Porta, 1908, 60, 96
bobrovoi Krotov & Delamure, 1952, 60
brevicolle (Malm, 1867) Porta, 1908, 60
caenoforme (Heitz, 1920) Meyer, 1932, 60
capitatum (Linstow, 1880) Porta, 1908, 60

Bolbosoma (cont.)
hamiltoni Baylis, 1929, 60
heteracanthis (Heitz, 1917) Meyer, 1932, 60
nipponicum Yamaguti, 1939, 60, 116
physeteris Gubanov, 1952, 60
porrigens (Rudolphi, 1814) Porta, 1908 – See *Bolbosoma balaenae* (Gmelin, 1790) Porta, 1908
scomberomori Wang, 1980, 60
thunni Harada, 1935, 60
tuberculata Skrjabin, 1970, 60
turbinella (Diesing, 1851) Porta, 1908, 60, 96, 97, 98, 99, 104, 105, 106
turbinella australis Skrjabin, 1972, 60
turbinella turbinella (Diesing, 1851) Porta, 1908, 60
vasculosum (Rudolphi, 1819) Porta, 1908, 61
Breizacanthus Golvan, 1969
chabaudi Golvan, 1969, 39
irenae Golva, 1969, 39
ligur Paggi, Orecchia & Della Seta, 1975, 39
Brentisentis Leotta, Schmidt & Kuntz, 1982
uncinus Leotta, Schmidt & Kuntz, 46

Caballerorhynchus Salgado-Maldonado, 1977
lamothei Salgado-Maldonado, 1977, 40
Cathayacanthus Golvan, 1969
exilis (Van Cleave, 1928) Golvan, 1969, 51
Cavisoma Van Cleave, 1931
chromitidis Cable & Quick, 1954 – See *Pseudocavisoma chromitidis* (Cable & Quick, 1954) Golvan & Houin, 1964
magnum (Southwell, 1927) Van Cleave, 1931, 40
Centrorhynchus Lühe, 1911
albensis Rengaraju & Das, 1975, 53
albidus Meyer, 1932, 53
aluconis (Müller, 1780) Lühe, 1911, 53, 310
amphibius Das, 1950, 53, 298
andamanensis Soota & Kansal, 1970 – See *Centrorhynchus spilornae* Schmidt & Kuntz, 1969
appendiculatum (Westrumb, 1821) Joyeux & Baer, 1937 – See *Centrorhynchus aluconis* (Müller, 1780) Lühe, 1911
asturinus (Johnston, 1912) Johnston, 1918, 53
atheni Gupta & Fatma, 1983, 53
bancrofti (Johnston & Best, 1943) Golvan, 1956, 53
batrachus Das, 1952, 53, 298
bazaleticus Kurachvilli, 1955, 53
bengalensis Datta & Soota, 1954, 53
bipartitus Solovieff, 1912 – See *Sphaerirostris pinguis* (Van Cleave, 1918) Golvan, 1956
bramae Rengaraju & Das, 1980, 53
breviacanthus Das, 1949, 53
brevicaudatus Das, 1949, 53
brumpti Golvan, 1965, 53
brygooi Golvan, 1965, 53
bubonis Yamaguti, 1939, 53
buckleyi Gupta & Fatma, 1983, 53.
buteonis (Schrank, 1788) Kostylev, 1914, 53, 310
californicus Millzner, 1924, 53
chabaudi Golvan, 1958, 53
cinctus (Rudolphi, 1819) Meyer, 1932 – See *Sphaerirostris lancea* (Westrumb, 1821) Golvan, 1956
clitorideus (Meyer, 1931) Golvan, 1956, 53
conspectus Van Cleave & Pratt, 1940, 53
corvi Fukui, 1929 – See *Sphaerirostris pinguis* (Van Cleave, 1918) Golvan, 1956
crocidurus Das, 1950, 53, 298
crotophagicola Schmidt & Neiland, 1966, 53
cylindraceus (Goeze, 1782) Kostylev, 1914 – See *Plagiorhynchus cylindraceus* (Goeze, 1782) Schmidt & Kuntz, 1966
dimorphocephalus (Westrumb, 1821) Meyer, 1932, 53
dipsadis (Linstow, 1888) Golvan, 1956, 53
elongatus Yamaguti, 1935, 54, 119, 240
embae Kolodkowski & Kostylev, 1916 – See *Sphaerirostris lancea* (Westrumb, 1821) Golvan, 1956
erraticus Chandler, 1925 – See *Sphaerirostris erraticus* (Chandler, 1925) Golvan, 1956
falconis (Johnston & Best, 1943) Golvan, 1956, 54, 298
fasciatus (Westrumb, 1821) Travassos, 1926, 54
freundi (Hartwick, 1953) Golvan, 1956, 54
galliardi Golvan, 1956, 54
gendrei (Golvan, 1957) Golvan, 1960, 54
giganteus Travassos, 1921, 54
globocaudatus (Zeder, 1800) Lühe, 1911, 54
golvani Anantaraman & Anantaraman, 1969, 54
grassei Golvan, 1965, 54
hagiangensis (Petrochenko & Fan, 1969) comb. n., 54
hargisi Gupta & Fatma, 1983, 54
horridus (Linstow, 1897) Meyer, 1932, 54

Centrorhynchus (cont.)
indicus Golvan, 1956, 54
insularis Tubangui, 1933, 54
itatsinis Fukui, 1929, 54
javanicans Rengaraju & Das, 1975, 54
knowlesi Datta & Soota, 1955, 54
kuntzi Schmidt & Neiland, 1966, 54
lancea (Westrumb, 1821) Skrjabin, 1913 – See *Sphaerirostris lancea* (Westrumb, 1821) Golvan, 1956
lanceoides Petrochenko, 1949 – See *Sphaerirostris lanceoides* (Petrochenko, 1949) Golvan, 1956
leguminosus Solovieff, 1912 – See *Sphaerirostris pinguis* (Van Cleave, 1918) Golvan, 1956
leptorhyncus Meyer, 1932, 54
lesiniformis (Molin, 1859) Bock, 1935, 54
longicephalus Das, 1950, 54, 298
lucknowensis Gupta & Fatma, 1983, 54
mabuiae (Linstow, 1908) Golvan, 1956, 54
macrorchis Das, 1949, 54
madagascariensis (Golvan, 1957) Golvan, 1960, 54
magnus Fukui, 1929, 54, 298
maryasis Datta, 1932 – See *Sphaerirostris maryasis* (Datta, 1932) Golvan, 1956
merulae Dollfus & Golvan, 1961, 54
microcephalus (Bravo-Hollis, 1947) Golvan, 1956, 54
microcerviacanthus Das, 1950, 54, 298
microrchis Fukui, 1929 – See *Centrorhynchus magnus* Fukui, 1929
migrans Zuberi & Farooq, 1974, 55
milvus Ward, 1956, 55, 242
mysentri Gupta & Fatma, 1983, 55, 298
narcissae Florescu, 1942, 55
nicaraguensis Schmidt & Neiland, 1966, 55
ninni (Stossich, 1891) Meyer, 1932, 55
olssoni Lundström, 1942 – See *Centrorhynchus aluconis* (Müller, 1780) Lühe, 1911
opimus Travassos, 1919 – See *Sphaerirostris opimus* (Travassos, 1919) Golvan, 1956
petrotschenkoi Kuraschvili, 1955, 55
picae Dollfus, 1953 – See *Sphaerirostris picae* (Rudolphi, 1819) Golvan, 1956
pinguis Van Cleave, 1918 – See *Sphaerirostris pinguis* (Van Cleave, 1918) Golvan, 1956
polemaeti Troncy, 1970, 55
ptyasus Gupta, 1950, 55, 298
renardi (Lindemann, 1865) Van Cleave, 1923, 55
reptans Bhalerao, 1931 – See *Sphaerirostris reptans* (Bhalerao, 1931) Golvan, 1956
scanensis Lundström, 1942 – See *Sphaerirostris lancea* (Westrumb, 1821) Golvan, 1956
sholapurensis Rengaraju & Das, 1975, 55
simplex Meyer, 1932, 55
skrjabini Petrochenko, 1949 – See *Sphaerirostris pinguis* (Van Cleave, 1918) Golvan, 1956
spilornae Schmidt & Kuntz, 1969, 55, 298
spinosus (Kaiser, 1893) Van Cleave, 1924, 55, 299
splendi Gupta & Gupta, 1970 – See *Centrorhynchus batrachus* Das, 1952
tenuicaudatus (Marotel, 1899) Lühe, 1911 – See *Sphaerirostris tenuicaudatus* (Marotel, 1899) comb. n.
teres (Westrumb, 1821) Travassos, 1926, 55, 299
tumidulus Neiva Cuhna & Travassos, 1914 – See *Centrorhynchus tumidulus* (Rudolphi, 1819) Travassos, 1923
tumidulus (Rudolphi, 1819) Travassos, 1923, 55
turdi Yamaguti, 1939 – See *Sphaerirostris turdi* (Yamaguti, 1939) Golvan, 1956
tyotensis Rengaraju & Das, 1975, 55
undulatus Dollfus, 1951, 55
wardae Holloway, 1958, 55
Chentrosoma Monticelli, 1905 – See *Corynosoma* Lühe, 1904
Chentrosoma Porta, 1906 – See *Centrorhynchus* Lühe, 1911
aluconis (Müller, 1780) Porta, 1909 – See *Centrorhynchus aluconis* (Müller, 1780) Lühe, 1911
zosteropis Porta, 1913 – See *Mediorhynchus zosteropis* (Porta, 1913) Meyer, 1932
Cleaveius Subramanian, 1927
circumspinifer Subramanian, 1927, 49
leiognathi Jain & Gupta, 1979, 49
mysti (Sahay & Sinha, 1971) comb. n., 49
portblairensis Jain & Gupta, 1979, 49
prashadi (Datta, 1940) Golvan, 1969, 49
secundus (Tripathi, 1959) Golvan, 1969, 49
Corynosoma Lühe, 1904
alaskensis Golvan, 1958, 61
antarcticum (Rennie, 1907) Leiper & Atkinson, 1914 – See *Corynosoma hamanni* (Linstow, 1892) Railliet & Henry, 1907
ambispigerinum Harada, 1935 – See

Mediorhynchus (cont.)
sharmai Gupta & Lata, 1967, 33
sipocotensis Tubangui, 1935, 33
taeniatus (Linstow, 1901) Dollfus, 1936, 33
tanagrae (Rudolphi, 1819) Travassos, 1921, 33
tenuis Meyer, 1931, 33
textori Barus, Sixl & Majumdar, 1978, 33
turnixena (Tubangui, 1931) Webster, 1948, 33
vaginatus (Diesing, 1851) Meyer, 1932, 33
vancleavei (Lundström, 1942) Golvan, 1962, 33
wardi Schmidt & Canaris, 1967, 33
zosteropis (Porta, 1913) Meyer, 1932, 33
Megapriapus Golvan, Gracia-Rodrigo & Diaz-Ungria, 1964
ungriai (Gracia-Rodrigo, 1960) Golvan, Gracia-Rodrigo & Diaz-Ungria, 1964, 40, 220
Megistacantha Golvan, 1960
horridum (Lühe, 1912) Golvan, 1960, 51
Mehrarhynchus Datta, 1940 – See *Cleaveius* Subramanian, 1927
mysti Sahay & Sinha, 1971 – See *Cleaveius mysti* (Sahay & Sinha, 1971) comb. n.
prashadi Datta, 1940 – See *Cleaveius prashadi* (Datta, 1940) Golvan, 1969
secundus Tripathi, 1959 – See *Cleaveius secundus* (Tripathi, 1959) Golvan, 1969
Metacanthocephaloides Yamaguti, 1959
zebrini Yamaguti, 1959, 50
Metacanthocephalus Yamaguti, 1959
campbelli (Leiper & Atkinson, 1914) Golvan, 1969, 50
ovicephalus (Zhukov, 1963) Golvan, 1969, 50
pleuronichthydis Yamaguti, 1959, 50
Metarhadinorhynchus Yamaguti, 1959
lateolabracis Yamaguti, 1959, 46
thapari Gupta & Gupta, 1975, 46
Metechinorhynchus Petrochenko, 1956 – See *Echinorhynchus* Zoega in Müller, 1776
alpinus (Linstow, 1901) Petrochenko, 1956 – See *Echinorhynchus salmonis* Müller, 1784
baeri (Kostylev, 1928) Petrochenko, 1956 – See *Echinorhynchus baeri* Kostylev, 1928
briconi (Machado, 1959) Golvan, 1969 – See *Echinorhynchus briconi* Machado, 1959
campbelli (Leiper & Atkinson, 1914) Petrochenko, 1956 – See *Metacanthocephalus campbelli* (Leiper & Atkinson, 1914) Golvan, 1969
cryophilus Sokolowskaja, 1962 – See *Echinorhynchus cryophilus* (Sokolowskaja, 1962) comb. n.
gomesi (Machado, 1948) Golvan, 1969 – See *Echinorhynchus gomesi* Machado, 1948
jucundum (Travassos, 1923) Petrochenko, 1956 – See *Echinorhynchus jucundus* Travassos, 1923
kushiroensis (Fujita, 1921) Golvan, 1969 – See *Echinorhynchus kushiroensis* Fujita, 1921
lageniformis (Ekbaum, 1938) Petrochenko, 1956 – See *Echinorhynchus lageniformis* Ekbaum, 1938
lateralis (Leidy, 1851) Golvan, 1969 – See *Echinorhynchus lateralis* Leidy, 1851
leidyi (Van Cleave, 1924) Golvan, 1969 – See *Echinorhynchus leidyi* Van Cleave, 1924
paranensis (Machado, 1959) Golvan, 1969 – See *Echinorhynchus paranensis* Machado, 1959
rhenanus Golvan, 1969 – See *Echinorhynchus rhenanus* (Golvan, 1969) comb. n.
salmonis (Müller, 1784) Petrochenko, 1956 – See *Echinorhynchus salmonis* Müller, 1784
salobrensis (Machado, 1948) Golvan, 1969 – See *Echinorhynchus salobrensis* Machado, 1948
truttae (Schrank, 1788) Petrochenko, 1956 – See *Echinorhynchus truttae* Schrank, 1788
Micracanthocephalus Harada, 1938 – See *Micracanthorhynchina* Strand, 1936
dakusuiensis Harada, 1938 – See *Micracanthorhynchina dakusuiensis* (Harada, 1938) Ward, 1951
hemirhamphi Baylis, 1944 – See *Micracanthorhynchina hemirhamphi* (Baylis, 1944) Ward, 1951
motomurai (Harada, 1935) Harada, 1938 – See *Micracanthorhynchina motomurai* (Harada, 1935) Ward, 1951
Micracanthorhynchina Strand, 1936
cynoglossi Wang, 1980, 50
dakusuiensis (Harada, 1938) Ward, 1951, 50
hemiculterus Demshin, 1965, 50
hemirhamphi (Baylis, 1944) Ward, 1951, 50

Subject index